Contributions to Economics

Brigitte Preissl • Justus Haucap • Peter Curwen
Editors

Telecommunication Markets

Drivers and Impediments

Physica-Verlag
A Springer Company

Editors
Dr. Brigitte Preissl
ZBW
Neuer Jungfernstieg 21
20354 Hamburg
Germany
b.preissl@zbw.eu

Professor Justus Haucap
University of Erlangen-Nuremberg
Deparment of Economics
Lange Gasse 30
90403 Nürnberg
Germany
justus.haucap@wiso.uni-erlangen.de

Professor Peter Curwen
Strathclyde Business School
Department of Management Science
Graham Hills Building
40 George Street
Glasgow G1 1QE
Scotland, UK
pjcurwen@hotmail.com

ISBN: 978-3-7908-2081-2 e-ISBN: 978-3-7908-2082-9

DOI: 10.1007/978-3-7908-2082-9

Springer Series in Contributions to Economics ISSN 1431-1933

Library of Congress Control Number: 2008943921

Cover design: WMXDesign GmbH, Heidelberg

Printed on acid-free paper

springer.com

Contents

Contributors

James H. Alleman
College of Engineering & Applied Science, University of Colorado, CB 530,
Boulder, CO 80309-0530, USA
james.alleman@Colorado.edu

Izak Atiyas
Faculty of Arts and Social Sciences, Sabanci University, Orhanli, Tuzla 34956,
Istanbul, Turkey
izak@sabanciuniv.edu

Leo Van Audenhove
IBBT-iLab.o & IBBT-SMIT, Vrije Universiteit Brussel, Pleinlaan 2,
1050 Elsene, Belgium
leo.vanaudenhove@vub.ac.be

Pieter Ballon
IBBT-SMIT, Vrije Universiteit Brussel, Pleinlaan 2, 1050 Elsene, Belgium
Pieter.Ballon@vub.ac.be

Johannes Bauer
Quello Center for Telecommunication Management and Law, Michigan State
University, 406 Communication Arts and Sciences, East Lansing,
Michigan 48824, USA
bauerj@msu.edu

Pierre-Jean Benghozi
Economics and Management Research Centre (PREG), CNRS 1,
Pôle de Recherche en Economie et Gestion de l'Ecole polytechnique,
1, rue Descartes, 75005 Paris, France

Harry Bouwman
Interim Chair Information and Communication Technology, Faculty
of Technology, Policy and Management, Delft University of Technology,
PO BOX 5015, 2600 GA Delft, The Netherlands

Mélisande Cardona
Ludwig-Maximilians-University Munich, Schackstr. 4/III, 80539 München, Germany
m.cardona@lmu.de

Leonel Cerno
Departamento de Economía, Universidad Carlos III de Madrid, C./ Madrid, 126, 28903 Getafe (Madrid), Spain
lcerno@eco.uc3m.es

Stéphane Ciriani
Orange Lab, 38 rue du Général Leclerc, 92130 Issy Les Moulineaux, France
stephane.ciriani@wanadoo.fr

Massimo G. Colombo
Department of Management, Economics and Industrial Engineering, Politecnico di Milano, P.za Leonardo da Vinci, 32, 20133 – Milan, Italy
massimo.colombo@polimi.it.

Ramón Compañó
European Commission, Directorate General Joint Research Centre, Institute for Prospective Technological Studies, Edificio EXPO C/ Inca Garcilaso, s/n, 41092 Sevilla, Spain
ramon.compano@ec.europa.eu

Peter Curwen
Department of Management Science, Strathclyde University, 40 George St, Glasgow, G1 1QE, Scotland
pjcurwen@hotmail.com

Toker Doganoglu
Department of Business and Economics, University of Southern Denmark, Campusvej 55, DK-5230 Odense M, Denmark and Faculty of Arts and Social Sciences, Sabanci University, Orhanli-Tuzla, Istanbul 34956, Turkey
tod@sam.sdu.dk

Morten Falch
CMI, Aalborg University, Lautrupvang 15, 2750 Ballerup, Denmark
falch@cmi.aau.dk

Laurent Gille
Département Sciences Economiques et Sociales, TELECOM ParisTech, 46,rue Barrault, 75013 Paris, France and TELECOM ParisTech, Département Sciences Economiques et Sociales, 46, rue Barrault, 75013 Paris Cedex 13, France

Luca Grilli
Department of Management, Economics and Industrial Engineering, Politecnico di Milano, P.za Leonardo da Vinci, 32, 20133 – Milan, Italy
luca.grilli@polimi.it.

Imsook Ha
Quello Center for Telecommunication Management and Law,
Michigan State University, 409 Communication Arts and Sciences,
East Lansing, Michigan 48824-1212, USA

Justus Haucap
University of Erlangen-Nuremberg, Department of Economics,
Lange Gasse 20, 90403 Nuremberg, Germany
justus.haucap@wiso.uni-erlangen.de

Anders Henten
CMI, Aalborg University, Lautrupvang 15, 2750 Ballerup, Denmark
henten@cmi.aau.dk

Guadalupe Flores Hernández
Paseo San Francisco Sales 9, 8 A, 28003 Madrid, Spain
mg.flores@alumnos.upm.es

Mitomo Hitoshi
Graduate School of Global Information and Telecommunication Studies (GITS)
Director, Waseda Institute for Digital Society, Waseda University,
Bldg. 29-7, 1-3-10, Nishiwaseda, Shinjuku-ku, Tokyo 169-0051, JAPAN
mitomo@waseda.jp

Nejc M. Jakopin
Arthur D. Little GmbH, Breite Strasse 27, 40213 Düsseldorf
jakopin.nejc@adlittle.com

Jörn Kruse
Helmut Schmidt Universität Hamburg, Institut für Wirtschaftspolitik,
Holstenhofweg 85, 22043 Hamburg
joern.kruse@hsu-hh.de

Patricia H. Longstaff
Newhouse School of Public Communications, Syracuse University,
215 University Place, Syracuse, New York 13244-2100, USA
phlongst@syr.edu

Gary Madden
Communications Economics & Electronic Markets Research Centre, Department of Economics, Curtin Business School, Curtin University of Technology, GPO Box U1987, Perth, WA 6845, Australia
maddeng@cbs.curtin.edu.au

Kiminori Nakaba
Consumer Marketing Department, Consumer Business Strategy Division KDDI CORPORATION GARDEN AIR TOWER, 3-10-10 Iidabashi, Chiyoda-ku, Tokyo 102-8460, Japan
Ki-nakaba@kddi.com

Koshiro Ota
Faculty of Economic Sciences, Hiroshima Shudo University, 1-1-1 Ozukahigashi, Asaminami-ku, Hiroshima 731-3195, Japan
k-ota@rf7.so-net.ne.jp

Tokio Otsuka
Institute for Digital Society, Waseda University, 29-7, 1-3-10 Nishi-Waseda, Shinjuku-ku, Tokyo 169-0051, Japan
t.otsuka@kurenai.waseda.jp

Teodosio Perez Amaral
Universidad Complutense de Madrid, Campus de Somosaguas, Edificio Prefabricado, N125, 28223 Madrid, Spain
teodosio@ccee.ucm.es

Svetlana Petukhova
Institute of Economy RAS, Novocheryomushkinskaya str. 42a, 117418 Moscow, Russia
petukhova@transecon.ru

Martijn Poel
TNO-ICT, Delft, Brassersplein 2, PO Box 5050, 2600 GB Delft, The Netherlands
martijn.poel@tno.nl

Brigitte Preissl
Intereconomics, ZBW, Neuer Jungfernstieg 21, 20354 Hamburg, Germany
b.preissl@zbw.eu

Bertrand Quélin
HEC Paris, 1, rue de la Libération, 78 351 Jouy-en-Josas, France
quelin@hec.fr

Paul N. Rappoport
Economics Department, School of Business and Management, Temple University, Philadelphia, PA 19122, USA
prapp@smb.temple.edu

Martin Reichhuber
LECG Ltd., Davidson Building, 5 Southampton Street, London WC2E 7HA, UK
mreichhuber@lecg.com

Mark de Reuver
Faculty of Technology, Policy and Management, Delft University of Technology,
PO BOX 5015, 2600 GA Delft, The Netherlands
G.A.dereuver@tudelft.nl

Delphine Riccardi
HEC Paris, 1, rue de la Libération, 78 351 Jouy-en-Josas, France
delphine.riccardi@gmail.com

Olaf Rieck
Nanyang Technological University, ITOM, S3-B1b Nanyang Ave,
Singapore 639798
orieck@ntu.edu.sg

Jeffrey H. Rohlfs
Analysys Mason, 818 Connecticut Ave NW, Suite 300, Washington DC 20006, USA
jeff.rohlfs@analysysmason.com

Boris Rotenberg[†]

Jonathan Sandbach
Vodafone Group, Vodafone House, The Connection, Newbury, Berkshire,
RG14 2FN, UK
jonathan.sandbach@vodafone.com

Dan Saugstrup
HiQ Copenhagen, Klampenborgvej 221, 2800 Kgs. Lyngby, Denmark
dan@saugstrup.net

Anton Schwarz
Austrian Regulatory Authority for Broadcasting and Telecommunications (RTR)
Mariahilfer Straße 77-79, 1060 Vienna, Austria
Anton.schwarz@rtr.at

Tomas Staelens
IBBT-iLab.o & IBBT-SMIT, Vrije Universiteit Brussel, Pleinlaan 2,
1050 Elsene, Belgium

Margarita Strepetova
Institute of Economy RAS, Novocheryomushkinskaya str. 42a,
117418 Moscow, Russia
strepetova@transecon.ru

Reza Tadayoni
CMI, Aalborg University, Lautrupvang 15, 2750 Ballerup, Denmark
tadayoni@cmi.aau.dk

[†]Deceased

Canh Thang Doan
535 Pierce Street, 3314 Albany, CA 94706, USA
Thang.Doan@gmail.com

Alain Vallée
TELECOM ParisTech, Département Sciences Economiques et Sociales,
46, rue Barrault, 75013 Paris Cedex 13, France

Margit A. Vanberg
Centre for European Economic Research (ZEW), Research Group Information and Communication Technologies, P.O. Box 103443, 68034 Mannheim, Germany
vanberg@zew.de

Nico van Eijk
Institute for Information Law (IViR), University of Amsterdam, Rokin 84,
1012 KX Amsterdam, The Netherlands
vaneijk@ivir.nl

Maartje de Visser
TILEC, Tilburg University, PO Box 90153, 5000 LE Tilburg, The Netherlands
Maartje.devisser@uvt.nl

Jason Whalley
Department of Management Science, Strathclyde University, 40 George St,
Glasgow, G1 1QE, Scotland
jason.whalley@stracth.ac.uk

B. Burcin Yurtoglu
Department of Economics, University of Vienna, Brünner Straße 72,
1210 Vienna, Austria
burcin.yurtoglu@univie.ac.at

Christine Zulehner
Department of Economics, University of Vienna, Brünner Straße 72,
1210 Vienna, Austria
christine.zulehner@univie.ac.at

Introduction

Brigitte Preissl, Justus Haucap, and Peter Curwen

Telecommunications markets are commonly assumed to have a high potential for innovation and growth. New technical developments offer opportunities to move into a world of exciting new communication-based services. The dynamics of this process depend on technological specifications (which standards are to be chosen?), on the generation of appropriate business models and business strategies (which service bundles will the market respond to?), on an optimal interaction of all players involved on the demand and on the supply side (which supplier configurations will be successful on the Internet?) and on regulatory decisions (how is spectrum allocated? which interconnection regime is most appropriate for next generation networks?). This list of questions is addressed in the chapters of this book. Attention is directed towards specific markets, technologies, strategies and regulatory measures, as well as towards examples of best practice or obvious mistakes. Thus, drivers of telecommunications markets can be identified that have promoted the development of product and service innovation and/or stimulated demand. Likewise other factors have impeded progress towards a differentiation of service supply and the growth of markets.

The question of what drives and what hinders the development of telecommunications markets has led to important strands of discussion in economics, and relatively new phenomena, such as Internet business models or mobile broadband technologies. It challenges theoretical explanations of market outcomes and models of regulatory governance. Medium- and long-term processes of telecommunications development are thus manifest in what are often very detailed questions. In this sense, this book also sheds light on the complexity and growing differentiation of markets, as it does not deal with market development in general but rather hints at selected problems which are discussed in detailed analyses.

The seven chapters of the book reflect the major challenges that are currently endemic in telecommunications markets across the globe.

B. Preissl(✉), J. Haucap, and P. Curwen
German National Library of Economics (ZBW), Hamburg
e-mail: B.preissl@zbw.eu

B. Preissl et al. (eds.), *Telecommunication Markets*, Contributions to Economics,
DOI: 10.1007/978-3-7908-2082-9_1, © Springer Physica-Verlag HD 2009

Theoretical Perspectives

In the opening paper, Izak Atiyas, Toker Doganoglu and Martin Reichhuber examine the properties of general one-way access payment mechanisms where access charges depend on the incumbent's retail prices (such as a retail-minus access charge). As the paper shows, regulators can, in theory, indirectly implement the optimal retail prices even if access charges linearly depend on retail prices. This contrasts with cost-based access price regulation which cannot lead to optimal retail prices in their model. Moreover, the authors show that linear access prices should decrease with the incumbent's retail price and increase with the entrant's own retail price, as an access pricing scheme of this kind induces a higher social welfare than schemes where the access price increases with the incumbent's retail price, such as the typical retail-minus scheme.

The second paper authored by Margit Vanberg addresses competition and cooperation in Internet backbone services. The paper analyzes the strong network externalities associated with Internet services from a competition policy perspective. Given the strength of the network effects, it is argued that all ISPs need to offer universal connectivity in order to be able to compete. Hence, new entrants to the Internet interconnectivity market need to establish a direct or indirect transit agreement with at least one Tier-1 ISP. This necessity may give rise to the concern that Tier-1 ISPs could abuse a dominant market position vis-à-vis lower level ISPs or collusively engage in price fixing with other Tier-1 ISPs. Margit Vanberg argues that these concerns are not borne out by economic analysis, as (a) there are strong competitive forces in the market for top-tier Internet interconnectivity, and (b) collusion between Tier-1 ISPs to collectively raise prices in the transit market is not likely to be stable given that market conditions are not conducive to stable collusion in the market for top-tier Internet interconnectivity services. Overall, the analysis supports the view that competitive forces in the transit market are working and can effectively hinder Tier-1 ISPs from discriminating against ISPs that are on lower levels of the Internet hierarchy.

The third and the fourth papers deal with charges for mobile telecommunications services, but the two papers take a quite different approach. Hitoshi Mitomo, Tokio Otsuka and Kiminori Nakaba study post-paid mobile telecommunications services and offer a behavioural economic interpretation of the apparent flat-rate bias, which can be explained by loss aversion, reference dependence, the shape of probability weighting function, mental accounting, ambiguity aversion and/or cognitive dissonance. The authors use non-parametric methods in their empirical analysis based on data collected for mobile phone services through an online survey in Japan. It is argued that the flat-rate preference mainly results from loss aversion and reference dependence, while the shape of the probability weight function is found to be insignificant.

The first section concludes with a paper by Morton Falch, Anders Henten and Reza Tadayoni on the regulation of international roaming charges in Europe. The paper begins by presenting a brief techno-economic analysis of the costs of providing international roaming. It then goes on to review the original proposal made by the European Commission and compares it with the final proposal adopted by the European Parliament and other proposals made during this process. As the authors argue, regulation of international roaming is more complicated than regulation of other telecommunications services for two reasons. In the first place, the market structures in mobile telecommunications are less concentrated than in markets for fixed-line telephony. Secondly, regulation of international roaming is difficult to implement at the national level as operators from more than one country are involved. The paper also shows that while a common price cap improves transparency for consumers, it may lead to situations (a) where operators in high-cost countries may have difficulties in covering their costs, and (b) where international roaming is cheaper than national roaming. Even though the authors criticize the heavy-handed regulation of international roaming, they conclude that the final agreement is a compromise, which is a considerable improvement compared to the previous situation from the point of view of consumers. It is, however, also a shift from regulation based on clear economic criteria and reasoning towards regulation based on political negotiations between parties with conflicting interests.

Internet Issues

In their paper, entitled "Substitution Between DSL, Cable, and Mobile Broadband Internet Services", Mélisande Cardona and her colleagues analyse residential demand for Internet access in Austria with a particular focus on broadband connectivity. Austria provides an interesting case study because there is both cable and DSL fixed-wire connectivity as well as extensive high-speed mobile connectivity via UMTS and HSDPA. The starting point of the paper is the requirement that national regulatory authorities (NRAs) in EU countries periodically review a number of electronic communications markets including that for wholesale broadband access – sometimes known as bitstream access. In this case as in others, the precise definition of the relevant market is a matter of controversy and opinion is divided as to whether cable and DSL provision should be considered to be a single market, in good part because wholesale provision via cable is uncommon in the EU and evidence about how the two access modes compete is scanty. Insofar that cable and DSL compete directly, this is claimed to be at the retail level – if a DSL monopolist raised its wholesale price the effect at the retail level would be to induce customers to switch to cable in order to avoid the consequent increase in the DSL retail price. Although this proposition has been tested, the results so far have been limited and inconclusive, so the authors set out to conduct a further test using a three-level nested logit discrete choice model. Through a comparison of estimated elasticities with critical elasticities, the authors are led to conclude that the demand for DSL is elastic, thereby indicating strongly that cable networks are part of the same market.

The other two papers in this section concern search engines. In the first of these, Boris Rotenberg and Ramón Compañó address "Search Engines for Audio-Visual Content: Copyright Law and Its Policy Relevance". Their starting point is the proposition that the explosion of Internet sites means that audio-visual (AV) search engines will become essential in order to deal with the wide variety of content finding its way onto the Internet. This is evidenced by the recent behaviour of the likes of Yahoo!, Google and Microsoft as well as public institutions. However, many policy challenges can be anticipated within Europe, so the authors seek in this paper to investigate in advance the legal, regulatory and policy dimensions of AV search. In particular, the paper analyses some recent case law relating to copyright litigation over deep linking, provision of snippets, cache copy, thumbnail images, news gathering and other aggregation services, while eschewing any analysis of illegal downloading. Given that copyright law is only partially harmonised within the EU, the paper does not conduct country case studies but rather aims to inform EU policy through legal analysis and an investigation of how copyright law could be a viable tool in achieving EU policy goals. At the heart of the paper lies the search for the appropriate regulatory balance: too much copyright protection and development of the AV search market will be stunted; too little protection and the creation of valuable content will be inhibited. Throughout, the paper sets out to draw appropriate comparisons between developments in the text-based sector and the AV sector of the search engine market.

Nico van Eijk is concerned not with legal issues but with the functionality and underlying business model of search engines in his paper entitled "Search Engines, the New Bottleneck for Content Access". The core functionality of a search engine consists of making content and information sources easily accessible. It is no longer an issue of whether the content is available – if anything, the exponential growth of content has itself become the problem – but rather of how to access it in a manner that is helpful to the searcher. As a result, search engines have become the key component of the search process, and insofar that they have proved to be reliable, they have high added value for society. However, search engines are themselves evolving with the result that the manner in which they aggregate content has a wider effect upon the content market. Hence the structure of the search engine market is itself a matter for debate, especially given the manner in which Google has achieved a position of dominance in the market. Since for the time being there is little regulatory control over the market, search engines with a significant market presence can in principle manipulate the market, thereby altering the business model compared to a mere 5 years ago and with implications for the presentation of search results. After analyzing these and other issues the author proceeds to examine a set of potential – albeit realistic and hence limited – remedies.

In their paper on "E-commerce Use in Spain" Leonel Cerno and Teodosio Peréz Amaral analyse one of the most popular adoptions of the Internet in private households: electronic commerce. Buying and selling online requires users to have Internet access and a fairly quick connection. In an econometric analysis the authors provide empirical evidence about what drives consumers to purchase goods and services on the internet. They use data from more than 5,000 internet users in Spain to create user profiles and to estimate expenditure for e-commerce. The analysis confirms earlier findings that younger people with a good education that live in medium-sized communities and have a median level of income are 'typical' e-commerce users. Access to credit and credit cards are other crucial factors.

Methodologically, the paper provides a new tool for policy-relevant forecasts by using the RETINA algorithm for data analysis.

Broadband Issues

The future of telecommunication services seems to rest on broadband technology. Once broadband is available at a sufficient number of customer access points, innovative applications seem to mushroom and existing services appear in an unprecedented quality. Hence, the construction of broadband infrastructure (in ever increasing bandwidths), and more recently the stimulation of demand for broadband connections are core questions in each country's ICT policy. In particular, the European Lisbon strategy relies on fast and widespread diffusion of the usage of broadband-based services. Three papers in this volume deal with the enhancing and inhibiting factors that can be observed in the deployment and diffusion of broadband technology.

Massimo Colombo and Luca Grilli have contributed a paper on "The Diffusion of Broadband-Based Applications Among Italian Small and Medium Enterprises". They address a problem which is common in innovative technologies, namely the reluctance of small- and medium-sized enterprises (SMEs) to adopt broadband applications. Although a vast literature on the issue already exists, the specific characteristics of broadband reveal new insights which are backed by the data presented in the paper. While broadband adoptions are said to compensate for disadvantages that SMEs face in their production technologies, the use of broadband technology is undoubtedly subject to economies of scale. Only empirical analysis can show how SMEs deal with this combination of opportunities and challenges. In an analysis which is based on a longitudinal survey of Italian firms, Colombo and Grilli find that a combination of intra-firm and extra-firm capabilities is required to be successful broadband adopters. Among the significant internal factors range the firm's labour productivity and the IT skills of personnel and managers. An external factor that proved to be highly relevant was the presence of a young and skilled labour force in the local labour market. Not surprisingly, the existence of a broadband connection alone had no impact on a firm's performance or competitive position. Advanced communication services (such as VoIP) and organisational changes that accompany the introduction of broadband connections are prerequisites of a positive impact.

Statistics that show the ranking of countries with respect to their broadband performance are very popular, in particular at the European level where they are used to confirm the success of regulatory policies. However, a closer look at these statistics reveals a differentiated picture. Thus, simple comparisons of broadband connections per number of inhabitants seem to be rather misleading. Nejc Jakopin analyses a whole set of factors that indicate whether there are deficiencies in a country's broadband performance in his paper on the "Drivers and Inhibitors of Countries' Broadband Performance – A European Snapshot". Correlation analyses for the EU 27 countries show that rankings can turn out to be quite diverse according to which indictors are used. Many of the existing dependencies, such as a direct correlation between regulatory measures and broadband penetration, turn out to be too simplistic. Jakopin uses a time lag of 2 years between independent and dependent variables in his correlations. He shows that initially, economic prosperity and computer literacy are decisive factors for quick adoption. Countries with good general English language proficiency show a quicker broadband diffusion than others, and a high level of teleworking among employees seems to be a ‘natural’ driver of broadband adoption. As expected, high levels of service sector employment correlate positively with broadband diffusion while unemployment and the resulting decline in consumers’ purchasing power correlate negatively. Existing analyses are confirmed for other variables: privatisation of the telecommunications sector, an independent regulator and early local loop unbundling influence broadband take-up positively, while market concentration has a negative influence. Contrary to common belief, however, inter-technology competition (for example, between cable and DSL access modes) does not show significant effects.

Japan’s telecommunication markets are always an interesting object of study. Whether it be the forerunner role that Japan has played with respect to many technologies, different approaches in regulation and industrial policies, or a culturally distinct reaction of consumers, an analysis of Japan provides surprising results. Thus, in presenting some important new insights, Koshiro Ota’s study of Japan’s broadband policies (“The Telecom Policy for Broadband Diffusion: A Case Study in Japan”) does not disappoint the reader. The distinctive features of the Japanese broadband market are very high capacities that are available at relatively low prices. At an earlier date than in other countries, fibre-to-the-home technology has overtaken DSL as the dominant access technology. However, how is this specific success story related to telecommunication policies? Ota finds that it is not only local loop unbundling as such which drives broadband markets, but the specific rate-setting methodology adopted by the regulator. He supports the view that a specific relationship between NTT, the incumbent operator and the regulator has created a smooth implementation of regulatory measures based on an attitude of consensus. This is due to the fact that the regulator considers NTT’s business situation in its own decision-making. Obviously, this is not seen as a policy which might harm competition and competitors in the Japanese market.

Mobile Drivers

Chapter 4 contains five papers on regulation and competition in mobile telecommunications markets, starting with a paper by Jörn Kruse on mobile termination carrier selection (MTCS). As he argues, the introduction of MTCS would turn mobile termination into a competitive market, thereby overcoming the current problems of regulating termination charges. After minor software changes to the GSM-standard, network operators would be able to deliver the termination service to any mobile device for all calls to mobile customers so that mobile termination services would be supplied competitively. The demand-side decisions involved in selecting the terminating networks would be made by either the calling parties themselves (through MTCS at the retail level, call-by-call or pre-selection) or by the originating networks (through MTCS at the wholesale level). The paper concludes that several arguments, including transaction cost economies, suggest that MTCS at the wholesale level would be most likely to prevail in the market so that efficient prices could be expected as a result of competition. Introducing MTCS would therefore dispense with any need for mobile termination rate regulation.

Jeff Rohlfs also addresses the regulation of mobile termination rates in his paper on "Countervailing Buyer Power and Mobile Termination". The paper discusses the analytical issues required to support public policies regarding mobile termination, including specification of the relevant product market, determination of market power, assessment of countervailing buyer power (CBP), regulatory intervention and regulatory restraint. The focus is largely on cost-based regulation of termination charges in the European Union, and recent challenges to this approach, which are based on the idea that buyers of mobile termination may have significant CBP. While in theory CBP could be exercised by refusing to interconnect, raising termination rates on the buyer's network, raising retail prices and/or withholding payment for mobile termination, the exercise of CBP is, in practice, restricted by regulatory mandates to interconnect, non-discrimination rules and enforcement of the obligation to make payments for mobile termination. As Rohlfs argues, CBP cannot therefore be used to ameliorate the terminating access monopoly, as buyers are not permitted to exercise their CBP. His analysis also shows that unrestrained exercise of CBP would likely harm consumers. More generally, the bilateral monopoly between buyers and sellers of mobile termination would, if unregulated, lead to perverse results for consumers. The paper concludes that existing policies of regulating mobile termination rates and restraining the exercise of CBP are beneficial. The alternative of deregulating mobile termination rates and allowing full scope for exercise of CBP would be much worse for consumers.

The third paper, "National Roaming Pricing in Mobile Networks", authored by Jonathan Sandbach, also deals with the (potential) regulation of mobile telecommunications. The paper develops a practical model of optimal and competitive neutral national roaming access prices. This method takes account of the geographical cost structure of the networks, and thus allows for the 'cream-skimming' effect whereby a new entrant will concentrate its own network build in low-cost (higher traffic density) urban areas, especially when it uses a technology that has a cost advantage in these areas. According to this model, both networks will invest in more geographic coverage when the national roaming access price is set higher. The incumbent will do so because of the extra revenue it will get from roaming charges, and the new entrant will do so in order to avoid paying roaming charges to the incumbent. The paper provides an illustration of how the method could be applied to a situation where the host incumbent is restricted to GSM 900 while a new entrant is deploying W-CDMA. The paper concludes that a competitively neutral national roaming access price will be about 38% above the average cost on the host incumbent's network, although this result will depend on the specific geographic distribution of traffic in the country concerned. An access price set at this level will ensure competitive neutrality between networks, and provide an efficient investment signal for the new-entrant network.

In the fourth paper of this chapter, Peter Curwen and Jason Whalley ask whether (more) competition can be introduced via the issue of new licences with specific reference to the experience of 3G licensing in Europe. In particular, the paper identifies those new entrants that emerged from the third-generation (3G) licensing process across Europe before examining how they have fared. Four operators seeking to become new entrants across a range of countries are identified, although only one of them – Hutchison Whampoa – is considered to be fully active as a 3G new entrant. It is argued that the success of Hutchison Whampoa to date has been patchy, and that a supportive parent company has significantly contributed to its ability to remain active.

The concluding paper in this chapter, authored by Delphine Riccardi, Stéphane Ciriani and Bertrand Quélin, provides an econometric analysis of market entry by mobile virtual network operators (MVNOs) in ten European countries between 1998 and 2005. The focus is on the question how differences in market structure and regulation affect the number of MVNOs that enter a market. As the authors show, the incumbents' strategy towards hosting MVNOs and the friendliness of the regulatory framework towards entry are key in explaining differences between countries. Of particular interest may be that the regulator's independence significantly contributes to explaining differences. With independent regulators MVNOs are more likely to enter, possibly due to the greater credibility stemming from better commitment possibilities.

Business Strategy

In an environment characterised by a technology with undetermined, but supposedly huge economic potential, high intensity of competition and often unpredictable reactions of consumers and regulators alike, business strategy is faced with unprecedented challenges. The success of individual firms, but also of entire markets, depends on the development of viable business models. Four papers deal with quite diverse aspects of business strategy in this volume. In their paper "Exploring Technology Design Issues for Mobile Web Services", Mark de Reuver, Harry Bouwman and Guadalupe Flores Hernández look at technological options for providing mobile internet services. Based on existing work on business models for mobile data services, they examine various options for the conception of mobile web services which differ with respect to industry structure and the division of labour along the value chain. It turns out that favouring one solution over another, that is, choosing an IP Multimedia Subsystem (IMS) model rather than a mobile web services model (MWS), not only has implications for the viability of business strategies, but it also affects the competitive situation in the market since the IMS model tends to support existing industry structures with a strong position of network operators.

In many cities, wireless broadband networks are being established as a quick solution for the realisation of comprehensive broadband coverage. The paper by Pieter Ballon, Leo van Audenhove, Martjin Poel and Tomas Staelens, "Business Models for Wireless City Networks in the EU and the US", analyses this phenomenon. Features of public goods and of infrastructure elements in these wireless networks have resulted in a strong public involvement in infrastructure building and in public–private-partnership approaches. Apart from the sometimes critical involvement of public money in a sector that once has been declared private and had to be privatised where it still was under public control, there are different ways of organising the partnerships. Ballon, van Audenhove, Poel and Staelens have analysed 15 cases of cities in the EU and in the United States that have established wireless networks with a public participation of varying intensity. The paper adopts the perspective of the public policy maker and attempts to identify an 'optimal mix' of public and private involvement in wireless city projects. Six 'typical' business models are identified that differ with respect to the conceptualisation of public–private partnerships. The comparison between EU and US cities adds a component that permits the analysis of the impact of differences in regulation and competition policies.

Pat Longstaff examines the reaction of firms to increasingly unpredictable environments in her paper "Managing Communications Firms in the New Unpredictable Environments: Watch the Movies!". Her thesis is that firms have to develop particular features of resilience that allow them to adapt to quickly changing conditions. Using the movie industry as an example, it can be shown how resilience underpins survival in an industry in which demand for the launched products is extremely unpredictable. Dramatic failures have to be faced by the industry, as it can never be said in advance whether a new film will be a success, not even if a predecessor in the same series was a 'mega-hit'. Management theory has created tools that allow the analysis of strategies that help complex adaptive systems to survive. As the scenario of unpredictable demand for services that require large upfront investments is typical not only of the movie industry, but also for many communication-based industries such as the mobile broadband industry, important lessons can be drawn from the analysis.

Mergers and Acquisitions (M&A) in the telecommunications sector provide a never-ending source of striking news about the sector. M&A activity is considered to be a key element of company strategy which merits the closest attention because of its implications for the destiny of the enterprise and because of its impact on competition in the markets concerned and on the internationalisation of markets. Olaf Rieck and Canh Thang Doan examine M&A from a shareholder's perspective in their paper "Shareholder Wealth Effects of Mergers and Acquisitions in the Telecommunications Industry". They study successful mergers, that is, those that have a positive shareholder value effect. In particular, the authors try to find out whether shareholder value effects of M&A depend on the company's motivation for engaging in the M&A activity. It might be argued that M&A activity driven by a service diversification strategy generates different effects from those driven by an internationalisation strategy. Indeed, according to the analysis presented, serviced diversification strategies generate more stable and more reliable returns than other strategies. M&A activity motivated by international expansion strategies shows mixed results: competition forces companies to diversify their geographic reach, but it is not guaranteed that this is accompanied by higher shareholder values. However, not doing it might harm international competitiveness.

Emerging Markets

New markets emerge from progress in communication technology, from new business models, but also from the opening up of geographical markets that were not part of an international market economy until recently. This chapter deals with all three phenomena.

Emerging markets imply a high degree of uncertainty about demand. Exciting though the technological opportunities arising from network investment might be, demand may be a limiting factor in generating the necessary revenue to justify this investment. If a large number of households will not be able to afford NGN services, a new digital divide lies around the corner. In "Next Generation Networks: The Demand Side Issues", Rappoport, Alleman and Madden analyse the patterns of expenditure across countries in order to determine the shares of income dedicated to communication services. This analysis has important policy implications. If an infrastructure project is intended to be successful, public subsidies may be needed to avoid a significant gap in demand which threatens business models of investors, as a critical mass of users may not be reached. Alternatively, increased regulation could impose NGN services as universal service obligations on network operators, thus creating affordable service prices. The analysis provides important background material for policy-making. This material reveals the expectation that the share of income spent on communication services in advanced countries also forms an upper limit for less-developed countries. Hence, if the resulting market volumes are not sufficient to justify the deployment of new infrastructure, economic policy measures might be necessary to avoid market failure.

Next generation networks (NGNs) will result in the emergence of new markets for very-high-speed transmission capacity, but also for a whole range of services relying on 'ultra-broadband' networks. Much discussion has taken place about how to finance NGNs, on technical specifications (fibre against cable) and about the implications for competition. Much less attention has been directed to the question of how demand will react to what NGNs have to offer. In their paper "A Cross-Country Assessment of the Digital Divide", James Alleman and Paul Rappoport argue that the demand for communication services ultimately determines the demand for network infrastructure. A forecast of market trends will shed light on the likely demand for NGNs. As this is also a crucial variable for the assessment of future market structures, important information for regulators will result from the analysis. The results of the Alleman/Rappoport paper suggest that the high probability of monopolistic or oligopolistic markets will require decisive regulatory measures.

Johannes Bauer, Imsook Ha and Dan Saugstrup examine a concrete example of an emerging market, that for mobile TV, in their paper "Technical, Business and

Policy Challenges of Mobile Television". The experience of South Korea reveals the complexity of the underlying business model, since the realisation of a mobile TV service requires the coordination of many players in an integrated value chain. Uncertainties about demand and the specific configurations that customers are likely to prefer, suggest that the actual solutions for mobile TV services in terms of technical access modes, content providers and price models are best determined by the market. Any attempts to manage markets by industrial policy measures are likely to generate less than optimal results as policy-makers will not be able to deal with these uncertainties satisfactorily. However, policy measures are required if obstacles exist to the development of mobile TV, for example in the form of inadequate spectrum allocation, and need to be removed to get the market going.

Russia is undoubtedly one of the most dynamically developing ICT markets in the world. In addition to a small group of individuals at the upper end of the income scale that represent demand for very high quality ICT services, a new middle class is slowly developing. It can be one of the pillars for the generation of a market for ICT goods and services which reflects the geographical dimension and the economic potential of the Russian Federation. Margarita Strepetova and Svetlana Petukhova Russian Information and Communication Technology in a Global Context describe the technological and economic factors which characterise the (admittedly slow) unfolding of this potential using the example of internet services. Technical standards, problems of qualification of engineers in ICT industries as well as the scarcity of qualified personnel in the adopting industries seem to be major problems. Economic structures which allow for the quick diffusion of knowledge and the adaptation of company managers to a new culture of openness and transparency are identified as major obstacles to market development.

New Perspectives on the Regulatory Framework

In his paper on "The Regulation Framework for European Telecommunication Markets: Between Subsidiarity and Centralisation", Justus Haucap discusses the pros and cons of centralising regulatory competences in the European telecommunications sector. He claims that the European Commission itself has not properly weighed the costs and benefits of further centralisation despite its many comments on the matter, and hence he has set out to develop clear economic criteria to underpin the allocation of executive competencies between the Commission and Member States. As he sees it, the benefits of further centralization such as the offsetting of negative externalities associated with national regulation and the risks of harmful regulatory competition need to be evaluated in relation to the downsides such as an increase in information costs and a failure to take account of significant differences in infrastructure development, taking into account the particularities of telecommunications markets. Based on his analysis, Haucap concludes that further centralization is not warranted in the EU at this point in time. Instead the risk of overregulation should be delimited through institutional safeguards, which make it easier for regulators to deregulate than to regulate a market. Therefore, the Commission's veto rights should be limited, in his view, to (a) markets, in which regulation creates significant cross-border externalities, and (b) cases where national regulators do not move into the direction of deregulation. If, however, a national regulator decides to deregulate a market, the Commission's ex ante veto right should be abandoned (but ex post intervention by the Commission still be possible) in order not to hamper deregulation which tends to be the desired result, in principle. In addition, the paper argues that the "insufficiency of competition law" criterion of the three-criteria-test needs to be taken more seriously than has been so far by national regulators.

The following paper by Maartje de Visser entitled "Surveying Regulatory Regimes for EC Communications Law", also addresses issues raised in the course of the Commission's 2006 Review and, like Haucap, she prefers evolution to revolution. The core of her argument is that the basic characteristics of the existing institutional model such as the need to achieve the Internal Market are sound and hence instead of institutionalization it would be preferable to move to a constitutionalization of the model through attention to such matters as effective judicial protection, accountability and legitimacy. The chapter begins with an evaluation of the core enforcement paradigm used by the Commission. Three models are examined: centralized enforcement, decentralized enforcement (via Member State regulators) and agency. An inherent trade-off between uniformity and subsidiarity is found to exist. A variety of policy options such as increased powers for the Commission or ERG are then assessed in the light of the strengths and weaknesses of the three models. It is concluded that the introduction of new institutions is likely to create more problems than it will solve, with a strengthening of the ERG proving to be the best choice among the available options because it provides consistency while maintaining the benefits of decentralized enforcement. However, if that route is followed then the downsides need to be addressed via amendments to the constitution of the EU in order to ensure optimal clarity and legal certainty for market operators, national authorities and the ERG itself.

The third paper authored by Laurent Gille and colleagues is entitled "Innovation and Regulation in the Digital Age: A Call for New Perspectives". It begins by making the point that innovative IT applications and networks involve new economic systems such as new markets, segmentation of the public, products, timing, transactions and pricing. In the process, frontiers between goods, services and segments of the market are increasingly blurred. As a result, there arise a new division of labour and the appearance of new actors in the value-added chain, and it follows that Internet innovations subvert systems of regulation based upon the creation of borders between stabilized institutional and physical spaces. However, by its very nature innovatory activity means that there is too much instability for any particular business model to be unanimously accepted. In particular, because the Internet creates huge difficulties in assigning market values to all of its activities, the majority scholarly opinion holds that cross-border activities should be exempted from government regulation. In practice, there are various potential methods of regulation including self-regulation by the parties to a given transaction which avoids the necessity for third-party intervention. Although there is a temptation simply to extend existing regulatory systems to innovative activities, the authors argue that this is unsatisfactory and that new approaches need to be introduced.

The editors of this book would like to thank first of all the authors of the 27 papers for their excellent contributions, their commitment and their patience. Special thanks go to Katharina Wetzel-Vandai, the Economics Editor at Physica-Springer who accepted this book for publication in the sure knowledge that it would be a difficult task to convert the draft material into a finished product. Her support and encouragement is very much appreciated. Saskia Wagner has helped with formatting and editing the chapters. She stepped in just when we needed her most, and she has done a great job. Ulrich Heimeshoff and Jürgen Kühling provided very helpful comments on the econometrics and the legal analysis in some of the papers. Last but not least, we are very grateful to the Alcatel-Lucent Stiftung for financial support, and in particular to Dieter Klumpp for enhancing and promoting this project in every possible way.

It is with deep regret that we bring to the attention of our readers the untimely death in a skiing accident of Boris Rotenberg, a contributor to this volume. Boris was a well-known and much respected contributor to the field of telecommunications, and the editors trust that his contribution in this volume will provide a worthy memorial to his talents.

Part I
Theoretical Perspectives

General Access Payment Mechanisms*

Izak Atiyas, Toker Doganoglu, and Martin Reichhuber

Abstract Despite the voluminous literature documenting their problems, per unit access pricing mechanisms are the most common ones used in practice. Interestingly, neither legal documents nor theoretical work on access payments provide any justifications for restricting access payments to per-unit charges. In this paper, we examine the properties of general one-way access payment mechanisms where payments from the entrants to the incumbent are expressed as functions of retail prices. We find that by imposing a linear access pricing mechanism the regulator can implement any pair of retail prices, including the first best. We also show that a per-unit access mechanism, including one which is cost-based, is incapable of implementing the first-best outcome. Moreover, we obtain a partial welfare ordering of payment mechanisms in that any linear access payment mechanism that depends negatively on the incumbent's price and positively on the entrant's price generates desirable outcomes with higher consumer welfare than payment mechanisms where parameters have the opposite signs.

Introduction

The last decade has seen a large wave of deregulation in telecommunications markets. Liberalization is also taking place in other network industries which were originally organized as monopolies, such as railway, electric power and natural gas. These industries share a common characteristic. An incumbent, which often is a formerly state owned company, is the sole proprietary of a network. When the industry is opened to competition, entrants buy access to the network infrastructure of the

*We would like to thank Justus Haucap, Brigitte Preissl and participants of the 18th Regional European ITS Conference for useful comments. All remaining errors are ours.

I. Atiyas (✉), T. Doganoglu, and M. Reichhuber
Sabanci University, Istanbul
e-mail: izak@sabanciuniv.edu

B. Preissl et al. (eds.), *Telecommunication Markets*, Contributions to Economics,
DOI: 10.1007/978-3-7908-2082-9_2,

incumbent and are therefore enabled to provide the same or a somewhat differentiated product in the market. Hence it is believed that competition in the market will increase, which should lower prices, and in turn, increase consumer surplus.

An important problem that regulators, incumbents and entrants are faced with is how the access of the entrants to the incumbent's network should be organized. Regulating access is a critical policy instrument that regulators try to use to ensure that the industry develops in both a sustainable and competitive manner. In most countries, regulation of access takes two main forms. In some cases the regulatory authority sets access charges directly. In others, the parties are free to negotiate access agreements. If the negotiation is successful, the agreement may need to be approved by the regulator. If negotiations are not successful, the regulator engages in dispute resolution and may end up imposing the terms of access. In most cases, regulator determined access charges are set on the basis of some measure of costs. With the introduction of some competition in retail sectors, regulators are also increasingly reluctant to regulate retail prices directly, with the hope that regulating access will be sufficient to generate socially desirable market outcomes.

There is now a voluminous literature on how access prices should be regulated (see, excellent surveys in Armstrong [2002] and Laffont and Tirole [2000] and Vogelsang [2003]).[1] A common feature of this literature is that access prices are treated on a per-unit basis. In a typical model of (one-way) access, the network (the essential facility) is owned by an incumbent. The incumbent sells access to the network to new entrants, which also compete with the incumbent in the retail market. In the basic setup of these models, profit functions of the incumbent and the entrants can be separated into two different parts. The first part is retail profit from providing the end service to consumers. The second part is the access payment, or the revenue/cost of providing and buying access to the incumbent's network. Inevitably, such an addition to the objectives of the firms crucially alter their retail pricing behavior.

In most models the access payment is simply a constant per-unit charge times the amount of access purchased. Given the wide-spread use of per-unit mechanisms in practice, this restricted focus may be justified. Most of the literature is concerned with the properties of this per-unit price and how it should be set in order to achieve certain social objectives. One of the general results is that setting the per-unit access charge to the marginal cost of access is optimal only if there is no need to undertake second-best corrections on the incumbent's retail prices. However simple cost-based pricing is not optimal when incumbent's retail prices are not cost based.[2] More generally, the access charge is often forced to correct too many imperfections. For example, when retail prices deviate from costs, optimal access prices can be above or below the cost of providing access, depending on the severity of imperfections. Under such conditions, as suggested in Armstrong (2002), supplementing the

[1]The access problem was initially analyzed by Rohlfs (1979) and Willig (1979). The seminal contributions include Laffont and Tirole (1994), Baumol and Sidak (1994) and Armstrong et al. (1996).

[2]Unless otherwise noted, the term "cost-based pricing" will refer to setting prices equal to marginal or incremental costs.

access charge with other policy instruments such as an output tax imposed on entrants improves total welfare.

Despite misgivings about the performance of per-unit access charges, it is interesting that the literature has not inquired about the welfare properties of other, more general mechanisms of access payments. Indeed, the literature does not provide any justification for restricting access payments to per-unit charges. In this paper, we examine the properties of one class of more general payment mechanisms, namely of payment mechanisms that are linear functions of retail prices. We argue that such a simple mechanism is capable of generating a wide variety of outcomes. The motivation behind this inquiry is simple: with per unit access charges, access payments are simply equal to the access charge times the quantity of access, and the quantity of access purchased is simply the volume of retail sales of the entrant, which, in a simple differentiated products framework, depends on the retail prices of the incumbent and the entrants. Hence, in most models, access payments are actually functions of the prices of the retail services provided by the incumbent and the entrants, but in a very restricted form. Our purpose is to examine the properties of access payment functions when they are expressed as more general functions of retail prices.

More specifically, we ask the following question: If access payment mechanisms are regulated but retail prices are not, what sort of outcomes would the regulator be able to implement? In a stylized model of retail competition between an incumbent and an entrant, we show that a simple access payment mechanism linear in retail prices is capable of implementing any pairs of retail prices, including the first best. By contrast, a per-unit access mechanism, including one which is cost-based, is incapable of implementing the first-best outcome.[3] We also show that we can obtain a partial welfare ranking of outcomes by simply focusing on the signs of the parameters of the linear access payment mechanism. Specifically, any linear access payment mechanism that depends negatively on the incumbent's price and positively on the entrant's price creates desirable outcomes with lower retail prices, and consequently higher consumer welfare, than payment mechanisms where parameters have the opposite signs. Finally, we refer to Doganoglu and Reichhuber (2007) who show that a desirable outcome can be achieved in a decentralized manner, in a procedure where the parameters of the mechanism are actually chosen by the operators themselves, requiring no cost or demand information on the part of the regulator. In this case, the linear access payment mechanism presents significant informational advantage over traditional ones.

The paper is organized as follows: In section "A Brief Review of Policy and Theory" we provide a brief review of the policy and theory of one-way access pricing. Section "The Linear Per-Unit Access Price Mechanism" provides a short historical

---

[3] Of course, per-unit access prices are also examples of linear access payment mechanisms, but linear in quantities rather than prices. In order to avoid confusion, we will use the term "per-unit access prices" to refer to payments which are expressed as charges per each unit of quantity (such as $a.q$ where a is the per-unit charge and q is quantity). As will be clear in section "A Stylized Model", we will use the term "linear access payment mechanisms" to refer to mechanisms which are expressed as linear functions of *retail prices*.

overview of the development of per-unit access prices and how cost-based access pricing came to dominate access policy. Section "Alternative Access Pricing Mechanisms" presents examples of alternative mechanisms that have been used in practice, such as capacity-based pricing. Section "Academic Literature on Interconnection Pricing" summarizes the theoretical literature that, inter alia, has underlined the shortcomings of cost-based per-unit access pricing. Section "A Stylized Model" introduces more general access pricing mechanisms and derives some of its properties. We conclude in section "Conclusions".

A Brief Review of Policy and Theory

The Linear Per-Unit Access Price Mechanism

Although interconnection of communication networks, and compensation mechanisms between different network operators has emerged as a topic of policy discussions after the Bell patent expired in 1894, it has not been in the forefront of debates until the early 1970s. This is due to the fact that the successful consolidation efforts of AT & T turned the US industry to one that is served by a regulated monopolist. On the other hand, telecommunication services were traditionally provided by state owned monopolies in the rest of the world. Such an industry structure was justifiable due to large fixed costs involved in providing telephone service making the industry a natural monopoly. Hence, for many years, interconnection of networks of different operators did not arise as a challenge to policy makers.

In the era of monopolistic national telecommunications operators, the only form of interconnection that was required was to complete international telephone calls. The operators serving each end of a telephone call were compensated via a bilateral revenue sharing arrangement,[4] the so called accounting rate system.[5] Essentially, this revenue sharing arrangement can be reinterpreted to be a linear per unit access charge for termination. With this view, when the traffic between two countries is balanced, there would be no payment required. On the other hand, when there is large imbalance in traffic, the country with lower originating minutes stands to earn a sizable amount.

FCC (1997) reports that the accounting rates used in practice were five to ten times the true cost of a call. There are a number of reasons why this was the case and these reasons are explained in great detail in Wright (1999) and Kahai et al. (2006).

---

[4]In its most basic form, two countries, A and B, negotiate an accounting rate, x, for per minute of a telephone call. Then for calls originating in country A and terminated in country B, country A pays country B a fraction α of the accounting rate. For calls in the reverse direction, country B pays country A a fraction $1-\alpha$ of the accounting rate. Most often, this sharing of the accounting rate was achieved by using a fraction of 0.5, that is accounting rate was shared equally among carriers.

[5]For a more thorough discussion of accounting rates, their determinants and their economic implications see Mason (1998), Wright(1999), Wallsten (2001) and Kahai et al. (2006).

A basic insight provided by Wright (1999) is that when there are income differences between countries, the high income country generates a larger number of calls directed to the low income country, and hence has more to loose in case of a disagreement. Therefore, at the end of, even cooperative, bargaining the negotiated accounting rates turn out to be above marginal cost of completing an international call. The important implication of such large markups over cost and technological progress is the emergence new firms which try to exploit these inefficiencies in the accounting rate system, for example by offering call-back or refiling services. In recent years, largely due to the unilateral efforts of FCC,[6] the accounting rate system has come under pressure. Moreover, increasing competitiveness of national markets are also driving international settlement rates to lower levels. Interestingly, even though national operators were able to search for a mechanism in a cooperative manner, they chose to use a linear per unit mechanism for compensating each other for exchanging international traffic.

Starting in the late 1960s, technological advances made it possible to provide long distance services through alternative means. Hence, interconnection and its pricing between providers of such services and the AT & T started to become an issue in the US. The main battle in this early phase was not on how to price interconnection however, but whether the incumbent would/should allow such interconnection to its the network in the first place (Vogelsang and Mitchell 1997). Nevertheless, in the late 1970s following the influential Execunet decisions of US Appeals Courts, interconnection of carriers such as MCI was negotiated to be priced at a discount from the charges for local interconnection assigned to AT & T's long distance service by the separations process. The discount was set rather large early on to facilitate new entry, and was scheduled to decrease as the revenues of new entrants increased (Temin 2003).

The emergence of heated debates on interconnection of networks and its pricing coincides with the break-up of AT & T in the USA and entry of Mercury in the U.K. These changes were largely fueled by the emergence of new technologies such as high capacity fiber optic cables, microwave and satellite communications that can be used to transmit signals over long distances relatively cheaply, rendering the provision of long distance telephone services no longer a natural monopoly. In the U.S. the organization of the industry was radically altered by vertically separating one of the largest American companies, AT&T. Seven local operating companies were awarded licenses which allowed them to provide local and intrastate telephone services, while interstate calls had to be made via a long distance carrier.[7] In the long distance market, there were two firms: the incumbent AT&T, and the new entrant, MCI. In the U.K., Oftel, the British regulator, did not alter the industry structure in such a dramatic manner. The new entrant Mercury had to compete with

[6]Wallsten (2001) reports that American telecommunications operators have paid $50 billion to non-US carriers as settlements for international traffic. This large sum is clearly more than enough of an incentive for such unilateral action.

[7]Some intrastate traffic also qualified as long distance, hence these firms could provide intrastate service as well.

the incumbent BT in essentially all possible segments of telephone service provision, although they mainly targeted highly profitable large business customers.

Clearly, in both countries, an interconnection regime was necessary to guarantee end-to-end operational telephone service. In 1982 in the U.S., the FCC approved an access charge regime to go into effect in 1984 with the divestiture of AT&T (Noam 2001, pp. 40–42) According to this regime, a fixed per line "customer access line charge" was collected directly from users. In addition, long distance companies had to pay local access carriers interconnection charges on a *per minute basis*. In the U.K., the negotiations between BT and Mercury failed to result in an agreement. Hence, the regulator, OFTEL, determined charges in 1985 in order to streamline entry of Mercury. The imposed settlement mechanism involved a fixed payment by Mercury to BT to cover fixed costs associated with interconnection as well as a *per minute* charge for the use of BT's local network. These per minute charges varied with time of day and distance of delivery (Valletti 1999).

New Zealand, serving as an experimental ground for the rest of the world, deregulated the telecommunications industry in April 1989. Subsequently, the liberalization move propagated as a wave across the globe throughout the 1990s. Not surprisingly, interconnection between the networks of different providers and its pricing turned out to be one of the hotly debated issues. Lawmakers, with the experience of the past 25 years, have tried to explicitly deal with the issue of interconnection in the legislation that opened telecommunications markets to competition. For example, one of the most important articles in the 1996 Telecommunications Act in the US addresses interconnection between networks. It asserts that interconnection should be provided on a nondiscriminatory manner to everyone who wishes; access to networks should be at a just and fair price; the access charges should be negotiated between interacting firms and binding agreements should be signed. These agreements are subject to the approval of FCC and Public Utility Commissions.[8] Like most laws, the 1996 bill uses vague language and it is subject to interpretation.

Intriguingly, we were not able to find a statement that restricts possible interconnection pricing mechanisms to a linear per-unit one. Nevertheless, in practice this seems to be the pricing mechanism that is considered most often. Thus, much of the discussion both in the industry and academic literature focuses on how to set these per-unit prices. On this issue, the first legal battle took place in New Zealand. As Armstrong et al. (1996) reports, interconnection negotiations between Telecom and Clear in New Zealand proved to be a rather lengthy and complicated process only to be resolved by the intervention of the Privy Council in London, which upheld the use of the Efficient Component Pricing Rule (ECPR). Despite its ease, ECPR generated a lot of discussion since it is efficient only under very strict assumptions, and seems to favor the incumbent monopolies.[9]

[8] See Telecommunications Act of 1996.

[9] Economides and White (1995) present a critique of ECPR, while Armstrong et al. (1996) reinterpret and extend the ECPR in the light of Laffont and Tirole (1994) as a Ramsey pricing rule for interconnection. See the discussion below.

Both in the US and Europe, the main evolution of interconnection pricing policy has been towards more widespread use of cost-orientation. Although cost orientation does not necessarily imply that access charges should be specified as per-unit charges, this has been the common practice. In some cases firms are allowed to negotiate compensation for interconnection between themselves, however any agreement they reach is subject to regulatory approval. Regulatory authorities also announce (or require dominant operators to announce) reference interconnection offers which are per-minute charges based on long run incremental cost and would be imposed in case of disagreements. Another important feature of global trends in interconnection policy is the widespread requirement of unbundled network elements which provides potential entrants with a variety of different business models regardless of the extent of their own facilities. Needless to say, traffic sensitive network elements are often charged on a per-unit basis.

With the advent of mobile telephony and internet, the interconnection issue is bound to remain an important policy problem. The types of problems that would emerge are numerous. An informative example, which incidentally highlights the arbitrariness of a linear per-minute access pricing regime, is the interconnection between a wireline and an internet telephony (VoIP) operator. These two services use fundamentally different technologies in their core to transmit voice over a network. A VoIP operator breaks voice signals into a number of smaller packets, and sends them through various available routes on the Internet. The seamless connection between two VoIP users are established by means of computers aggregating these packets at the terminating end of the call.[10] Thus, a natural unit of service for a VoIP call is a packet. On the other hand, although becoming increasingly digital, a landline telephony network forms a live circuit between originating and terminating ends of a call for the duration of a call. Hence, a unit described in terms of time seems to be appropriate way to measure the provided service. Clearly, there is no apparent reason to use a linear per-unit access price for one operator to compensate the other in this particular case.

Alternative Access Pricing Mechanisms

In this subsection, we want to report on a few mechanisms for interconnection pricing that are not linear or per-minute. The earliest example we can find goes back to the early days of telephony. After Bell Patent expired in 1894, a number of independent carriers flourished in the USA. Gabel (2002) reports that, before state or federal regulation, interconnection between these operators and the Bell system were often based on a revenue sharing mechanism. Typically 15–25% of the originating revenue of a call would be paid to the terminating local exchange carrier.

[10] Mobile telephony works in a similar fashion in that it also sends digital packets. Although routing of these packets are controlled more centrally by the operator.

Furthermore, whenever traffic between operators were balanced, a bill-and-keep mechanism was also frequently used. One of the earliest examples of a dispute ending with regulation of interconnection prices took place in 1910s in Wisconsin, U.S.A. In this particular case, The Wisconsin Railroad Commission set the rate for interconnection at "*5 cents per local telephone call, 10 cents for calls between 50 and 100 miles, and 15 cents for calls over longer distances.*" (Noam 2001, p. 19) It is interesting to note that in this case the unit which the access payment is based on is a *call* and *not a minute of a call.*

According to a report by WIK and EAC prepared for the European Commission in 1994 (WIK-EAC), the most common one-way or two-way access regime adopted at the time was one of pricing on a per-unit basis, generally quantity being measured in terms of minutes or pulses, sometimes as part of a two-part tariff that also included a fixed charge. However, some countries developed their idiosyncratic mechanisms in dealing with the interconnection issue. In Greece, the interconnection agreement between two mobile operators was a bill and keep arrangement: "*Each of the two operators keeps what he gets*" (WIK EAC 1994, p. 171). Revenue sharing agreements was another possibility. For example, in the case of Italy, prior to 1994 there were four telecommunications companies, one of which (SIP) operated the local network and provided voice telephony services, another (IRITEL) which owned 60% of the trunk network and a third that provided international telephone services (Italcable). In this arrangement, the local company collected the revenues and distributed part of that revenue to Iritel, if the Iritel network was used, and to Italcable, if the call was an international call. Revenue sharing agreements were also used in Turkey, in the period 1994–1998, between newly entering mobile operators and the fixed line incumbent.

In the UK, the first determination of Oftel regarding interconnection was in 1985 between Mercury and BT, whereby Oftel required that mercury would pay for direct costs and a per-minute charge. Valletti (1999) also states: "In practice, access charges were felt by all commentators as being discounts on the incumbent's retail prices thus providing a signal to the entrant about the profitability of entry in different segments." Interestingly, even back then capacity based interconnection pricing was one of the alternatives being considered. Specifically, a document published by the Director General of Telecommunications in the U.K. stated a desire to investigate feasibility of such a payment mechanism as an alternative to the standard per-minute charges (WIK EAC 1994, p. 196). Mercury itself proposed capacity-based pricing of interconnection in early 1990s. In particular, Mercury proposed that BT's interconnection charges should consist of a fixed element, that would be paid upfront or through a recurrent rental arrangement, and a variable component that would depend on the number of call attempts, capturing thereby BT's call set-up charges (see OECD , pp. 99–100 for a summary).

This proposal of Mercury was not accepted but capacity based interconnection pricing continued to attract interest. Payments based on usage of capacity was supported by the idea that network costs are actually fixed and sunk in nature.

In a capacity payment regime, the entrant would rent capacity and payments would not depend on actual usage. There was some discussion of capacity type mechanisms in the WIK-EAC study. Although the main proposal of the study was based on average incremental costs (per-minute), it also suggested that charges for capacity could be applied in the spirit of peak-load pricing.

In Europe, the first country to introduce a complementary interconnection regime based on capacity based pricing in voice telephony was Spain (in 2001). According to the 12th Implementation Report of the European Commission (European Commission 2007) in 2006, about half of fixed access and termination interconnection in Spain was capacity-based. In 2006, Portugal and Poland also introduced capacity-based interconnection in voice telephony and have requested their respective incumbents to revise their reference interconnection offers accordingly. The European Commission states (2007, p. 26): "This interconnection model allows an operator to contract certain capacity for interconnection services from the dominant operator at a specific point of interconnection, paying a fixed cost, regardless of the traffic minutes actually routed. It also gives an incentive to alternative operators to increase traffic volumes, allowing the routing of a higher number of minutes at lower unit costs. The increased flexibility offered by capacity-based interconnections makes it easier for alternative operators to provide a varied range of retail offers, such as flat rates for voice or data, bundled offers, free or discounted calls."

Capacity-based pricing is more widespread in internet services, and many countries in Europe have introduced flat rate internet access call origination (FRAICO). Some countries (such as UK) have required internet operators to introduce flat rate internet access call origination (for a discussion, see OECD 2004).

Another approach that many regulators use primarily in wholesale broadband access markets is the "retail-minus" method which resembles the simple version of ECPR employed in practice. Under this method, the access charge of the wholesale product is linked directly to the retail price of the corresponding retail service and is determined so as to leave a margin of profitability to new entrants. The discount of the retail price is often set as a percentage of the retail price. As of September 2007, Austria, Germany, Ireland, Portugal and Spain used the "retail-minus" approach to price bitstream or DSL resale services (www.cullen-international.com). While the retail-minus approach is not common for access charges in voice telephony, OECD (2004, p. 114) reports that Australia prices access on a retail-minus basis for local calls. However, it is interesting to note that local calls are priced on a per-call rather than per-minute basis. The retail-minus method is a variant of per-unit access pricing with a specific link to retail prices.

Finally, it can be mentioned that two-part tariffs, in the form of a call setup charge and a per-minute charge, have also found some use in access pricing. According to data from Cullen International (www.cullen-international.com), as of May 2008 call termination and/or call origination charges of incumbent operators have entailed two-part tariffs in countries such as Belgium, Denmark, Finland, France, Portugal, Sweden and Czech Republic.

Academic Literature on Interconnection Pricing

The one way access problem focuses on the situation where an incumbent operator provides access to an entrant with whom it competes in the downstream retail market. Almost all theoretical discussions of the problem of how payments between the provider and user of access should be designed treat access charges on a per-unit basis, units measured most frequently in terms of minutes. As discussed above, the approach that has been most popular among regulators is to set access charges on the basis of some estimates of incremental costs of access, sometimes supplemented by a mark-up intended to capture a share of common costs. Whatever properties cost-based per-unit interconnection charges may have, optimality in terms of economic efficiency is not necessarily one of them. Economic efficiency often entails two components: allocative efficiency and cost efficiency. Allocative efficiency means that a good or a service should be produced as long as the social valuation of that good is above its marginal cost. Cost based per-unit interconnection pricing is optimal in the sense of allocative efficiency only when retail prices are cost based as well. This is most often not the case in the telecommunications industry. Final prices of the incumbent operator often have not been cost-based because of market power, or universal service obligations. Furthermore, in many instances new entrants may possess market power as well, especially if they are able to provide differentiated services. Under these circumstances, in accordance with the theory of the second best, it is optimal for access prices to deviate from costs.

The main determinant of productive efficiency is the cost structure of the industry. Costs critically depend on the entry decisions of potential entrants, their costs, and their decisions regarding bypass of the incumbents' infrastructure. Cost efficiency requires that overall costs are minimized in the industry. Hence achieving productive efficiency requires that new entry occurs only when the new entrant's cost of providing the retail service is lower than that of the incumbent and that bypass occurs only when the new entrant can generate the essential facility more cheaply than the incumbent. Strictly speaking, these considerations would require that access prices are set equal to marginal or incremental costs. However, given other distortions in the industry, especially when retail prices are not based on costs, then optimality requires that access prices deviate from costs as well. In fact, the presence of fixed costs of access is sufficient to create a tension between allocative and productive efficiency: If access charges are determined according to a "cost-plus-mark up", then the mark up will provide distorted signals for bypass.

The non-optimality of cost-based pricing of access can also be phrased in the following terms: One of the important principles of economic policy is that one needs as many instruments as targets to reach optimality. As emphasized by Laffont and Tirole (2000), Armstrong (2002) and Vogelsang (2003), access policy is often required to address more objectives than it can handle. In such cases, it is advocated that policy makers resort to additional instruments such as output taxes or subsidies on entrants. To the extent that such additional instruments are available, then access prices can be set closer to incremental costs. Often regulators are unwilling or are unauthorized to use such additional instruments.[11]

One alternative to the cost-based approach, the ECPR, has been used by a few regulators but has been subject of an extensive debate in the theoretical literature. The ECPR approach also specifies the access charge on a per-unit basis. The ECPR states that the access charge should be set equal to the marginal cost of providing access plus the "opportunity cost" of access. In its simple form, the second term captures the lost profits that the provider of access suffers when the entrant uses access to steal customers from the incumbent in the retail market. In this version, the access charge is simply set to marginal cost of access plus the retail margin of the incumbent (the so-called "margin rule"). The initial attractiveness of the ECPR was due to the simplicity of the margin rule. However, the literature has shown that the simple ECPR rule is optimal only under very special circumstances (namely when the downstream services of the incumbent and the new entrant are perfect substitutes, the entrant does not have bypass opportunities, retail prices are at desired levels [say, set by regulators] and downstream industry produces at constant returns to scale). These conditions may be met, for example, when the entrant simply re-sells the incumbent's retail products. When these conditions do not hold, Armstrong et al. (1996) have shown that the term reflecting the opportunity cost of access becomes a much more complicated term involving cross elasticities of final demand and technical substitution possibilities as well as the nature of competition downstream.

The generalization provided by Armstrong et al. (1996) is an instance of the Ramsey approach, which derives second best optimal access charges on a per-unit basis as well. The Ramsey approach was initially developed to examine optimal deviations of final goods prices from marginal costs when costs contain non-convexities (and therefore marginal cost pricing would result in losses). In the context of access pricing, the Ramsey approach implies that optimal access prices are determined jointly with retail prices as solutions to a constrained optimization problem, the constraint being the zero-profit condition of the incumbent. With imperfect competition downstream, the pricing of access needs to strike a balance between lowering downstream markups through lower access charges (allocative efficiency) and discouraging inefficient entry (cost efficiency). When the zero profit constraint is binding, the solution typically can be represented as the summation of an ECPR term (with opportunity costs correctly defined) and an additional term that reflects the budget constraint of the incumbent: Hence, with binding profit constraints the access price is higher than the ECPR level, thus allowing a decrease in the price of the retail service and thereby reducing the role of retail margin's contribution towards balancing the incumbent's budget. Again, with more instruments,

[11]One important exception is the area of universal service in the EU, where the current approach relies on the formation of a universal service fund.

higher levels of welfare may be reached. In that case, the Ramsey access price gets closer to incremental costs. The Ramsey approach to access pricing (or the "correct" version of the ECPR, for that matter) has not been implemented in practice. Vogelsang (2003) suggests that this may be due to pressures from interest groups opposed to mark-ups as well as due to the fact that the resulting equations are complex and informationally demanding. Instead, as discussed above, cost-based pricing has been the norm.

Why cost-based pricing has been so popular is somewhat of an enigma. The most-often cited reason is that it is simple. However, we find this characterization misleading because measuring costs is not easy at all.[12]An even more intriguing aspect of the literature briefly surveyed above is its preoccupation with per-unit prices. Nowhere in the literature is it shown that the determination of access payments on a per-unit or even more specifically per-minute basis is optimal. Hence neither legal documents nor theory suggests that access prices should be determined on a per-unit basis, but this is what has been done both in theory and in practice. Given especially the fact that cost-based per-unit access prices have few desirable properties from a welfare point of view, it is interesting that the literature has not inquired about alternative and perhaps more general specifications of access payment mechanisms.

One exception could be the case of two-part access charges, that is, access charges that consist of a fixed fee and a per-unit price charged for each unit of access (Gautier 2006; Valletti 1998). In a model where the entry decision of the potential competitor is endogenous, the entrant's marginal cost is not known by the regulator, and public transfers are costly, Gautier shows that the choice between uniform (the term used in the article is "single") and two-part access charges depends on tradeoffs between financing network costs, productive (entry) efficiency and allocative efficiency. When the incumbent is relatively efficient, the regulator uses the uniform access tariff because a two-part tariff would bar entry. By contrast, when the probability of entry is high enough (i.e. when the incumbent is relatively less efficient), the regulator uses a two-part access charge because this results in smaller allocative inefficiency. In this paper we take the market structure as given and do not endogenize the entry decision.

Even though not a major focus of the present paper, one can also mention the academic literature on the impact of access pricing regimes on investment behavior (for example Gans and King 2004; Gans and Williams 1999). Gans (2006) provides an overview of this literature. One of the interesting findings is that while under unregulated access the timing of investment may deviate from the social optimum, optimal regulation of access charges may help the regulator to induce socially optimal investment timing.

[12]In many countries operators with significant market power are required to adopt the practices of accounting separation and cost accounting, themselves highly costly, with the purpose of generating the necessary cost information on which the determination of access charges can be based.

A Stylized Model

We have in mind an environment as depicted in Fig. 1.

An incumbent I and an entrant E compete in the retail market. Services produced by the two firms are imperfect substitutes with demand given by

$$q_i = q_i(p_i, p_j)\ \forall i \in \{I, E\}\ i \neq j.$$

and the demand derivatives satisfy

$$\frac{\partial q_i(p_i, p_j)}{\partial p_i} < 0 \quad \frac{\partial^2 q_i(p_i, p_j)}{\partial p_i^2} \geq 0 \quad \frac{\partial^2 q_i(p_i, p_j)}{\partial p_i \partial p_j} \geq 0 \quad \forall\ i \in \{I, E\}.$$

Both firms produce the retail product at marginal cost c_s. In the upstream market the incumbent I owns a physical network which is an essential input for the end service in the downstream market. Each unit of end service sold requires exactly one unit of input. Firm I can provide a unit of capacity to the entrant at marginal cost c_I. We assume that the cost of setting up a new fully fledged network is prohibitively high, so that the upstream market remains a monopoly. We also assume away any possibility of bypass.

We consider a rather general access payment mechanism, $A(p_I,p_E)$, which determines the payments E needs to make to I in order to become operational. Although we represent the mechanism in terms of retail prices, it clearly is able to describe all the mechanisms we have described above, as well as a host of others.

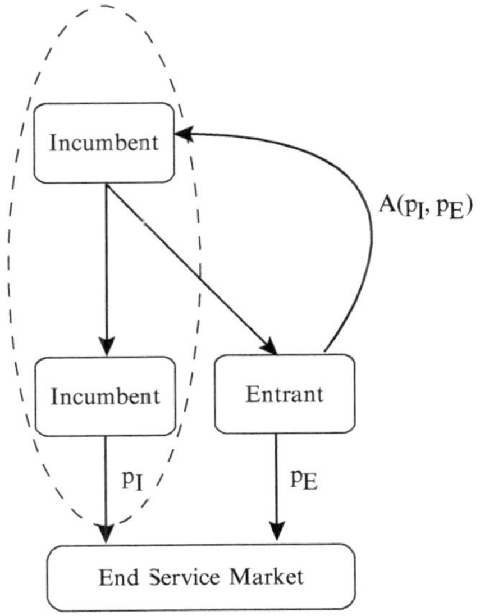

Fig. 1 Up and downstream market

For example, a mechanism based on a per unit access price a is a special case of the general access payment mechanism considered here:

$$A(p_i, p_E) = \alpha q_E(p_I, p_E).$$

Similarly, a fixed payment mechanism can be viewed as a special case as well where $A(p_I, p_E)=K$ and when $K=0$ this would be the bill and keep mechanism. Also, the traditionally used simple version of ECPR is given by $A(p_I, p_E) = (p_I - c_S)q_E(p_I, p_E)$.

Given the access payment mechanism, the profit functions of I and E are as follows:

$$\Pi^I = (p_I - c_T - c_S)q_1(p_I, p_E) - c_T q_E(p_I, p_E) + A(p_I, p_E) \quad (1)$$

$$\Pi^E = (p_E - c_S)q_E(p_I, p_E) - A(p_I, p_E) \quad (2)$$

Consistent with scenario imagined in most laws regulating interconnection between firms, we consider a model with two stages. In the first stage, firms or the regulator select a payment mechanism, and in the second stage firms compete in prices given the payment scheme. In most of what follows, we will consider the case where the access payment mechanism is set by the regulator in order to implement an outcome with a given pair of retail prices. In particular, we will examine the extent to which a regulator can implement socially desirable market outcomes in the retail market by designing the access mechanism appropriately. In stage 2, firms set their prices simultaneously given the access payment mechanism, $A(p_I,p_E)$.

Assuming for the moment that solution to the first order conditions describe a Nash equilibrium in the second stage pricing game,[13] the equilibrium will be characterized by the simultaneous solution of the system

$$(p_I - c_T - c_S)\frac{\partial q_I(p_I, p_E)}{\partial p_I} + q_I(p_I, p_E) - c_T\frac{\partial q_E(p_I, p_E)}{\partial p_I} + \frac{\partial A(p_I, p_E)}{\partial p_I} = 0 \quad (3)$$

$$(p_E - c_S)\frac{\partial q_E(p_I, p_E)}{\partial p_E} + q_E(p_I, p_E) - \frac{\partial A(p_I, p_E)}{\partial p_E} = 0 \quad (4)$$

It is immediate that by choosing an access payment mechanism with appropriate slopes in terms of retail prices, any desired price pair, (p_I^*, p_E^*), can be implemented in equilibrium. We would like to emphasize the fact that these slope restrictions are required to hold only at the desired vector of prices. That is, as long as $\frac{\partial A(p_I^*, p_E^*)}{\partial p_I}$ and $\frac{\partial A(p_I^*, p_E^*)}{\partial p_E}$ provide the *right* values, two firms noncooperatively selecting p_I and p_E will choose the desired prices in equilibrium.

[13] We will state specific conditions for this assumption to hold below for a special class of access payment mechanisms.

The easiest way to fulfill these slope conditions is to imagine a mechanism with constant slopes in both retail prices. With this in mind, we claim that any pair of retail prices can be implemented in equilibrium by the following mechanism:

$$A_L(p_I, p_E) = K + a_I p_I + a_E p_E. \tag{5}$$

Here, by choosing $a_I = \frac{\partial A(p_I^*, p_E^*)}{\partial p_I}$ and $a_E = \frac{\partial A(p_I^*, p_E^*)}{\partial p_E}$, firms can be induced to choose the retail prices p_I^* and p_E^*. The additional instrument K then can be used to distribute industry profits in order to fulfill financial viability commitments to the firms. The argument behind this claim is provided below.

Using the mechanism in (5), the first order conditions (3) and (4) reduce to:

$$(p_I - c_T - c_S)\frac{\partial q_I(p_I, p_E)}{\partial p_I} + q_I(p_I, p_E) - c_T \frac{\partial q_E(p_I, p_E)}{\partial p_I} + a_I = 0 \tag{6}$$

$$(p_E - c_S)\frac{\partial q_E(p_I, p_E)}{\partial p_E} + q_E(p_I, p_E) - a_E = 0 \tag{7}$$

It is cumbersome but straightforward to obtain the slopes of the best response function of each firm by using implicit function theorem. These are given by[14]

$$\frac{dp_I}{dp_E}\Big|_{R_I} = -\frac{(p_I - c_T - c_S)\frac{\partial^2 q_I}{\partial p_I \partial p_E} + \frac{\partial p_I}{\partial p_E} - c_T \frac{\partial^2 q_E}{\partial p_I \partial p_E}}{(p_I - c_T - c_S)\frac{\partial^2 q_I}{\partial p_I^2} + 2\frac{\partial q_I}{\partial p_I} - c_T \frac{\partial^2 q_E}{\partial p_I^2}} \tag{8}$$

$$\frac{dp_E}{dp_I}\Big|_{R_E} = -\frac{(p_E - c_S)\frac{\partial^2 q_E}{\partial p_I \partial p_E} + \frac{\partial q_E}{\partial p_I}}{(p_E - c_S)\frac{\partial^2 q_E}{\partial p_E^2} + 2\frac{\partial q_E}{\partial p_E}} \tag{9}$$

It is useful to note a few implications of (8) and (9). First of all, the slopes of the best functions are independent of the parameters of the linear access payment mechanism, a_I and a_E. These two parameters effect the level of the best response curves for both firms, but they have no effect on their slopes. Second, the denominators of these two expressions in (8) and (9) are the second derivatives of the profits of the incumbent and the entrant, respectively, with respect to their own choice variable. If these two expressions are negative,[15] a necessary condition for first order conditions to characterize

[14] We suppressed dependency of demand functions on retail prices, p_I and p_E, for ease of exposition.

[15] A sufficient condition for this to occur is that the demand functions are not too convex.

an equilibrium, then the sign of the slope of each firm's best response function depends on the expression in the numerator. Hence, provided that demand cross partials are positive as we have already assumed, and the cost of termination is sufficiently small, both best responses will have a positive slope, and increase in their rival's price. That is, if these conditions were satisfied, then the prices are strategic complements.

If, in addition, both best responses have a slope that is less than one in absolute value, then they cross only once implying a unique Nash equilibrium for all reasonable values of a_I and a_E.[16] In particular, it is useful to evaluate conditions (6) and (7) at the point $a_I = a_E = 0$. This will be called "bill-and-keep" point. Clearly, if there exists a unique equilibrium under the "bill-and-keep" regime – a relatively simpler condition to check, then there would be a unique equilibrium for all values of a_I and a_E.

All the conditions we mentioned in the last two paragraphs are satisfied for linear demands, for example. Hence, for any demand system not too far from linear, any price pair p_I and p_E can be implemented as the unique equilibrium outcome of a game where a regulator selects the parameters of an access payment mechanism, and given this, firms compete in prices. We would like to summarize this finding as our first result.

Result 1. By choosing the parameters of the access payment mechanism in Equation (5) appropriately, the regulator can implement any desired pair of retail prices as the unique equilibrium of the price competition between the incumbent and the entrant.

This result states that even when the regulator does not directly regulate retail prices, it can in fact regulate them indirectly, by regulating the access payment mechanism.

Special Cases

First-Best Retail Prices

Suppose that the regulator wishes to implement marginal cost pricing at the retail level. This may be desirable objective if there are no fixed costs. That is, we wish that the operators will choose cost based retail prices, i.e. $p_i^s = c_T + c_s$. This would be the case, if the following first order conditions are satisfied:

$$(p_I^S - c_S - c_T)\frac{\partial q_I(p_I^s, p_E^S)}{\partial q_I} = 0 \tag{10}$$

$$(p_E^S - c_S - c_T)\frac{\partial q_E(p_I^s, p_E^S)}{\partial p_E} = 0 \tag{11}$$

---

[16] A sufficient condition for is that demand functions have larger derivatives with respect to own price as compared with the derivatives with respect to rival's price – a reasonable assumption to make.

Comparing these with the first order conditions (6) and (7) it is clear that in order to induce operators to choose these prices, then the parameters of the access mechanism need to satisfy:

$$a_I = -q_I(c_T + c_S, c_T + c_S) + c_T \frac{\partial q_E(c_T + c_S, c_T + c_S)}{\partial p_I} \tag{12}$$

and

$$a_E = c_T \frac{\partial q_E(c_T + c_S, c_T + c_S)}{\partial p_E} + q_E(c_T + c_S, c_T + c_S) \tag{13}$$

In general the signs of the parameters are ambiguous. However when c_T is small then it can be concluded that a necessary conditions for implementing first best retail prices is $c_I > 0$ and $a_E < 0$. In other words, to induce marginal cost pricing, and assuming the cost of providing access is small, then the access mechanism is arranged so that both operators are penalized for increasing their retail prices. That makes sense: When the cost of access is small, then the main focus of the payment mechanism is to correct for distortions in allocative efficiency that arise due to market power. To summarize:

Result 2. When the cost of access is low, then inducing operators to choose first best retail prices requires $a_I < 0$ *and* $a_E > 0$.

Result 2 can be compared to the situation where the regulator sticks to a per-unit access price and leaves the retail prices unregulated, i.e. $A(p_I, p_E) = aq_E(p_I, p_E)$. Notice that with this regime we have $\frac{\partial A(p_I, p_E)}{\partial p_I} > 0$ and $\frac{\partial A(p_I, p_E)}{\partial p_I} < 0$, given our assumptions regarding the demand derivatives. If one wishes to implement the outcome induced by any linear per unit mechanism with a linear access payment mechanism, (p_I^D, p_E^D), the choices of a_I must be positive and a_E must be negative. In fact, a mechanism with $a_I = c_T \frac{\partial q_E(p_I^D, p_E^D)}{\partial p_I} > 0$, and $a_E = \frac{c_T \partial q_E(p_I^D, p_E^D)}{\partial p_E} < 0$ would implement the retail prices that would prevail under a cost based per-unit access pricing regime as an equilibrium outcome with our linear payment mechanism.

With linear per unit access prices, the first order conditions (3) and (4) become:

$$(p_I - c_T - c_S)\frac{\partial q_I(p_I, p_E)}{\partial p_I} + q_I(p_I, p_E) - c_T \frac{\partial q_E(p_I, p_E)}{\partial p_I} + \frac{a\partial q_E(p_I, p_E)}{\partial p_I} = 0$$

$$(p_E - c_S)\frac{\partial q_E(p_I, p_E)}{\partial p_E} + q_E(p_I, p_E) - \frac{a\partial q_E(p_I, p_E)}{\partial p_E} = 0$$

Comparing these with (10) and (11) we find that to implement the first best under the per-unit access mechanism, the following equations need to be satisfied:

$$a = \left[-q_I(c_T + c_S, c_T + c_S) + c_T \frac{\partial q_E(c_T + c_S, c_T + c_S)}{\partial p_I} \right] / \left[\frac{\partial q_E(p_I, p_E)}{\partial p_I} \right]$$

$$a = \left[c_T \frac{\partial p_E(c_T + c_S, c_T + c_S)}{\partial p_E} + q_E(c_T + c_S, c_T + c_S) \right] / \left[\frac{\partial p_E(p_I, p_E)}{\partial p_E} \right]$$

Unless the right hand sides of these equations are equal by chance, there is no single a that satisfies both equations. Hence we have a corollary:

Corollary 1. It is not possible to implement the first best through a constant per-unit access mechanism.

Integrated Monopoly or the Collusive Outcome

The first-best solution can be compared to the polar opposite case, that of an integrated monopoly, or the case when the operators choose the payment mechanism so as to maximize joint profits. In that case the payment mechanism can be seen as a commitment device that will ensure that the operators will collude in the retail market. Now the joint profits can be written as

$$\Pi^M = (p_I - c_T - c_S)q_I(p_I, p_E) + (p_E - c_S - c_T)q_E(p_I, p_E) \tag{14}$$

The prices that maximize joint profits, p_I^M and p_E^M, have to satisfy the first order conditions:

$$(p_I^M - c_T - c_S)\frac{\partial q_I(p_I^M, p_E^M)}{\partial p_I^M} + q_I(p_I^M, p_E^M) + (p_E^M - c_S - c_T)\frac{\partial q_E(p_I^M, p_E^M)}{\partial p_I^M} = 0 \tag{15}$$

$$(p_E^M - c_T - c_S)\frac{\partial q_E(p_I^M, p_E^M)}{\partial p_E^M} + q_E(p_I^M, p_E^M) + (p_I^M - c_S - c_T)\frac{\partial q_I(p_I^M, p_E^M)}{\partial p_E^M} = 0 \tag{16}$$

Comparing these with the first order conditions (6) and (7) it is clear that p_I^M and p_E^M can be implemented if the parameters of the payment functions are chosen according to:

$$a_I = (P_E^M - c_S)\frac{\partial q_E(P_I^M, P_E^M)}{\partial P_I} > 0$$

$$a_E = c_T \frac{\partial q_E(p_I^M, p_E^M)}{\partial p_E} - \left(p_I^M - c_S - c_T\right)\frac{\partial q_I(p_I^M, p_E^M)}{\partial p_E} < 0$$

Result 3. A payment mechanism that commits the operators to collude in the retail market has $a_I > 0$ *and* $a_E < 0$.

It is interesting to note the difference in the signs of the access payment mechanism parameters which implement a collusive outcome, and those which implement the first best. They are completely opposite. Furthermore, the signs of parameters required to implement any outcome that can be induced by a linear per-unit mechanism coincide with that of a mechanism which implements the integrated monopoly outcome.

A Further Characterisation

As we have seen in the previous two subsections, finding the parameters of the linear access payment mechanism for a given pair of retail prices is a matter of picking the right a_I and a_E so that the first order conditions for the both firms are satisfied at the desired pair of prices. There are a few more noteworthy points. A close observation of (6) and (7) suggests that holding $a_E(a_I)$fixed and changing $a_I(a_E)$ has no impact on the best response curve of the entrant(incumbent), while it just merely causes a shift in the best response curve of the incumbent(entrant). This is only a shift since as can be from Equations (8) and (9), the parameters of the linear access payment mechanism have no effect on the slopes of the best responses of both firms. These arguments suggest the following. Suppose the curve BR_I^0 in Fig. 2 represents the best response curve for the incumbent when a_I=0 for some value of a_E. Then decreasing $a_I < 0$ yields a new best response curve that is shifted left for the incumbent represented by BR_I^1, while an increase in $a_I > 0$ yields a right shift yielding the new best response curve to move to BR_I^2. Movements of the best response curve of the entrant can be tracked in a similar similar fashion for changes in a_E.

This simple effect of the access payment mechanism on best response curves in fact allows a further characterization. For example, the retail prices in Region *I* in the left panel in Fig. 3 can be implemented with corresponding (a_I,a_E)combinations in the right panel of the figure. That is, to obtain a retail price combination in Region *I*, it is needed that $a_I < 0$ and $a_E > 0$. For example the first best outcome is in this region. Region *I*, in fact, provides a set of desirable prices, since at each of the price combinations in this region, welfare is higher than the welfare under the bill and keep regime, and moreover, welfare in Region *II*. In Region *II*, we have $a_I > 0$ and $a_E < 0$ – the signs exactly as in the integrated monopolist outcome as well as the cost based per unit access pricing outcome.

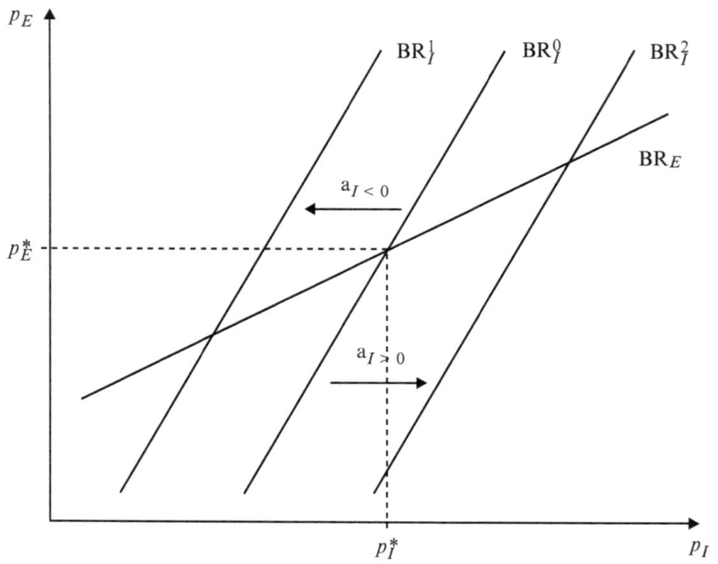

Fig. 2 Best response curves with changing a_I while holding a_E constant

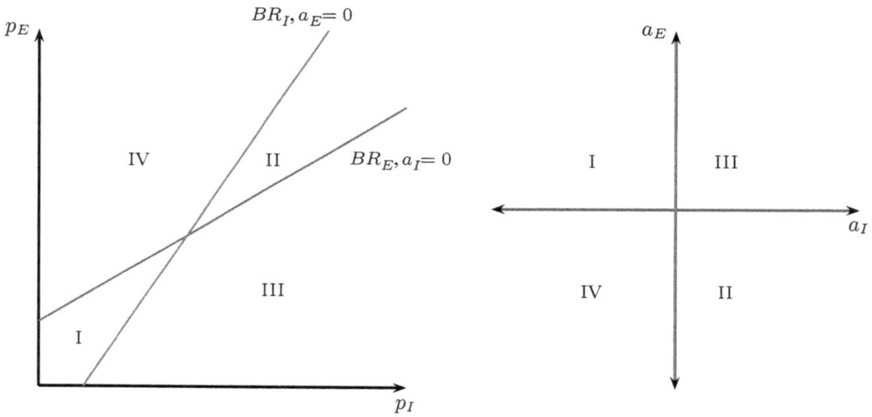

Fig. 3 Regions of retail prices, and regions of corresponding (a_I, a_E) values which can be used to implement them as equilibrium outcomes

Note that determining the correct a_I and a_E is not an easy task and informationally very demanding. The regulator would need to have perfect information on demand and costs. However, the insight developed above suggests that there may be very simple strategy for a completely uninformed regulator which would achieve a desirable outcome: This entails requiring the operators to negotiate the parameters of a linear access payment mechanism with the restriction that a_I is negative and a_E is positive and retail prices are non-negative. Under these restrictions the results in Fig. 3 would imply that operators would set both parameters equal to zero and essentially would have to agree on a transfer payment to ensure non-negative profits. Full collusion would result in a bill-and-keep type agreement with a fixed side payment and would yield an outcome that is more desirable than the one that would be obtained by a cost-based per-unit access payment mechanism. The next section summarizes results from another paper which shows that in fact the regulator can do even better.

A Decentralized Solution to the One-Way Access Pricing Problem

As can be seen in examples with access payment mechanisms implementing the first best or the integrated monopoly outcome, the values of a_I and a_E depend on demand and cost parameters in a non-trivial manner. In fact, the informational requirements are very similar to those required for selecting the Ramsey access charge. Naturally, then, a question arises: "If we need so much information to set a_I and a_E optimally, why worry about a new payment mechanism and why not just regulate the per-unit access price at its Ramsey level?"[17] Therefore, one needs to question whether there is any further advantage of an access payment mechanism like the one we studied above.

In a recent paper, Doganoglu and Reichhuber (2007) provide an affirmative answer to the question. In a stylized model, they present an access payment mechanism linear in prices, similar to the one described above, and show that a regulator can induce retail prices below those that would obtain under cost based per unit access prices. Moreover, in order to achieve this the regulator does not need information on costs and demands; only the firms are required to have perfect information about them. In their model, the regulator designs a three stage game, and informs the potential players with its structure. Most importantly, the players are informed about the fact that they will use a linear access payment mechanism whose parameters are going to be selected by the incumbent and the entrant firm. The slope parameters of the payment mechanism are actually chosen in a decentralized manner by the operators themselves.

In the first stage, a license to operate in the market as a rival to the incumbent is auctioned off. The winner of the auction, the entrant, then selects a_I, while the incumbent selects a_E. Notice that each firm selects the parameter that interacts with their rival's retail price. In the third stage, given a_I and a_E each firm selects their own retail price. In the subgame perfect equilibrium of this game, the entrant selects $a_I < 0$ and the incumbent selects $a_E > 0$. Following our arguments in the previous subsection, the retail prices these parameters yield in equilibrium are in Region *I* of the left panel of the Fig. 3. Clearly, these retail prices are more desirable from a welfare perspective to all price pairs in Region *II*. In particular, the outcome is more desirable when compared to retail prices that obtain under cost based per unit access prices (which are in Region *II*). Notice that in this equilibrium, both firms select access payment mechanism parameters such that they punish their rival's higher retail prices. This provides both firms with incentives to chose lower retail prices in the third stage. In this equilibrium, it turns out that the access

[17] One does not need to stop questions at this point. A more crucial question would be the following: "If all the necessary information is available, why do we not think of simply setting retail prices at their Ramsey levels?"

revenues of the incumbent does not cover the costs of termination it needs to incur by providing interconnection to the entrant. However, by transferring some or all of the auction revenues raised in the first stage to the incumbent, the regulator can keep the incumbent financially viable.

Conclusions

The pricing of access services has emerged as an important policy issue with the liberalization of telecommunications markets. With a few exceptions, both in theory and practice, access payments are treated as per-unit charges on the amount of the volume of access services purchased by new entrants. In most jurisdictions, regulated access prices are treated in a cost-oriented manner. Even though the theoretical literature has underlined a number of shortcomings of per-unit access charges, little intellectual effort has been spent on examining other, more general forms of access payments. Moreover, neither legal documents nor theoretical work on access payments provide any justifications for restricting access payments to per-unit charges. The purpose of this paper is to suggest that more general alternatives may prove useful. We introduce one very simple such alternative mechanism and examine some of its properties.

The suggested mechanism treats access payments as linear functions of retail prices of the incumbent and the entrant. In a stylized model, we examine what sort of outcomes a regulator can implement when it regulates access in this manner and allows complete freedom to operators to choose their retail prices. We find that by imposing a linear access pricing mechanism the regulator can implement any pair of retail prices, including the first best. We also show that a per-unit access mechanism, including one which is cost-based (i.e. where the access charge is set equal to the marginal cost of access), is incapable of implementing the first-best outcome. Moreover, we obtain a partial welfare ordering of payment mechanisms in that any linear access payment mechanism that depends negatively on the incumbent's price and positively on the entrant's price generates desirable outcomes with higher consumer welfare than payment mechanisms where parameters have the opposite signs. We also refer to Doganoglu and Reichhuber (2007) who show that similar desirable outcomes can be obtained through a decentralized procedure whereby the parameters of the linear payment mechanism are determined by the operators themselves.

References

Armstrong M (2002) The theory of access pricing and interconnection. In: Caves ME, Majumdar SK, Vogelsang I (eds), The Handbook of Telecommunication Economics. Elsevier, Boston, MA, Vol. 1, pp. 295–384

Armstrong M, Doyle C, Vickers J (1996) The access pricing problem: a synthesis. Journal of Industrial Economics, 44(1), 131–150

Baumol W, Sidak J (1994) The pricing of inputs sold to competitors. Yale Journal on Regulation, 11(1), 171–202

Doganoglu T, Reichhuber M (2007) An Interconnection Settlement Based on Retail Prices. Mimeo. Sabanci University/University of Munich

Economides N, White L (1995) Access and interconnection pricing: how efficient is the: "Efficient Component Pricing Rule"? Antitrust Bulletin, XL(3), 557–579

European Commission (2007) Commission Staff Working Document – Annex to the Communication from the Commission to the European Parliament, The Council, The European Economic and Social Committee and the Committee of the Regions European Electronic Communications Regulation And Markets 2006 (12th Report),' Volume 1 [COM(2007) 155]

FCC US Federal Communications Commission (1997) Report and Order in the Matter of International Settlement Rates. FCC 97-280, IB Docket No. 96-261

Gabel DJ (2002) A competitive market approach to interconnection payments in the US. In Mansell R, Samarajiva R, Mahan A (eds), Networking Knowledge for Information Societies: Institutions and Intervention. Delft University Press, The Netherlands, pp. 132–140

Gans JS (2006) Access pricing and infrastructure investment. In Haucap J, Dewenter R (eds), Access Pricing: Theory and Practice. Elsevier Science, Amsterdam, pp. 41–64

Gans JS, King SP (2004) Access holidays and the timing of infrastructure investment. Economic Record, 80(248), 89–100

Gans JS, Williams PL (1999) Access regulation and the timing of infrastructure investment. Economic Record, 79(229), 127–138

Gautier A (2006) Network financing with two-part and single tariffs. In Haucap J, Dewenter R (eds), Access Pricing: Theory and Practice. Elsevier, Amsterdam, pp. 65–90

Haucap J, Dewenter R (eds) (2006) Access Pricing: Theory and Practice, Elsevier, Amsterdam

Kahai SK, Kahai PS, Leigh A (2006) Traditional and non-traditional determinants of accounting rates in international telecommunications. International Advances in Economic Research, 12, 505–522

Laffont JJ, Tirole J (1994) Access pricing and competition. European Economic Review, 38(2), 1673–1710

Laffont JJ, Tirole J (2000) Competition in telecommunications. Cambridge, MA: MIT press

Mason R (1998) Internet telephony and the international accounting rate system. Telecommunications Policy, 22(11), 931–944

Noam E (2001) Interconnecting the network of networks, MIT Press, Cambridge, MA

OECD (2004) Access pricing in telecommunications. OECD, Paris

Rohlfs J (1979) Economically-efficient bell-system pricing. Bell Laboratories Discussion Paper No. 138

Temin P (2003) Continuing confusion: entry prices in telecommunications. In Guinnane T, Sundstrom VA, Whately W (eds.), History Matters. Stanford University Press, Stanford, pp. 163–186

Valletti TM (1998) Two-part access prices and imperfect competition. Information Economics and Policy, 10(3), 305–323

Valletti TM (1999) The practice of access pricing: telecommunications in the United Kingdom. Utilities Policy, 8, 83–98

Vogelsang I (2003) Price regulation of access to telecommunications networks. Journal of Economic Literature, XLI, 830–862

Vogelsang I, Mitchell BM (1997): Telecommunications Competition: the Last Ten Miles. AEI Press, Cambridge/London

Wallsten S (2001) Telecommunications investment and traffic in developing countries: the effects of international settlement rate reforms. Journal of Regulatory Economics, 20(3), 307–323

WIK EAC (1994) Network interconnection in the domain of ONP. (Study for DG XIII of the European Commission)

Willig R (1979) The theory of network access pricing. In Trebing H (ed), Issues in Public Utility Regulation. Michigan State University Press, East Lansing, MI, pp. 109–152

Wright J (1999) International telecommunications, settlement rates, and the FCC. Journal of Regulatory Economics, 15(3), 267–292

Competition and Cooperation in Internet Backbone Services*

Margit A. Vanberg

Abstract This paper analyzes the strong network externalities associated with Internet services from a competition policy perspective. In the market for Internet services, network effects are so important that an ISP needs to be able to offer universal connectivity in order to survive in this market. To reach universal connectivity, new entrants to the Internet interconnectivity market need to establish a direct or indirect transit agreement with at least one Tier-1 ISP. The fear that a single Tier-1 ISP could abuse a dominant market position in a transit agreement with lower level ISPs is not substantiated by the analysis. Competitive forces in the market for top-tier Internet interconnectivity are strong. A collusion between Tier-1 ISPs to collectively raise prices in the transit market is also not likely to be stable because the prerequisites for a stable collusion are not fulfilled in the market for top-tier Internet interconnectivity services. The analysis supports the view that competitive forces in the transit market are working and can effectively hinder Tier-1 ISPs from discriminating ISPs that are on lower levels of the Internet hierarchy.

Introduction

This paper discusses the effect of the strong network externalities that are associated with Internet service provision on competition in the market for Internet backbone services. In Internet services network effects are so important that an Internet service

M.A. Vanberg
Centre for European Economic Research (ZEW)
Research Group Information and Communication Technologies
e-mail: vanberg@zew.de

*This paper is based on Chapter 7 of Vanberg (2009).

B. Preissl et al. (eds.), *Telecommunication Markets*, Contributions to Economics,
DOI: 10.1007/978-3-7908-2082-9_3, © Springer Physica-Verlag HD 2009

provider (ISP) needs to be able to offer universal connectivity in order to survive in this market. To reach universal connectivity, new ISPs need to establish a direct or indirect transit agreement with at least one Tier-1 ISP. The focus in this paper is on understanding the consequences of network externalities on market structure in the Internet backbone services market from a competition policy perspective. U.S. and European competition authorities have studied the effects of network externalities on competition in Internet backbone services extensively.[1]At the focus of their analysis were the proposed mergers of large telecommunications companies (MCI and Worldcom and later of MCIWorldcom and Sprint) with notable market shares in the Internet backbone services market. The question concerning the competition authorities was whether a larger provider of Internet backbone services would have an incentive and the means to discriminate against smaller rivals because of network externalities in the market?

Based on the *disaggregated regulatory approach* (Knieps 1997 and 2006), the logical layer of Internet service provision is analyzed in isolation from the vertically related upstream market for physical network infrastructure (the physical layer) and the downstream market for Internet applications services (the applications layer). The main services provided on the logical layer of Internet service provision are Internet traffic services: *Internet access services*, which are provided on top of local communications infrastructure and serve to transmit Internet traffic between the end-users premises and a point of presence of an ISP's network and *Internet backbone services*, which are provided over long-distance communications infrastructure and serve to transmit data within an ISP's networks and between ISPs' networks. The main network elements of the logical layer are routers and switches which are combined with software and Internet-addressing standards. Furthermore, network management functions and the negotiation of interconnection agreements belong to the logical layer. The communication lines over which Internet traffic is transmitted are part of the physical layer of Internet service provision.

The paper is structured as follows: Section "Network Effects in Internet Service Provision" introduces the specifics of network externalities in the applications layer of Internet service provision and how they relate to the logical layer of Internet service provision. Section "Terms of Interconnection Among ISPs in a Competitive Environment" reviews the terms of interconnection between ISPs which are observable in today's unregulated Internet interconnection markets. Section "Dominance at the Tier-1 Level" reviews the literature on interconnection incentives of ISPs with a focus on the single-dominance case. Section "Collusion on the Tier-1 Level" analyzes whether the Tier-1 ISPs as a group could successfully form a stable collusion on the market for transit services and thereby collectively discriminate ISPs on lower hierarchy levels (collective dominance). Section "Conclusions" concludes the paper.

---

[1] See European Commission (1998, 2000).

Network Effects in Internet Service Provision

The Internet, as a classical communications network, belongs to the class of goods which exhibit positive external benefits in consumption. Direct external effects are due to the fact that the utility of belonging to the Internet community is directly related to the number of other people and services that can be reached via the Internet. Indirect network effects result from the fact that the more people use Internet services, the more applications and complementary products are offered to Internet users.

The utility derived from the consumption of any network good can be decomposed into a so-called network effect, resulting from the number of people reachable via the network, and a so-called technology effect, resulting from the technological characteristics of the network the user is connected to (Blankart and Knieps 1992: 80). In the context of Internet service provision the network effect can be expected to dominate the technology effect because users are more likely to give up benefits from a preferred technology for a wider reach in the Internet.

One way of maximizing the benefits from the network effect is to have only one network supply its services to all users. This would, however, imply that consumers can derive no benefits from competition over price, product or service quality. As an alternative to a single large network, network interconnection among otherwise independent network operators can allow that users enjoy the positive network externalities associated with a single network while benefiting from product diversity in product dimensions other than network size.

Indeed, the principal attraction of the Internet is that because of interconnection among ISPs anyone connected to the Internet is reachable by all other users of the public Internet, irrespective of the home ISPs these users subscribe to. Internet users expect this universal connectivity from their ISP, that is, the ability to reach all destinations reachable on the public Internet. For universal connectivity all networks need to be either directly or indirectly connected to one another. The strong network effects experienced on the retail level of Internet services provision therefore translate into a demand for Internet interconnection by ISPs on the logical layer of Internet service provision. Still, an ISP's incentives for interconnection may be contradictory, when on the one hand an ISP wants to offer universal connectivity to its customers and therefore will seek to interconnect with rival networks, but on the other hand it could try to gain a competitive advantage by refusing to interconnect with some ISPs, thereby keeping them out of the market and trying to lure the customers of these ISPs to its own network instead.

Terms of Interconnection Among ISPs in a Competitive Environment

The interconnection of networks has three aspects. Firstly, a logical interconnection of the networks needs to define which services are to function across the network boundaries and at which quality. Secondly, a physical interconnection between the network infrastructures needs to be established. Lastly, the ISPs need to negotiate how the costs of the physical interconnection and the costs for the traffic transmission via this interconnection ought to be split.

The advantage of the Transmission Control Protocol/Internet Protocol (TCP/IP) standard is that two IP-based networks can agree to use the TCP/IP protocol and thereby define much of what the logical interconnection parameters will be. ISPs can negotiate further quality of service parameters which they want to guarantee across network boundaries. Advanced services, such as real-time Voice over Internet Protocol (VoIP) capabilities or Television over Internet Protocol (IP-TV) services can, for instance, be offered only to users within one and the same network by running additional protocols on top of the standard TCP/IP protocols.[2] They can however, also be offered across network boundaries, if the ISPs agree to guarantee the required quality parameters.

Negotiations over physical interconnection as well as the financial terms of network interconnection need to address the following questions: (1) where to establish the location of the interconnection, (2) how to cover the costs of the network infrastructure which physically connects the two networks and (3) how the two networks ought to split the costs for traffic transmission to and from the other's network. The following subsections present the typical financial agreements for Internet interconnection services today.

Costing and Pricing of Internet Traffic Services

Early interconnection of IP-based networks in the NSFNET era[3] functioned basically without monetary compensation between the connecting parties. The rationale may have been that traffic flows could be expected to be roughly symmetrical. More importantly, however, the funding for the network infrastructure at this time was in most cases provided by the government. Network administrators therefore considered the effort to install complex traffic metering dispensable. This situation changed fundamentally, when the National Science Foundation (NSF) reduced funding and networks had to become self-supporting, this being the case even more so when commercial ISPs took up business. The need arose to recover network costs according

[2] See, for instance, Buccirossi et al. (2005). According to Marcus (2006: 34) these technologies are already widely deployed for controlling the quality of service within networks.

[3] When computer-networking was increasingly used in the 1970s the U.S. National Science Foundation (NSF) played an important role in the development of network interconnection. The NSF initially funded regional networks in the academic community. In 1986, the NSF build the NSFNET, a long-distance network connecting five sites at which NSF funded supercomputers could be accessed. The NSFNET was a network of high-capacity links spanning the entire United Sates and connecting the supercomputer sites (Rogers 1998). This network was open to interconnection by previously existing regional networks in support of research and communication (Jennings et al. 1986). The NSFNET was therefore the first common backbone, or "network of networks".

to some cost-causation principle. It is no coincidence that interconnection agreements changed dramatically at the time of the privatization of the Internet, and that at the same time concerns regarding the possibility of anti-competitive interconnection agreements started to be intensely analyzed by competition authorities and competition economists.

The costs of providing Internet traffic services include the access costs to network resources of the physical layer as well as the costs of switches and routers, the costs for transmission software and the costs for employed staff. These costs are driven by the geographic extent of the network as well as by the bandwidth of the links making up the network.[4] Most of these costs are long-run variable costs. The short-run marginal costs for any particular product or service provided over a given infrastructure are close to zero. As is typical for network services, most of the costs involved in Internet traffic services are also overhead costs, meaning that they cannot be allocated to the incremental costs of particular products and services. The pricing for Internet backbone services therefore necessarily does not reflect short-run marginal costs or even long-run incremental costs of the service.

In general, the price of a particular product must cover at least the long-run incremental costs of this product. If these are not covered then, from an economic point of view, the product should not be produced. In addition, the entire set of products and services offered must cover all overhead costs of production, that is, all costs which cannot be allotted to the incremental costs of a particular product or service. To cover their considerable overhead costs, network operators use pricing strategies that calculate mark-ups on the incremental costs, which allocate the overhead costs to particular products and services according to the price elasticity of demand for these products and services.

The elasticity of demand for Internet backbone services depends on the possibilities for substitution. To offer universal connectivity, a network provider can combine the components (1) own network services, (2) network services from peering partners, and (3) network services from transit partners. These components are interchangeable to a degree and the amount used will depend on the costs of each of these services. With network interconnection, an ISP can avoid building out its own network to particular regions and customer groups, instead profiting from the network investments made by the interconnection partners. The following two subsections look at the pricing of peering and transit interconnection respectively.

The Implicit Price of Peered Interconnection

The main difference between interconnection by a transit contract and interconnection by peering is the degree of coverage of the Internet offered by either transit

[4] Transmission links can be leased. Leased lines are priced by their length and by the capacity of the pipe. The larger the extent of the network, the more switches and routers are needed. The costs for employees also rise with the geographical extent of the network.

(complete coverage) or peering (only the direct customers and transit customers of the peering partner are reached).[5] Furthermore, peering generally involves no monetary compensation for using the peering partner's network while in a transit relationship one party pays the other party for delivery of its data traffic from and to the rest of the Internet.

There is, however, an "implicit price for peered interconnection" (Elixmann and Scanlan 2002: 47), namely the cost of providing the reciprocal service for one's peering partner. In order to understand which interconnection services ISPs consider equal, one must understand how traffic exchange among peering partners is organized. The practice in question tellingly has been called "hot potato routing" (Kende 2000: 5ff.). Peering partners generally interconnect their networks at several dispersed geographic locations. For any data transmission, traffic is passed on to the peering partner at the nearest point of exchange to the origin of the communication.[6] The bits of data are then transported to the receiving user on the receiving user's network.

When the geographic extent of the networks of two ISPs are comparable, and when the end-users connected to the ISPs are similar with respect to the data flows they initiate and receive, then ISP 1 and ISP 2 will carry roughly the same amount of traffic for roughly the same distances as a result of a peering agreement. It is interesting to note, that under these circumstances the number of users connected to the ISPs is irrelevant.[7] If, however, ISP 2 had a network of smaller geographic coverage than ISP 1, then ISP 1 would have to carry the traffic further on its own network before having the opportunity to hand the traffic off to ISP 2. ISP 2 would then profit disproportionately from the peering agreement. Furthermore, if ISP 2's customers had more outbound than inbound traffic flow, for instance if ISP 2 had many content servers on its network which receive only small packages containing content requests but send out large volumes of data, then ISP 1 would carry a larger data volume on its network on the return trip than ISP 2 had carried for the content requests. ISP 1 would then need to invest more into the bandwidth of its network without compensation by ISP 2. Again, ISP 2 would profit disproportionately from a peering agreement.

---

[5] For an overview of transit and peering see also Laffont et al. (2001: 287ff.).

[6] This convention also makes sense, considering that the physical geographic location of the receiving host is known only to the home network of the receiving host.

[7] If ISP 1 had more Internet-users than ISP 2, then traffic flows between the two networks would still be balanced, when the probability of communication between all users is the same, and when the geographic extent of the networks is the same (Economides 2005: 381). Consider, for instance, the following example: Suppose a network with 1,000 attached users interconnects with a network with 100 attached users. If every user corresponds once with every other user, then the smaller network transmits 100 × 1000 contacts to the larger network, amounting to 100,000 contacts. The larger network transmits 1000 × 100 contacts to the smaller network, therefore also 100,000 contacts. Thus, if the data volume that the users send to one another is roughly equal, then the traffic carried by the large and the small network is the same, as long as the types of users are the same across the networks and as long as the operators have networks of similar geographic extent.

These examples illustrate that a change in the relative geographic extent of the networks or in the product portfolio of the peering partners (which would attract different types of customers) can result in an unequal distribution of the advantages from a peering contract and lead the party which profits less by the arrangement to terminate the contract. This shows that the observation that an ISP is terminating peering agreements does not suffice as evidence of anti-competitive behavior. If termination of a contract were not allowed (as some ISPs have demanded from the competition authorities), infrastructure investments would degenerate at the rate at which some ISPs would practice "backbone-free-riding"[8] at the costs of other ISPs. If competition policy forbade positive settlement fees in interconnection contracts, then this would lead to under-investment in network infrastructure (Little and Wright 2000). In conclusion, ISPs will enter into peering agreements only if their prospective peering partners have a network of similar geographic extension and have invested into comparable network bandwidth that can guarantee an equivalent level of quality of service. Furthermore, ISPs generally require traffic flows to be roughly similar. For this it is not important to have the same number of customers, only the same type of customers.

The Price for Transit Interconnection

Transit can be bought from transit givers at any available point of network interconnection with transit fees covering at least the costs of the network resources into which a transit provider has invested to be able to offer transit services and the interconnection fees. In addition, a transit giver will try to cover some of its overhead costs by a mark-up on the incremental costs of providing the transit service. In practice, transit fees are typically two-part tariffs. A flat-fee is charged, which varies depending on the bandwidth of the pipe connecting the two networks and the arranged peak throughput of data on this pipe. A variable fee is charged for traffic in excess of this agreed level, generally charged on a Mbit/s basis. The transit giver therefore has the opportunity to price-differentiate in the market for Internet backbone services. A transit taker will pay a lower average price if more traffic is sent via a particular interconnection and if the amount of traffic sent over this interconnection is correctly predicted beforehand. For inelastic demand, often characterized by a short-term need to shift traffic to a new transit provider, the average price paid will be higher. Yet, such price differences cannot be taken as evidence of significant market power by the transit giver. The need to cover the substantial overhead costs in this market force the transit giver to find ways of implementing surcharges on marginal costs, that can cover the overhead costs of production.

The above analysis shows that a transit interconnection requires far less investments into network infrastructure as well as human resources than peering does.

[8]This term was coined by Baake and Wichmann (1998: 2).

Since a transit contract also offers universal connectivity, whereas peering offers only limited coverage of the Internet, a smaller ISP will often find it less costly to pay for transit services in order to reach universal connectivity than to meet the network requirements necessary to peer with several ISPs of higher hierarchy levels. Peering is therefore not always preferred to transit interconnection, even though it generally involves no monetary compensation for the exchange of traffic. Transit fees are justified by the fact that transit givers invest more into their network infrastructure than transit takers.

Dominance at the Tier-1 Level

The preceding section focused exclusively on the decision on whether to interconnect via a peering or transit agreement. It was shown that the differences in the terms for peering or transit do not necessarily reflect discrimination between ISPs operating on different levels of the network hierarchy. The decision to interconnect either via a peering or a transit agreement is not driven by the number of IP-addresses an interconnection partner offers access to. Rather, factors such as the type of customer mix and the relative geographic extent of the two networks were shown to be important. In contrast, the focus of the following analysis is the decision on whether to interconnect at all. In the course of this decision the network reach provided by a potential interconnection partner is of fundamental importance, because the ultimate goal of network interconnection is to provide universal connectivity. All ISPs not active on the highest level of the Internet hierarchy need at least one transit agreement with a Tier-1 ISP or with an ISP that has such a transit interconnection. Therefore the question arises whether it is likely that a merger on the Tier-1 level of the Internet hierarchy could negatively impact competition in Internet backbone services in the sense that a Tier-1 ISP may have an incentive to discriminate lower level ISPs.

As was discussed above, the demand for Internet backbone services on the logical layer of Internet service provision is a derived demand from the end-user demand for universal connectivity on the retail level of Internet service provision. In the retail market, universal connectivity signifies that all other end-users and content providers on the Internet can be reached via one's home ISP. In the Internet backbone services market, universal connectivity signifies that an ISP can send and receive data to and from all IP-addresses allocated to public uses in the Internet.

The literature on Internet backbone services does not differentiate clearly between universal connectivity on the applications layer and universal connectivity on the logical layer of Internet service provision. The difference is, however, of importance when, as is often the case, the number of "customers" attached to ISPs is used as the measure for the Internet coverage the ISP provides. This is a concept relevant on the applications layer of Internet service provision. On the logical layer a customer of an ISP can, however, be either an end-user, representing only one of millions of Internet-Protocol addresses (IP-addresses) or another ISP, representing an important fraction of all registered IP-addresses. For the purposes of measuring Internet coverage on the logical layer of Internet service provision it is therefore more meaningful to speak of the coverage of IP-addresses which this ISP can offer as a peering partner. Transit services, by definition, offer universal connectivity.

Economists have developed models that try to capture the interconnection incentives of ISPs. Theoretical models are of particular relevance in the context of merger policy because competition authorities cannot look at actual market conduct for their analysis. Policy makers depend on predictions derived from economic modeling to understand whether efficiency considerations or attempted exclusionary conduct are at the core of proposed mergers. The model that was influential in the merger proceedings surrounding the MCI and Worldcom merger in 1998 and the attempted merger of the resulting firm MCI/Worldcom and Sprint in 2000 offered initial interesting insights into the interconnection incentives of ISPs with asymmetric installed customer bases. Since then, the literature on interconnection incentives of ISPs has refined this model considerably. The following two subsections shall review the theoretical debate on the interconnection incentives of ISPs in more detail.

The Crémer, Rey and Tirole Model

The reasoning that led the competition authorities to impose severe conditions on the merger of MCI and Worldcom in 1998[9] was based to a great extent on one of the earliest theoretical models, which tried to capture the strategic interconnection decision of ISPs. From this model by Crémer et al. (2000) the conclusion was drawn that an ISP that is dominant in terms of attached customer base in the retail market, would have the means to dominate the market for Internet backbone services. It would either refuse to peer with smaller rivals or price-squeeze them out of the market (Kende 2000: 22–23).[10]

The model by Crémer, Rey and Tirole builds on the Katz and Shapiro (1985) model of network externalities. As Katz and Shapiro, Crémer, Rey and Tirole model the number of firms in the market as exogenously given and assume that there is no product differentiation. Consumers exhibit different basic willingness to pay for the service but show no technology preferences and express the same evaluation of the network effect.

In a first scenario, Crémer, Rey and Tirole focus on interconnection decisions in an asymmetric duopoly situation. The existing users of the two networks are assumed to be locked-in. The networks compete à la Cournot over the addition of new customers to their networks. The choice of the quality of interconnection between the networks is introduced as a strategic variable. In the first stage of the game the

[9] MCI had to divest it's Internet operations before a merger with Worldcom was approved (European Commission 1998).

[10] Crémer, Rey and Tirole argue that a customer in this model can be either an end-user or an ISP. They do not differentiate between the two.

quality of interconnection is determined by the network which sets the lower quality level. Given the interconnection quality, the networks then choose their capacity and prices. In equilibrium, the network with the larger installed customer base prefers a lower level of interconnectivity than the smaller rival because it can expect to dominate the market for new customers.

Two effects determine the equilibrium outcome. Firstly, lower connectivity levels lead to an overall demand reduction in the market, which negatively impacts all firms. Secondly, reduced interconnectivity introduces an element of quality differentiation between the firms, which in this model, can only differentiate among themselves along the dimension of network size. The network with the initially larger locked-in customer base profits from this quality-differentiation effect because it can offer more benefits from network externalities to new users. The bigger network trades off the negative effect of the demand reduction against the positive effect of the quality differentiation. The incentive to choose a lower level of interconnection quality is the more positive the stronger the network externalities are and the greater the difference in installed bases is. A differential analysis shows that the incentive to increase the level of interconnection quality may rise when the number of locked-in customers is already very large, because then the demand expansion effect triggered by a larger network becomes so important, that good quality interconnection is preferred. This equilibrium solution to the model has been the basis for arguing that a dominant Tier-1 ISP would have an incentive to refuse or degrade interconnection with rivals, especially in dynamic markets with high growth potential.

In a second scenario, Crémer, Rey and Tirole (ibid. 456ff.) analyze a market initially consisting of four equal sized ISPs. As long as all four have the same size, all are interested in a good quality interconnection because all profit equally from a demand expansion effect. The elicitor of a quality degradation would suffer the same negative demand reduction as its three rivals without a compensatory gain from a positive quality-differentiation effect. The authors then show how the incentives to interconnect change when two of the ISPs merge and the resulting market of three ISPs then includes one firm with an installed base of at least the size of the combined installed bases of the other two firms. In this scenario the largest firm is generally not interested in deteriorating the quality of interconnection with both of the rival networks, although, in some circumstances, it can profit from a targeted degradation strategy, in which it refuses good quality interconnection with one of the smaller rivals while it continues good quality interconnection with the other rival. This conclusion depends on the non-targeted firm not offering transit services to the targeted firm.[11] The positive quality-differentiation effect will then result in the targeted firm not attracting any new customers while the dominant firm and the non-targeted firm gain more customers (even though the non-targeted rival profits more from the quality-differentiation

[11] Crémer, Rey and Tirole (ibid. 458) argue that the dominant firm can limit the capacity of the interface with the non-targeted network to such an extent that the capacity is only sufficient to provide good quality interconnection for the traffic of the non-targeted network but would result in very bad interconnection quality if the traffic should grow to encompass also the traffic of the targeted network.

effect). It was especially this result that competition authorities relied upon in their decision on the merger by MCI and Worldcom in 1998.

Critique of the Crémer, Rey and Tirole Model and Alternative Modeling

The results of the model by Crémer, Rey and Tirole depend critically on the additional assumptions besides the network effects included in the modeling set-up. It is these assumptions which lead to the result that the largest firm prefers a lower level of interconnection quality compared to its smaller rivals. Below it is discussed whether these assumptions are relevant for the market for Internet backbone services.

Market Entry Conditions

First, consider the assumption of a fixed number of firms in the market. This assumption does not correspond well to the thousands of active ISPs observable in reality. If at all, then this assumption may apply to the market for Tier-1 ISP services in which only five to ten ISPs are active. But whether this market has structural barriers to entry, which justify the assumption of a fixed number of firms, is what is trying to be proved. To start with this assumption distorts the analysis of the effects of network externalities on competition in this market.

It can be shown that the equilibrium results of the model by Crémer, Rey and Tirole change dramatically when the number of firms in the market is endogenized (Malueg and Schwartz 2006). Consumers do not necessarily choose the firm with the initially larger installed base. When this firm chooses not to be compatible with it's smaller rivals,[12] and when smaller rivals in sum have a minimum initial market share and choose to remain compatible among themselves, then, for a large set of parameter values, new consumers will sign on to this network of smaller compatible firms in the expectation that in a dynamic market setting this network will eventually incorporate more contacts than the single-firm network of the initial market leader.[13] If payments for interconnection were introduced, the parameter values for which the initially larger

[12] The targeted degradation scenario is not considered by Malueg and Schwartz. In a related working paper (Malueg and Schwartz 2002: 37) the authors argue that the parameter values that make targeted degradation profitable to the dominant firm imply unrealistic values for price relative to marginal cost and the consumer surplus of the median subscriber.

[13] Even when the dominant network's installed customer base is larger than the combined installed customer bases of its rivals, there are parameter regions in which the rivals will be more successful in adding new customers to their networks (Malueg and Schwartz 2006: 9). This is due to the customer's expectations of market evolvement in dynamic market settings, in which networks are expected to have a high growth potential. This conclusion is comparable to the results by Economides (1996) for a monopolist that prefers inviting market entry.

firm would choose autarky would be even more limited because smaller firms could share their gain from increased connectivity by offering payments to the larger firm.

That the smaller rivals will remain compatible amongst one another and will have a significant network reach through the interconnection of their networks is very realistic for the Internet backbone services market. The presence of many ISPs at Internet exchange points and the availability of standardized contracts together with the fact that market conditions for transit services are transparent facilitate interconnection agreements. The subscribers of the interconnected networks on the lower hierarchy levels can reach all users of these networks. Considering that many subscribers of Internet services are multi-homed (i.e. subscribe to several networks) and that all those customers of the dominant firm that are multi-homed can be reached via an alternative network, it becomes clear that the Internet reach provided to the customers of the lower-level ISPs can be increased significantly by coordination on the lower hierarchy levels.

Product Differentiation

Secondly, consider the assumption that customers do not have individual preferences for technology characteristics of the network they subscribe to. This assumption does not correspond well to the reality of a large degree of product differentiation observable among ISPs. On the Internet backbone services market, ISPs offer their services to other ISPs, to web-hosting services, to large business users and to private end-users. They offer different service levels according to their customers' needs and they offer their services at diverse locations, again according to their customers' needs. An ISP that would hope to make the market tip in its favor would have to cater to all customers in the market. This may not be the most profitable market strategy in a world of customer heterogeneity. ISPs that focus on particular customer groups have comparative advantages in supplying the types of services that these customers prefer. In this case, the proper theoretical reference model may be that ISPs are supplying components of systems rather than competing systems. In such markets, compatible products (as, for instance, interconnected networks) cater to the needs of particular customers. Competition between the products is not as strong as in a market of competing systems because the possibility to make profits is often increased by compatibility (see Economides 1989; Einhorn 1992). When product differentiation is introduced into the model by Crémer, Rey and Tirole it can be shown that in any shared market equilibrium both firms profit from a higher interconnection quality because competition becomes less aggressive when the firms can offer the same positive network externality effect to their customers (Foros and Hansen 2001).[14] General analysis on the compatibility incentives of providers of differentiated network goods come to comparable results (Doganoglu and Wright 2006).

[14] In this model there is also no installed customer base. This fact of course also has an important impact on the results of the model. This aspect is in the focus of a model structure by Economides (2005) which is discussed below.

Switching-Costs

There are other critical assumptions in the Crémer, Rey and Tirole model which do not correspond to the characteristics of the Internet backbone services market. Firstly, consider the assumption that installed bases are locked-in. In reality, switching ISPs is not difficult for end-users or ISPs. Only the cancellation period of their contract may delay the reaction for some weeks. Larger customers such as firms and ISPs are often multi-homed, that is, they connect to more than one ISP at any given time. This is important for the ISP to be able to guarantee its contractual service level vis-à-vis its customers. It is also a signal that traffic can be diverted quickly from one ISP to another without large transaction costs involved. The fact that switching is relatively easy increases the competition between Internet backbone service providers. When the assumption of a locked-in customer base is relaxed, it can be shown that the initially dominant network has an incentive to keep up a high quality of interconnection (Economides 2005, Appendix). A degradation of interconnection quality with one of the smaller rivals would lead to a loss of universal connectivity that would result in a severe demand response by the installed customer base as well and therefore to revenue and profit loss.

Collusion on the Tier-1 Level

Only Tier-1 ISPs can guarantee universal connectivity without relying on a transit offer. The preceding section showed that one Tier-1 alone cannot successfully refuse interconnection with other ISPs or raise interconnection prices in the hopes of ousting competitors from the market. The transit offers of Tier-1 ISPs are perfect substitutes. Absent any collusive practices there is intense competition in this market. This fact provides the Tier-1 ISPs with a motive to collude on the market for transit services. If all Tier-1 ISPs acted simultaneously in increasing prices for transit services, then lower level ISPs would have no alternative transit provider from whom to buy universal connectivity services. And no new provider of universal connectivity could enter the market as long as the Tier-1 ISPs would successfully foreclose this market by not entering into any new peering agreements. The question analyzed in the present section is whether Tier-1 ISPs can organize a stable collusion in the wholesale market in order to collectively raise the price of transit services?

There is a literature on two-way access in telecommunications markets which analyzes whether cooperation on the wholesale level can help enforce collusion on the retail level.[15] A two-way access scenario is given when customers connect to only one network, such that the two networks reciprocally need access to each other's customers on the wholesale level. Termination in this scenario is comparable to a monopolistic bottleneck. The application of this literature has mostly been to

[15] The seminal articles in this research are Laffont et al. (1998a, b).

voice telephony markets, for instance, mobile telephony or reciprocal international termination. Considering that ISPs have a termination monopoly whenever customers exclusively connect to their network only, the models may, however, also be applicable to the market for Internet backbone services. If a large fraction of end-users are connected to only one network, then ISPs may have the possibility to collude on the retail market.

The assumptions that are necessary for successful collusion in a market with reciprocal termination are:

- There is no free market entry.
- There are no capacity limitations.
- Every customer connects to only one network.
- The calling party pays for the connection. The receiving party does not care about the price the caller has to pay to reach him.
- There is no price differential for calling customers on the same network (on-net calls) or on another network (off-net calls).
- Access charges for call termination are set reciprocally.
- Both networks have the same costs of production.
- The probability of a call is independent of the home-network of the calling parties. This implies that given same marginal prices for on-net and off-net calls, the share of calls that originates with network 1 and terminates with network 2 will be equivalent to the market share of network 2.[16]

It can be shown, that when the reciprocal access charge set by the firms is not too high compared to the marginal costs of termination, and when the substitutability between the two networks is also not too high, then there exists a unique equilibrium to this model (Laffont et al. 1998a: 10). In this equilibrium, the retail price is increasing in the access charge for termination. The firms can therefore use the access charge to enforce a higher retail price than would be the outcome of competition.

The intuition behind this result is that if access charges are set at the level of the actual marginal costs of terminating a call, then the marginal costs of producing either an on-net call or an off-net call are the same for the originating network. If the access charges are above the marginal costs of termination, then the costs of producing an off-net call are higher than those of producing an on-net call. The higher the access charge, the higher the marginal costs of producing an off-net call. This mechanism can be used to raise the rival's costs of production and put pressure on retail prices.

For the collusion to be stable the access charge must not be too far above the marginal costs of termination and the substitutability between the networks must not be too high. When the access charge is set well above the marginal costs of termination, a firm has an incentive to increase its market share and avoid paying

[16] This so-called "balanced calling pattern assumption" has important implications for the model. It implies that "…for equal marginal prices, flows in and out of the network are balanced – even if market shares are not." (Laffont et al. 1998a: 3). When wholesale access charges are set reciprocally, this assumption implies that the wholesale interconnection payments cancel each other out.

termination fees.[17] When the substitutability between the networks is high, attempts to increase one's own market share by luring the customers of the other network to switch networks will more likely be successful.

The incentives to compete rather than collude in the retail market are further intensified by allowing for more complex price structures in the retail market besides identical linear prices for on-net and off-net calls. Firstly, consider the possibilities offered by non-linear pricing structures. When charging two-part tariffs the firms can use a lower fixed fee to increase market share while keeping the unit price on the collusive level such as not to induce a quantity expansion effect. As a result of the higher market share, the firm will have less off-net traffic and less termination charges to be paid. With non-linear pricing in the retail market competition is intensified and collusion, again, becomes more difficult (Laffont et al. 1998a: 20ff.). Secondly, consider price discrimination in the retail market. In a companion article, Laffont, Rey and Tirole show that collusion is destabilized when retail prices differentiate between on-net and off-net calls (Laffont et al. 1998b). A defecting firm can use low on-net prices to increase its market share but keep off-net prices on the collusion level such as not to induce a quantity expansion effect which could produce an access-deficit.

Application to the Market for Internet Backbone Services

The model above shows that while collusion via wholesale access charges is possible it is only stable under very restrictive assumptions. Given this information, what can be learned with respect to the market for top-tier Internet backbone services? Is it likely that Tier-1 ISPs can use their wholesale agreements to stabilize higher transit prices?

Some of the assumptions of the model set-up by Laffont, Rey and Tirole fit relatively well with the market characteristics of the Internet backbone services market, at least when only the highest level of the Internet hierarchy is in focus. For instance, for Internet interconnection via peering it is true that there is no price differential between on-net and off-net connections. Furthermore, Tier-1 ISPs, as peering partners, generally set their access charges reciprocally (albeit to the level of zero). Tier-1 ISPs can also be considered to have a similar cost-structure for terminating each others connections. Lastly, the assumption of a balanced calling pattern between Tier-1 ISPs is fitting, given that they are peering partners and can therefore be assumed to have a similar customer structure.

Other assumptions of the model by Laffont, Rey and Tirole, however, do not correspond as well to the market for Internet backbone services on the highest hierarchy level. As these assumptions are essential to the stability of the collusion

[17] Even when the net payments between the two networks are zero with reciprocal access charges and balanced calling patterns, they perceive the access charge as a marginal cost of production and will want to avoid them.

equilibrium the fact that they do not correspond to the market in question is an indication that collusion in the market for top-tier transit services is difficult to maintain. Firstly, consider the assumption that every customer is connected to only one network as prerequisite for the termination monopoly. This assumption is too strong for the market for Internet backbone services, as many small ISPs and many business customers are multi-homed. Therefore, the termination monopoly in Internet interconnection is not as stable as assumed in the model by Laffont, Rey and Tirole. Next, consider the number of players in the market. It can be argued that market entry into Tier-1 Internet service provision is not free because any new entrant must reach a peering agreement with all other Tier-1 ISPs. None the less, there are already several active firms on the Tier-1 level of Internet backbone services which increases the number of potential substitutes and destabilizes any collusive agreement.

Furthermore, the assumption that the receiving party of a connection does not care about the costs the calling party has to pay for the connection is not appropriate in the context of Internet interconnection. Businesses offering content and information on the Internet care very much about the costs their targeted customers face for reaching this content. The costs of being reached are a significant factor in their decision where to place their content on the Internet. The access charge is therefore not only indirectly but also directly a strategic element in the competition over end-users.

Decisive for the stability of any collusion are the level of the access charge and the substitutability of the network offers. Between Tier-1 ISPs the access charge is generally set to the level of zero. It therefore corresponds to the prerequisite that is should not be too far above marginal costs of termination. However, for collusive purposes a termination fee would need to be introduced where there was none before. This may be more difficult than an incremental increase of an existing termination charge. Furthermore, the degree of substitutability between the transit offers of Tier-1 ISPs can be considered to be very high. This fact makes collusion interesting, but at the same time it represents a high risk of instability of any collusion because any of the Tier-1 ISPs could hope to increase its market share by offering a lower transit charge than its competitors.

Lastly, consider the price structures in the market for transit services provided by Tier-1 ISPs. Transit prices generally are not differentiated according to the destination network. However, non-linear prices for transit services are the norm in the transit market. In general a transit taker will pay a fixed fee that depends on the bandwidth by which the two networks are connected plus a variable fee for traffic exceeding a previously defined threshold. The ability to compete in two-part tariffs is a further hindrance to stable collusion in the transit market. To summarize, the prerequisites for a stable collusion are not fulfilled in the market for Tier-1 backbone services.

Conclusions

The purpose of this paper was to analyze the strong network externalities associated with Internet services from a competition-policy perspective. It was argued that in the market for Internet services network effects are so important that an ISP needs to be able to offer universal connectivity in order to survive in this market. The demand for universal connectivity on the logical layer is a derived demand from the demand for universal connectivity on the applications layer. To reach universal connectivity, new entrants to the Internet backbone services market will need to establish a direct or indirect transit agreement with at least one Tier-1 ISP.

Tier-1 ISPs enter into peering agreements only when the benefits from the interconnection are roughly similar to both parties. The fear that a single Tier-1 ISP could be able to abuse a dominant market position in a transit agreement with lower-level ISPs was not substantiated by the analysis. Competitive forces in the market for top-tier Internet backbone services are strong. Tier-1 ISPs compete with product differentiation tactics. Customers frequently multi-home and can relatively conveniently switch their home network. As a result, Tier-1 ISPs cannot benefit from refusing to interconnect respectively from deteriorating interconnection quality with lower-level networks.

In principle, some market constellations are conducive to collusion on the retail level, stabilized via cooperation on the wholesale level. A collusion between Tier-1 ISPs to collectively raise prices in the transit market is not likely to be stable because the prerequisites for a stable collusion are not fulfilled in the market for top-tier Internet backbone services. Most importantly, the assumption of a termination monopoly is not fulfilled. To summarize, the discussion in this paper has provided strong support that competitive forces in the transit market are working and can effectively hinder Tier-1 ISPs to discriminate ISPs on lower levels of the Internet hierarchy.

References

Baake P, Wichmann T (1998) On the economics of internet peering. Netnomics 1:89–105

Blankart C, Knieps G (1992) Netzökonomik. Jahrbücher für neue politische Ökonomie 11:73–87

Buccirossi P, Ferrari Bravo L, Siciliani P (2005) Competition in the internet backbone market. World Competition 28(2):235–254

Crémer J, Rey P, Tirole J (2000) Connectivity in the commercial internet market. J Ind Econ XLVIII:433–472

Doganoglu T, Wright J (2006) Multihoming and compatibility. Int J Ind Organ 24:45–67

Economides N (1989) Desirability of compatibility in the absence of network externalities. Am Econ Rev 79:1165–1181

Economides N (1996) Network externalities, complementarities, and invitations to enter. Eur J Polit Econ 12:211–233

Economides, N (2005) The economics of the internet backbone. Majumdar S et al. Handbook of Telecommunications Economics, Vol. 2. Amsterdam: North Holland

Einhorn M (1992) Mix and match: compatibility with vertical product dimensions. RAND J Econ 23: 535–547

Elixmann D, Scanlan M (2002) The Economics of IP Networks-Market, Technical and Public Policy Issues Relating to Internet Traffic Exchange. wik-Consult Final Report, Bad Honnef

European Commission (1998) Commission decision of 8 July 1998 declaring a concentration to be compatible with the common market and the functioning of the EEA agreement (Case IV/M.1069 '– WorldCom/MCI). Official Journal of the European Commission L116:1–35

European Commission (2000) Commission decision of 28 June 2000 declaring a concentration incompatible with the common market and the EEA agreement (Case COMP/M.1741 – MCI WorldCom/Sprint)

Foros O, Hansen B (2001) Competition and Compatibility Among Internet Service Providers. Info Econ Pol 13(4):411–425

Jennings M, Landweber LH, Fuchs IH, Farber DJ, Adrion WR (1986) Computer networking for scientists. Science 231:943–950

Katz M, Shapiro C (1985) Network externalities, competition, and compatibility. Am Econ Rev 75:424–440

Kende M (2000) The Digital Handshake: Connecting Internet Backbones. OPP Working Paper 32. FCC, Washington, DC

Knieps G (1997) Phasing out sector-specific regulation in competitive telecommunications. Kyklos 50:325–339

Knieps G (2006) Sector-specific market power regulation versus general competition law: Criteria for judging competitive versus regulated markets. In: Sioshansi FP and Pfaffenberger W (eds.), Electricity Market Reform: An International Perspective. Amsterdam: Elsevier

Laffont JJ, Rey P, Tirole J (1998a) Network competition: I. Overview and nondiscriminatory pricing. RAND J Econ 29:1–37

Laffont JJ, Rey P, Tirole J (1998b) Network competition: II. Price discrimination. RAND J Econ 29:38–56

Little I, Wright J (2000) Peering and settlements in the internet: An economic analysis. J Regul Econ 18:151–173

Malueg D, Schwartz M (2002) Interconnection Incentives of a Large Network. Georgetown University Department of Economics Working Paper 01-05, August 2001, revised January 2002

Malueg D, Schwartz M (2006) Compatibility incentives of a large network facing multiple rivals. J Ind Econ 54:527–567

Marcus S (2006) Framework for Interconnection of IP-Based Networks: Accounting Systems and Interconnection Regimes in the USA and the UK. wik-Consult Report, Bad Honnef

Rogers JD (1998) Internetworking and the politics of science: NSFNET in internet history. Inf Soc 14(3):213–228

Vanberg M (2009) Competition and Cooperation among Internet Service Providers: A Network Economic Analysis. Baden-Baden: Nomos

A Behavioral Economic Interpretation of the Preference for Flat Rates: The Case of Post-paid Mobile Phone Services

Hitoshi Mitomo, Tokio Otsuka, and Kiminori Nakaba

Abstract This paper aims to empirically test the existence of a biased preference for flat rate service plans related to mobile phones, and to examine how psychological factors can affect such preferences. We define such preference as "flat-rate preference" and interpret it in terms of behavioral economic concepts. Behavioral economics, in spite of its limitations in empirical analysis, provides deeper insights into human behavior than traditional economic models since it considers psychological factors within decision-making processes and allows for irrational choices by consumers. By applying several important concepts from behavioral economics, we seek to investigate a more reasonable explanation for mobile users' flat-rate preference. Loss aversion, reference dependence, the shape of probability weighting function, mental accounting, ambiguity aversion and cognitive dissonance are employed to examine such preference. Non-parametric methods are applied in the empirical analysis to data that was collected through an online survey in Japan. We successfully show the existence of the flat-rate preference in terms of loss aversion and reference dependence although we failed to identify the influences of the shape of the probability weighting function. The other three concepts could also be recognized as factors conducive to preference behaviors.

Introduction

Flat rates have been recognized as having positive influence upon the usage of telecommunications services. Selective tariffs are often applied to such services and subscribing to a flat rate plan is considered an attractive option for many consumers.

H. Mitomo (✉), T. Otsuka, and K. Nakaba
Graduate School of Global Information and Telecommunication Studies,
Waseda University, Japan
e-mail: mitomo@waseda.jp

B. Preissl et al. (eds.), *Telecommunication Markets*, Contributions to Economics,
DOI: 10.1007/978-3-7908-2082-9_4, © Springer Physica-Verlag HD 2009

In most markets, a user can choose a service plan which is suitable to his/her intended usage patterns. In many cases, a flat rate is preferred to a measured rate. Various services ranging from basic telephony to broadband Internet access have been offered at flat rates. A major reason for this would be that users want to avoid uncertain bill payments. Fluctuations in phone bill payments under measured rates create many uncomfortable issues for modern consumers. For local telephone services, Train et al. (1987) found that users tended to choose a fixed charge system rather than a measured system even in the case where they paid the same total amount for their phone bill. Train defined this phenomenon as a "flat-rate bias".

A flat-rate bias is found in a wide range of studies such as Train (1991), Train et al. (1987), Mitchell and Vogelsang (1991), Taylor (1994), Kling and van der Ploeg (1990), Kridel et al. (1993), Lambrecht and Skiera (2006), and others. However, Miravete (2002) finds evidence for rational consumer choice, suggesting that no flat rate bias exists. Similarly, Narayanan et al. (2006) focus on consumers' learning about their own usage and find that they learn faster with a metered plan than a fixed plan.

On the other hand, traditional economics has emphasized the importance of measured rates since they are believed to provide consumers with economic efficiencies. When flat rates are applied, user demand becomes insensitive to the price, and usage is likely to increase. From the perspective of the service provider, supply increases in response to increased demand while the revenue does not increase in a similar manner. Flat rates therefore do not achieve efficient resource utilization. In the analysis of a two-part tariff, the relative importance between the fixed fee and the usage fee has been discussed. For example, Dominance of lump-sum fees over usage-sensitive prices is discussed by Oi (1971), in the context of the Disneyland economy. On the other hand, Mitomo (2002) advocates that even a monopoly supplier should arrange pricing plans in such a way that the fixed fee is set below the per capita cost to attract more subscribers while the usage fee is higher than the associated marginal cost.

However, we will focus more on what is behind users' biased choices for flat rates, so that we define users' inclinations to prefer flat rates as *flat-rate preference*. We investigate the reasons for such biased preference, which traditional economics has failed to explain, by employing several important concepts established in behavioral economics. We will show the results of our empirical tests which illustrate how the concepts from behavioral economics can successfully explain the existence of flat-rate preference. This paper is organized as follows: Section "Telecommunications Services in Japan and Consumers' Flat-Rate Preference" provides an overview of how flat rate services have been applied in the Japanese telecommunications market. Section "Interpreting Flat-Rate Preference Through Behavioral Economics" outlines how concepts from behavioral economic can be applied to explain the flat-rate preference. Section "The Empirical Study" shows the results of our empirical tests. Post-paid mobile phone services are selected for the tests since they have been supplied under both flat and measured rates. Section "Conclusion" concludes the paper.

Telecommunications Services in Japan and Consumers' Flat-Rate Preference

A variety of pricing rules have been applied in telecommunications markets around the world. The two most popular models are (1) measured rates, and (2) flat rates. In Japan, plain old telephony services (POTS), both local and long-distance, have been supplied using measured rates, more specifically with two-part tariffs. Measured rates have also been applied to most mobile phone services, but flat rates are now included in selective pricing plans. On the other hand, broadband access services such as DSL and FTTH are typically flat-rate services (see Fig. 1). Measured rates used to be offered for narrowband access services such as dial-up connections. There is a natural inclination of telecommunications consumers to favor flat-rate service rather than a measured-rate service.

Consequently, application of a flat-rate model has been regarded as an important measure for telecommunications service providers to attract more users. So far, however, these companies have been very cautious in their introduction of flat-rate tariff systems. This is because with flat rates their revenues will not increase proportionally with system usage, but will remain fixed with the number of subscribers. At the same time, it is most likely to increase system usage because flat rates allow for unlimited use by consumers. This increased usage will require increased facility investment and management, which will increase the financial pressures on telecommunications providers.

Besides the increasing consumer demand for flat-rate services, service providers have begun to realize that flat-rate plans do not necessarily bring about negative consequences. In many cases, revenues have not fallen drastically and sometimes the financial benefits have outweighed the costs. This is because (i) flat-rate services can attract more users, but many of them do not use as much as they pay, (ii) revenues are constant and stable, (iii) management, calculation and billing are simplified due to the standardized pricing, and (iv) the creation of business plans is easier than in the case of measured rates. Furthermore, flat-rate services yield greater customer satisfaction. Examples of successful flat-rate mobile services in Japan include WILLCOM's flat-rate voice communication service for their PHS users; NTT

Measured Rate (Incl. two-part tariff)	Measured+Flat	Flat Rate
POTS	Dial-up Mobile (Voice) Mobile (packet) PHS IP phone	Broadband

Fig. 1 Measured and flat rates as applied to telecom services

DoCoMo's flat-rate data packet communication service called "Pake-houdai"; and the KDDI's "au" two-stage flat-rate tariff called "Double-Teigaku".[1] These services have attracted many users. DoCoMo's "Pake-houdai" has been especially successful in making many mobile users aware of the convenience of flat-rate services. The monthly charge for DoCoMo's "Pake-houdai" is JPY4,095 (including tax), with unlimited mobile web access and emails available. Since its introduction, the number of users subscribing to the service continues to grow rapidly.

Flat-rate pricing can also solve a number of problems that have emerged from measured pricing plans. For example, youngsters addicted to mobile communications rely on packet communication such as i-mode Internet access and e-mail services for their daily communications needs. They do not realize how much they have used these mobile phone services within a given month until after they receive their phone bill, as post-paid billing is common in Japan. Most manage to pay their bill on time, but some of them fall behind in their payments. This problem called "pake-shi" has become a social concern in Japan. Such problems will quickly disappear with the introduction of flat-rate services.

Interpreting Flat-Rate Preference through Behavioral Economics

According to traditional economic theories, only a measured rate (a single price) can work as a parameter and achieve economic efficiency through market mechanisms. Flat rates are not believed to attain higher economic efficiencies. Consumer preference for flat-rate pricing has been regarded as a consequence of consumers' *risk-averse behavior.*

However, we have realized from our experience that human behaviour is not necessarily as rational as traditional economic theory assumes. As such, the hypothesis that consumers always work to maximize their utility is too simplistic. Our actual decisions or choices often violate this expected utility hypothesis. An interpretation of such behaviors based on risk avoidance is useful but not sufficient for explaining the preference for flat rates. Therefore, the "rationality" of consumer behavior should be re-defined by incorporating psychological factors which describe these more realistic processes in consumer decision-making behaviors.

Behavioral economics, which was initiated by Kahneman and Tversky (1979) and others, have integrated insights from psychological research into economic science, especially concerning human judgment and decision-making under uncertainty. Prospect theory, which is one of the important theories in behavioral economics, provides a

[1] Regarding their pricing plans, see the following websites (cited 23 April 2008):
WILLCOM: http://www.willcom-inc.com/en/price/voice/index.html
NTT DoCoMo: http://www.nttdocomo.co.jp/english/charge/discount/pake_houdai/index.html
KDDI: http://www.au.kddi.com/english/charge_discount/discount/double_flat_rates/index.html

framework for explaining how people prospect for and behave towards a risk, and contains several important concepts including *loss aversion*, *reference dependence* and *the shape of the probability weighting function*.

Figures 2 and 3 illustrate the two functions which characterize prospect theory: the *value function* which replaces the utility function from traditional economic theory and indicates that a loss is more serious than the gain of the same size; and the *probability weighting function* which represents modified expected value weighted by the subjective evaluation of the probability that a phenomenon is likely to occur, and assumes that a lower probability is overestimated.

Loss aversion explains the consumer tendency to place substantially more weight on avoiding losses than obtaining objectively commensurate gains in the evaluation of prospects and trades (Kahneman and Tversky 1979). This can be

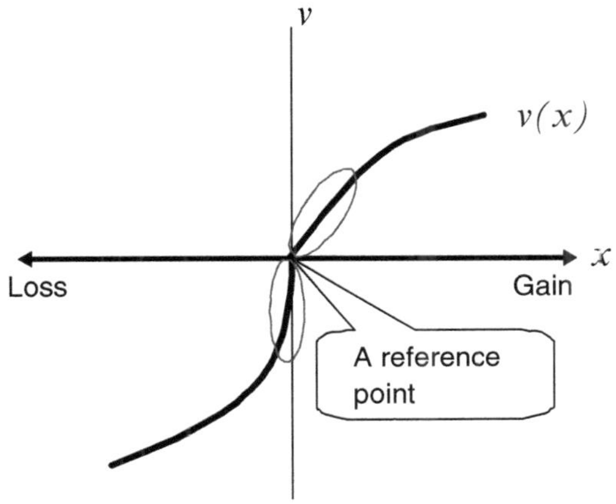

Fig. 2 The shape of a value function

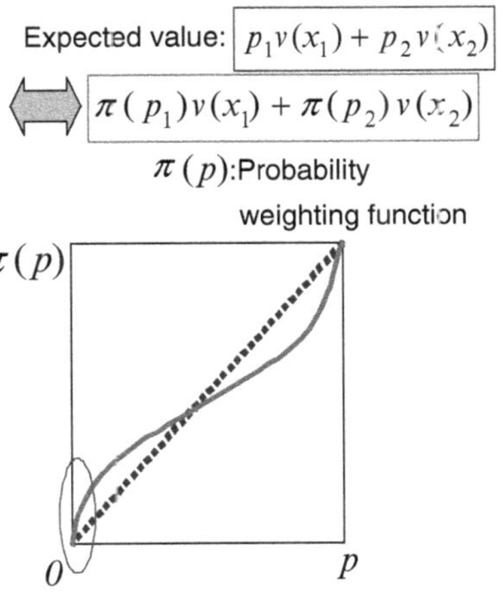

Fig. 3 The shape of the probability weighting function

represented in Fig. 2 by a steeper value function curve in the loss direction than in the gain direction near the reference point. Reference dependence represents a dependence of preference with respect to one's reference point, which is shown as the origin in Fig. 2. The shape of the probability weighting function reveals the tendency that low probabilities are over-weighted while high probabilities are under-weighted relative to the objective probabilities in Fig. 3.

In addition to these concepts from prospect theory, we also employ three additional concepts of *mental accounting*, *ambiguity aversion* and *cognitive dissonance*. Thaler (1980) first introduced the concept of mental accounting, proposing that people set up mental accounts for outcomes that are psychologically separate, and described the rules that govern the integration of gains and losses from such outcomes. Ambiguity aversion describes the preference for known risks over unknown risks (Camerer et al. 2004, Chapter 1). The key finding from such studies suggests that measures of certainty affect decisions and people tend to avoid decision-making in uncertain situations. Cognitive dissonance is defined as the psychological tension which results from behaviors that conflict with one's own beliefs. These concepts can be applied to interpret consumer preferences related to flat versus measured rate pricing as follows:

Loss aversion

If a monthly payment is larger than the reference point (i.e. the average monthly bill payment), users tend to overestimate the loss. To avoid such losses, they prefer flat rates.

Reference dependence

If the reference point represents the total phone bill amount that a user is accustomed to paying, the payment level affects the subjective evaluation of the shift to a flat rate. User preferences for flat rates do not depend on the absolute level of payment but on the digression from the reference point.

The shape of the probability weighting function

With a typical probability weighting function, users with a low probability of overuse tend to overestimate the probability. They tend to avoid an extraordinary payment and will choose a flat rate.

Mental accounting

This factor represents the psychological impact of expenditures (Thaler 1980). Under a measured rate, users are aware of their monthly bill payment amounts, while, under flat rates, users are not psychologically burdened with such concerns.

Ambiguity aversion

This denotes the behavior to avoid uncertainty. Under measured rates, the monthly payment amount is uncertain, while under flat rates the amount of payment is constant. Because of this, consumers prefer the latter than the former.

Cognitive dissonance

Once a flat rate is chosen, users will not change to a measured rate plan, even if it is reasonable.

The Empirical Study

The Survey and Basic Demographics

An online survey was fielded in order to examine the existence of flat-rate preference and to interpret this preference in terms of the above six concepts. The survey was fielded in February 2006 and included 232 mobile users ranging from their teens to forties who each answered the entire survey. Respondents were selected randomly from a pre-registered consumer panel. The survey included 28 questions focused upon flat-rate preferences in addition to ten questions which collected demographic information. Table 1 and Fig. 4 outline some of the key demographic features of the sample.

Figure 5 represents the percent share of each of the mobile phone operators used by respondents. NTT DoCoMo held a 43% share (including 25% and 18% for 3G and 2G, respectively). *au* by KDDI held second place with an approximately 28%

Table 1 Demographic information of the respondents

The number of samples		232
Gender	Male	114
	Female	118
Average monthly disposable money		JPY31,254

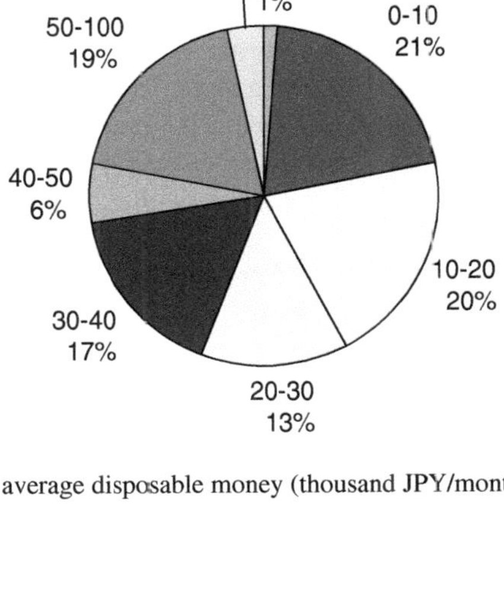

Fig. 4 The respondents' average disposable money (thousand JPY/month)

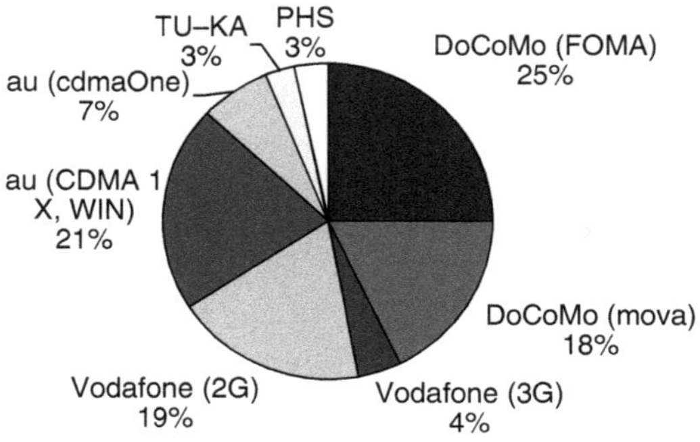

Fig. 5 Mobile phones used by the respondents

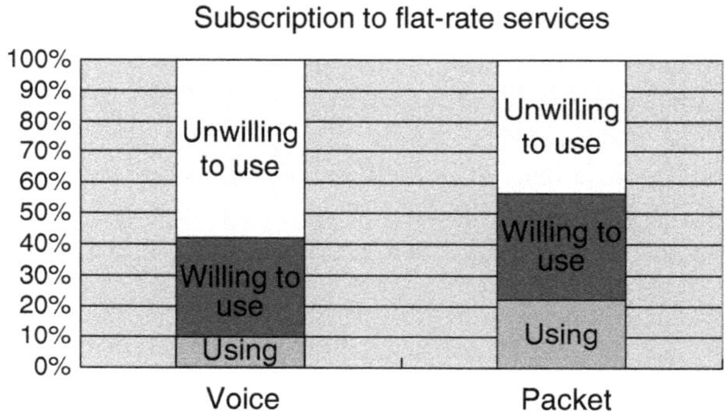

Fig. 6 Subscriptions to flat-rate services

share (21% and 7%). Vodafone, which was purchased by Softbank Mobile, had a 23% share (19% and 4%). Subscriptions to flat-rate services are shown in Fig. 6. With regard to voice communication services, 7.8%, 34.1% and 58.1% were using, willing to use and unwilling to use flat-rate services, respectively. As for packet communication services, 22.0%, 34.5% and 43.5% were using, willing to use and unwilling to use the services, respectively. This means that more than 34% of the respondents were willing to use flat-rate services if they were available.

Interpretation of the Flat-Rate Preference Based on Behavioral Economic Concepts

This section examines whether psychological factors can explain mobile users' flat-rate preference in terms of the six concepts quoted from behavioral economics. Statistical methods were applied to test the hypotheses.

Loss Aversion

In order to interpret the flat-rate preference in terms of loss aversion, we asked two questions shown below. While these two questions ask the same thing in principal, Q.1 emphasizes the gain of Plan A while Q.2 emphasizes the loss.

Q.1 Plan A: JPY7,000/month + JPY6,000 if you exceed the communications allowance

Plan B: JPY9,000/month + no extra charge

According to your previous experience, you will not spend your allowance with a 67% probability when you choose Plan A.

Q.2 Plan A: JPY7,000/month + JPY6,000 if you exceed the communications allowance

Plan B: JPY9,000/month + no extra charge

According to your previous experience, you will exceed your allowance with a 33% probability when you choose Plan B.

The answers given by respondents to these paired questions are shown in Fig. 7. The horizontal axis categorizes the degree of preference between Plan A and Plan B, and the vertical axis shows percentage shares of the respondents. Whether the two distributions are identical or not can be tested by applying the Wilcoxon signed rank sum test. The null hypothesis to be tested is that the median of a distribution is equal to some value. The results indicate that the answers to the two questions have significantly different distributions. The choice of the flat rate (Plan B) with emphasis on the loss in Q.2 is significantly greater than with the emphasis on the gain in Q.1 ($p < 0.01$). When the loss is emphasized in choosing a mobile phone tariff, users are more likely to prefer flat-rate services, taking the loss more serious than the gain.

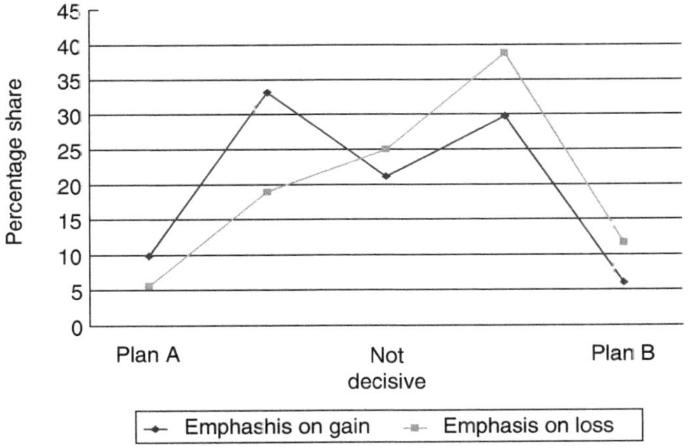

Fig. 7 Distributions of the choices of the mobile users over the preferred plans in the two questions

Reference Dependence

To examine whether the concept of reference dependence can explain flat-rate preference, respondents were asked to accept the shift from either one of hypothetical average bill levels (as the reference point). Measured rates were set at JPY1,000 through 9,000 with fluctuations between −50% and +50%, to one of the flat-rates ranging from the same amount to +60%.

The Settings of the Question:

Your average monthly tariff last year was [A], while the highest tariff was [B] and the lowest [C]. Do you hope to change the fixed tariff system which costs [D]?

[A]: JPY1,000, 3,000, 5,000, 7,000 and 9,000

[B]: +50% amount of [A]

[C]: −50% amount of [A]

[D]: the same amount, +20%, +40% and +60% of [A]

The results are illustrated in Fig. 8. The horizontal axis represents the current payment under the measured rate, and the vertical axis shows the percentage of users choosing flat rates. If there is no reference dependency, each distribution should be a uniform distribution, because those who choose a flat rate should be indifferent to the levels of the current bill payment. To estimate how closely an observed distribution matches an expected distribution, Chi square tests were applied to test the hypothesis that the distributions are uniform. The results indicate that a uniform distribution could not be applied to the shape of the curves ($p < 0.01$). We must conclude then that different bill payment levels result in different willingness to adopt a flat rate, and thus flat-rate preference depends on the reference point.

The Shape of the Probability Weighting Function

Tendencies that mobile users overweight lower probabilities was examined through a question which asked respondents to choose between measured rates with

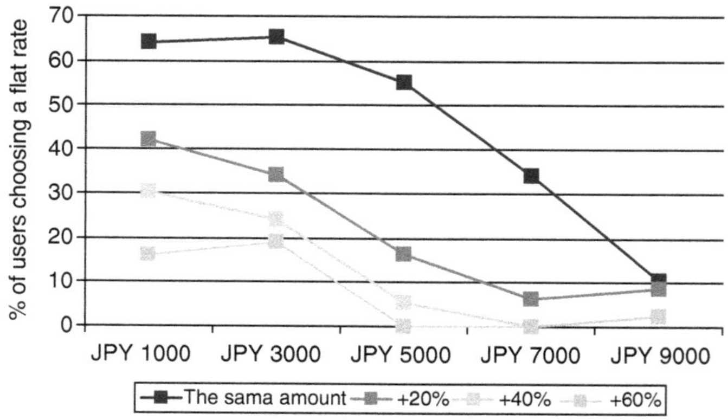

Fig. 8 Dependencies on the reference points

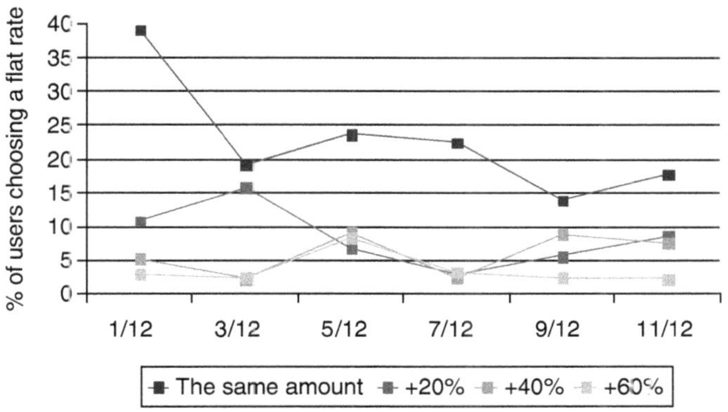

Fig. 9 Stochastic fluctuations in phone bills and the choice of flat rates

stochastic fluctuations and a flat rate. Our hypothesis was that users with lower probabilities of fluctuations would be more likely to choose flat rates than those with higher probabilities.

The Settings of the Question: To ask the willingness to accept the shift from a measured rate with which the respondent has to pay the double of the average with probabilities ranging from 1/12 (once in a year) to 11/12 (11 times in a year) to a flat rates of the same amount, +20%, +40% and +60%.

The results are plotted in Fig. 9. The horizontal axis represents the probabilities of fluctuation in the phone bills ranging from once to 11 times a year, and the vertical axis shows the percentage of users who chose flat rates. If the tendencies exist, the left hand side of the curves in the graph should be downward-sloping. In that case, the null hypothesis that the distributions are uniform should be rejected. We applied chi square tests to test this hypothesis, and our results show that the null hypothesis cannot be rejected except for the case of respondents shifting to flat rate plans of the same amount ($p < 0.01$). With the data collected through our survey, we could not find firm evidence that overweighting lower probabilities exists.

Mental Accounting

Mental accounting is the psychological recognition of gains verses losses, and respondents were asked whether flat rates were perceived as reducing the psychological costs of mobile phone subscriptions:

Question: Do flat rate services influence your level of comfort relative to using mobile phone services?

Respondents were given five ordinal scale choices for their answers. Figure 10 illustrates the distribution of their answers. If respondents were aware of mental accounting, that is, if they felt the exact amount of payment was psychologically cheaper than the actual amount, the distribution of the answers would be diverted

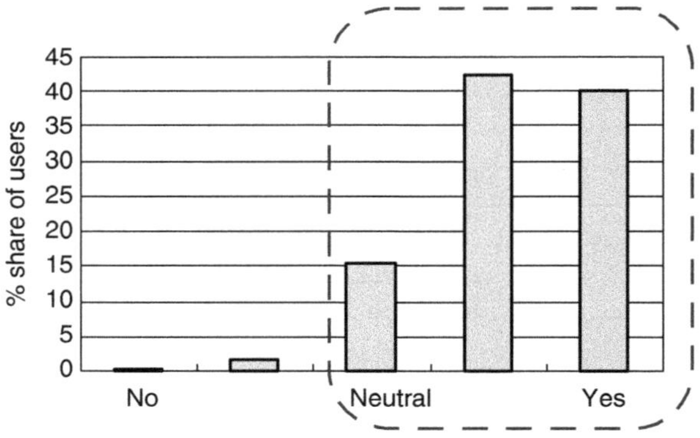

Fig. 10 Flat rates reduce mental costs

from a symmetric distribution such as a normal or uniform distribution. The Kolmogolov-Smirnov (K-S) test was applied to check whether an underlying probability distribution differed from a hypothesized distribution. The null hypothesis which assumed that the distribution was either a normal or uniform distribution was rejected ($p < 0.01$). The result shows that flat rates relieve the psychological burden of mobile phone bills and suggests that mental accounting can be embedded in the flat-rate preference.

Ambiguity (Uncertainty) Aversion

Ambiguity aversion is a behavior to avoid uncertainty. In uncertain situations, people do not know the probability that a certain incident will occur, while *risk* suggests that the probability of such incidents occurring is known (Epstein 1999). The influence of ambiguity aversion on the choice of flat rates was examined by asking the following question:

Question: Is it a merit of flat rates that the monthly payment is fixed?

The distribution of answers is shown in Fig. 11. As in the case of mental accounting, the answers were selected from five ordinal scale choices and the K-S test was applied to examine whether or not the distribution was diverted from symmetric distributions. The results indicate that the null hypothesis of having a normal or uniform distribution was rejected ($p < 0.01$). The concept of ambiguity aversion therefore is also embedded in the concept of flat-rate preference. Note that there might be some laziness or comfortableness effect (Garbarino and Edell 1997) in this survey response.

Cognitive Dissonance

Cognitive dissonance represents psychological tension resulting from behaviors that conflict with an individual's internal perceptions of appropriate decisions.

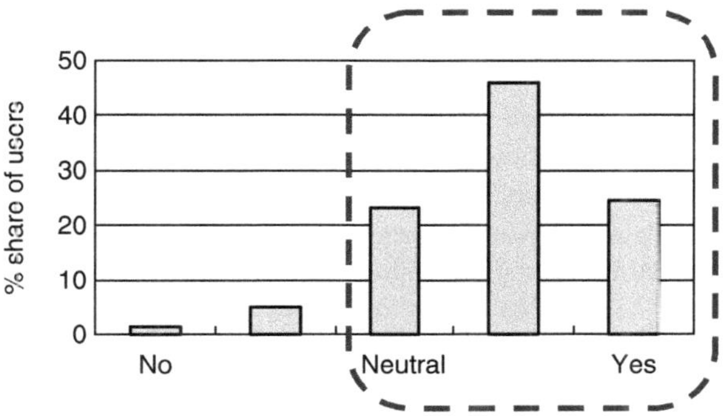

Fig. 11 Uncertainty avoidance was seen in the users' flat-rate preference

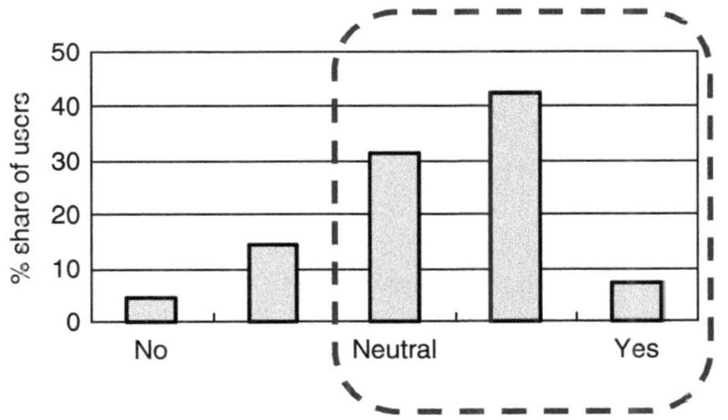

Fig. 12 Conflicts with beliefs can cause flat-rate preference

More concretely, it can be defined as unwillingness to accept an inconvenient truth. After a flat rate is selected, users believe it is the best payment choice and don't want to change to other options even if they are more efficient. This can be examined by asking the following question:

Question: Would you continue to use a flat-rate billing plan even if it did not appropriately reflect the amount of your mobile phone usage?

The K-S test was applied to the distribution shown in Fig. 12 in the same way as the above two cases, and the hypothesis of having a normal or uniform distribution was rejected ($p < 0.01$). This indicates that many of the respondents stick to their decision once they choose a flat rate even though users have noticed that their choice is no longer appropriate. The result suggests that cognitive dissonance can also be used to explain the existence of the flat-rate preference. Note that there is some room to explain this phenomenon within the context of switching costs. Although this question ignores the existence of switching costs, respondents may have considered it implicitly and responded accordingly.

Conclusion

In this paper, we have proposed the application of concepts from behavioral economics to more reasonably explain the flat rate preferences of mobile subscribers. We have examined how three fundamental concepts from Prospect Theory including loss aversion, dependency of reference points and the characteristics of the probability weighting function can explain the consumer inclination to prefer flat rates. The other three important factors, i.e., mental accounting, ambiguity aversion and cognitive dissonance have also been employed to explain such preferences. Non-parametric statistical tests were applied and the results show that except for the shape of the probability weighting function, these concepts can be recognized as important factors of flat-rate preference. Table 2 summarizes the results.

Flat rates have been gaining increasing attention as a means to promote the usage of ICT services. If ICT services are supplied at flat rates, our results suggest that overall ICT usage will increase drastically. The framework adopted in this analysis can provide mobile operators and policy makers with initial insights into the reasons underlying consumer preferences for flat rate plans. However, this study is just the first step toward building a detailed understanding of this phenomenon.

Further extension and elaboration is necessary to deepen and widen our collective understanding of the consequences of flat-rate applications. For example, although non-parametric approaches are useful in investigating the significance of each behavioral economic concept, their relative importance cannot be identified. Parametric approaches will be able to overcome the shortcomings of the non-parametric approaches because they deal with the factors affecting decisions within a single framework and can specify relative importance. Comparison with other services will also provide more profound insights into the impact of flat-rate applications.

Acknowledgment The authors are indebted to the reviewer for his helpful comments on an earlier version of this paper.

Table 2 Summary of the results from the empirical tests

The concepts conducive to flat-rate preference	Concepts that failed to explain flat-rate preference
Loss aversion	The shape of the probability weighting function
Reference dependence	
Mental accounting	
Ambiguity aversion	
Cognitive dissonance	

References

Camerer CF, Loewenstein G, Rabin M (2004) Advances in Behavioral Economics. Princeton University Press, New Jersey, NJ

Epstein LG (1999) A definition of uncertainty aversion. The Review of Economic Studies 66: 579–608

Garbarino EC, Edell JA (1997) Cognitive effort, affect, and choice. Journal of Consumer Research 24(2): 147–158

Kahneman D, Tversky A (1979) Prospect theory: An analysis of decision under risk. Econometrica 47(2): 263–291

Kling, Van der Ploeg (1990) Estimating local telephone call elasticities with a model of stochastic class of services and usage choice. In de Fontenay A, Shugard MA, Sibley DS (Eds.), Telecommunications Demand Modeling: An Integrated View. North-Holland, Amsterdam

Kridel DJ, Lehman DE, Weisman DL (1993) Option value, telecommunications demand, and policy. Information Economics and Policy 5: 125–144

Lambrecht A, Skiera B (2006) Paying too much and being happy about it: Existence, causes and consequences of tariff-choice biases. Journal of Marketing 43: 212–223

Miravete EJ (2002) Choosing the wrong calling pattern? Ignorance and learning. The American Economic Review 93(1): 297–310

Mitchell BM, Vogelsang I (1991) Telecommunications Pricing: Theory and Practice. Cambridge University Press, Cambridge

Mitomo H (2002) Heterogeneous subscribers and the optimal two-part tariff of telecommunications service. Journal of the Operations Research Society of Japan 35(2): 194–214

Narayanan S, Chintagunta PK, Miravete EJ (2007) The role of self selection, usage uncertainty and learning in the demand for local telephone service. Quantitative Marketing and Economics 5: 1–34

Taylor LD (1994) Telecommunications Demand in Theory and Practice. Kluwer, Dordrecht

Thaler R (1980) Toward a positive theory of consumer choice. Journal of Economic Behavior and Organization 1(1): 39–60

Train KE (1991) Optimal Regulation. MIT Press, Cambridge, MA

Train KE, McFadden DL, Ben-Akiva M (1987) The demand for local telephone service: A fully discrete model of residential calling patterns and service choices. RAND Journal of Economics 1(1): 109–123

Oi WY (1971) A disneyland dilemma: Two-part tariffs for a Mickey Mouse monopoly. Quarterly Journal of Economics 85: 77–96

Regulation of International Roaming Charges – The Way to Cost-Based Prices?

Morten Falch, Anders Henten, and Reza Tadayoni

Abstract This paper reviews EU regulation on international roaming and discusses whether this will lead to more cost orientation when setting roaming charges in Europe. First, a cost analysis for providing international roaming is presented. Then, the various proposals put forward during the debate in the EU Parliament are discussed. Finally, the issue of cost orientation is discussed.

Introduction

EU roaming regulation entered into force on 30 June 2007. On 23 May 2007 the European Parliament voted for a text on EU regulation of International roaming charges within Europe, which later was endorsed by the EU Ministers at their meeting on 7 June (European Parliament 2007c). Following the proposal, international roaming charges are now subject to a price cap, which was to be fully implemented in September 2007 and which will last for 3 years.

This intervention has led to reductions of 57% and 60% in charges for outgoing and incoming roaming calls, and is the result of a lengthy process, which began mid-1999 when the European Commission decided to carry out a sector inquiry covering national and international roaming services (CEC 2006a). This paper discusses whether European regulation of international roaming charges will lead to more cost orientation in international roaming charges, and what impact this will have on competition.

First, a brief techno-economic analysis of the costs of providing international roaming is presented. Second, the original proposal made by the Commission is reviewed and compared to the final proposal adopted by the European Parliament, as well as other proposals made during this process. In particular, the following proposals are addressed:

M. Falch(✉), A. Henten, and R. Tadayoni
CMI, Aalborg University
e-mail: falch@cmi.aau.dk

B. Preissl et al. (eds.), *Telecommunication Markets*, Contributions to Economics,
DOI: 10.1007/978-3-7908-2082-9_5, © Springer Physica-Verlag HD 2009

- The final proposal from the EU Commission, published 12 July 2006 (CEC 2006b)
- Draft opinion of the Committee on the Internal Market and Consumer Protection for the Committee on Industry, Research and Energy (ITRE), published 9 February 2007
- Report from the Committee on Industry, Research and Energy (ITRE), published 20 April 2007
- The final text adopted by the European Parliament 23 May 2007 (European Parliament 2007b)

The final text adopted the European Parliament is in substance identical to the final legislation. On this background, it is discussed how the changes proposed by the European Parliament will affect cost orientation as the proposals include two interventions in separate (although interrelated) markets. The paper treats the issues of wholesale and retail regulation in two separate sections.

Roaming Technology

The most important components used when international roaming is required are the Home Location Register (HLR), the Visiting Location Register (VLR), and the Mobile Switching Center (MSC). They provide the call-routing and roaming capabilities of the GSM network. The signalling system which is used for communication between these intelligent network components in the GSM network is the Signalling System 7 (SS7), which is widely used also in PSTN and ISDN networks. Other components in the mobile network system are the Equipment Identity Register (EIR), the Authentication Center (AUC), and the Gateway Mobile Switching Center (GMSC).

When a mobile terminal is turned on or moved to a new location area, it will register its location information to the VLR.[1] The VLR sends the location information of the mobile station to the HLR. In this way the HLR is always updated with regard to location information of subscribers registered in the network. The information sent to the HLR is normally the SS7 address of the new VLR, although it may be a routing number. A routing number is not normally assigned, even though it would reduce signalling, because there are only a limited number of routing numbers available in the new MSC/VLR and they are allocated on demand for incoming calls. If the subscriber is entitled to service, the HLR sends a subset of the subscriber information, needed for call control, to the new MSC/VLR, and sends a message to the old MSC/VLR to cancel the old registration. Call routing is based on the dialled mobile number, which is an E.164 number starting with country code, etc. If the dialled number is a local number, the connection is set up locally, otherwise the call is transmitted to the country to which the number belongs.

Depending on the usage scenario, different routing modes can be used for international roaming calls. At least four different scenarios can be distinguished:

---

[1] In the event that the user is in an area where there is no coverage from his/her home network, e.g. in another country, the precondition for registration to the VLR is that there is a roaming contract between the visiting network and the user's home network.

- Scenario 1: Calls inside a visited country.
- Scenario 2: Calls from a visited country to the user's home country.
- Scenario 3: Calls from a visited country to a third country.
- Scenario 4: Calls received in a visited country.
- It follows that three different countries may be involved in the handling of an international roaming call.
- 'Home Country': The country, where the user has his/her subscription. We have chosen Denmark as the home country.
- 'Visited Country': The country visited by the user. We have chosen France to be the country visited.
- 'Third Country': The country to which the call is directed, if different from the home country and the visited country. We have chosen Germany as the third country.

Scenario 1: Calls Inside a Visited Country

Different variations of scenario 1 are depicted in Fig. 1.

*1a) A **Danish user** travelling in France calls a **French user** staying in France.*

As seen, the call is routed locally in the visited country (France). The call set-up and switching are performed and maintained in France. However, even though the

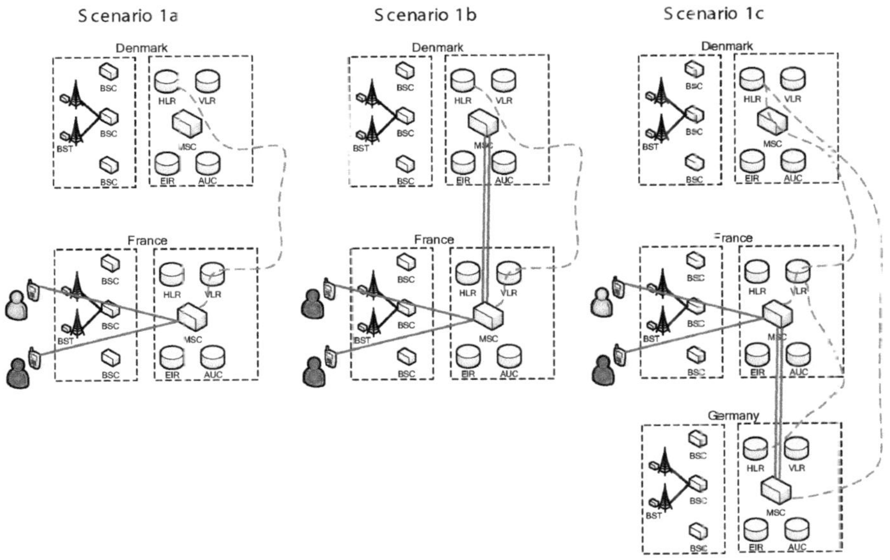

Fig. 1 Scenario 1: Calls inside a visited country
Note: In the figures, the Danish user is Red, the French user is Blue and the German user is Yellow. Dashed red lines indicate signalling channels, and blue bold lines indicate voice channels

call is routed locally, there are signalling communications between Denmark and France. For the voice connection, one origination and one termination are deployed.

*1b) A **Danish user** travelling in France calls another **Danish user** travelling in France.*

The call is routed to Denmark and the switching and call set-up are performed in Denmark. So apart from the origination and termination there are two international transits between France and Denmark included in order to maintain the connection. This routing method is called 'tromboning'[2] in the literature, which indicates that the voice channel is sent to the home network and back. This method is the common practice, but there are technologies which can eliminate the 'tromboning' part and maintain a local termination in this scenario.[3] This requires standardisation and agreement between the operators, and the cost reduction incentives are not very high, which has resulted in relatively limited use of these technologies.

*1c) A **Danish user** travelling in France calls a **German user** travelling in France.*

This is like the 1b scenario, but here the call is sent to Germany. Also additional signalling is needed.

Scenario 2: Calls from a Visited Country to the Home Country

Different variations of scenario 2 are depicted in Fig. 2.

*2a) A **Danish user** travelling in France calls a **Danish user** staying in Denmark.*

The call is sent to Denmark. The call set up is performed in Denmark. There is one origination, one termination and one transit.

*2b) A **Danish user** travelling in France calls a **French user** travelling in Denmark.*

The call set up is maintained in France. There is one origination, one termination and one transit. There is additional signalling between VLR in Denmark and HLR in France.

*2c) A **Danish user** travelling in France calls a **German user** travelling in Denmark.*

The call is sent to Germany. The call set-up is performed in Germany. There is one origination, one termination, one transit between France and Germany and one transit between Germany and Denmark. There is additional signalling between Denmark and Germany.

---

[2] See for example The Mobile Application Part (MAP) of GSM, Jan A. Audestad, Telektronikk 3, 2004.

[3] Ibid.

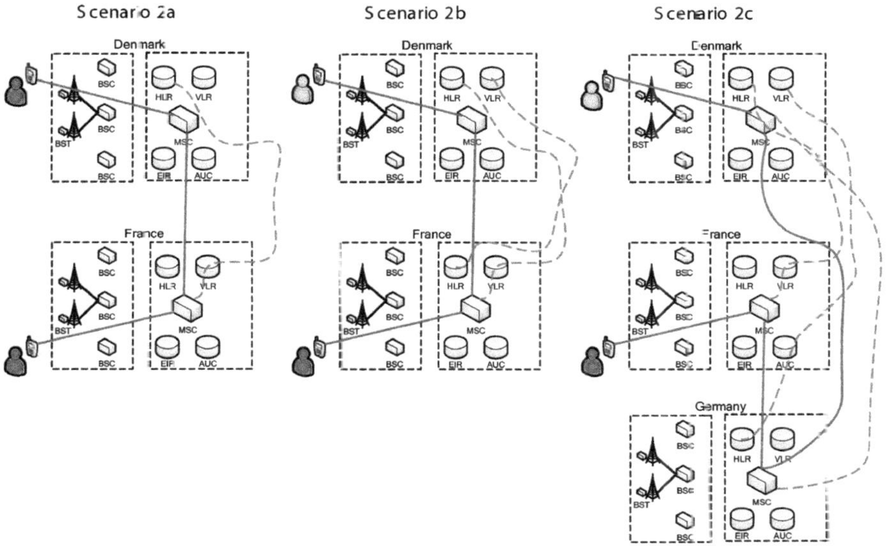

Fig. 2 Scenario 2: Calls from a visited country to the home country

Scenario 3: Calls from a Visited Country to a Third Country

Different variations of scenario 3 are depicted in Fig. 3.

3a) A ***Danish user*** *travelling in France calls a* ***German user*** *staying in Germany.*

The call is sent to Germany. The call set-up is performed in Germany. There is one origination, one termination and one transit.

3b) A ***Danish user*** *travelling in France calls a* ***Danish user*** *travelling in Germany.*

The call is sent to Denmark. The call set up is performed and maintained in Denmark. There is one origination, one termination, one transit between France and Denmark and one transit between Denmark and Germany.

3c) A ***Danish user*** *travelling in France calls a* ***French user*** *travelling in Germany.*

The call set up is performed in France. There is one origination, one termination and one transit between France and Germany.

Scenario 4: Receiving Calls in a Visited Country

This applies to all the above-mentioned scenarios, however with the difference that here, the Danish user travelling in France receives a call. This will in all cases involve one termination.

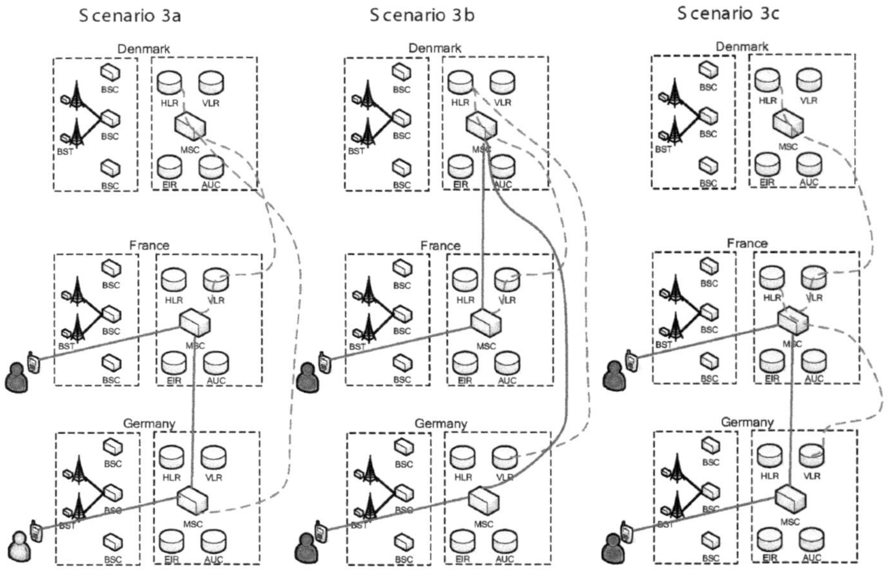

Fig. 3 Scenario 3: Calls from a visited country to a third country

All scenarios assume that calls are terminated in a mobile network. Scenarios similar to scenarios 1–3 could be made for calls with fixed termination. In scenario 4, calls can be originated either in the fixed or the mobile network. This is however not relevant in this context, as the roaming charge paid in this scenario does not include call origination (this is paid by the caller).

Techno-economic Analysis of Roaming Costs

Basically, international roaming involves the following functions:

- Mobile origination (MO)
- Mobile/Fixed termination (MT/FT)
- International Transit (IT)
- Roaming specific costs (RSC)

The costs of mobile origination are comparable to those of mobile termination. Mobile termination rates are subject to regulation within the EU and are in principle cost-based. Mobile termination rates per minute varied in October 2006 between 0.0225 in Cyprus and 0.1640 in Estonia (Fig. 5). However in most countries the rates are close to the EU average of 0.114. The European average for local fixed termination is 0.0057. It may be argued that it is more appropriate to use the double transit charge of 0.0125 (European average), as the calls to be terminated are international.

International transit costs depend on the inter-operator tariffs agreed between operators. These tariffs are confidential, but some information on these has been provided to the Commission. According to Copenhagen Economics, international transit costs vary between 0.01 and 0.025/min (Jervelund et al. 2007). They use in their calculations 0.02/min as a high estimate for international transit costs. INTUG, for instance, estimates that the wholesale cost for international calls between EU countries is of the order of 0.01/min (INTUG 2006). In the report from Copenhagen Economics, roaming specific costs are estimated to account for 0.01–0.02/min. The costs used in the calculations are summarized below (Table 1).

Using the cost estimates from Fig. 1, roaming costs for each scenario can be calculated as depicted in Table 2. The results are in line with the wholesale cost estimated in the impact assessment report prepared by the Commission. In this report the average international roaming costs are estimated to be slightly below 0.2/min (CEC 2006a). It follows from the table that the major cost components are origination and termination of a call. These two components add up to 0.1265 or 0.228 depending on the kind of termination. In spite of this, retail charges for international roaming call are almost four times higher than for national mobile calls (Fig. 4). This indicates that the charges currently paid by international roaming customers are way above the underlying costs, and that the Commission therefore has a strong case for suggesting regulatory intervention.

Table 1 Cost estimates of key network functions in international roaming

	/min
Mobile origination/termination	0.114
Fixed termination	0.0125
International transit	0.02
Roaming specific costs	0.02

Table 2 Roaming costs per scenario (/min)

Scenario	Mobile termination		Fixed termination	
0	2 * MT	0.228	FT + MT	0.1265
1a	2 * MT + RSC	0.248	FT + MT + RSC	0.1465
1b	2 * MT + RSC + 2 * IT	0.288	FT + MT + RSC + 2 * IT	0.1865
1c	2 * MT + 2 * RSC + 2 * IT	0.308	FT + MT + 2 * RSC + 2 * IT	0.2065
2a	2 * MT + RSC	0.248	FT + MT + RSC	0.1465
2b	2 * MT + RSC + IT	0.268	FT + MT + RSC + IT	0.1565
2c	2 * MT + 2 * RSC + 2 * IT	0.308	FT + MT + 2 * RSC + 2 * IT	0.2065
3a	2 * MT + RSC + IT	0.268	FT + MT + RSC + IT	0.1665
3b	2 * MT + RSC + 2 * IT	0.288	FT + MT + RSC + 2 * IT	0.1865
3c	2 * MT – RSC + IT	0.268	FT + MT + RSC + IT	0.1665
4	IT + RSC	4	–	–

Note: 2 Scenario 0 includes the costs for a national call without roaming. RSC are included for 2c as this type of call involves more complicated call handling than the other scenarios. Scenario 4 includes costs incurred in addition to those paid by the calling party only.

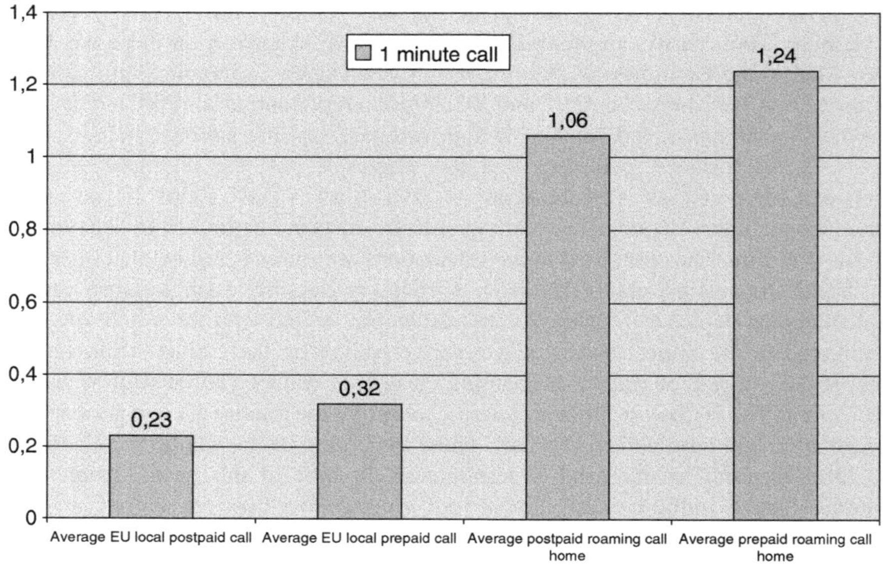

Fig. 4 Prices of local and roaming calls (From CEC 2006a)

Regulation of Wholesale Prices

The first proposal from the Commission linked the prices paid for international roaming to prices paid by customers for ordinary mobile calls in their home country. This home pricing principle was replaced by a "European Home Market Approach" in the revised proposal, in which the same maximum price limits are applied in all the EU member states. In the final proposal adopted by the Parliament, the concept Eurotariff is used for the regulated price that the operator charges its customers for international roaming calls within the EU area. The Europe-wide maximum tariffs are defined both for wholesale and retail charges.

The proposal from the European Commission with respect to wholesale international roaming prices includes the following elements:

- Wholesale price ceilings for initiating roaming calls is set with reference to multiples of the average per-minute mobile termination rate (MTR) for operators with significant market power (SMP).
- In the case of a regulated roaming call to a number assigned to a public telephone network in the Member State in which the visited network is located (scenario 1a), the maximum wholesale price is set at two times the average per-minute MTR (2 × MTR).
- In the case of a regulated roaming call to a number assigned to a public network in a Member State other than that in which the visited network is located, the maximum wholesale price is set at 3 × MTR.

In the case of receiving calls when roaming in other countries, the wholesale charge payable to the operator of the network on which the visiting customer is roaming is not subject to regulation according to the Commission proposal for wholesale international roaming. The wholesale price is the specific mobile termination rate of the operator used in the visited country.

The approach taken in the proposal raises a number of issues with regard to how a price limit on international roaming charges can be set. Is MTR the best basis for setting ceilings for international wholesale roaming prices, and are multiples of MTR the best way of setting maximum wholesale prices for the different kinds of roaming calls? The list of issues debated includes:

- Should not only MTRs be used but also origination and transit rates?
- Should the same maximum cross-Community MTR be used in all countries?
- Should there be common rules for cost calculation methods when setting national maximum MTRs?
- Should a 75 percentile be used when calculating the cross-Community MTR?
- Should peak or average rate MTRs be used?
- Should there be a common wholesale cap for all roaming scenarios including the case of receiving calls when roaming?
- Should MTRs be calculated on a per-minute basis of each international call or on an average operator basis?

MTR is a reasonable element when setting wholesale roaming prices. In technical terms, roaming services basically consist of origination, termination, and transit. In the proposal from the Commission, the costs of each of these three elements are set at one MTR. This helps to create a pricing scheme, which is both simple and consistent. But it is debatable whether this pricing scheme reflects the underlying cost structure.

With respect to the termination rate as a proxy for an origination rate, termination rates (market 16) could be said to have the advantage, from a regulatory point of view, that they are (or will be) regulated and that rates, therefore, are known to the regulators. Origination is only regulated in a limited number of countries, as the wholesale origination markets (market 15) mostly are considered as competitive and, therefore, not subject to ex ante regulation. Furthermore, although termination rates vary between countries (from 2.25/min in Cyprus to 16.49/min in Poland (CEC 2007), Fig. 25). Most termination rates are in the vicinity of the average EU termination rate.

Wholesale origination rates are not public and therefore difficult to compare with termination rates. The underlying costs for origination and termination are almost identical and rates would be expected to be the same. Origination is subject to competition in most countries and the market will therefore, in theory, ensure cost-based prices. Also termination charges should be cost-based – not because of market forces but due to regulation. However, the indication is that origination rates are lower than termination rates. For the users this can be seen in the fact that a fixed-to-mobile call often is more expensive than a mobile-to-fixed call. The indication is therefore that 2 × MTR for a roaming call to a local customer within a visited country is above the wholesale cost of termination + origination. (Also because termination may be on a fixed network, where termination rates are considerably lower than on mobile networks). But then again, MTR can be considered as a reasonable proxy for origination. It should be noted that regulation of termination rates has just been implemented, and that it is likely that MTR will decrease (further) in the coming years. Termination rates constitute an area, which is subject to increasing regulatory attention, as MTRs are considered as being too high.

With respect to transit, the 3 × MTR is a rather favourable proposition for the operators. As indicated above, wholesale costs of transit are far below termination tariffs. It could, therefore, be argued that 2 × MTR in all cases of roaming call origination should be used as the maximum wholesale price. The overall conclusion must be that using MTR as the basis for wholesale prices is a reasonable solution, but that the most debatable issue is whether MTR is a reasonable proxy for the transit element. The argument for using MTR as a proxy for transit is that it results in a consistent and fairly simple pricing structure.

Secondly, with respect to calculating the MTR, different questions have been discussed. The first question is whether it is reasonable to use the same MTR in all countries. Are the costs of roaming different from country to country to the extent that an average EU MTR poses grave problems? Indeed, termination rates vary from country to country – in a few cases considerably, as indicated. This, however to some extent, reflects different pricing strategies more than differences in costs. The costs of providing roaming services do not differ substantially between countries, except for the unusual cases of popular tourist sites, which require networks with huge over-capacity in non-tourist seasons. Also, it is true that it is more expensive to build mobile networks in mountainous countries than in flat countries. However, these differences in costs are not necessarily reflected in present MTRs. Therefore, the advantages for users of a common cross-Community MTR outweigh the difficulties created by differences in network costs in different countries.

Thirdly, one might ask whether there should be common rules for setting MTR costs. In an increasing number of countries, mobile LRIC prices are being introduced. In general this leads to lower MTRs. In Sweden, for instance, the calculated 2007 LRIC MTR price is set at SEK 0.5425 (app. 0.06)/min, while the actual MTRs charged by Tele2, Vodafone and TeliaSonera are respectively SEK 0.99, SEK 1.35 and SEK 0.80 (app. 0.11, 0.15 and 0.09) per minute (Nordic NRA 2006). With respect to a common EU costing method for calculating MTR, this is not likely. More and more countries are implementing mobile LRIC prices, but other costing methods are also used. There is no tradition for imposing one specific costing method for specific services on Member States. LRIC will generally lower MTR prices, but it is not realistic to implement common cost calculation methods for specific services.

Fourth, there is a discussion regarding the use of a 75 percentile for calculating a European MTR. This is a proposal that ERG introduced. The Commission proposal is an average MTR, and the 75 percentile will in most cases be above the average and is in the MTR case slightly higher than the average MTR (app. 0.02). The MTR will then be higher and so will customer prices. The argument for using a 75 percentile is that it helps ensure that only a few operators will witness a lower cross-Community MTR than their own MTRs. But a 75 percentile based calculation will take rates further away from a most efficient operator approach, which is the traditional basis for cost calculations. An average MTR is therefore more reasonable than a 75 percentile based MTR.

A fifth issue concerns the use of peak or average rates for MTRs. Peak rates are obviously higher than average rates and, therefore, an advantage for (most) operators and more costly for customers. The argument for peak rates could be that they help avoid situations where the discrepancy between a cross-Community MTR and the local MTR in peak situations is too large. This situation can be avoided by the use of maximum instead of average costs as a benchmark for cost determination. It can be added that the peak and off-peak MTRs are the same in most countries in the EU (see Fig. 1). The use of peak rates instead of average rates, therefore, has a limited impact on the level of roaming charges. A peak-rate-based cross-Community MTR will, of course, be higher than an average-based MTR – but not substantially and therefore it is acceptable.

A sixth issue is whether it is reasonable to charge the same for receiving calls as for initiating calls. Wholesale charges for receiving calls are not included in the Commission proposal for regulation of international roaming. For receiving calls, the wholesale price charged by the operator handling termination of the call will not be the EU average MTR, but the MTR of the local operator in the visited country. An argument for using the average EU MTR also in this case is that operator specific MTRs may be relatively high and consume a large share of the wholesale price received by the operator in the visited country in the case, for instance, where a visiting customer calls a mobile customer in a third country (scenario 3). The problem with high retail prices for receiving calls are discussed further in the section on retail prices below.

A seventh issue is whether per minute or average operator MTR should be used. An average operator MTR is clearly the most flexible solution for operators and could, therefore, be the preference for operators. However, for reasons of transparency for customers, and for reasons of lowering regulatory burdens, a per minute MTR is preferable. The advantages of price transparency for the users with a per-minute charge outweigh the advantages of flexibility with an average operator MTR solution.

The proposal from the Commission has been subject to intensive negotiations and discussions within the Parliament, with the Ministry Council, and with the industry. And substantial revisions have been made before the final adoption by the Parliament.

The Committee on the Internal Market and Consumer Protection under the European Parliament issued a draft opinion on 9 February 2007 suggesting a number of amendments to the proposal from the Commission (CEC 2006b). With regard to determination of the level of charges, this proposal was largely built on recommendations from a report prepared for the committee by Copenhagen Economics (Jervelund et al. 2007). The most important changes were:

- Instead of using peak MTRs national average MTRs are used for calculating the EU average MTR. This results in a decrease in the EU MTR of 0.013.
- Instead of the average MTR applied in the Commission proposal, the 75th percentile should be used for calculating the price cap. This results in an increase in the MTR of 0.0278.
- A maximum wholesale charge is set at two times MTR for all roaming calls.

The argument for using two MTRs for all types of roaming is as noted above that one MTR is a very generous price limit for handling of international transit. The argument for using the 75th percentile as suggested by ERG is to protect high cost operators. But it also renders the reduction from three to two MTRs more acceptable to operators. In the subsequent draft report the proposal concerning using the 75th percentile has been replaced by a special clause regarding operators located in high cost regions. In total the amendments imply a further tightening of the wholesale regulation. The amendments must however be seen within the context of a suggested increase in the maximum profit margin for the retail market (see next section).

The changes proposed by the Parliament better reflect the cost structure for international roaming, as the costs of transit are negligible compared to the costs of origination and termination. However it is not clear how the changes made in this proposal will impact the revenue, as this will depend on the call pattern. The price limit for a call charged by three MTR in the original proposal has been reduced by about 0.10, while the limit for other calls has been increased by about 0.015. However, it can be expected that calls, which involve two countries only, represent an overwhelming majority of the calls.

The text adopted by the Parliament on 23 May still recognizes the use of the concept of MTR, but only a as benchmark. The cap on wholesale charges is set at 0.30 the first year, and then reduced to 0.28 after 1 year and 0.26 after 2 years. After 3 years, the regulation may be extended or amended following a review by the Commission. This approach is clearly more beneficial for the operators, as the minute charge is increased from 0.2468 to 0.30. Reductions are built into the system, but also the use of the approaches previously suggested would result in further reductions if national MTRs are reduced. The major reason for this change is to reach agreement with the Telecom Ministry Council, who suggested even higher charges (first 0.50 and later 0.60) (Table 3).

Regulation of Retail Prices

The introduction of price regulation at retail level is certainly more controversial than price regulation at wholesale level. It is generally acknowledged within the EU that the best way to ensure competition and bring down retail prices is to ensure open access to network facilities provided at cost-based prices. Therefore, the EU Commission

Table 3 Wholesale charges allowed in the three proposals (€/min) (From Jervelund et al. 2007), (European Parliament 2007b)

Commission proposal	0.2320 (2 × MTR)		0.3480 (3 × MTR)
European Parliament draft opinion (9 February 2007)		0.2468	
European Parliament draft report (20 April 2007)		0.2180	
European Parliament (23 May 2007)	30 (1st year)	28 (2nd year)	26 (3rd year)

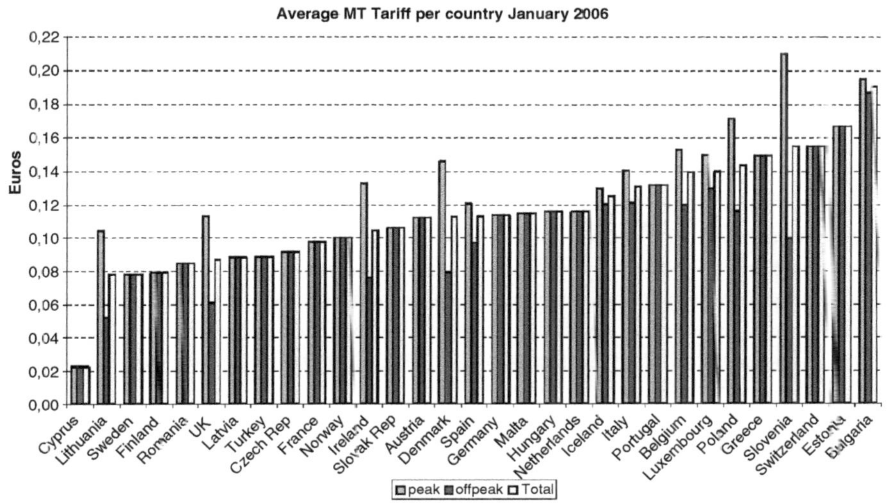

Fig. 5 Average MT Tariff per country January 2006 (From ERG)

recommends applying price regulation mainly at wholesale level. During the first-phase consultation preceding the proposal from the Commission, the great majority of the respondents in favour of regulation preferred regulation at wholesale level only. For instance, ERG favoured introduction of regulation at wholesale market first, and adoption of a 'wait-and-see' approach to regulation of the retail market.

The argument concerning retail regulation from the Commission is that there is no 'guarantee that lower wholesale prices will be passed through to retail roaming customers, given the lack of competitive pressures on operators to do so.' This argument could be used for regulation of retail prices for any service provided in markets with limited competition. A relevant question is, therefore, whether there are any special reasons for allowing more tight regulation of international roaming services than other retail telecom services.

In the impact assessment report, it is argued that there is no clear relationship between costs and end user prices for roaming services. Some European operators have entered into mutual agreements with foreign operators and have in this way been able to buy roaming services at reduced prices. However, these operators do not yet offer cheaper retail roaming services than others.

But this lack of relationship can also be observed for fixed services. In many countries reductions in charges for switched interconnection prices have not been followed by similar reductions in retail prices, but have instead led to increased margins between wholesale and retail prices.

In addition, the market for international roaming services is not without competition. Mobile operators may use low charges on international roaming services as a competitive parameter in order to attract more customers. Also, mobile service providers can offer cheap international calls, if the operators choose to maintain an excessive profit margin in this market.

The main argument for retail price regulation in this field is to ensure a fast and Community-wide lowering of end-user prices in a field that has been plagued with prices that are far too high. Experience shows that wholesale price deductions are not transferred to retail price reductions. However, the same argument could be made in other cases where there has been no regulation.

In the proposal from the Commission, retail price regulation is used in combination with regulation of wholesale prices. It could be argued that a price cap on retail prices is sufficient to ensure low charges for users. However, this might lead to a profit squeeze, where service providers would be unable to cover their costs and it would, therefore, harm competition. This argument is used both in the proposal and in the impact assessment report. A combination of retail and wholesale regulation is the most appropriate solution – or at least, if there is retail regulation, there should also be wholesale regulation.

The proposal from the European Commission with respect to regulation of retail international roaming prices is as follows:

- A uniform price cap of 130% of the wholesale price is introduced for all calls within the EU made by a roaming customer.
- A price cap of 130% of the average mobile termination rate on charges paid for receiving calls while roaming within the EU.

The suggested price cap implies a margin of 30% of the wholesale price. The question here is whether this is sufficient to cover the costs of customer handling and other retail services, and also to ensure a reasonable profit for service providers. AT Kearney argues that retail costs alone constitute more than 30% for at least some operators (Kearney 2006). According to the impact assessment report, the current margin is 46% of the wholesale price (average wholesale and retail charges are 0.75 and 1.10/min respectively) (CEC 2006a).

On the other hand a margin of 30% of the wholesale price (equivalent to 23% of the retail price cap) is in line with the margin used for price setting of wholesale prices using the 'retail minus' principle used for some wholesale telecom services. For instance, mobile service operators in Denmark are offered wholesale products at a price equivalent to the end user price minus 21%. This seems to be sufficient to cover both retail costs and some profit, as a number of service providers are able to operate on these terms.

The impact assessment report compares the 30% with the EBIT of European mobile operators. It is, however, not obvious that this is a relevant comparison. The EBIT margin of mobile operators is a measure of the profitability and does not relate to the size of the retail costs. A 30% mark-up is in line with, for instance, the 'retail minus' rate used for mobile service operators and can, consequently, be deemed reasonable.

This margin has however been criticized by several parties. AT Kearney argues that retail costs are independent of whole costs and that a percentage mark-up is not appropriate. It is certainly correct that there is no direct relationship between these two types of costs. Substantial reductions in wholesale prices could, therefore, lead to a profit squeeze for service providers. The question is, however, if a manageable alternative to a percentage mark-up is available. Percentage mark-ups are used in many different contexts, e.g. cost studies and price regulation, if more exact data are unavailable. Determination of a cost-based absolute mark-up would require that empirical studies documenting retail costs were carried out at regular intervals. Furthermore, the profit margin should still be calculated as a percentage mark up. A percentage mark-up instead of an absolute mark-up is therefore the most manageable solution in a cost based pricing regime.

Copenhagen Economics supports the suggestion by AT Kearney and has recommended that the Parliament suggests an absolute mark-up of 0.14. This amount is founded on the cost analysis made by AT Kearney. It should however be noted that the AT Kearney study was commissioned by the GSM Association serving the operators' interests. It would therefore be obvious to make an independent assessment of retail costs before cost data are imported directly into new legislation. Nevertheless the absolute mark-up of 0.14 has been maintained in the Draft Opinion prepared by the Parliament.

In their opinion issued on 22 March 2007, the Committee on Economic and Monetary Affairs states that the proposed margin of 13% is 'excessively low'. They suggest retail charges of 0.50 and 0.25 accordingly for making and receiving calls. This reflects a profit margin of 150% (Losco 2007). The same charges have been suggested by the Committee on the Internal Market and Consumer Protection.

A second issue is whether MTR is an appropriate benchmark for the costs of receiving a call. It should be noted that the caller pays for both origination and termination of the call. The charge for receiving an international roaming call should therefore cover the costs of international transit and of roaming specific costs. These costs add up to no more than 0.04/min plus the costs of retail operations (customer-handling, billing, etc.). Furthermore it should be noted that some of these costs have been covered already, if the calling party makes use of international roaming as well.

In the final text adopted by the Parliament, the charges have been increased to 0.49 and 0.24 the first year. The final text does not reveal how these prices have been decided, but the charges are seen as a compromise with the Telecom Ministry Council, who first proposed retail charges of 0.50 and 0.25 and later 0.60 and 0.30 (Table 4) (European Parliament 2007a).

Table 4 Retail charges allowed in the four proposals (€/min) (From Jervelund et al. 2007), (European Parliament 2007b)

	Making a local call	Making a call home/ to third country	Receiving a call
Commission proposal	0.3016	0.4524	0.1508
European Parliament draft (9 February 2007)	0.3868		0.2634
European Parliament Report (20 April 2007)	40		15
European Parliament (23 May 2007)	49 1st year		24 1st year
	46 2nd year		22 2nd year
	43 3rd year		19 3rd year

Discussion

Regulation of international roaming is more complicated than regulation of other telecom services for two reasons. First the market structures on mobile markets are different than on markets for fixed services. Markets for fixed services are dominated by incumbent operators who have their own fixed infrastructures. Regulatory intervention demanding open access to these networks will benefit new entrants and promote competition at least in the short term. As regards mobile markets, the situation is slightly different as there are more mobile infrastructures on each market. It is therefore less obvious what the market implications will be if a similar kind of obligation is imposed on mobile networks. Second, regulation of international roaming is difficult to implement at national level as operators from more than one country are involved.

For these reasons, a common framework for regulation has not yet been adopted at EU level before now. International roaming was defined as a separate market in the market definitions applied in the EU regulatory framework. But the implementation of the new telecom regulation package has not led to any intervention on this market at national level. Although in August 2006 market analyses for other telecom services had more or less been completed in most countries, only Finland had made a decision on international roaming; here the conclusion was that the market was competitive. Thus regulation of the market for international roaming seems to be more difficult for national regulators to handle than regulation of markets for other telecom services.

The proposal for regulation of international roaming put forward by the EU Commission suggests the introduction of price caps in both retail and wholesale markets for international roaming. The major argument for such heavy-handed regulation is that at present international roaming prices are much higher than cost-based prices, and that roaming charges represent a major barrier towards growth in international mobile communication within the EU.[4]

An interesting aspect of the proposal from the Commission is the use of a European home-market approach, which implies use of common price caps for all EU member states. This implies that determination of price caps are moved from national to European level. This may therefore be seen as a step towards decreasing the power of national telecom authorities and strengthening regulation at EU level. A common price cap will improve transparency for consumers, but it may create a situation where operators in high cost countries may have difficulties in covering their costs in full. It may also create strange pricing schemes, where international roaming becomes cheaper than national roaming.

The price caps suggested by the EU Commission are based on mobile termination rates (MTRs). The arguments for using MTRs are that these are already subject to regulation, and that in principle they are therefore cost based. MTRs are used not

---

[4]Documented by Special Eurobarometer on Roaming published March 2007 http://ec.europa.eu/information_society/newsroom/cf/document.cfm?action = display&doc_id = 250

only as a proxy for wholesale costs of termination, but also for wholesale costs of origination in a foreign network. In order to keep regulation as simple as possible, MTRs are also used as an approximation for the cost of transmission of a call from one country to another. This is obviously problematic as this is a completely different service, and the cost of this service is only a fraction of the costs of mobile termination. This part of the proposal has been disputed by the European Parliament. The first ITRE report suggests a price cap of 2 × MTR for wholesale international roaming, while the Commission suggest a price cap of 3 × MTR, if the call goes to another country within the EU.

This suggestion has been maintained in the later draft from April 20 also, but in the final proposal the wholesale charges have been increased, without any justification in costs, but as a compromise with the Telecom Ministry Council.

The suggested price cap on retail prices is the most controversial part of the proposal, as regulation of wholesale prices as a means to bring down retail prices is preferred in other telecom markets. ERG and ten member states have all suggested that regulation of retail charges is delayed until the impact of wholesale regulation can be observed. The Commission acknowledges that regulation of retail prices should only be made in exceptional cases, and that international roaming represents such a case.[5]

Also in this case the price caps have been changed in the various proposals. The Parliament first suggested an absolute price cap of 0.14 was used, but returned to the 30% mark-up as suggested by the Commission. The 30% mark-up was considered too low by several parties, and also in this case a compromise was made with the Telecom Ministry Council who wanted substantially higher rates.

Both wholesale charges and retail charges have been subject to intensive debate. From the beginning operators were very much against any form of regulation, in particular at the retail level. In spite of amble documentation proving excessive rates without any relationship to costs, it is claimed that there is effective competition on the international roaming market. The proposal "smacks of planned economy-style approach to the market" according to a spokesman for the GSM Association (*Herald Tribune* 15 May 2007).

Also some Governments have been very reluctant towards regulation. In particular in tourist destinations in Southern Europe, international roaming has proved to be an important source of income. This was reflected in the negotiations between the Telecom Ministry Council and the European Parliament, where the Parliament had to renounce on the cost-based pricing principle in order to achieve an agreement.

The final agreement is a compromise, but seen from the consumers' point of view, it is a considerable improvement compared to the present situation, and it has been implemented with an impressive speed (less than 1 year after the proposal from the Commission was published). It is also a move away from regulation based on more or less objective economic evidence towards regulation based on political negotiations between parties with conflicting interests.

---

[5]http://ec.europa.eu/information_society/newsroom/cf/itemlongdetail.cfm?item_id = 3309.

The new legislation brings international roaming charges closer to costs, but it is less clear whether it will lead to more cost orientation. The proposal from the Commission is that legislation defines guidelines on how to determine international roaming charges by use of MTRs, which again are subject to cost-based regulation. The intermediate proposals put forward by the Parliament propose various changes justified by economic arguments. The final legislation takes a completely different approach. Here, price caps are defined in nominal terms and MTRs are used as benchmarks only. It seems that it has been too complicated to invent a pricing principle which could be used to justify the rates agreed upon with the Telecom Ministry Council.

References

CEC (2006a) Impact assessment of policy options in relation to a commission proposal for a regulation of the European Parliament and of the council on roaming on public mobile networks within the community.

CEC (2006b) Proposal for a regulation of the European Parliament and of the council on roaming on public mobile networks within the community and amending directive 2002/21/EC on a common regulatory framework for electronic communications networks and services. COM(2006) 382 final. Brussels.

CEC (2007) 12th report on the implementation of the telecommunications regulatory package – 2006. Vol. 2. Brussels.

European Parliament (2007a) European Parliament – news press service – MEPs deliver on cheaper roaming: calling rates to drop by the summer holidays.

European Parliament (2007b) European Parliament legislative resolution of 23 May 2007 on the proposal for a regulation of the European Parliament and of the council on roaming on public mobile networks within the community and amending directive 2002/21/EC on a common regulatory framework for electronic communications networks and services (COM(2006)0382C6-0244/2006/0133(COD)), European Parliament.

European Commission (2007c). The new EU Regulation to reduce roaming charges ahead of the European Parliament plenary vote: Frequently asked questions, European Commission. MEMO/07/158 Brussels, 23 May 2007.

INTUG (2006) European Commission – international mobile roaming an INTUG response to the DG information society second phase consultation on roaming charges, INTUG.

Jervelund C, Karlsen S, et al. (2007) Roaming – an assessment of the commission proposal on roaming. Brussels, European Parliament.

Kearney AT (2006). International roaming regulation – proposed retail mark-up and allocation of actual industry average retail costs, GSM Association.

Losco A (2007) Opinion of the committee on economic and monetary affairs, European Parliament.

Part II
Internet Issues

Substitution Between DSL, Cable, and Mobile Broadband Internet Services*

Mélisande Cardona, Anton Schwarz, B. Burcin Yurtoglu, and Christine Zulehner

Abstract This article reviews substitution patterns on the market for broadband internet services in Austria. We present survey evidence and demand estimations which suggest that DSL and cable are part of the same market at the retail and at the wholesale level. Survey and estimation results from most other countries point in the same direction. We also consider substitution to mobile broadband via UMTS/HSDPA and describe recent developments in Austria which is one of the leading countries in the adoption of mobile broadband.

Introduction

Broadband internet services are usually not only considered as of great importance for society, but are also important for sector specific regulation in telecommunications. In the US as well as in the EU, there have been intense debates about how to properly define broadband markets and if there is a need for regulation.[1] One of the main questions was whether broadband internet delivered via (upgraded) cable TV networks is part of the same market as broadband internet delivered via copper twisted pairs by means of DSL technology.

This article presents survey evidence and results from a demand estimation for Austria which deals with this question. The evidence suggests that DSL and cable broadband internet services are part of the same market at the retail level. This is supported by evidence from other countries such as the UK, the US, Portugal and Malta.

A. Schwarz (✉), M. Cardona, B.B. Yurtoglu and C. Zulehner
Regulatory Authority for Broadcasting and Telecommunications (RTR)
e-mail: Anton.Schwarz@rtr.at

*All views expressed are solely the author's and do not bind RTR or the Telekom-Control-Kommission (TKK) in any way nor are they official position of RTR or TKK.

[1] For the US see, e.g., Crandall et al (2002), for the EU see, e.g., European Commission (2004), Schwarz (2007) and Inderst and Valletti (2007).

B. Preissl et al. (eds.), *Telecommunication Markets*, Contributions to Economics,
DOI: 10.1007/978-3-7908-2082-9_6, © Springer Physica-Verlag HD 2009

Evidence for Austria and some other countries also suggests that DSL and cable are part of the same market at the wholesale level.[2] Estimates for Japan show, however, that DSL may also form a separate market under particular circumstances. A detailed analysis on a case-by-case basis is therefore necessary before concluding on the appropriate market definition.

We also consider the question whether broadband delivered via mobile networks by means of UMTS and HSDPA is part of the same market as DSL and cable. Survey evidence from end of 2006 suggests that this is not the case. However, recent developments suggest that there is an increased competitive pressure from mobile on fixed network broadband connections. Despite this evidence we conclude that mobile broadband is in a too early stage of development to draw firm conclusions on market definition.

The rest of the article is structured as follows: section "The Austrian Market for Broadband Internet Services" gives a brief overview of the Austrian market for broadband internet services. The next two sections present empirical evidence on consumer behavior: section "Consumer Survey Results" describes evidence from a consumer survey and section "Estimation Results" discusses results from a nested logit demand estimation. The results are compared to results from other countries. Section "The Development of Mobile Broadband" discusses recent developments of mobile broadband in Austria and the effects on fixed broadband. Section "Conclusions" concludes.

The Austrian Market for Broadband Internet Services

Broadband internet via cable networks became available in Austria in 1996 and DSL followed in 1999. By the end of June 2007 there were 1.54 million fixed network and about 350,000 mobile broadband connections. This corresponds to a fixed network broadband penetration rate of 44% of all households which is almost exactly the OECD average.[3] While Austria has been falling behind other countries in the past years with regard to fixed broadband connections, it seems to be a leading country with regard to mobile broadband.[4]

---

[2] According to the 2003 regulatory framework, national regulatory authorities in the EU are required to periodically analyse the state of competition on the market for wholesale broadband access which is defined as "bit-stream' access that permit the transmission of broadband data in both directions and other wholesale access provided over other infrastructures, if and when they offer facilities equivalent to bit-stream access.' (see Commission Recommendation of 11 February 2003 on relevant product and service markets within the electronic communications sector susceptible to ex ante regulation in accordance with Directive 2002/21/EC of the European Parliament and of the Council on a common regulatory framework for electronic communication networks and services, OJ L 114/45).

[3] See OECD (2007).

[4] See Berg Insight (2007) or Analysys (2007) and the discussion in section "The Development of Mobile Broadband".

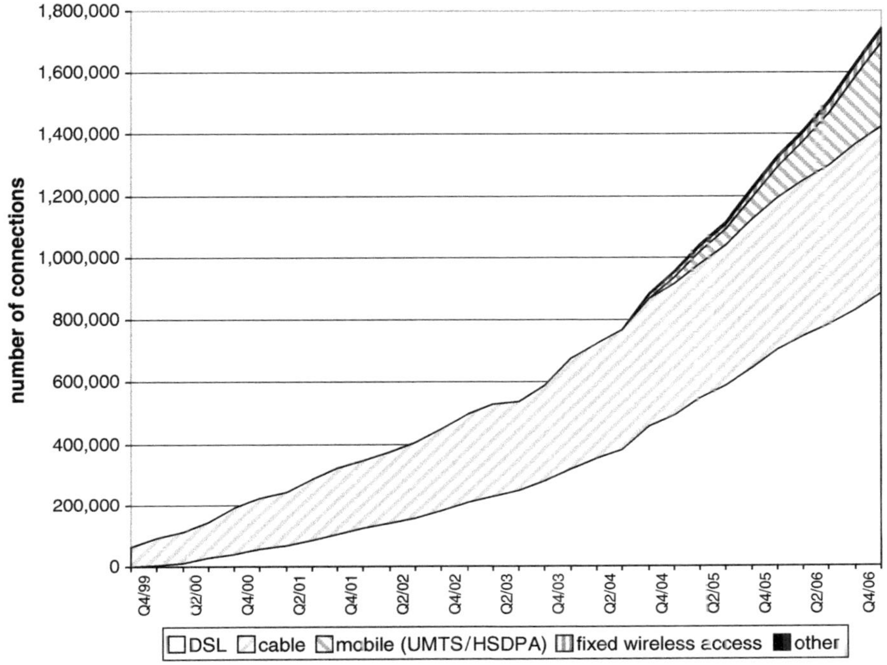

Fig. 1 Development of broadband connections (From RTR 2007a).

Figure 1 shows the development of broadband connections by technology.

The cable network coverage is approximately 50% of all households and relatively high compared to most other EU countries.[5] There are more than 100 cable network operators which offer broadband services in different regions of Austria (cable networks usually do not overlap), however, almost 90% of all cable connections are offered by six bigger operators. The DSL coverage is, like in most other EU countries, above 95% of all households. DSL services are offered by the former fixed network monopolist Telekom Austria as well as by alternative operators using the unbundled local loop (ULL) or Telekom Austria's 'bitstream' wholesale product.

Since 2003, mobile broadband via UMTS and since 2006, mobile broadband via HSDPA is available. HSDPA allows download rates of (theoretically) up to 7.2 Mbit/s. Since its introduction, mobile broadband has continuously grown much stronger than broadband delivered via fixed networks. Mobile broadband via HSDPA is usually available in all bigger cities (more than 5,000 inhabitants) and operators are continuing to roll out their networks.

[5] Exceptions are the Netherlands, Belgium, Luxembourg and Switzerland with almost full cable coverage.

Broadband delivered via fixed wireless access (W-LAN, WLL/WiMax or Wifi) and other technologies (Satellite, Power Line, Fibre to the home) only has a very small share of the market.

Consumer Survey Results

The data presented here is from a survey commissioned by RTR (the Austrian National Regulatory Authority) which was conducted in November 2006. Four-thousand-and-twenty-nine households and 1,510 businesses were interviewed about the type and characteristics of the internet connection they use, their monthly expenses and on their past and potential switching behavior. For households, individual specific data such as age, education and household size were also collected.

Looking at product characteristics such as price, download rate, included download volume and speed in November 2006 reveals significant differences among the access types on average. While users on average spend around 10.0 less on DSL than on cable connections (around 30.0 for DSL and 40.0 for cable), DSL connections come – on average – with lower speed and volume. Cable connections are also much more frequently bought with flat rate (58% of all cable connections) than DSL products (8%). The included volume for mobile broadband is much lower than for fixed broadband connections while the average expenses on such products are close to those of DSL. Nevertheless, the product portfolio of DSL and cable operators is such that for most DSL profiles there exists a comparable cable profile and the other way around.

Past Switching Behavior

Questioned on their past switching behavior, 22% of households claim to have changed their connection type at least once. Not surprisingly the biggest movement has taken place from narrowband (dial-up and ISDN) to broadband connections. But following this, the switching between cable and DSL is the next notable movement (see Fig. 2). Switching to mobile broadband has been far less intense. A similar pattern can be observed for business users although switching between DSL and cable occurs to a somewhat smaller extent (5.9% of all business users who have switched from DSL to cable and 3.2% in the other direction).

Potential Substitution Between DSL and Cable

To investigate potential future switching behavior in response to a (small but significant) price increase, households and businesses with a DSL connection which are aware of cable availability in their area were directly asked whether they

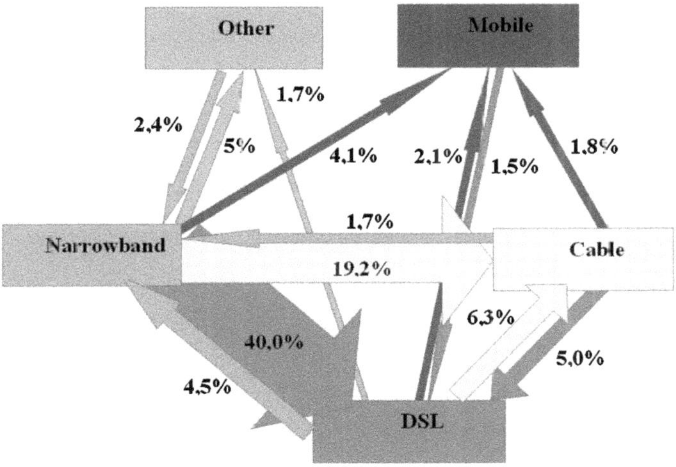

Fig. 2 Switching between broadband connection types (in percentage of all households who have switched)

consider cable as an "appropriate substitute"[6] for DSL.[7] Respondents who agreed with this were furthermore asked to assess the effort related to change the connection type. Results from these questions can be seen in Fig. 3 for households. The 10.4% of households who consider cable as an appropriate substitute and also consider switching costs as low[8] can be classified as the part of the population that are likely to substitute in case of a price increase. The corresponding figure for businesses is 6.2%. This allows us to calculate elasticities for this sub-sample of the survey. Assuming a 5% price increase this yields a DSL own-price elasticity of −2.08 for households and −1.24 for businesses.[9] The stronger preference for DSL of businesses is also reflected in a higher market share of this connection type for business users compared to households. Cable network operators are more focused on households (as they started their business selling TV services) and therefore

[6] The original expression used in the German interviews was "guter Ersatz".

[7] There may be households which are not even aware of cable availability in their area. These households may be less likely to switch in case of price increase. Our data on cable availability and household location is not sufficiently detailed, however, to allow for a good estimate of the share of households not aware of cable availability. One could also argue that in case of a price increase households would get informed about alternative access types and then may on average react like the households who are already aware of their alternatives.

[8] A rather high share of households which consider cable as an appropriate substitute to DSL also consider switching costs as low (65%). This is not implausible since the monetary switching costs are low in most cases due to promotions (no installation/set-up fee).

[9] This is the elasticity which results from substitution to cable only. The proper own-price elasticity of DSL would – as in the studies presented in section "Estimation Results" – also consider switching to other access types such as mobile and narrowband.

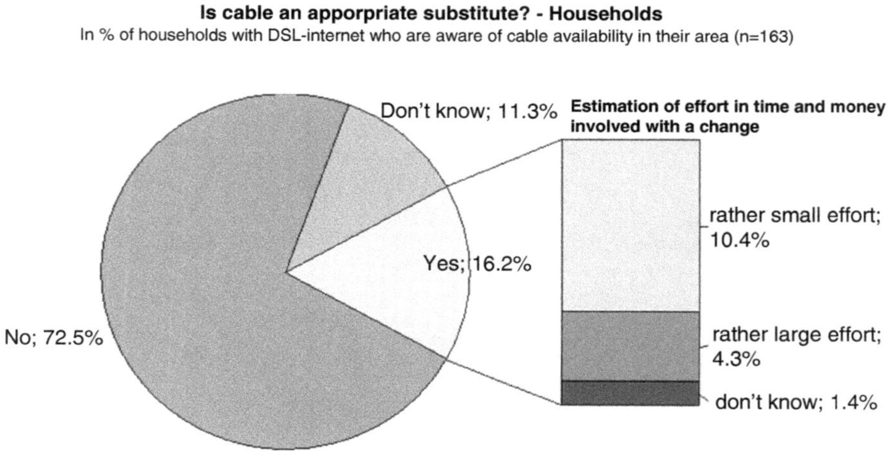

Fig. 3 Cable an appropriate substitute – households

might not be considered as a good alternative by many businesses. Nevertheless, these results as well as the past switching behavior indicate that DSL and cable are substitutes for a significant share of households and businesses and that the price elasticity of demand for DSL is elastic. Of course, these elasticities have to be interpreted with caution since they are based on the abovementioned assumptions and on stated (and not revealed) preferences.

Research to address the question of substitutability between cable and DSL access has also been carried out by the Malta Communications Authority MCA (2007) and the UK national regulatory authority Ofcom (2006). In Malta a survey showed that more than 30% of respondents with internet subscription regard cable an appropriate substitute for ADSL and more than 40% the other way around. Further, 53% of all internet households consider it as not difficult to switch from ADSL to cable or vice versa.

In the UK, 25% of respondents with an ADSL connection and 28% with a cable connection claim that they would switch following a 10% price increase. This compares to 16% claiming to switch connection following a price increase in both ADSL and cable. Business consumers show overall less willingness to switch, with 17% claiming to switch if the price for only DSL rose by 10%, and 8% claiming to switch if the price increase occurred across all types of broadband access. Ofcom interprets this as an upper bound since it is likely that there are consumers who claim that they would switch but would not actually do so.

Similar to the results form the Austrian survey, both studies indicate the existence of significant competitive pressure from cable on DSL services at the retail level. All three authorities concluded that DSL and cable are part of the same market at the retail as well as at the wholesale level (i.e., the wholesale broadband access market).[10]

[10] See Ofcom (2007), MCA (2006) and RTR (2007b).

Substitution Between Mobile and Fixed Broadband

In Austria households and businesses were also asked on their actual and potential use of mobile broadband via UMTS/HSDPA. At the time of the survey, 4% of households with internet connections used mobile broadband compared to 15% of businesses. While businesses use mobile broadband predominantly complementary to fixed broadband, private customers usually either have a fixed or a mobile connection, but not both (see Fig. 4).

The effect of the adoption of mobile internet on the fixed internet access for businesses is shown in Fig. 5. Sixty-eight percent of businesses with a mobile

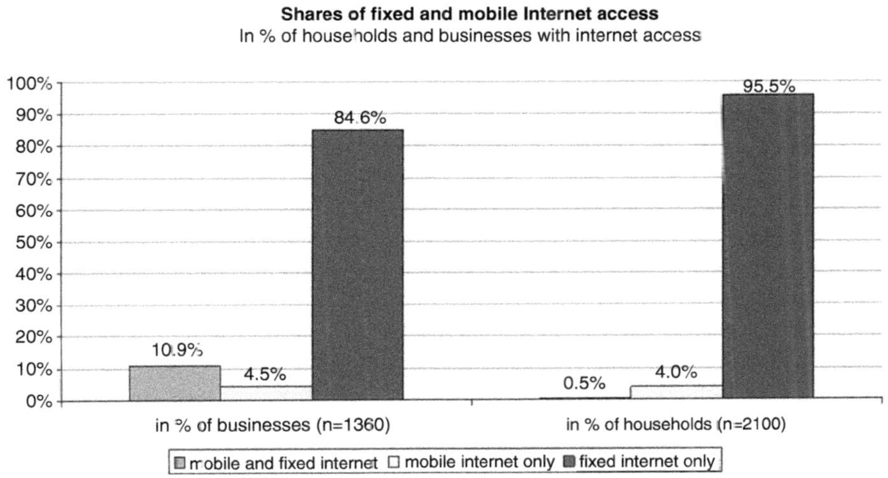

Fig. 4 Shares of fixed and mobile Internet – households and businesses

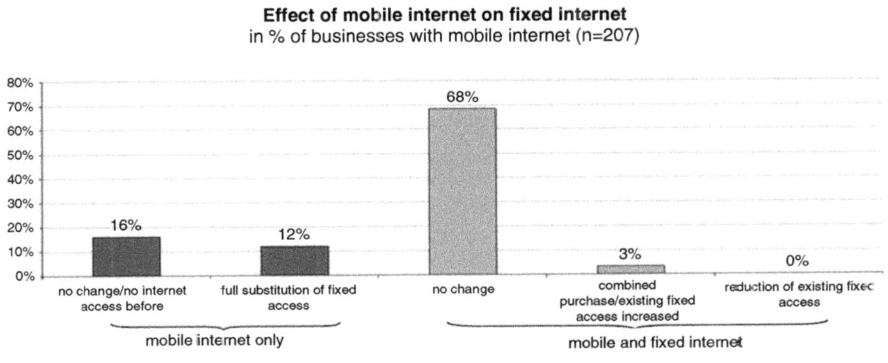

Fig. 5 Effect of mobile Internet on fixed Internet – businesses

connection said that purchasing a mobile connection did not affect their fixed line access. For these users, mobile and fixed access obviously are complements. Three percent even said that they expanded their fixed line access together with the adoption of mobile access or that they purchased both at the same time. On the other hand, 12% said that they cancelled their fixed line connection, i.e., switched from fixed to mobile access. Another 16% subscribed to mobile access without having had a fixed access before. Assuming that at least a part of these users would have bought a fixed broadband access if mobile had not been available; this can also be regarded as substitution to some extent.

To assess potential future switching behavior, DSL and cable households were also asked whether they consider mobile broadband an appropriate substitute and how they perceive switching costs. Households (8.1%) with a DSL or cable connection say that they consider mobile broadband as a good substitute and regard switching costs as small (see Fig. 6) – a figure somewhat lower than the 10.4% for DSL and cable.

While there appears to be some potential for substitution with regard to private users, business users are more likely to continue their complementary use of both fixed and mobile broadband access products. Eighty-four percent of all business users who plan to buy a mobile connection within the next year plan to do so in addition to their fixed line connection while only a very small share is planning to totally give up their fixed network access (see Fig. 7).

Concluding, the results show a certain acceptance of mobile broadband as a further access alternative. In particular private customers may substitute their fixed line connection for mobile connections. Business customers appear to use both types of connections more in a complementary way. Compared to the (actual and potential) substitution from DSL to cable, substitution from fixed to mobile connections is more limited. The Austrian regulatory authority therefore concluded that mobile broadband should not be included in the relevant market (see RTR 2007b). Of course these data are only indicative as they were collected in November 2006 when introduction of mobile broadband to the market had been rather recent and

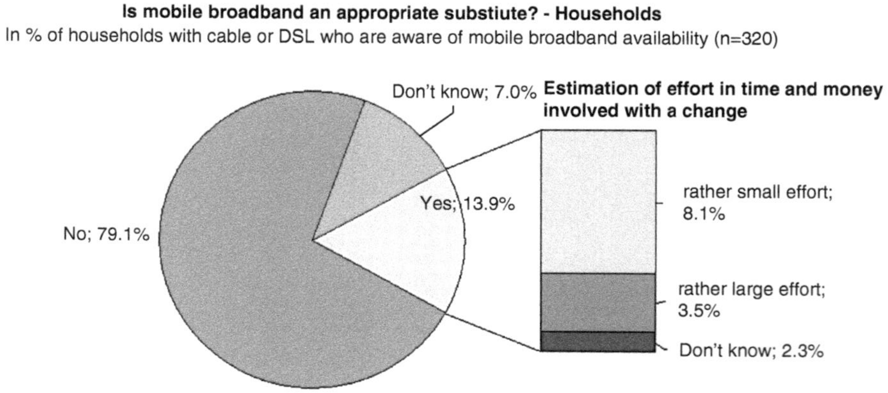

Fig. 6 Is mobile broadband an appropriate substitute? – Households

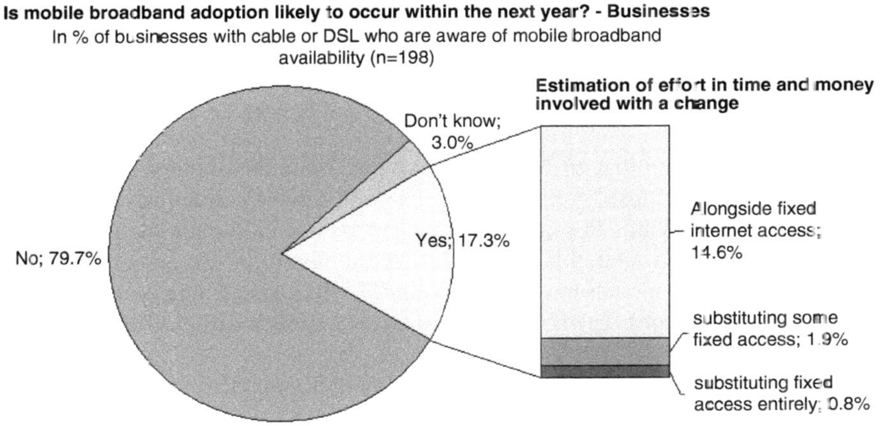

Fig. 7 Future mobile broadband adoption – Businesses

mobile penetration rates were still low. The development of mobile broadband in 2007 is described in section "The Development of Mobile Broadband".

The data collected in the survey has also been used to estimate price elasticities of demand for different types of internet access services.

Estimation Results

Cardona et al. (2007) use a nested logit discrete choice model to estimate price elasticities of demand for DSL and cable internet access. This section briefly describes the methodology used, the main results from the estimation, and compares the results to results from other studies.

The analysis in Cardona et al. (as well as in the other studies discussed in this section) is based on a random utility model where consumers choose from a set of choices. The utility a consumer derives from a particular product depends on characteristics of that consumer and on the characteristics of the product. To account for characteristics that are unobserved by the econometrician, the utility of consumer i for product j is of the form

$$U_{ij} = V_{ij} + \varepsilon_{ij}, \tag{1}$$

where i and j are the indices for consumer i, $i = 1, \ldots, I$, and product j, $j = 1, \ldots, J$, and where the term Vij reflects the deterministic part of consumers' utility. The error εij is a residual that captures for example, the effects of unmeasured variables or personal idiosyncrasies. It is assumed to follow an extreme value distribution of type I.

Consumers are assumed to purchase that product that gives them the highest utility. The probability Pij that consumer i purchases product j is equal to the probability that *Uij* is larger than the utility consumer *i* experiences from any other product, i.e., *Uij* > *Uij'* for all *j'*≠*j*. This probability is equal to

$$P_{ij} = P[U_{ij} > U_{ij'} \forall j' \neq j] = P[\varepsilon_{ij'} - \varepsilon_{ij} \leq V_{ij} - V_{ij'} \forall j' \neq j]. \quad (2)$$

Under the assumption that e*ij* follows an extreme value distribution of type I, the probability *Pij* has a closed form solution: The well-known conditional logit model (McFadden 1974). Within this model, we have to assume independence of irrelevant alternatives (IIA). To relax this assumption and to allow for correlations between choices, nested logit models have been developed.[11] In a nested logit model, choices are grouped in branches. The IIA property then only applies within a branch but not across branches.

Figure 8, depicts the preferred nested logit models considered in Cardona et al. Consumers are assumed to first decide whether they want to be connected to the internet or not. Then they decide between getting a narrowband or a broadband connection. If they decide for broadband, they have to choose between DSL, cable, and mobile access.

Cardona et al. use sequential maximum likelihood estimation to obtain estimates for the price elasticity. In doing that they consider the product characteristics like download rate and download volume, and consumer characteristics like age, household size, education, gender and whether the consumers are located in the capital city Vienna.

The price elasticities for broadband services are in a range of −2.6 to −2.4. The elasticity of DSL services is −2.55 indicating that 1% increase in price yields a 2.5% decrease in the demand for DSL services. The corresponding figures for mobile and cable services are −2.48 and −2.62. The elasticity for narrowband services is equal to −1.68. The results indicate that demand for all services is elastic with broadband services appearing to be more elastic than narrowband services. Different broadband services (in particular DSL and cable) constrain each other more than narrowband services. The elasticities for DSL and cable from the nested logit model are somewhat more elastic than those directly derived from survey questions (see section "Estimation Results"). One explanation for this might be that the nested logit model does not allow for switching costs while the survey questions

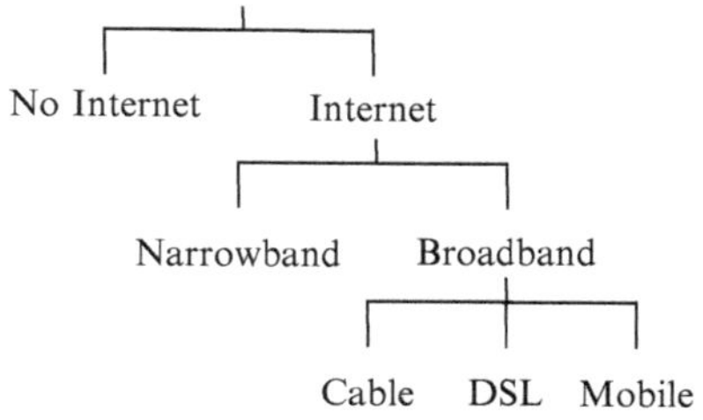

Fig. 8 Decision tree for the nested choice model

[11] See, for example, Maddala (1983), Greene (2003) or Train (2002).

explicitly took such costs into account. In addition, the discrete choice approach allows for substitution from DSL not only to cable but also to mobile, narrowband and no internet.

Applying a hypothetical monopolist test (HM-test, also called SSNIP-test)[12] for market definition, the results for the Austrian market further show that a 5–10% price increase from the competitive level would not be profitable for a hypothetical monopolist of DSL lines. Cardona et al. therefore conclude that cable services have to be included into the relevant market. They also conclude that the extent of substitution between DSL and cable is high enough so that both products are also part of the same market at the wholesale level. Since the penetration rate of mobile broadband was still very low in 2006, the authors do not investigate whether DSL and cable taken together would be constrained by mobile broadband.

These results can be compared to estimates from other countries. Despite the importance of the topic there seem to be only a small number of studies that have estimated price elasticities of demand for different types of broadband access services.[13]

Pereira and Ribeiro (2006) estimate demand elasticities for broadband access to the internet in Portugal, where the incumbent operator offers broadband access to the internet both via DSL and cable modem. The authors' main aim is to analyze the welfare implications of the structural separation of these two businesses. They use a random effects mixed logit model to estimate price elasticities of demand for different types of broadband services with panel data from April 2003 to March 2004 (1,650 households). The results suggest that households are very sensitive to price variations in Internet access services. More specifically, the demand for broadband access is more elastic than the demand for narrowband access, with an estimate of −2.836 and −1.156, respectively. They conclude that broadband and narrowband access are substitutes, however, the demand for broadband access is less sensitive to the price of narrowband access than the demand for narrowband access to the price of broadband access, with cross price elasticities being 0.503 and 0.876, respectively. Considering DSL and cable individually yields even higher elasticities of −3.195 and −3.130 respectively – a magnitude comparable to Cardona et al. (2007).

Crandall et al. (2002) use a nested logit model to estimate the elasticity of demand for broadband access to the internet in the USA. They use survey data gathered from the first quarter of 2000 to the fourth quarter of 2001. The survey is conducted with 3,500 respondents and covers information on broadband access availability, prices, and also socio-economic characteristics including income, race, occupation, education and age. The estimates are obtained by modeling the demand using a two layer nested logit model with no internet, narrowband and broadband at the first level and DSL and cable in the broadband nest. The own-price elasticities of demand for

[12] SSNIP is the acronym for small but significant non-transitory increase in prices.

[13] Other studies of demand for broadband internet services are Madden and Simpson (1997), Varian (2000) (experimental study), Goel et al. (2006) and Goolsbee (2006) (both using aggregate data). However, none of these studies investigates the demand elasticities for DSL or cable individually.

broadband access to the Internet through DSL reported is −1.184 compared to −1.220 for cable modem access. The cross-price elasticity of demand for cable modem access with respect to the price of the DSL is equal to 0.591. The authors consider this as evidence for DSL and cable being part of the same market (without going through the details of a SSNIP-Test).

A similar study for the USA making use of a discrete choice model is Rappoport et al. (2003). They also employ survey data from more than 20,000 randomly selected households over the January–March 2000 period. The survey data contains information on household size, income, education, age and gender of the respondents. For the area where all types of internet access (DSL, cable, narrowband) are available, a three level nested logit as in Fig. 8 is estimated. The estimated own-price elasticity of demand for DSL is elastic (−1.462) while the elasticity for cable is inelastic (−0.587). The estimate of the cross-price elasticity of demand for cable with respect to the price of DSL is 0.618.

Ida and Kuroda (2006) estimate several versions of conditional logit and nested logit models for broadband services in Japan. Their choice set includes five internet access alternatives: narrowband (dial-up and ISDN), DSL, cable, and fiber to the home (FTTH) – a rapidly growing access technology in Japan – in their choice set. Their data are from a survey, which was carried out using a web questionnaire. The dataset from 2003 with around 800 observations contains data on average expenditures (price), access speed, type of internet access line provider or service providers and individual characteristics such as gender, age, income, occupation and type of residence. Two thirds of the sample is made up by households that have chosen the DSL alternative. Their model allows the choice between narrowband and broadband access in the first layer and the choice among the three broadband alternatives, DSL, cable and FTTH in the second. Ida and Kuroda conclude that demand for DSL (at this time the main access technology with a share of 75%) is inelastic with an own price elasticity of −0.846. On the other hand, the demand for cable and FTTH is elastic with an estimate of the own price elasticity in the range of −3.150 and −2.500. They conclude that the DSL market is independent of other services. However, they also find that the upper and lower ends of the DSL market (i.e., very high and low bandwidths) are highly elastic (elasticities between −9 and −11) as they directly compete with FTTH and cable on the high end and dial-up and ISDN (narrowband) on the low end.

Summing up, most studies indicate that the pricing of DSL is significantly constrained by cable services where such services are available and that both products are likely to be part of the same market. Evidence from Japan (with a very high share of DSL-users) however shows that this is not always the case and that a detailed analysis of consumer preferences is necessary before concluding on the appropriate market definition. A limitation to these models is, of course, that they are static, and therefore switching costs are not allowed for. In future estimates it might also be useful to consider demand for service bundles (e.g., broadband with voice and/or TV) since such products are likely to gain importance. A relevant (future) question is also whether fixed network broadband services such as DSL and cable are constrained by mobile broadband services.

The Development of Mobile Broadband

This section looks at the development of mobile broadband delivered via UMTS/HSDPA in Austria and how it affected the demand for fixed broadband connections.

While Austria is only around the OECD average with regard to fixed broadband penetration, it seems to be among the leading countries with regard to mobile broadband. Analysys (2007), for example, compares Austria, Singapore, Sydney and Germany and finds that Austria has the highest share of mobile broadband connections (21% compared to less than 10% for the others in 2007). According to Berg Insight (2007), Mobilkom Austria was among the top five European operators with regard to the number of subscribers by end of 2006 (despite the relatively small size of the country) and Austria is one of the countries with the lowest prices for mobile broadband connections.

Mobile Broadband via UMTS is available in Austria since 2003. However, UMTS only allows bandwidths up to 384 kbit/s. With the introduction of HSDPA (basically a software-upgrade to UMTS), bandwidth of up to 3.6 Mbit/s and later 7.2 Mbit/s became possible. Although in practice usually only up to around 1 Mbit/s can be reached, these bandwidths are comparable to those of some fixed network broadband connections by means of DSL or cable.

Until the beginning of 2007, however, prices of mobile connections were significantly higher than prices of fixed network connections. As pointed out in section "The Austrian Market for Broadband Internet Services" this lead to a situation, where mobile connections were used to a large extent by business users in addition to their fixed connection while private consumers used mobile broadband only to a limited extent. This changed in February 2007, when one of the mobile operators significantly reduced prices and the other operators followed within weeks. Table 1 reports the development of the price per GB of monthly included download volume for a product of the largest mobile operator, Mobilkom Austria. A similar development could be observed for products of the other three operators.

These developments made the prices of mobile connections comparable to those of fixed network connections.[14] The share of mobile connections in total connections increased from 3.6% to 23.5% from Q2/05 to Q2/07, while the growth of fixed

Table 1 Development of mobile broadband prices (Mobilkom Austria product)

	November 2006	April 2007	July 2007	November 2007
Price	59.0	25.0	25.0	20.0
Included GB	1.2	1.5	3	3
Price/GB	49.2	16.7	8.3	6.7

[14] Actually, prices for fixed and mobile connections are hard to compare since prices of fixed connections vary with download speed and volume (with an increasing number of flat rate products) while prices for mobile connections only vary with volume. It seems reasonable, however, to compare mobile broadband prices to prices of fixed connections with 1 Mbit/s or 2 Mbit/s which were between 20 and 40 in July 2007.

broadband connections decelerated significantly in Q2/07 (see Fig. 9). Although there is still need for further investigation, this indicates that competitive pressure from mobile on fixed network connections increased considerably after the price cuts in February/March 2007. As the fixed network broadband penetration is only around OECD average, and given the timely coincidence of the two developments a pure saturation effect for fixed network connections appears unlikely.

This poses the question of whether mobile connections constrain fixed connections sufficiently in order to be included into the same market. As the development of mobile broadband is still in its beginnings and detailed up to date analysis of consumer behaviour is still missing, it seems too early to draw a firm conclusion. As the speed of mobile connections decreases with the number of users (if the network is not upgraded and/or enlarged sufficiently),[15] the further development of the 'mobile hype' cannot be predicted with any certainty. But certainly the development of mobile broadband connections and its effects on fixed connections warrants highest attention.

Another interesting question is why the development of mobile broadband in Austria is much faster than in other countries. There appear to be two reasons for this[16]:

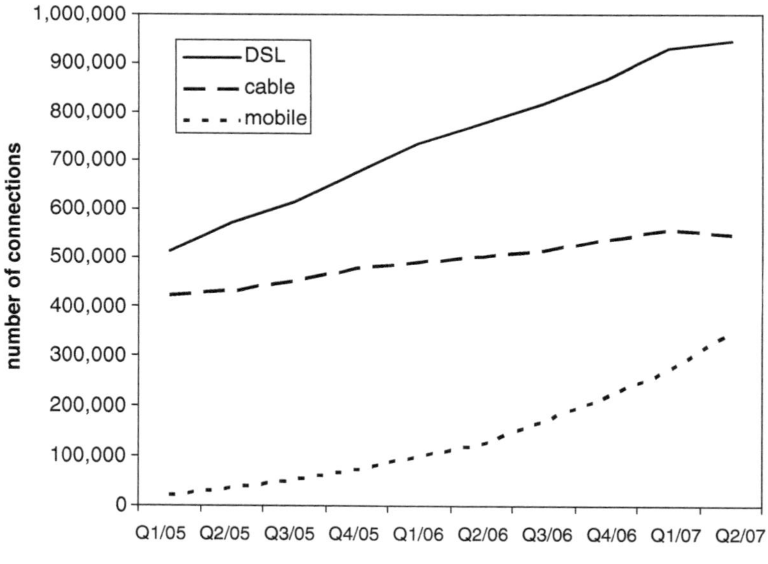

Fig. 9 Development of fixed and mobile broadband connections (From RTR 2007a)
The number of mobile broadband connections is the number of mobile broadband contracts including ≥250 MB per month. Contracts with less than 250 MB per month are unlikely to be used as substitutes to fixed broadband connections

[15] See, for example, a test in Konsument (2007) which finds that mobile connections only deliver 1/7 of the advertised maximum bandwidth of 7.2 Mbit/s while fixed network connections in general are much closer to their advertised maximum bandwidths.

[16] See also Analysys (2007) and Willmer (2007).

One is the relatively high prices for fixed broadband connections in Austria. Several international comparisons show that prices for DSL and cable connections were above the European average or even among the highest in Europe (see, e.g., Anacom 2007 or Kopf 2007). Another driving factor seems to be the high extent of fixed-mobile substitution. Austria is among the countries with the highest share of "mobile-only" households in Europe (see Elixmann et al. 2007; Kruse 2007). A combination of these factors together with a competitive mobile sector and spare capacities in the operators' UMTS/HSDPA networks seem to be the main causes for the developments observed. However, an in-depth analysis is still missing.

Conclusions

Survey evidence and demand estimation (Cardona et al. 2007) indicates that DSL and cable broadband internet access are likely to be part of the same market at the retail as well as at the wholesale level in Austria. Evidence from other countries like the UK (Ofcom 2006), Malta (MCA 2007), Portugal (Pereira and Ribeiro 2006), and the US (Crandall et al. 2002; Rappoport et al. 2003) points in the same direction. Evidence from Austria also suggests that competitive pressure from mobile broadband via UMTS/HSDPA on fixed broadband connections is also significant. While businesses seem to use mobile broadband mainly in addition to their fixed connection, more and more private users appear to have switched from fixed to mobile in 2007. National regulatory authorities therefore should closely examine the impact of cable and mobile broadband on DSL connections either at the level of market definition or at the level of market analysis.[17] Since cable networks and sometimes also high speed mobile broadband connections may not be available throughout the territory, a geographic differentiation of regulation (if necessary) such as, for example, in Ofcom (2007) might be justified if the competitive pressure from the other platform(s) is strong enough.

References

Anacom (2007) International Comparison of Broadband Prices. http://www.anacom.pt/txt/template12.jsp?categoryId=234442. Cited 12 December 2007.

Analysys (2007) Has Wireless Broadband Become Mainstream? http://www.analysys.com/default_acl.asp?Mode=article&iLeftArticle=2473&m=&n. Cited 12 December 2007.

Berg Insight (2007) The European Mobile Broadband Market. VAS Res Ser. http://www.berginsight.com

[17] Whereas some regulatory authorities opted fort the inclusion of cable networks in the wholesale broadband access market at the stage of market definition, the European Commission is in favour of considering the "indirect" constraints from cable on DSL at the stage of market analysis, see European Commission (2004).

Cardona M, Schwarz A, Yurtoglu BB, et al. (2007) Demand Estimation and Market Definition for Broadband Internet Services. Working Paper. http://homepage.univie.ac.at/Christine.Zulehner/broadband.pdf. Cited 12 December 2007.
Crandall RW, Sidak JG, Singer HJ (2002) The Empirical Case Against Asymmetric Regulation of Broadband Internet Access. Berkeley Law and Technol J, Vol. 17(1):953–987.
Elixmann D, Schäfer RG, Schöbel A (2007) Internationaler Vergleich der Sektorperfomance in der Telekommunikation und ihr Bestimmungsgründe. WIK-Diskuss Nr. 289, February 2007.
European Commission (2004) Notifications Received Under Article 7 of the Framework Directive – Wholesale Broadband Access – Commission Briefing Paper to ERG. 20 September 2004.
Goel RK, Hsieh ET, Nelson MA, et al. (2006) Demand Elasticities For Internet Services. Appl Econ, Vol. 38(9):975–980.
Goolsbee A (2006) The Value of Broadband and the Deadweight Loss of Taxing New Technologies. Contrib to Econ Anal and Policy, Vol. 5(1):1505–1505.
Greene WH (2003) Econometric Analysis. Fifth Edition. Prentice Hall, Upper Saddle River, NJ.
Ida T, Kuroda T (2006) Discrete Choice Analysis of Demand for Broadband in Japan. J of Regul Econ, Vol. 29(1):5–22.
Inderst R, Valletti T (2007) Market Analysis in the Presence of Indirect Constraints and Captive Sales. J of Compet Law and Econ, published online on 21 May 2007, http://www3.imperial.ac.uk/portal/pls/portallive/docs/1/15263697.PDF. Cited 12 December 2007.
Kopf W (2007) VDSL and NGN Access Strategies. WIK Conference "VDSL – The Way to Next Generation Networks". Königswinter, 21/22 March 2007.
Konsument (2007) Breitband-Internet. Das Blaue vom Himmel. Issue 10, 2007.
Kruse J (2007) 10 Jahre Telekommunikations-Liberalisierung in Österreich. Schriftenreihe der Rundfunk-und Telekom Regulierungs-GmbH, Vol. 2, 2007. www.rtr.at/de/komp/Schriftenreihe Nr22007/Band2-2007.pdf. Cited 12 December 2007.
Maddala GS (1983) Limited-Dependent and Qualitative Variables in Econometrics. Cambridge University Press, Cambridge.
Madden G, Simpson M (1997) Residential Broadband Subscription Demand: An Econometric Analysis of Australian Choice Experiment Data. Appl Econ, Vol. 29(8):1073–1078.
MCA (2006) Wholesale Broadband Access Market. Identification and Analysis of Markets, Determination of Market Power and Setting of Remedies. Consultation Document. http://www.mca.org.mt/infocentre/openarticle.asp?id=869&pref=6. Cited 12 December 2007.
MCA (2007) End-users Perceptions Survey – Broadband Services. http://www.mca.org.mt/infocentre/openarticle.asp?id=1079&pref=48. Cited 12 December 2007.
McFadden D (1974) Conditional Logit Analysis of Qualitative Choice Behaviour. In: Zarembka P (ed) Frontiers in Econometrics, Academic Press, New York, pp. 105–142.
OECD (2007) OECD Broadband Statistics to June 2007. http://www.oecd.org/document/60/0,3343,en_2.1825_495656_39574076_1_1_1_1,00.html. Cited 12 December 2007.
Ofcom (2006) Consumer Research to Inform Market Definition and Market Power Assessments in the Review of the Wholesale Broadband Access Markets 2006/07. http://www.ofcom.org.uk/consult/condocs/wbamr/research.pdf. Cited 12 December 2007.
Ofcom (2007) Review of the Wholesale Broadband Access Markets 2006/07. http://www.ofcom.org.uk/consult/condocs/wbamr07/wbamr07.pdf. Cited 12 December 2007.
Pereira P, Ribeiro T (2006) The Impact on Broadband Access to the Internet of the Dual Ownership of Telephone and Cable Networks. NET Institute Working Paper No. 06–10.
Rappoport P, Kridel D, Taylor L, et al. (2003) Residential Demand for Access to the Internet. In: Madden G (ed) Int Handbook of Telecommun Econ, Volume II. Edward Elgar, Cheltenham, UK.
RTR (2007a) RTR Telekom Monitor. 3. Quartal 2007. http://www.rtr.at/de/komp/TKMonitor_Q32007/TM3-2007.pdf. Cited 12 December 2007.
RTR (2007b) Abgrenzung des Marktes für breitbandigen Zugang auf Vorleistungsebene. http://www.rtr.at/de/komp/KonsultationBBMarkt2007/Untersuchung_Breitbandmarkt.pdf. Cited 12 December 2007.
Schwarz A (2007) Wholesale Market Definition in Telecommunications: The Issue of Wholesale Broadband Access. Telecommun Policy, Vol. 31:251–264.

Train KE (2002) Discrete Choice Methods with Simulation. Cambridge University Press, Cambridge. Available online at http://elsa.berkeley.edu/books/choice2.html.
Varian H (2000) Estimating the Demand for Bandwidth. Discussion Paper, University of California, Berkeley, CA.
Willmer G (2007) Growing HSPA Impact Helps Drive Data Acceleration Across Europe. Mobile Commun Europe, 2 October 2007: 3–4.

Search Engines for Audio-Visual Content: Copyright Law and Its Policy Relevance

Boris Rotenberg and Ramón Compañó

Abstract The first generation of search engines caused relatively few legal problems in terms of copyright. They merely retrieved text data from the web and displayed short text-snippets in reply to a specific user query. Over time, search engines have become efficient retrieval tools, which have shifted from a reactive response mode ('user pull') to pro-actively proposing options ('user push'). Moreover, they will soon be organising and categorising of all sorts of audio-visual information. Due to these transformations, search engines are becoming fully-fledged information portals, rivalling traditional media. This will cause tensions with traditional media and content owners. As premium audiovisual content is generally more costly to produce and commercially more valuable than text-based content, one may expect copyright litigation problems to arise in the future. Given this perspective, this article briefly introduces search engine technology and business rationale and then summarizes the nature of current copyright litigation. The copyright debate is then put in the audiovisual context with a view to discussing elements for future policies.

In Memoriam: Boris Rotenberg passed away on 23rd December 2007 in an unfortunate ski-accident at the age of 31. This is the last article he wrote. His colleagues from the Institute for Prospective Technological Studies will always remember him for his professional achievements and his personal life which will remain an example for many.

Disclaimer: The views expressed in this article are those of the authors and do not necessarily reflect the official view of the European Commission on the subject. Neither the European Commission nor any person acting on behalf of the European Commission can be made responsible for the content of this article.

R. Compañó
European Commission, Joint Research Centre, Institute for Prospective Technological Studies, Sevilla (Spain)
e-mail: ramon.compano@ec.europa.eu

B. Preissl et al. (eds.), *Telecommunication Markets*, Contributions to Economics,
DOI: 10.1007/978-3-7908-2082-9_7, © Springer Physica-Verlag HD 2009

Introduction

We are currently witnessing a trend of data explosion. The amount of information created, stored and replicated in 2006 is thought to be about 161 billion gigabytes – equivalent to 3 million times the information in all books ever written. That figure is expected to reach 988 billion gigabytes by 2010.[1] This data comes in a variety of formats, and content has evolved far beyond pure text description. It can be assumed that search engines, in order to cope with this increased creation of audiovisual (or multimedia) content, will increasingly become audio-visual (AV) search engines.

By their nature, audio-visual search engines promise to become key tools in the audio-visual world, as did text search in the current text-based digital environment. Clearly, AV search applications would be necessary in order to reliably index, sift through, and 'accredit' (or give relevance to) any form of audiovisual (individual or collaborative) creations. AV search moreover becomes central to predominantly audiovisual file-sharing applications. AV search also leads to innovative ways of handling digital information. For instance, pattern recognition technology will enable users to search for categories of images or film excerpts. Likewise, AV search could be used for gathering all the past voice-over-IP conversations in which a certain keyword was used. However, if these applications are to emerge, search technology must transform rapidly in scale and type. There will be a growing need to investigate novel audio-visual search techniques built, for instance, around user behaviour. Therefore, AV search is listed as one of the top priorities of the three major US-based search engine operators – Google, Yahoo! and Microsoft. The French Quaero initiative, for the development of a top-notch AV search portal, or the German Theseus research programme on AV search, provide further evidence of the important policy dimension.

This paper focuses on some policy challenges for European content industries emanating from the development, marketing and use of AV search applications. As AV search engines are still in their technological infancy, drawing attention to likely future prospects and legal concerns at an early stage may contribute to improving their development. The paper will thus start with a brief overview of trends in AV search technology and market structure.

The central argument of this paper concerns the legal, regulatory and policy dimension of AV search. Specifically, the paper analyses copyright law.[2] With its

[1] *See* Andy McCue, Businesses face data 'explosion', *ZDNet*, 23rd May 2007, *at* http://news.zdnet.co.uk/itmanagement/0,1000000308,39287196,00.htm (last visited: 18th December 2007), referring to IDC/EMC Study *The expanding Digital Universe*.

[2] It is acknowledged that a number of other legal instruments deserve a closer look when studying search engines from a legal point of view. However, data protection law, competition law, trademark law, etc. are beyond the scope of this paper.

dual economic and cultural objectives, copyright law is a critical policy tool in the information society because it takes into account the complex nature of information goods. It seeks to strike a delicate balance at the stage of information creation. Copyright law affects search engines in a number of different ways, and determines the ability of search engine portals to return relevant "organic" results.[3] Courts across the globe are increasingly called on to consider copyright issues in relation to search engines. This paper analyses some recent case law relating to copyright litigation over deep linking, provision of snippets, cache copy, thumbnail images, and news gathering services (e.g. Google Print). However, the issue of secondary copyright liability, i.e. whether search engines may be liable for facilitating the downloading of illegal copies of copyright content by users, is beyond the scope of this paper.

Copyright law is not the same for the whole of Europe. Though it is harmonized to a certain extent, there are differences across EU Member States, due to the fact that, following Article 295 of the EC Treaty, the EU does not interfere into the national regulation on property ownership of the Member States. It is not the intention of this paper to address particular legal questions from the perspective of a particular jurisdiction or legal order, but rather to tackle important questions from the higher perspective of European policy. The aim is to inform European policy in regard to AV search through legal analysis, and to investigate how copyright law could be a viable tool in achieving EU policy goals.

This paper argues that finding the proper regulatory balance as regards copyright law will play a pivotal role in fostering the creation, marketing and use of AV search engines. Too strong copyright protection for right-holders may affect both the creation and availability of content and the source of income of AV search engine operators and, thus, hamper the development of innovations for AV search. Conversely, copyright laws which are unduly lenient for AV search engine operators may inhibit the creation of novel content. The paper will refer each time to relevant developments in the *text* search engine sector, and will consider to what extent the specificities of AV search warrant a different approach.

The second section briefly describes the functioning of web search engines and highlights some of the key steps in the information retrieval process that raise copyright issues. Section "Copyright in the Search Engine Context: Business Rationale and Legal" reviews the business rationale and main legal arguments voiced by content providers and search engine operators respectively. Section "Policy Dimension: Placing the Copyright Debate in the Audio-Visual Context" places these debates in the wider policy context and Section "Conclusion" offers some conclusions.

---

[3]Organic (or natural) results are not paid for by third parties, and must be distinguished from sponsored results or advertising displayed on the search engine portal. The main legal problem regarding sponsored results concerns trademark law, not copyright law.

Search Engine Technology

For the purposes of this paper, the term 'web search engine' refers to a service available on the public Internet that helps users find and retrieve content or information from the publicly accessible Internet.[4] The best known examples of web search engines are Google, Yahoo!, Microsoft and AOL's search engine services. Web search engines may be distinguished from search engines that retrieve information from non-publicly accessible sources. Examples of the latter include those that only retrieve information from companies' large internal proprietary databases (e.g. those that look for products in eBay or Amazon, or search for information inside Wikipedia), or search engines that retrieve information which, for some reason, cannot be accessed by web search engines.[5] Similarly, we also exclude from the definition those search engines that retrieve data from closed peer-to-peer networks or applications which are not publicly accessible and do not retrieve information from the publicly accessible Internet.

Likewise, it is more adequate to refer to search results as "content" or "information", rather than web pages, because a number of search engines retrieve other information than web pages. Examples include search engines for music files, digital books, software code, and other information goods.[6]

In essence, a search engine is basically composed of three essential technical components: the crawlers or spiders, the (frequently updated) index or database of information gathered by the spiders, and the query algorithm that is the 'soul' of the search engine. This algorithm has two parts: the first part defines the matching process between the user's query and the content of the index; the second (related) part of this algorithm sorts and ranks the various hits. The process of searching can roughly be broken down into four basic information processes, or exchanges of information: (a) information gathering, (b) user querying, (c) information provision, and (d) user information access. As shall be seen below, some of the steps or services offered in this process raise copyright issues.[7]

---

[4]See for a similar definition, James Grimmelmann, The Structure of Search Engine Law (draft), October 13, 2006, p. 3, at / (last visited: 18th December 2007). It is acknowledged that many of the findings of this paper may be applicable to different kinds of search engines.

[5]Part of the publicly accessible web cannot be detected by web search engines, because the search engines' automated programmes that index the web, crawlers or spiders, cannot access them due to the dynamic nature of the link, or because the information is protected by security measures. Although search engine technology is improving with time, the number of web pages increases drastically too, rendering it unlikely that the 'invisible' or 'deep' web will disappear in the near future. As of March 2007, the web is believed to contain 15 to 30 billion pages (as opposed to sites), of which one fourth to one fifth is estimated to accessible by search engines. See and compare http://www.pandia.com/sew/383-web-size.html (last visited: 18th December 2007) and http://technology.guardian.co.uk/online/story/0,,547140,00.html (last visited: 18th December 2007).

[6]Search engines might soon be available for locating objects in the real world. *See* John Battelle, *The Search: How Google and its rivals rewrote the rules of business and transformed our culture (2005)*, p. 176. See James Grimmelmann, *supra*.

[7]*See* James Grimmelmann, *Ibid*.

Technological aspects of no interest for the subject of this article, such as the search algorithm for the ranking, will not be discussed, although they are important elements for the search engine functioning.

Indexing

The web search process of gathering information is driven primarily by automated software agents called robots, spiders, or crawlers that have become central to successful search engines.[8] Once the crawler has downloaded a page and stored it on the search engine's own server, a second programme, known as the indexer, extracts various bits of information regarding the page. Important factors include the words the web page or content contains, where these key words are located (e.g. title) and the weight that may be accorded to specific words, as well as any or all links the page contains.

A search engine index is like a big spreadsheet of the web. The index breaks the various web pages and content into segments. It reflects where the words were located, what other words were near them, and analyses the use of words and their logical structure. Importantly, the index is therefore not an actual reproduction of the page or something a user would want to read.

The index is further analysed and cross-referenced to form the runtime index that is used in the interaction with the user. By clicking on the links provided in the engine's search results, the user may retrieve from the content provider's server the actual version of the page.

Caching

Most of the major search engines now provide "cache" versions of the web pages that are indexed. The search engine's cache is, in fact, more like a temporary archive. Search engines routinely store for a long period of time, a copy of the indexed content on their own server. When clicking on the "cache version", the user retrieves the page as it looked the last time the search engine's crawler visited the page in question. This may be useful for the user if the server is down and the page is temporarily unavailable, or if the user intends to find out what were the latest amendments to the web page.

---

[8]There are also non- or semi-automated alternatives on the market, such as the open directory project whereby the web is catalogued by users, or search engines that tap into the wisdom of crowds to deliver relevant information to their users, such as Wiki Search, the wikipedia search engine initiative (http://search.wikia.com/wiki/Search_Wikia) (last visited: 18th December 2007), or ChaCha (http://www.chacha.com/) (last visited: 18th December 2007). *See* Wade Roush, New Search Tool Uses Human Guides, *Technology Review*, 2nd February 2007, *at*http://www.techreview.com/Infotech/18132 (last visited: 18th December 2007).

Robots Exclusion Protocols

Before embarking on legal considerations, it is worth recalling the regulatory effects of technology or code. Technology or 'code' plays a key role in creating contract-like agreements between content providers and search engines. For instance, since 1994 robots exclusion standards or protocols have allowed content providers to prevent search engine crawlers from indexing or caching certain content. Web site operators can do the same by simply making use of standardised html code. Add '/robots.txt' to the end of any site's web address and it will indicate the site's instructions for search engine crawlers. Similarly, by inserting NOARCHIVE in the code of a given page, web site operators can prevent caching. Content providers can also use special HTML `<META>` tags to tell robots not to index the content of a page, and/or not scan it for links to follow.[9]

Standardising bodies are currently working on improving standards to go beyond binary options (e.g. to index or not to index). Right now content providers may opt-in or opt-out, and robots exclusion protocols also work for keeping out images, specific pages (as opposed to entire web sites). Though methods are now increasingly fine-grained, allowing particular pages, directories, entire sites, or cached copies to be removed,[10] many of the intermediate solutions are technologically still hard to achieve. There are currently no standardised ways – for instance – to indicate that text can be copied, but not the pictures. Fine grained technology could allow content owners to decide that pictures can be taken on condition that photographer's name also appear. Indicating payment conditions for the indexing of specific content might also be made possible with the improved robots exclusion protocol.[11] This way, technology could enable copyright holders to determine the conditions in which their content can be indexed, cached, or even presented to the user.

Automated Content Access Protocol (ACAP) is such a standardized way of describing some of the more fine-grained intermediate permissions, which can be applied to web sites so that they can be decoded by the crawler. The ACAP initiative supported mainly by the publishing and content industries have just launched version 2.0 of its robots.txt standard on 29th November, 2007. The ACAP protocol emphasises granting permissions and blocking, and supports time-based inclusion and exclusion (i.e. include or exclude until). The ACAP standard includes among others (i) a special "crawl function" that determines whether the search engine is allowed to crawl the page (as opposed to indexing it) for determining the relevance of a site, and (ii) a "present function" that governs the search engine's ability to

---

[9] For precise information, the robots.txt and meta robots standards can be found at http://www.robotstxt.org/ (last visited: 18th December 2007).

[10] *See* for a detailed overview Danny Sullivan, Google releases improved Content Removal Tools, *at* http://searchengineland.com/070417-213813.php (last visited: 18th December 2007).

[11] *See* Struan Robertson, Is Google Legal?, *OUT-LAW News*, October 27, 2006, *at* http://www.out-law.com/page-7427 (last visited: 18th December 2007).

display content as well as the specific manner in which the content would be displayed (e.g. the size of thumbnails).[12]

However, it must be noted that none of the major search engines currently support ACAP.[13] Each search engine provides its own extensions to the standardised robots exclusion protocols. These extensions enable detailed ways of excluding content from particular search engines' index and/or cache. When it suited them, search engines have united around a common standard – for instance, with respect to the sitemaps standard.[14] In the long term, one may expect some elements of ACAP to enter the new version of the robots exclusion protocols.

From Displaying Text Snippets and Image Thumbnails to 'Pro-active Information Portals'

Common user queries follow a 'pull'-type scheme. The search engines react to keywords introduced by the user and then submit potentially relevant content.[15] Current search engines return a series of text snippets of the source pages enabling the user to select among the proposed list of hits. For visual information, it is equally common practice to provide thumbnails (or smaller versions) of pictures.

However, search engines are changing from a reactive to a more proactive mode. One trend is to provide more personalized search results, tailored to the particular profile and search history of each individual user.[16] To offer more specialized

---

[12] *See* Danny Sullivan, ACAP Launches, Robots.txt 2.0 For Blocking Search Engines?, *Search Engine Land*, 29th November 2007, *at* http://searchengineland.com/071129-120258.php (last visited: 18th December 2007).

[13] *See* Danny Sullivan, ACAP Launches, Robots.txt 2.0 For Blocking Search Engines?, *Search Engine Land*, 29th November 2007, *at* http://searchengineland.com/071129-120258.php (last visited: 18th December 2007).

[14] *See* http://www.sitemaps.org/ (last visited: 18th December 2007). The sitemaps standard is an easy way for webmasters to inform search engines about pages on their sites that are available for crawling. In its simplest form, a Sitemap is an XML file that lists URLs for a site along with additional metadata about each URL (when it was last updated, how often it usually changes, and how important it is, relative to other URLs in the site) so that search engines can more intelligently crawl the site.

[15] A number of new search engines are being developed at the moment that propose query formulation in full sentences (e.g. Hakia, Powerset), or in audio, video, picture format.

[16] *See* Your Google Search Results Are Personalised, February 5, 2007, *at*http://www.seroundtable.com/archives/007384.html (last visited: 18th December 2007). *See* also Kate Greene, A More Personalized Internet?, *Technology Review*, February 14, 2007, *at* www.technologyreview.com/Infotech/18185/ (last visited: 18th December 2007). This raises intricate data protection issues. *See* Boris Rotenberg, Towards Personalised Search: EU Data Protection Law and its Implications for Media Pluralism. In Machill, M.; M. Beiler (eds.): *Die Macht der Suchmaschinen/The Power of Search Engines*. Cologne [Herbert von Halem] 2007, pp. 87–104. Profiling will become an increasingly important way for the identification of individuals, raising concerns in terms of privacy and data protection. This interesting topic is however beyond the scope of this paper.

results, search engines need to record (or log) the user's information. Another major trend is news syndication, whereby search engines collect, filter and package news, and other types of information. At the intersection of these trends lies the development of proactive search engines that crawl the web and 'push' information towards the user, according to this user's search history and profile.

Audio-Visual Search

Current search engines are predominantly text-based, even for AV content. This means that non-textual content like image, audio, and video files are indexed, matched and ranked according to textual clues such as filenames, tags, text near images or audio files (e.g. captions) and even anchor text of links that point directly at AV content. Truveo is an example of this for video clips,[17] and SingingFish for audio content.[18]

While text-based search is efficient for text-only files, this technology and methodology for retrieving digital information has important drawbacks when faced with other formats than text. For instance, images that are very relevant for the subject of enquiry will not be listed by the search engine if the file is not accompanied with the relevant tags or textual clues. For instance, although a video may contain a red mountain, the search engine will not retrieve this video when a user inserts the words "red mountain" in his search box. The same is true for any other information that is produced in formats other than text. In other words, a lot of relevant information is systematically left out of the search engine rankings, and is inaccessible to the user. This in turn affects the production of all sorts of new information.[19]

There is thus a huge gap in our information retrieval process. This gap is growing with the amount of non-textual information that is being produced at the moment. Researchers across the globe are currently seeking to bridge this gap. One strand of technological developments revolves around improving the production of meta-data that describes the AV content in text format. A solution could be found by, for instance, developing "intelligent" software that automatically tags audio-visual content.[20]

See Clements, B, et al., Security and Privacy for the Citizen in the Post-September 11 Digital Age: A Prospective Overview, 2003, EUR 20823, *at* http://cybersecurity.jrc.es/docs/LIBE%20STUDY/LIBEstudy%20eur20823%20.pdf (last visited: 18th December 2007).

[17] http://www.truveo.com (last visited: 18th December 2007).

[18] SingingFish was acquired by AOL in 2003, and has ceased to exist as a separate service as of 2007. *See* http://en.wikipedia.org/wiki/Singingfish (last visited: 18th December 2007).

[19] *See* Matt Rand, Google Video's Achilles' heel, *Forbes.com*, March 10, 2006, *at* http://www.forbes.com/2006/03/10/google-video-search-tveyes-in_mr_bow0313_inl.html (last visited: 18th December 2007).

[20] *See* about this James Lee, Software Learns to Tag Photos, *Technology Review*, November 9, 2006, *at* http://www.technologyreview.com/Infotech/17772/.*See* Chris Sherman, Teaching Google to See Images, *Search Engine Land*, April 5, 2007, *at* http://searchengineland.com/070405-172235.php (last visited: 18th December 2007).

However, though technology is improving, automatic tagging is still very inefficient due to complex algorithms and high processing or computational requirements. Another possibility is to create a system that tags pictures using a combination of computer vision and user-inputs.[21] However, manual tagging by professionals is cumbersome and extremely costly. A cheaper option is to make use of collective user tagging as performed in online social networks sites. This Web 2.0-option does not yet comply with high-quality standards in terms of key word accuracy and consistency for high-value applications, but such bottom-up approach may become a viable solution for other applications.

AV search often refers specifically to new techniques better known as content-based retrieval. These search engines retrieve audio-visual content relying mainly on pattern or speech recognition technology to find similar patterns across different pictures or audio files.[22] Pattern or speech recognition techniques make it possible to consider the characteristics of the image itself (for example, its shape and colour), or of the audio content. In the future, such search engines would be able to retrieve and recognise the words "red mountain" in a song, or determine whether a picture or video file contains a "red mountain," despite the fact that no textual tag attached to the files indicate this.[23]

The search engine sector is currently thriving, and examples of beta versions across those various strands abound, both for visual and audio information. Tiltomo[24] and Riya[25] provide state-of-the-art content-based image retrieval tools that retrieve matches from their indexes based on the colours and shapes of the query picture. Pixsy[26] collects visual content from thousands of providers across the web and makes these pictures and videos searchable on the basis of their visual characteristics. Using sophisticated speech recognition technology to create a spoken word index, TVEyes[27] and Audioclipping[28] allow users to search radio, podcasts,

---

[21] *See* Michael Arrington, Polar Rose: Europe's Entrant Into Facial Recognition, *Techcrunch*, December 19, 2006, *at* http://www.techcrunch.com/2006/12/19/polar-rose-europes-entrant-into-facial-recognition (last visited: 18th December 2007).

[22] Pattern or speech recognition technology may also provide for a cogent way to identify content, and prevent the posting of copyrighted content. See Anick Jesdanun, Myspace Launches Pilot To Filter Copyright Video Clips, Using System From Audible Magic, *Associated Press Newswires*, February 12, 2007.

[23] *See* Dr. Fuhui Long, Dr. Hongjiang Zhang and Prof. David Dagan Feng, Fundamentals of Content-Based Image Retrieval, *at* http://research.microsoft.com/asia/dload_files/group/mcomputing/2003P/ch01_Long_v40-proof.pdf (last visited: 18th December 2007).

[24] http://www.tiltomo.com (last visited: 18th December 2007).

[25] http://www.riya.com (last visited: 18th December 2007).

[26] http://www.pixsy.com (last visited: 18th December 2007).

[27] http://www.tveyes.com (last visited: 18th December 2007); TVEyes powers a service called Podscope (http://www.podscope.com

[28] http://www.audioclipping.de (last visited: 18th December 2007).) (last visited: 18th December 2007) that allows users to search the content of podcasts posted on the Web.

and TV programmes by keyword.[29] Blinkx[30] and Podzinger[31] use visual analysis and speech recognition to better index rich media content in audio as well as video format. However, the most likely scenario for the near future is a convergence and combination of text-based search and search technology that also indexes audio and visual information.[32] For instance, Pixlogic[33] offers the ability to search not only metadata of a given image but also portions of an image that may be used as a search query.

Two preliminary conclusions may be drawn with respect to AV search. First, the deployment of AV search technology is likely to reinforce the trends discussed above. Given that the provision of relevant results in AV search is more complex than in text-based search, it is self-evident that these will need to rely even more on user information to retrieve pertinent results. As a consequence, it also seems likely that we will witness an increasing trend towards AV content 'push', rather than merely content 'pull'. Second, the key to efficient AV search is the development of better methods for producing accurate meta-data for describing and organising AV content. This makes it possible for search engines to organise the AV content optimally (e.g. in the run-time index) for efficient retrieval. One important factor in this regard is the ability of search engines to have access to a wide number of AV content sources on which to test their methods. Another major factor is the degree of competition in the market for the production of better meta-data for AV content. Both these factors (access to content, market entry) are intimately connected with copyright law.

The next section will briefly consider some high profile copyright cases that have arisen. It will discuss the positions of content owners and search engines on copyright issues, and provide an initial assessment of the strengths of the arguments on either side.

Copyright in the Search Engine Context: Business Rationale and Legal Arguments

Introduction

Traditional copyright law strikes a delicate balance between an author's control of original material and society's interest in the free flow of ideas, information, and commerce.

---

[29]*See* Gary Price, Searching Television News, *SearchEngineWatch*, February 6, 2006, *at* http://searchenginewatch.com/showPage.html?page=3582981 (last visited: 18th December 2007).

[30]http://www.blinkx.com (last visited: 18th December 2007).

[31]http://www.podzinger.com (last visited: 18th December 2007).

[32]*See* Brendan Borrell, Video Searching by Sight and Script, *Technology Review*, October 11, 2006, *at* http://www.technologyreview.com/read_article.aspx?ch=specialsections&sc=personal&id=17604 (last visited: 18th December 2007).

[33]http://www.pixlogic.com (last visited: 18th December 2007).

Such a balance is enshrined in the idea/expression dichotomy which states that only particular expressions may be covered by copyright, and not the underlying idea.

In US law, the balance is moreover struck through the application of the "fair use" doctrine. This doctrine allows use of copyrighted material without prior permission from the rights holders, under a balancing test.[34] Key criteria determining whether the use is "fair" include questions as to whether it is transformative (i.e. used for a work that does not compete with the work that is copied), whether it is used for commercial purposes (i.e. for profit), whether the amount copied is substantial, and whether the specific use of the work has significantly harmed the copyright owner's market or might harm the potential market of the original. This balancing test may be applied to any use of a work, including the use by search engines.

By contrast, there is no such broad catch-all provision in the EU. The exceptions and limitations are specifically listed in the various implementing EU legislations. They only apply provided that they do not conflict with the normal exploitation of the work, and do not unreasonably prejudice the legitimate interests of the right-holder.[35] Specific exemptions may be in place for libraries, news reporting, quotation, or educational purposes, depending on the EU Member State. At the moment, there are no specific provisions for search engines, and there is some debate as to whether the list provided in the EU copyright directive is exhaustive or open-ended.[36] In view of this uncertainty, it is worth analysing specific copyright issues at each stage of the search engines' working.

The last few years have seen a rising number of copyright cases, where leading search engines have been in dispute with major content providers. Google was sued by the US Authors' Guild for copyright infringement in relation to its Google Book Search service. Agence France Presse filed a suit against Google's News service in March 2005. In February 2006, the *Copiepresse* association (representing French and German-language newspapers in Belgium) filed a similar lawsuit against Google News Belgium.

As search engines' interests conflict with those of copyright holders, copyright law potentially constrains search engines in two respects. First, at the information gathering stage, the act of indexing or caching may, in itself, be considered to infringe the *right of reproduction*, i.e. the content owners' exclusive right "

[34] A balancing test is any judicial test in which the importance of multiple factors is weighed against one another. Such test allows a deeper consideration of complex issues.

[35] See Art.5.5, Directive 2001/29/EC of the European Parliament and of the Council of 22 May 2001 on the Harmonisation of Certain Aspects of Copyright and Related Rights in the Information Society (hereinafter: EUCD), OJ L 167, 22.6.2001.

[36] *See* Institute for Information Law Report, The Recasting of Copyright & Related Rights for the Knowledge Economy, November 2006, pp. 64–65, *at* www.ivir.nl/publications/other/IViR_Recast_Final_Report_2006.pdf (last visited: 18th December 2007). Note, however, that Recital 32 of the EUCD provides that this list is exhaustive.

to authorise or prohibit direct or indirect, temporary or permanent reproduction by any means and in any form, in whole or in part" of their works.[37] Second, at the information provision stage, some search engine practices may be considered to be in breach of the *right of communication to the public*, that is, the content owners' exclusive right to authorise or prohibit any communication to the public of the originals and copies of their works. This includes making their works available to the public in such a way that members of the public may access them from a place and at a time individually chosen by them.[38]

Right of Reproduction

Indexing

Indexing renders a page or content searchable, but the index itself is not a reproduction in the strict sense of the word. However, the search engine's spidering process requires at least one initial reproduction of the content in order to be able to index the information. The question therefore arises whether the act of making that initial copy constitutes, in itself, a copyright infringement.

Copyright holders may argue that this initial copy infringes the law if it is not authorized. However, the initial copy is necessary in order to index the content. Without indexing the content, no search results can be returned to the user. Hence it appears search engine operators have a strong legal argument in their favour. The initial copy made by the indexer presents some similarities with the reproduction made in the act of browsing, in the sense that it forms an integral part of the technological process of providing a certain result.

In this respect, the EU Copyright Directive states in its preamble that browsing and caching ought to be considered legal exceptions to the reproduction right. The conditions for this provision to apply are, among others, that the provider does not modify the information and that the provider complies with the access conditions.[39]

The next section considers these arguments with respect to the search engine's cache copy of content.

Caching

The legal issues relating to the inclusion of content in search engine caches are amongst the most contentious. Caching is different from indexing, as it allows the users to retrieve the actual content directly from the search engines' servers. The first issues in regard to caching relate to the reproduction right.

[37] *See* Art. 2 EUCD, supra,, *OJ L 167*, 22.6.2001.

[38] *See* Art. 3 EUCD.

[39] *See* EUCD, *supra*, Recital 33.

The question arises as to whether the legal provision in the EU Copyright Directive's preamble would really apply to search engines. One problem relates to the ambiguity of the term 'cache'. The provision was originally foreseen for Internet Service Providers (ISPs), whose cache enabled them to speed up their information provision process. The use of the word "cache" by search engines may give the impression that content is only temporarily stored on an engine's servers for more efficient information transmission. Search engines may argue that the copyright law exception for cache copies also applies to search engines. Their cache copy makes information accessible even if the original site is down, and it allows users to compare between live and cached pages. However, cache copies used by search engines fulfil a slightly different function. They are more permanent than the ones used by ISPs and can, in fact, resemble an archive. Moreover, the cache copy stored by a search engine may not be the latest version of the content in question.

In US law, the legal status under copyright law of this initial or intermediate copy is the subject of fierce debate at the moment.[40] For instance, in the on-going litigation against Google Book Search, publishers are arguing that the actual scanning of copyrighted books without prior permission constitutes a clear copyright infringement.[41]

In the EU, an issue appears to relate to the use of particular content, or whether and how it is communicated to the public. In the *Copiepresse* case, the Belgium Court made clear that it is not the initial copy made for the mere purpose of temporarily storing content that is under discussion, but rather the rendering accessible of this cached content to the public at large.[42]

Right of Communication to the Public

Indexed Information

Text Snippets

It is common practice for search engines to provide short snippets of text from a web page, when returning relevant results. The recent Belgian *Copiepresse* case

[40]*See*, for instance, Frank Pasquale, Copyright in an Era of Information Overload: Toward the Privileging of Categorizers, *Vanderbilt Law Review*, 2007, p. 151, *at* http://ssrn.com/abstract=888410 (last visited: 18th December 2007); Emily Anne Proskine, Google Technicolor Dreamcoat: A Copyright Analysis of the Google Book Search Library Project, 21 *Berkeley Technology Law Journal* (2006), p. 213.

[41]Note that this is essentially an information security argument. One of the concerns of the publishers is that, once the entire copy is available on the search engines' servers, the risk exists that the book becomes widely available in digital format if the security measures are insufficient.

[42]*See* Google v. Copiepresse, Brussels Court of First Instance, February 13, 2007, at p. 38.

focused on Google's news aggregation service, which automatically scans online versions of newspapers and extracts snippets of text from each story.[43] Google News then displays these snippets along with links to the full stories on the source site. Copiepresse considered that this aggregation infringed their copyright.[44] The argument is that their members – the newspapers – have not been asked whether they consent to the inclusion of their materials in the aggregation service offered by the Google News site.[45]

Though it has always been common practice for search engines to provide short snippets of text, this issue had not raised copyright issues before. However, this may be a matter of degree and the provision of such snippets may become problematic, from a copyright point of view, when they are pro-actively and systematically provided by the search engines. One could argue either way. Search engines may argue that thousands of snippets from thousands of different works should not be considered copyright infringement, because they do not amount to one work. On the other hand, one may argue that, rather than the amount or quantity of information disclosed, it is the quality of the information that matters. Publishers have argued that a snippet can be substantial in nature – especially so if it is the title and the first paragraph – and therefore communicating this snippet to the public may constitute copyright infringement. One might also argue that thousands of snippets amount to substantial copying in the qualitative sense.

The legality of this practice has not yet been fully resolved. On 28th June 2006, a German publisher dropped its petition for a preliminary injunction against the Google Book Search service after a regional Hamburg Court had opined that the practice of providing snippets of books under copyright did not infringe German copyright because the snippets were not substantial and original enough to meet the copyright threshold.[46]

---

[43]*See* Google v. Copiepresse, Brussels Court of First Instance, February 13, 2007, at p. 36. The *Copiepresse* Judgment is available at http://www.copiepresse.be/copiepresse_google.pdf (last visited: 18th December 2007). *See* Thomas Crampton, Google Said to Violate Copyright Laws, *New York Times*, February 14, 2007, *at* http://www.nytimes.com/2007/02/14/business/14google.html?ex=1329109200&en=7c4fe210cddd59dd&ei=5088&partner=rssnyt&emc=rss (last visited: 18th December 2007).

[44]As explained above, Copiepresse is an association that represents the leading Belgian newspapers in French and German.

[45]*See* Latest Developments: Belgian Copyright Group Warns Yahoo, ZDNet News, January 19, 2007, at http://news.zdnet.com/2100-9595_22-6151609.html (last visited: 18th December 2007); Belgian Newspapers To Challenge Yahoo Over Copyright Issues, *at* http://ecommercetimes.com/story/55249.html (last visited: 18th December 2007). A group representing french- and german-language belgian newspaper publishers has sent legal warnings to yahoo about its display of archived news articles, the search company has confirmed. (They complain that the search engine's "cached" links offered free access to archived articles that the papers usually sell on a subscription basis.) See also Yahoo Denies Violating Belgian Copyright Law, *Wall Street Journal*, January 19, 2007, *at* http://online.wsj.com/

[46]*See* Germany and the Google Books Library Project, Google Blog, June 2006, *at* http://googleblog.blogspot.com/2006/06/germany-and-google-books-library.html (last visited: 18th December 2007).

By contrast, in the above mentioned *Copiepresse* case, the Belgian court ruled that providing the titles and the first few lines of news articles constituted a breach of the right of communication to the public. In the court's view, some titles of newspaper articles could be sufficiently original to be covered by copyright. Similarly, short snippets of text could be sufficiently original and substantial to meet the 'copyrightability' threshold. The length of the snippets or titles was considered irrelevant in this respect, especially given that the first few lines of articles are often meant to be sufficiently original to catch the reader's attention. The Belgian court was moreover of the opinion that Google's syndication service did not fall within the scope of exceptions to copyright, since these exceptions have to be narrowly construed. In view of the lack of human intervention and fully automated nature of the news gathering, and the lack of criticism or opinion, this could not be considered news reporting or quotation. Google News's failure to mention the writers' name was also considered in breach of the moral rights of authors. If upheld on appeal, the repercussions of that decision across Europe may be significant for search engine providers.

Image Thumbnails

A related issue is whether the provision by search engines of copyrighted pictures in thumbnail format or with lower resolution breaches copyright law. In *Arriba Soft v. Kelly*,[47] a US court ruled that the use of images as thumbnails constitutes 'fair use' and was consequently not in breach of copyright law. Although the thumbnails were used for commercial purposes, this did not amount to copyright infringement because the use of the pictures was considered transformative. This is because Arriba's use of Kelly's images in the form of thumbnails did not harm Kelly's market or reduce the picture's value. On the contrary, the thumbnails were considered ideal for guiding people to Kelly's work rather than away from it, while the size of the thumbnails makes using these versions, instead of the original, unattractive. In the *Perfect 10* case, the US court first considered that the provision of thumbnails of images was likely to constitute direct copyright infringement. This view was partly based on the fact that the applicant was selling reduced-size images like the thumbnails for use on cell phones.[48] However, in 2007 this ruling was reversed by the Appeals Court, in line with the ruling on the previous *Arriba Soft* case. The appeals

---

[47]*See* Kelly v. Arriba Soft, 77 F.Supp.2d 1116 (C.D. Call 1999). *See* Urs Gasser, Regulating Search Engines: Taking Stock and Looking Ahead, 9 *Yale Journal of Law & Technology* (2006) 124, p. 210; *at* http://ssrn.com/abstract=908996 (last visited: 18th December 2007).

[48]The court was of the view that the claim was unlikely to succeed as regards vicarious and contributory copyright infringement. *See* Perfect 10 v. Google, 78 U.S.P.Q.2d 1072 (C.D. Cal. 2006).

court judges ruled that "Perfect 10 is unlikely to be able to overcome Google's fair use defense."[49] The reason for this ruling is the highly transformative nature of the search engine's use of the works, which outweighed the other factors. There was no evidence of downloading of thumbnail pictures to cell phones, nor of substantial direct commercial advantage gained by search engines from the thumbnails.[50]

By contrast, a German Court reached the opposite conclusion on this very issue in 2003. It ruled that the provision of thumbnail pictures to illustrate some short news stories on the Google News Germany site did breach German copyright law.[51] The fact that the thumbnail pictures were much smaller than the originals, and had much lower resolution in terms of pixels, which ensured that enlarging the pictures would not give users pictures of similar quality, did not alter these findings.[52] The court was also of the view that the content could have been made accessible to users without showing thumbnails – for instance, indicating in words that a picture was available. Finally, the retrieving of pictures occurred in a fully automated manner and search engines did not create new original works on the basis of the original picture through some form of human intervention.[53]

The German Court stated that it could not translate flexible US fair doctrine principles and balancing into German law. As German law does not have a fair use-type balancing test, the Court concentrated mainly on whether the works in question were covered or not by copyright.[54] Contrary to text, images are shown in their entirety, and consequently copying images is more likely to reach the substantiality threshold, and be copyright infringing.[55] It may therefore be foreseen that AV search engines are more likely to be in breach of German copyright law than mere text search engines.

A related argument focuses on robots exclusion protocols. The question arises as to whether not using them can be considered by search engines as a tacit consent to their indexing the content. The Courts' reaction to these arguments in relation to caching is significant here.

Cached Information

The second set of issues related to the caching of content revolves around the right of communication to the public. When displaying the cache copy, the search engine

---

[49] *See* Perfect 10, Inc. v. Amazon.com, Inc., (9th Cir. May 16, 2007), judgment available *at* http://lawgeek.typepad.com/LegalDocs/p10vgoogle.pdf (last visited: 18th December 2007).

[50] *See* p. 5782 of the judgment.

[51] *See* the judgment of the Hamburg regional court, available at http://www.jurpc.de/rechtspr/20040146.htm (last visited: 18th December 2007), in particular on pp. 15–16. See on this issue: http://www.linksandlaw.com/news-update16.htm (last visited: 18th December 2007).

[52] *Ibid.*, p. 14

[53] .*Ibid.*, p. 15.

[54] *Ibid.*, p. 19.

[55] *Ibid.*, p. 16.

returns the full page and consequently users may no longer visit the actual web site. This may affect the advertising income of the content provider if, for instance, the advertising is not reproduced on the cache copy. Furthermore, Copiepresse publishers argue that the search engine's cache copy undermines their sales of archived news, which is an important part of their business model. The communication to the public of their content by search engines may thus constitute a breach of copyright law.

The arguments have gone either way. Search engines consider, that information on technical standards (e.g. robots exclusion protocols), as with indexing, is publicly available and well known and that this enables content providers to prevent search engines from caching their content. But one may equally argue the reverse. If search engines are really beneficial for content owners because of the traffic they bring them, then an opt-in approach might also be a workable solution since content owners, who depend on traffic, would quickly opt-in.

Courts on either side of the Atlantic have reached diametrically opposed conclusions. In the US, courts have decided on an opt-out approach whereby content owners need to tell search engines not to index or cache their content. Failure to do so by a site operator, who knows about these protocols and chooses to ignore them, amounts to granting a license for indexing and caching to the search engines In *Field v Google*,[56] for instance, a US court held that the user (as opposed to the search engine) was the infringer, since the search engine remained passive and mainly responded to the user's requests for material. The cache copy itself was not considered to directly infringe the copyright, since the plaintiff knew and wanted his content in the search engine's cache in order to be visible. Otherwise, the Court opined, the plaintiff should have taken the necessary steps to remove it from cache. Thus the use of copyrighted materials in this case was permissible under the fair use exception to copyright. In *Parker v Google*,[57] a US court came to the same conclusion. It found that no direct copyright infringement could be imputed to the search engine, given that the archiving was automated. There was, in other words, no direct intention to infringe. The result has been that, according to US case law, search engines are allowed to cache freely accessible material on the Internet unless the content owners specifically forbid, by code and/or by means of a clear notice on their site, the copying and archiving of their online content (Miller 2006).[58]

In the EU, by contrast, the trend seems to be towards an opt-in approach whereby content owners are expected to specifically permit the caching or indexing of content over which they hold the copyright. In the *Copiepresse* case, for instance, the Belgian

[56]*See* Field v. Google, F.Supp.2d, 77 U.S.P.Q.2d 1738 (D.Nev. 2006); judgment available at http://www.eff.org/IP/blake_v_google/google_nevada_order.pdf (last visited: 18th December 2007).

[57]*See* Parker v. Google, Inc., No. 04 CV 3918 (E.D. Pa. 2006); judgment available at http://www.paed.uscourts.gov/documents/opinions/06D0306P.pdf (last visited: 18th December 2007)

[58].*See* David Miller, Cache as Cache Can for Google, March 17, 2006, at http://www.internetnews.com/bus-news/article.php/3592251 (last visited: 18th December 2007).

Court opined that one could not deduce from the absence of robots exclusion files on their sites that content owners agreed to the indexing of their material or to its caching.[59] Search engines should ask permission first. As a result, the provision without prior permission of news articles from the cache constituted copyright infringement.[60]

Conclusion

The view of content providers is straightforward. They argue that search engines are making money out of their creations, without paying any of the costs involved in their production. The content generated by the providers is used by search engines in two distinct ways. First, search engines use the content providers' content as a bait to derive their (sometimes future) advertisement income.[61] Second, search engines can become fully-fledged information portals, directly competing with the content providers that provide their very content. Therefore, content providers are increasingly unwilling to allow search engines to derive benefits from listing or showing their content without remuneration. In addition, they argue that not using robots exclusion protocols in their websites cannot be considered as an implicit permission to use their content, since robots exclusion protocols cannot be regarded as law. And there is currently no legal regulation in force stating that the non-use of robots exclusion protocols is equal to implicitly accepting indexing and caching.

Search engines have a diametrically opposed view. They emphasise their complementary role as search engines (as opposed to information portals) in directing web-traffic to content providers. A recent report by the consulting company Hitwise shows that US newspapers' web sites receive 25% of their traffic from search engines.[62] Consequently, the search engines' view is that the commercial relationship is mutually beneficial, in that search engines indirectly pay content providers through the traffic they channel to them. Further they argue that if content providers prefer not to be included in the index or cache, they simply have to include the robots exclusion protocols in their website, while asking all content providers for their prior permission one by one would be unfeasible in practice. On the other

---

[59] *See* Google v. Copiepresse, Brussels Court of First Instance, February 13, 2007, at p. 35; *see* also the judgment of the Hamburg regional court, *at* http://www.jurpc.de/rechtspr/20040146.htm (last visited: 18th December 2007), p. 20.

[60] *See* Struan Robertson, Why the Belgian Court Ruled Against Google, *OUT-LAW News*, February 13, 2007, *at* http://out-law.com/page-7759 (last visited: 18th December 2007).

[61] *See* Google v. Copiepresse, Brussels Court of First Instance, February 13, 2007, supra note 46, at p. 22.

[62] *See* Tameka Kee, Nearly 25% of Newspaper Visits Driven by Search, *Online Media Daily*, Thursday, May 3, 2007, *at* http://publications.mediapost.com/index.cfm?fuseaction=Articles.showArticleHomePage&art_aid=59741 (last visited: 18th December 2007).

hand, automation is inherent to the Internet's functioning: permission and agreement should, in their view, be automated.

Copyright infringement ultimately depends on the facts. Search engines may retrieve and display picture thumbnails as a result of image search, or they may do so proactively on portal-type sites such as Google news to illustrate the news stories. The copyright analysis might differ depending on particular circumstances. The analysis shows how US courts have tended to be more favourable towards search engine activities in copyright litigation. This can be seen, for instance, in the litigation on caching, the displaying of thumbnails, and the use of standardised robots exclusion protocols. The open-ended 'fair use' provision has enabled US courts to balance the pros and cons of search engine activities case by case. However, the balancing test does not confer that much legal certainty.

European case law shows that European courts have been rather reluctant to modify their approaches in the wake of fast-paced technological changes in the search engine sector. For instance, EU courts have stuck more to the letter of the law, requiring express prior permission from right-holders for the caching and displaying of text and visual content. This is partly because European copyright laws do not include catch-all fair use provisions. The result is, however, that while US courts have some leeway to adapt copyright to the changing circumstances, the application of copyright law by European Courts is more predictable and confers greater legal certainty.

The paper finds, first, that different courts have reached diametrically opposed conclusions on a number of issues. Second, case law appears to indicate that the closer search engines come to behaving like classic media players, the more likely it is that copyright laws will hamper their activities. Likewise, it appears that the current EU copyright laws make it hard for EU courts to account for the specificities and importance of search engines in the information economy (for instance, increased automation and data proliferation).

The question thus arises whether current copyright law is in accord with European audio-visual policy. We should also ask whether copyright law can be used as a policy lever for advancing European policy goals, and if so, how.

Policy Dimension: Placing the Copyright Debate in the Audio-Visual Context

Copyright Law Is a Key Policy Lever

Search engines are gradually emerging as key intermediaries in the digital world, but it is no easy task to determine from a copyright point of view whether their automated gathering and displaying of content in all sorts of formats constitute copyright infringement. Due to their inherent modus operandi, search engines are pushing the boundaries of existing copyright law. Issues are arising which demand a reassessment of some of the fundamentals of copyright law. For example, does scanning books constitute an infringement of copyright, if those materials were scanned with the sole aim of making them searchable? When do text snippets become substantial enough to break copyright law if they are reproduced without the content owners' prior permission?

The paper has shown some tensions regarding the application of copyright law in the search engine context. Comparing EU and US copyright laws in general terms, we can say that EU laws tend to provide a higher degree of legal certainty but its application to search engines may be considered more rigid. US law, on the other hand, is more flexible but may not confer as much legal certainty. Both approaches are not mutually exclusive and a key question for policy makers is how to find a balance between conferring rather rigid legal certainty and a forward-looking more flexible approach in such a fast-paced digital environment.

The importance of copyright is visible in the increasing amount of litigation expected. Its role as a key policy lever in the AV era can be inferred from the twin axioms underpinning it. First, copyright law has an economic dimension. It aims at promoting the creation and marketing of valuable works, by offering a framework for licensing agreements between market players regarding this content. Second, copyright law has a cultural dimension. It is widely considered to be the 'engine of free expression'[63] *par excellence* in that copyright law incentivises the creation of other cultural expressions. The tuning of the boundaries of copyright law, by defining what is covered or not, and by balancing different interests through exceptions to copyright, makes it a key policy lever.

Copyright Law Impacts Other Regulatory Modalities

However, copyright law is not the only policy lever. There are other regulatory, technical and economic means of advancing the interests of the European AV content and AV search industry. However, these regulatory means are influenced by copyright law and this determines the permissible uses of certain content by search engines. Specifically, copyright law may have an impact on the use of certain technologies and technological standards; and copyright law may influence the conclusion of licensing agreements between market players.

Technology

The first dimension of the copyright issue is technological. A solution to copyright-related problems arising from fast-paced technological change may come from

[63]*See* for the use of this well-known metaphor in US law, *Harper & Row Publishers, Inc. v. Nation Enterprises*, 471 U.S. 539, 558 (1985).

technology itself. Technology determines behaviour as it allows or curtails certain actions. The search engine context is yet another area of the digital environment where this assertion is relevant. The increased standardisation of robots exclusion tools may give content owners fine-grained control over their content, and enable technologically determined contracts between content owners and information organisers (such as search engines). This is reminiscent of the debate on digital rights management (DRM) where technology enables fine-grained technological contracts between content owners and users.

On the one hand, developments which aim to increase flexibility are welcome, because there is probably no one-size-fits-all solution to the copyright problem. Technology may fill a legal vacuum, by allowing parties at distinct levels of the value chain to reach agreement on the use of particular content. This approach has the advantage of being fully automated.

On the other hand, the question arises as to whether society wants content providers to exert, through technological standards, total control over the use of their content by players such as search engines. Such total control over information could indeed run counter to the aims of copyright law, as it could impede many new forms of creation or use of information. This is a recurrent debate. For example in the DRM debate, many commentators are skeptical about technology alone being capable of providing the solution. Just like with DRM, it is doubtful that fair use or exceptions to copyright could be technologically calculated or mathematically compute. Fair use and exceptions to copyright are essential means for striking the appropriate balance for our ‘information ecosystem’. Providing content owners with technological control over the use of their content by search engines in terms of indexing, caching and displaying risks may change this delicate balance.

Market

Another regulatory modality is the market, or contractual deals amongst market players. As mentioned before, copyright law’s uncertain application in the search engine context has sparked a series of litigations, and seems to point to conflicts. At the same time, however, there have been a number of market deals between major content providers and major search engines. In August 2006, Google signed a licensing agreement with Associated Press. Google also signed agreements with SOFAM, which represents 4,000 photographers in Belgium, and SCAM, an audio-visual content association. Initially, both SOFAM and SCAM were also involved in the Copiepresse litigation. On 3 May 2007, the Belgian newspapers represented by Copiepresse were put back on Google news. Google agreed to use the no-archive tag so that the newspapers’ material was not cached On 6 April 2007, Google and Agence France Presse reached an agreement concerning licensing.

Consequently, as regards policy, the question arises as to whether there ought to be any legal intervention at all, since the market may already be sorting out its own problems. A German Court supported this view in its decision on thumbnails.[64] As it is a non-consolidated business and information is scarce, it is currently difficult to judge whether there is a market dysfunction or not. One of the salient facts here is that the exact terms of the deals were not rendered public, but in each one Google was careful to ensure that the deal was not regarded as a licence for the indexing of content. Google emphasised the fact that each deal will allow new use of the provider's content for a future product.

Some commentators see the risk that, while larger corporations may have plenty of bargaining power to make deals with content owners for the organisation of their content, the legal vacuum in copyright law may erect substantial barriers to entry for smaller players who might want to engage in the organisation and categorisation of content. "In a world in which categorizers need licenses for all the content they sample, only the wealthiest and most established entities will be able to get the permissions necessary to run a categorizing site."[65]

This prospect may become particularly worrying for emerging methods for categorizing and giving relevance to certain content, like the decentralised categorisation by user-participation. Although automated, search engines are also dependent on (direct or indirect) user input. The leading search engines observe and rely heavily on user behaviour and categorisation. A famous example is Google's PageRank algorithm for sorting entries by relevance which considers the number clicks, and ranks the most popular URLs according to the link structure. There is a multitude of other sites and services emerging, whose main added value is not the creation of content but categorising it. This categorisation may involve communicating to the public content produced by other market players. Examples include shared bookmarks and web pages,[66] tag engines, tagging and searching blogs and RSS feeds,[67] collaborative directories,[68] personalized verticals or collaborative search engines,[69] collaborative harvesters,[70] and social Q&A sites.[71] This emerging market for the user-driven creation of meta-data (data about the data) may be highly creative, but may nonetheless be hampered by an increasing reliance on licensing contracts for the categorisation of content.

When compared to pure text-based search, copyright litigation in the AV search environment may be expected to increase for two reasons. First, AV content is on

---

[64] *See* Judgment 308 O 449/03 (Hamburg Regional Court), 05.09.2003, p. 20, *at* http://www.jurpc.de/rechtspr/20040146.htm (last visited: 18th December 2007).

[65] Frank Pasquale, *supra*, pp. 180–181.

[66] For instance, Del.icio.us, Shadows, Furl.

[67] For instance, Technorati, Bloglines.

[68] For instance, ODP, Prefound, Zimbio and Wikipedia.

[69] For instance, Google Custom Search, Eurekster, Rollyo.

[70] For instance, Digg, Netscape, Reddit and Popurl.

[71] For instance, Yahoo Answers, Answerbag.

average more costly to produce and also commercially more valuable. Content owners will therefore be more likely to seek to keep control over this source of income against search engines. Second, effective AV search will depend on gathering user data, i.e. carrying out user profiling. Search engines will use profile data in a pro-active manner in order to push relevant content to the user. Search engines are increasingly taking over some of the key functions of traditional media players while using their content, increasing the likelihood that these classic players will contest through copyright litigation the search engines' use of their content.

The next section focuses on the effect of copyright law on the creation of meta-data for efficient AV content retrieval and search.

EU Copyright Law and the Creation of Meta-Data for AV Search

The discussion above indicates a number of unresolved issues in applying copyright law to search engines. One important issue with respect to AV search engines relates to the copyright status of producers of meta-data, i.e. information (data) about particular content (data).[72] In an audio-visual environment, metadata will become increasingly important to facilitate the understanding, use and management of data – in other words, to organize the massive flow of audio-visual information.[73] Two issues arise with respect to the role of meta-data producers. First, it is worth clarifying the scope of the right to reproduction with respect to 'organisers' of digital data. For their operation, organisers, such as search engines, need to make an initial (temporary) reproduction in order to organise the content. A possibility would be to distinguish more clearly this action from the right to communicate the data to the public. An extensive right to reproduction can hardly coexist with a broad right of communication to the public. One option might be to adopt a more normative stance by taking into account the purpose of the initial copying to determine whether there is reproduction or not.[74]

[72]Meta-tags or meta-data *sensu stricto* vary with the type of data and context of use. In a film, – for instance – the metadata might include the date and the place the video was taken, the details of the camera setting, the digital rights of songs, the name of the owner, etc. The metadata may either be automatically generated or manually introduced, like tagging of pictures in online social networks (e.g. Flickr). For the purposes of this paper, meta-data is considered more broadly. For instance, in organising and indexing all sorts of information, search engines also massively produce meta-data.

[73]Legal questions concerning the ownership of particular meta-data are beyond the scope of this paper, though it is acknowledged that this is bound to become a key policy issue in view of the proliferation of unstructured AV content and the importance to have access to meta-data to organise and make sense of AV content.

[74]*See* Chapter II, IVIR Study, The Recasting of Copyright & Related Rights for the Knowledge Economy, November 2006, pp. 64–65, *at* www.ivir.nl/publications/other/IViR_Recast_Final_Report_2006.pdf (last visited: 18th December 2007).

Second, search engines have become indispensable organisers and categorizers of data. They enable users to filter huge amounts of data and thus play an increasingly pivotal role in the information society. Search engines' main contribution is producing meta-data. However, this may raise questions about some of the fundamental assumptions of copyright law in the light of data proliferation. How should we consider, from a copyright point of view, the creativity and inventiveness of search engines in their organising of data or producing of meta-data?

Copyright law originates from the 'analogue era' with rather limited amounts of data. In those times, obtaining prior permission to reproduce materials or to communicate them to the public was still a viable option. Nowadays with huge amounts of data, automation is the only efficient way of enabling creation in the digital era. Automation raises intricate and unforeseen problems for copyright law. In addition, the automatic collection and categorisation of information by search engines and other meta-data producers is all-encompassing. Search engine crawlers collect any information they can find, irrespective of its creative value. They do this in a fully automated manner. The result may eventually be that search engines are forced to comply with the strictest copyright standard, even for less creative content.

Changing (slightly) the focus of EU copyright law could have positive *economic* effects. Today's main exceptions to copyright law are the right to quotation, review, or the special status granted to libraries. Automatic organization and filtering of data are not the focus of current copyright law. The above view suggests, however, that there is value in an efficient and competitive market for the production of meta-data, where the organisation of information is becoming increasingly critical in environments characterised by data proliferation. Some commentators consider that it would be beneficial to give incentives not only for the creation of end-user information, but also for the creation of meta-data. This could be achieved by including a legal provision in the copyright laws that take into account new methods for categorising content (e.g. the use of snippets of text, thumbnail images, and samples of audiovisual and musical works), some of which even as additional exceptions or limitations of copyright.[75] Increasing clarity on these practices might ease the entry of smaller players into the emerging market for meta-data.

Similar arguments also apply to the *cultural or social* dimension, where copyright can be regarded as a driver of freedom of expression through its incentives to people to express their intellectual work. Again, given today's information overload, categorizers of information appear to be important from a social point of view. First, the right to freedom of expression includes the right to receive information or ideas.[76] One may argue that, in the presence of vast amounts of data, the right to receive information can only be achieved through the organization of information. Second, categorisations – such as the ones provided by search engines – are also expressions of information or ideas. Indeed, the act of giving relevance or accrediting

---

[75]*See* Frank Pasquale, supra, p. 179 (referring to Amazon's "look inside the book" application)

[76].*See* among other legal provisions Art. 10, paragraph 1, European Convention on Human Rights.

certain content over other content through, for instance, ranking, is also an expression of opinion.[77] Third, the creation or expression of new information or ideas is itself dependent on both the finding of available information and the efficient categorisation of existing information or ideas.

Conclusions

1. The first generation of search engines caused relatively few problems in terms of copyright litigation. They merely retrieved text data from the web, and displayed short snippets of text in reply to a specific user query. Over time, one has witnessed a steady transformation. Storage, bandwidth and processing power have increased dramatically, and automation has become more efficient. Search engines have gradually shifted from a reactive response to the user ('pull') to pro-actively proposing options to the user ('push'). Future search will require increasing organisation and categorisation of all sorts of information, particularly in audio-visual (AV) format. Due to this shift from pure retrievers to categorisers, search engines are in the process of becoming fully-fledged information portals, rivalling traditional media players.
2. Much of the information collected and provided to the public is commercially valuable and content owners find that search engines are taking advantage of their content without prior permission, and without paying. As a result copyright litigation has come to the forefront, raising a set of completely new legal issues, including those surrounding the caching of content, or the scanning of books with a view to making them searchable. New legal issues arise due to search engines' unique functionality (retrieving, archiving, organising and displaying). The paper makes two points in this regard.
3. First, EU and US courts appear to have drawn markedly different conclusions on the same issues. Comparing EU and US copyright law in general terms, we can say that EU law tends to provide a higher degree of legal certainty but its application to search engines may be considered more rigid. US law, on the other hand, is more flexible but may not confer as much legal certainty.
4. The second point relates to the AV search context. The more audio-visual – rather than solely text-based – content is put on the Internet, the more one may expect copyright litigation problems to arise with respect to AV search engines. The reason is that premium AV content is generally more costly to produce and commercially more valuable than text-based content. Moreover, given that it is already difficult to return pertinent results for text-based content, AV search

[77] *See* for a US case where such arguments were made: KinderStart.com LLC v. Google, Inc. C 06-2057 JF (N.D. Cal. March 16, 2007). *See for comments*: Eric Goldman, KinderStart v. Google Dismissed – With Sanctions Against KinderStart's Counsel, March 20, 2007, at http://blog.ericgoldman.org/archives/2007/03/kinderstart_v_g_2.htm (last visited: 18th December 2007).

engines will have to rely even more on user profiling. By the same token, user profiles will enable search engines to target users directly and thereby compete with traditional media and content owners.

5. Copyright law is a key policy lever with regard to search engines. The wording of the law, and its application by courts, has a major influence on whether a thriving market will emerge for search engines, including the future AV search engines. This paper argues that the shift towards more audio-visual search offers the opportunity to rethink copyright law in a digital environment, characterised by increased automation and categorisation. The paper makes the following two considerations.
6. Copyright law is only one of several possible regulatory modalities which could determine whether the most appropriate balance is struck between giving incentives to the creation of digital content on the one hand, and – on the other hand – the categorisation and organisation of this content by a wide range of players such as search engines. Other essential elements in this debate are technological standardisation (e.g. robots exclusion protocols), and commercial agreements between market players. Far from being independent from one another, these regulatory modalities impact each other. For instance, copyright law determines the use of robots exclusion protocols. Similarly, the way copyright law is applied may increase or decrease the pressure on search engines to conclude licensing agreements with content owners.
7. A basic goal of copyright law is to give incentives for the creation of content. Given the proliferation of digital content, it becomes more difficult to locate specific content. It becomes comparatively more important to promote the development of methods for accurate labelling and indexing or organisation of AV content than to incentivise creation. This is particularly true in the AV search context, where describing and organising AV content for efficient retrieval is a major challenge. Many players are currently competing to provide the leading technology or method for producing accurate meta-data (data about the data).
8. The paper claims that copyright's policy relevance lies in its possible effects on the emerging market for meta-data production (i.e. meta-tags, and indexing/organisation of content). Strong copyright law will force AV search engines to conclude licensing agreements over the organising of content. It supports technology's role in creating an environment of total control whereby content owners are able to enforce licences over snippets of text, images and the way they are used and categorised. By contrast, a more relaxed application of copyright law might take into account the growing importance of creating a market for AV meta-data production and meta-data technologies in an environment characterised by data proliferation. This approach would give incentives for the creation of content, while allowing the development of technologies for producing meta-data.
9. One should consider if a slight refocusing of copyright law may be necessary. Today's copyright exceptions include the use of copyright content for quotation, review, or the special status granted to libraries. The use of copyright content for automatic organization and filtering of data, for the production of meta-data, or for the categorization of content (e.g. by means of snippets of text, thumbnail

images, and samples of audiovisual and musical works) are currently not foreseen as exceptions to copyright. Given the increasing importance of AV content retrieval in an age of AV content proliferation, it is worth debating whether new types of copyright exceptions should be introduced. More clarity might ease the entry of new players into the vital market for meta-data provision. One subsequent policy question will then concern the legal status of meta-data in terms of ownership, disclosure and use.

Search Engines, the New Bottleneck for Content Access*

Nico van Eijk

Abstract The core function of a search engine is to make content and sources of information easily accessible (although the search results themselves may actually include parts of the underlying information). In an environment with unlimited amounts of information available on open platforms such as the internet, the availability or accessibility of content is no longer a major issue. The real question is how to find the information. Search engines are becoming the most important gateway used to find content: research shows that the average user considers them to be the most important intermediary in their search for content. They also believe that search engines are reliable. The high social impact of search engines is now evident. This contribution discusses the functionality of search engines and their underlying business model – which is changing to include the aggregation of content as well as access to it, hence making search engines a new player on the content market. The biased structure of and manipulation by search engines is also explored. The regulatory environment is assessed – at present, search engines largely fall outside the scope of (tele)communications regulation – and possible remedies are proposed.

Search Engines: We Cannot Do Without Them

Search engines have become an essential part of the way in which access to digital information is made easier. They are used by virtually all internet users (in February 2007, US internet users conducted 6.9 billion searches), who moreover believe that searching through search engines is reliable and the best way of

N. van Eijk
Institute for Information Law (IViR, University of Amsterdam),
e-mail: vaneijk@ivir.nl

* Nico van Eijk is professor by special appointment of Media and Telecommunications Law. This paper contains updated parts of his inaugural address, of which an edited version was published in English ("Search engines: Seek and ye shall find? The position of search engines in law", *IRIS plus* [supplement to *IRIS - Legal observations of the European Audiovisual Observatory*], 2006-2).

B. Preissl et al. (eds.), *Telecommunication Markets*, Contributions to Economics,
DOI: 10.1007/978-3-7908-2082-9_8, © Springer Physica-Verlag HD 2009

finding websites.[1] "Googling" has become an autonomous concept and an independent form of leisure activity, similar to zapping through television channels. Anybody who cannot be found via a search engine does not exist: "To exist is to be indexed by a search engine."[2] Because of its prominent position, Google is often used as an example in the following paragraphs (Table 1).

How a Search Engine Works

The main function of a search engine is that of enabling access; it is a gateway to possibly relevant information on the internet. However, it is a two-directional gateway: from the information provider to the user and from the user to the information provider. A search engine determines which information provided by an information provider can be found by the end-user as well as what information the end-user will ultimately find. The search facility provided and the underlying search algorithm thus control supply and demand. Or to put it more simply: it is a bottle-neck with two bottles attached to it. How does a search engine work? Most search engines use more or less the same method to achieve search results.[3]

The process starts with searching the internet for information. This automated process uses intelligent "sleuths" called spiders, bots or crawlers. These sleuths surf the internet using criteria set previously by the search-engine provider.

The information found is thus made uniform and structured, laying the basis for its traceability. Then the information is indexed. This indexing determines the criteria for what are considered relevant words or combinations of words. Irrelevant information, such as fillers and punctuation marks, is deleted. At this stage the information is also streamlined in such a way that, for example, differences between singular and plural forms of words or variations due to declensions produce identical search results. Certain recognisable words, such as people's names and basic concepts, are possibly identified. The rest of the information is then "weighted", based on the frequency of words in a text and the contextual relevance or significance (or otherwise). This enriched information forms the ultimate basic material for the search engine.

Table 1 Number of searches in the United States (From comScore Networks)

Search engine	01/2006	02/2007
Google	2.3 billion	3.3 billion
Yahoo	1.6 billion	2 billion
MSN	752.5 million	730 million
Others	827.5 million	870 million
Total	5.48 billion	6.9 billion

[1] See, inter alia: Rainie and Shermak (2005).

[2] Introna and Nissenbaum (2000, p. 171).

[3] Liddy (2002, pp. 197–208).

When a search engine is consulted, a process is used that is largely the opposite of the indexing process. The end-user formulates a search question that is broken down and analysed by the search engine. In this process, non-relevant elements (such as fillers) are deleted, the relationships between the search terms are looked at (this can be indicated in the search query i.e. by using Boolean operators, e.g. AND, OR, NOT), and the relative importance of the search terms entered is charted. This leads to several search results, which are displayed on the end user's screen.

The search engine process						
Searching the inter net	Structuring collected data	Indexing data	Search query	Analysis query	Linking with index	Search result

It is by no means true that all information that is present on the internet is found and indexed by search engines. In the literature, there are claims that individual search engines index only 16% of all the information present on the internet, and all the search engines together cover no more than 42% of all available information.[4] Other estimations contradict these low numbers, but the observation that only a limited amount of the information is present, or can be, indexed, remains valid.

There are various reasons for this. Some of the information is hidden in files that cannot be indexed, such as text in graphics files. However, search engines are becoming increasingly intelligent and are increasingly capable of analysing more and more formats of information (e.g. Word, PDF and JPG-files). There is also information that the providers do not want to have included in search engines. News information that is rapidly refreshed, for example, is not suitable for inclusion in search engines, as the information quickly becomes obsolete (sometimes months pass before a spider attempts to re-index the site). There is also information that is accessible via the internet but that is not itself present on the internet, such as information stored in external databases. Moreover, the internet is still constantly growing and changing.

The model of collecting and ordering information and making information available is only one reflection of reality. What actually happens before a search result is made available is very complex and is characterised in an important way by the many subjective elements woven into the process (also see Paragraph 5).

The Search Engine Market

Not so long ago, at the beginning of the century, a lot of search engines were active, and it was the general assumption that competition between search engines would discipline the market. Both information providers and users would be able to benefit

[4] Lawrence and Giles (1999, pp. 107–109).

from this. Although the number of search engines is still significant, this cannot be said about their market shares.

Recent statistics on the US market show that Google, Yahoo, MSN/Livesearch and ASK together have a market share of 92%. All the other search engines account for the remaining 8% of the market. Google is clearly the market leader (Table 2).

There is an interesting difference between the US and Europe. Although an American company, Google is even more dominant in Europe. Recent figures about the Dutch market speak for themselves. Google has reached a 96% market share, whereas the second player, Ilse (Dutch), has a share of only 2%. The Dutch figures are extraordinary, but Google dominates in many European countries with a market share above 80% (Table 3).

Where Does the Money Come from?

Search engines generate income mainly from one source: advertising. Again, we take Google as an example. Google generates almost all of its income from advertising. This income is generated mainly by "Google AdWords". AdWords enables advertisers to create their own advertisements and state how much money they are willing to spend. They are then charged on the basis of the number of times that the advertisement is clicked on. The advertisements appear on the Google web site next to the results of a search request. Google decides which advertisement appears when and does this mainly in relation to the search request.

Table 2 Percentage of US searches among leading search engine providers (From Hitwise)

Domain	Mar 2007	Feb 2007	Mar 2006
www.google.com	64.13%	63.90%	58.33%
search.yahoo.com	21.26%	21.47%	22.30%
search.msn.com	9.15%[a]	9.30%*	13.09%
www.ask.com	3.48%	3.52%	3.99%

[a]Includes executed searches on Live.com and MSN Search.

Table 3 Market share of search engines in the Netherlands

Table	02/02	05/02	01/03	08/03	02/04	10/04	01/05	04/05	01/06	10/06	02/07
Google	32	40	52	65	68	74	84	85	91	90	94
Ilse	19	16	14	17	19	14	9	8	5	4	2
Livesearch	4	3	5	6	4	3	2	3	2	1	1
Yahoo	3	2	1	1	1	4	1	1	0	0	0
Lycos	2	2	2	1	1	0	0	0	0	0	0

The second source of income consists of placing the advertisements on third parties' websites. This is done via the AdSense program, which has two variations: "AdSense for search" and "AdSense for content". With "AdSense for search", advertisements are placed in relation to search requests on third parties' web sites. With "AdSense for content", advertisements are linked to the content of websites. For AdSense, Google has a revenue-sharing model, with some of the advertising income generated going to the information providers. These providers are thus in a position to take this into account when putting together the content of their website and to "optimise" the content.

Just to illustrate the financial impact: according to industry data for 2005, the four largest search engines/portals had captured more than half that year's US internet ad spending of $12.5 billion. In 2007, projections suggest that two-thirds of the $19.5 billion spent online will go to Google, Yahoo, AOL and MSN. Google alone reported a total advertising income for 2006 of almost $10.5 billion.

Syndicating ad space (related to search results and other available data) is now being extended to become a more general mechanism to allocate advertising slots in other media like radio, TV and print. Again, Google is an active market player in this respect.

US Online Advertising Spending Growth, 2002–2011 (% increase/decrease vs. prior year) (From eMarketeer February 2007 (www.emarketeer.com))	
2002	–15.8%
2003	20.9%
2004	32.5%
2005	30.3%
2006	30.8%
2007	18.9%
2008	22.1%
2009	18.1%
2010	14.9%
211	13.0%

US Online Advertising Revenues at Top Four Portals as a Percent of Total Online Advertising Spending 2004–2007 (From EMarketeer, February 2007 (Company reports, 2004–2007; eMarketeer calculations, www.emarketeer.com))

	2004	2005	2006	2007
Google	13.1%	19.2%	25%	32.1%
Yahoo!	18.4%	19.4%	18.3%	18.7%
AOL	6.8%	7.2%	7.5%	9.1%
MSN	9.4%	7.8%	6.7%	6.8%
Total top 4	47.8%	53.7%	57.4%	66.6%

It offers the possibility to use the Adwords mechanism to sell airtime to radio advertisers ("AdioAds"). Already 1,600 radio stations – including the 675 Clear Channel stations – use the service. More recently, Google announced the acquisition of DoubleClick, one of the leading companies in digital marketing. The announcement caused quite some reactions about the possible negative effects on the market and with regards to privacy. It is a sign on the wall that companies like Microsoft and AT&T where amongst those who expressed their concerns. This horizontal extension of its market should generate further advertising-related income and contribute to the diversification of revenue resources. The transaction is still under review by the (US and EU) competition authorities.

Certain search engines (i.e. Yahoo) offer the possibility to influence search results and/or ranking positions. This is not a dominant activity, but remains often unclear for the user.[5]

Manipulation of Search Results

The manipulation of search results takes at least two forms: manipulation by the search engine and manipulation by information providers by boosting their ranking in the search results.

Search Engines

The first form of manipulation is carried out by search-engine providers. They draw up the criteria on the basis of which the information present on the internet is collected, ordered and made available. Information that is not searched for is not found. If a spider is instructed to ignore certain information, this information will never appear as the result of a search action. The analysis of a search query and the answer to be given are determined by the algorithm that the search engine uses. This algorithm is the true secret to the way the process works, and it is the ultimate manipulation tool. It resembles to some extent the secret recipe for Coca-Cola.

Here are a few examples from practice to illustrate the manipulation by search engines.

Some search engines offer the opportunity of "buying" a high position on the list of search results. There are different variations of this. The simplest method involves literally selling the position. Other search engines priority-index the pages of paying parties, so that they rank higher in the list of search results.

---

[5]See: Nicholson (2005). Also: http://blogoscoped.com/archive/2007-07-16-n41.html en http://www.accc.gov.au/content/index.phtml/itemId/792088/fromItemId/142.

For commercial or policy reasons, some search engines – using filters – deliberately do not reproduce any certain results. For example, it is claimed that Google does not make certain search results available in the case of search queries from specific countries.[6]

Furthermore, search engines can be under legal obligations not to provide certain search results. Criteria for exclusion can originate from legislation or be based on jurisprudence. For example, in Germany and France restrictions exist on the portrayal/promotion of Nazi-related material (the famous Yahoo case). Courts regularly interfere based on trademark, copyright or unfair business practices regulation. Research shows that the results of search requests differ, not only depending on the search engine used, but also depending on whether Google.com, Google.de or Google.fr is used.[7]

There are search engines that, in addition to automated systems, also use a human factor: search results are manually adjusted by their own employees on the basis of more detailed criteria that have been formulated, both subjectively and otherwise.

Finally, the relationship between search and advertising income has already been mentioned in Paragraph 4. The need to optimize revenues causes search engines to take this relationship into account.

Information Providers

The second form of manipulation is manipulation by information providers. They can do this by paying for a higher ranking in some cases or by exercising direct influence on the search-engine provider, but more often it is a matter of cleverly designing the information provider's own web information to create a profile in such a way that the information is placed high up on the list of search results by the search engines. In doing this, they attempt to anticipate the search engine's algorithm (to the extent that this is actually known). A classic example is the manipulation of one's own metatags by adding attractive search words that have nothing to do with one's own service provision (such as football, pornography or the brand names of competitors).

However, search engines are becoming increasingly clever and are often capable of "neutralising" the effects of manipulated metatagging. More advanced methods are therefore currently used to attract greater attention. Fake sites are being set up, for example, that contain a lot of references to one's own site in order to influence page-ranking systems. Popular sites are being copied and included invisibly in one's own site so that unsuspecting users end up at other sites than those they intended to access.

[6]Zittrain and Edelman (2003).

[7]See, inter alia: Zittrain and Edelman (2003, pp. 137–148).

These and other forms of manipulation or deception are known as spamdexing, cloaking, link farming, webring, redirects, doorway pages, page-jacking, etc. All these methods aim to improve the ranking in the search results.

These manipulation techniques are combated by the search engines but not always successfully. At Google, the ultimate sanction is the exclusion of the offender, whose pages are then no longer indexed. The party concerned can then no longer be found via the search engine. The offenders are not just shady characters: they include governments and reputable companies, which use agencies to optimise the search results. An entire industry has emerged around this optimisation of search results. Under the name "search engine marketing" companies offer services aimed at improving rankings. They are also called SEOs, "search engine optimisers", a nice euphemism. Search engines in general do have policies on optimisation and "allow" certain types of manipulation by information providers.

Data Retention and Content Aggregation

The functionality of search engines is to a large extent determined by the nature and extent of the underlying data. The systems not only gather information about the data available on the internet, they also link that to what they know about the people submitting search queries. It means that the query itself plays an additional but crucial role.

This paragraph also looks at the fact that, in certain cases, search engines are developing a vertical relationship in respect of the content they are processing and analysing.

Data Retention

In the first instance, a search engine is dependent upon data generated by third parties. That is the information available on the internet, in the form of websites and the associated data, such as metatags. The engines interpret that information, which results in the recording of a large amount of selected data. That is then saved so that, amongst other things, a more accurate interpretation can be provided and hence a better search result generated. This process is described in section "How a Search Engine Works" above.

Information is not only gathered from the internet, user data is also generated. This consists of data made available by users themselves. It may come from submitted information specifying personal preferences, but it can also be derived from user-authorised analysis of personal documents such as e-mails (as is the case with Gmail, Google's e-mail service) or the use of online or offline applications like Google Desktop, Picasi and Google Docs & Spreadsheets.[8]

Thirdly, there is the data generated by the search queries themselves. In principle, these provide information about both the user – such as personal preferences, possibly combined with personal data – and what they are looking for.

If all the data mentioned are recorded, it creates a vast database. The size of that is determined by such factors as:

(a) When data recording began
(b) What data is selected
(c) How long the data is retained
(d) How and when data is re-evaluated and
(e) When aggregated data is deleted

Although the phenomenon as such is not unfamiliar – data warehousing and data mining are well-known terms, after all – relatively little is known about the data recorded by search engines. They are very coy about this aspect of their activities. We shall return to the sensitivities associated with data retention when discussing the regulatory aspects of the issue.

Content Aggregation

Several search engines are seeking vertical integration. This trend is reflected in their efforts to own, acquire or otherwise control content or its associated exploitation rights.

[8]From the privacy notice of Google Docs & Spreadsheets: "*Account activity.* You need a Google Account to use Google Docs & Spreadsheets. Google asks for some personal information when you create a Google Account, including your e-mail address and a password, which is used to protect your account from unauthorized access. Google's servers automatically record certain information about your use of Google Docs & Spreadsheets. Similar to other web services, Google records information such as account activity (e.g., storage usage, number of log-ins, actions taken), data displayed or clicked on (e.g., UI elements, links), and other log information (e.g., browser type, IP address, date and time of access, cookie ID, referrer URL); *Content.* Google Docs & Spreadsheets stores, processes and maintains your documents and previous versions of those documents in order to provide the service to you… We use this information internally to deliver the best possible service to you, such as improving the Google Docs & Spreadsheets user interface and maintaining a consistent and reliable user experience. Files you create with Google Docs & Spreadsheets may, if you choose, be read, copied, used and redistributed by people you know or, again if you choose, by people you do not know. Information you disclose using the chat function of Google Docs & Spreadsheets may be read, copied, used and redistributed by people participating in the chat. Use care when including sensitive personal information in documents you share or in chat sessions, such as social security numbers, financial account information, home addresses or phone numbers. You may terminate your use of Google Docs & Spreadsheets at any time. You may permanently delete any files you create in Google Docs & Spreadsheets. Because of the way we maintain this service, residual copies of your files and other information associated with your account may remain on our servers for three weeks."

In this respect, Google is a striking example. It is building a database of world literature, Google Books, by digitising the contents of libraries. Out-of-copyright works are being made available online in their entirety; in the case of books still subject to copyright protection, only an excerpt known as a "snippet" can be viewed. Another case in point is the company's acquisition of YouTube, the website on which companies and individuals can post videos for viewing by other internet users. And a third example is Google's activities in the field of mapping and geographical information.

As well as acquiring content directly in this way, search engines are also entering into special or preferential relationships with information providers. These can be based either upon the "manipulation" model described earlier – privileging certain providers in return for payment – or upon some form of revenue sharing (see section "Manipulation of Search Results").

Other Search Engine Involvement

Search engines have activities in many other areas inside and outside the vertical value chain. For example, search engines actively participate in the debate about network neutrality. They clearly seek control over the underlying (tele)communications infrastructure as was recently illustrated again by the interest of Google in acquiring frequencies. (This aspect will not be further discussed here.)

Regulatory Aspects

With the growing role of search engines, the question increasingly arises as to where to position them in law.[9] The myth of the self-regulating internet, the idea that it is "different", seems to have been exploded. The next-generation internet, the much-hyped "Web 2.0" will definitely bridge the gap between the "old" and the "new" worlds as far as its regulatory aspects are concerned. It might be somewhat controversial to put it this way, but the internet is becoming embedded into the day-to-day business of regulation. This is a sign of the internet's maturity and of its growing social and economic importance.[10]

Nevertheless, search engines are still largely "lost in law". The applicability of existing legal concepts needs further testing, while sector-specific rules such as European media regulation or the European regulatory framework for the communications sector have not been written with the phenomenon of the search engine in mind.

A myriad of topics could be discussed under the heading "regulatory aspects". Within the framework of this paper, however, only a limited number of aspects will be

[9]On the legal aspects of search engines, see, inter alia: Elkin-Koren (2001) Schulz et al. (2005), Grimmelmann (2007).

[10]See: Van Eijk (2004).

looked into – with an emphasis on the European regulatory perspective.[11] First of all, the question can be raised as to whether or not generic regulation might be or become relevant. We will look briefly at two aspects of this: freedom of expression and competition. Secondly, does sector-specific regulation come into play? And more particularly, do existing regulatory frameworks such as the European directives on audiovisual media services, the communications sector or privacy apply to search engines?

Freedom of Expression

Given their role in the information society, it goes without saying that freedom of expression as a fundamental value is at the heart of the legal context pertaining to search engines. However,, in particular as laid down in Article 10 of the European Convention on Human Rights (and Article 11 of the EU Charter of Fundamental Rights), freedom of expression does not directly cover the core activity of search engines. This has to do with the fact that Article 10 deals with the freedom to hold opinions and to receive and impart information. Search engines are primarily making information accessible which is already available. None the less, in my view this *making information accessible* is so closely linked with the basic aspects of freedom of expression that it should be treated similarly.[12]

Competition Law

It goes without saying that the generic national and European rules on competition apply to search engines. Abuse of a dominant position is prohibited, and the European Commission has specific powers to control mergers. However, it is also quite clear that, under the present market conditions as described above, the position of one search engine in particular has begun to draw attention in that respect: Google.

It is difficult to say whether Google is abusing its market power at the present time. Before that can be done, we first need to establish what market search engines are actually operating in. More research is then going to be needed to reveal whether there is any abuse of power. Nevertheless, we can confidently identify some market areas in which there is a potential for abuse.

(a) *Inclusion in search results*. Information providers could object to the fact that they are being excluded from or incorrectly included in the results generated by searches. Thus far, no European case law exists to establish whether or not there is any entitlement to such inclusion. Under US law, search engines have

---

[11]To mention some of the legal issues which fall outside the scope of this paper: general liability issues, copyright, trademark, unfair business practices, criminal law aspects (including data retention) and e-commerce. We also overlook the issue of jurisdiction and assume that search engines – although mostly of US origin – have to comply with European regulation.

[12]Van Eijk (2006, p. 5).

successfully claimed that obligations to include specific search results infringe their freedom of expression (i.e. the famous Kinderstart case).

(b) *Preferential treatment for in-house information services.* Quite apart from the issue of whether other providers of information services are disadvantaged, it may be that the search engine's own services are given preferential treatment. Such a situation seems more likely the greater a search engine's interest in specific content becomes. One specific example is Google searches for video files, where results on Google Video and YouTube are – allegedly – given a preferred position.[13]

(c) *Access to the advertising market.* The business model adopted by search engines is driven predominantly by advertising. Large shares of the search market imply a concentration of so-called "eyeballs" – a phenomenon already familiar in the broadcasting market. This entails the risk that prices will be driven up, bias in the selection process will occur and intransparancies become part of the advertising model.

Viewed from a merger's point of view, these three examples give rise to a number of pertinent questions. Competition in the marketplace could be affected adversely if, for example, (a) other search machines were taken over, (b) there were a takeover within the vertical business column (content) or (c) there were a horizontal takeover in the advertising brokerage market.

Within competition law, there is also the issue of whether search engines qualify as an "essential facility" (the term "natural monopoly" has even been used!). Essential facilities are primarily a feature of network-related sectors; whether a service counts as one depends in part upon whether substitution is possible. And one important factor in determining that is how high the barriers to entry are. In the case of search engines, it can be stated that in principle those barriers are very low indeed and that setting up a new service is by no means a complicated procedure. This is a point of view I have adopted in the past, but it has to be said now that there is good reason to review that opinion. In particular, Google's dominant position raises the question of whether relevant substitution really is possible. Let me give just one example. If the database built up by Google is indeed significant in its own right, then we have to ask whether other market players are still in any position to put together comparable databases of their own.

Sector-Specific Regulation

What about the applicability of sector-specific regulation? The present European involvement with both the media and the telecommunications sector does not really take search engines into account.

Both the Television without Frontiers Directive and its successor, the Audiovisual Media Services (AVMS) Directive regulate primarily traditional television

[13]See: Louderback (2007). "Although there are thousands of useful video sources on the Net, Google delivers results only from its own YouTube and Google Video – along with third party MetaCafe. That's just wrong, and..."

broadcasting and explicitly exclude services like search engines.[14] The framework for the communications sector has a similar handicap. Under the definitions in its core "Framework"-directive,[15] only electronic communication services are covered. This means services which consist "wholly or mainly in the conveyance of signals on electronic communications networks". Service providing or exercising editorial control over content are excluded.

In my view, search engines have characteristics of both information and communications services. They are a good example of convergence in the information society. But the information service aspects dominate: it is an understatement to see search engines as a mere directory service.

Privacy

The same applies to privacy as to freedom of expression. It is a right which enjoys constitutional protection under Article 8 of the European Convention on Human Rights and Articles 7 and 8 of the EU Charter. European law on this matter is further defined in a general privacy directive and a special directive applicable to the telecommunications sector.[16]

In general terms, the European privacy rules are easy to describe. They are based upon the principle that a minimum of personal data should be stored and processed, and that there must exist a direct relationship between what is done with data and the reason why it has been collected. Moreover, permission is required to gather data and the person involved must be able to verify and correct the information held. In all cases, proportionality is required. And compliance is regulated.

[14]EC Council Directive 89/552/EEC on the co-ordination of certain provisions laid down by law, regulation or administrative action in Member States concerning the pursuit of television broadcasting activities, adopted on 3 October 1989, OJ L 298, 17 October 1989, p.23, as amended by Directive 97/36/EC of the European Parliament and of the Council of 30 June 1997 amending Council Directive 89/552/EEC on the co-ordination of certain provisions laid down by law, regulation or administrative action in Member States concerning the pursuit of television broadcasting activities, adopted on 30 June 1997, OJ L 202, 30 July 1997, p. 60. The "AVMS"-directive: directive 2007/65/EC of the European Parliament and of the Council of 11 December 2007 amending Council Directive 89/552/EEC on the coordination of certain provisions laid down by law, regulation or administrative action in Member States concerning the pursuit of television broadcasting activities, OJ L 332/27, 18 December 2007.

[15]Directive 2002/21/EC of the European Parliament and of the Council of 7 March 2002 on a common regulatory framework for electronic communications networks and services (Framework Directive) OJ L 108/33 (24.04.2002).

[16]Directive 95/46/EC of the European Parliament and of the Council of 24 October 1995 on the protection of individuals with regard to the processing of personal data and on the free movement of such data, OJ L 281, 23/11/1995 pp. 0031-0050; Directive 2002/58/EC of the European Parliament and of the Council of 12 July 2002 concerning the processing of personal data and the protection of privacy in the electronic communications sector (Directive on privacy and electronic communications) OJ L 201/37 (31.07.2002).

The national regulators in Europe are members of an official working party,[17] which has recently launched an investigation into Google's observance of the European privacy regulations. This has prompted a correspondence[18] with the company, including a reference by the working party to the Resolution on Privacy Protection and Search Engines adopted at the 28th International Data Protection and Privacy Commissioners'.[19] This resolution more or less transposes the general characteristics mentioned above into conditions relevant to the situation of search engines. The agenda has thus been set, with the working party indicating that it has now begun a general investigation of search engines. "Taking into account the current situation initiated by the 'Google case'," it says, "the Working Party will deal with search engines in general, because this issue affects an ever growing number of users."[20]

The privacy directive for the communications sector contains more detailed rules, specifically covering the service level. As well as upholding the confidentiality of communications, it regulates such matters as the use of traffic and location data. As mentioned earlier, these rules are not specifically tailored to the search-engine industry either and it is quite uncertain whether the directive applies to them at all. As far as I am ware of, no regulator has yet issued an opinion on that applicability. What is certain is that some other services frequently provided by search engine operators – such as e-mail – are governed by the directive. However, in this respect search engine operators do not substantially differ from traditional internet service providers.

Analysis

As stated at the beginning of this paper, search engines represent an essential part of the way in which digital information is made easily accessible. However, they have also become a bottleneck in access to information, with both its providers and users depending upon the engine's intermediary function. At the same time, the way in which search engines work poses quite a few challenges. Nevertheless, they are able to generate serious revenues, primarily through advertising. But new elements are now being added, covering both vertical and horizontal issues – control over content, expansion into other advertising markets and marketing areas, and so on. Meanwhile, Google's dominant position in the market cannot be ignored. Policy makers and regulators are becoming increasingly aware of the role played by search engines in society, and the possible effects of reduced competition in the sector.[21]

---

[17]http://ec.europa.eu/justice_home/fsj/privacy/workinggroup/index_en.htm.

[18]See: http://ec.europa.eu/justice_home/fsj/privacy/news/docs/pr_google_16_05_07_en.pdf

[19]d.d. 2/3 November 2006. Text of the resolution: http://ec.europa.eu/justice_home/fsj/privacy/news/docs/pr_google_annex_16_05_07_en.pdf.

[20]Article 29 Data Protection Working Party, press release, Brussels, 21 June 2007.

[21]Which has lead to new support for creating European alternatives (The German Theseus and French Quaero-initiatives).

The interests at stake are huge, certainly in a situation where market dominance is a factor. It is possible that there may eventually be some role for competition law here, but more pressing and increasingly relevant is the question of whether sector-specific regulation is needed for search engines.

From a European perspective, that could take its lead from the industry-specific frameworks applied to the telecommunications sector.[22] However, the rules as they currently stand simply do not take into account a phenomenon like the search engine. Despite that, it is quite possible to investigate whether existing legal concepts like "significant market power" should be applied in this domain. Search engines with significant market power could be required to comply with obligations in respect of such matters as access, non-discrimination, transparency and accountability. Even where processes of a commercially confidential nature are at issue, that should not stand in the way of independent audits. They could, for example, establish whether search results are indeed generated in an objective way. They could also investigate whether recorded data is being stored and processed correctly. (The existing privacy regulations might in fact be sufficient for this to be done already, but so far they have never been invoked to justify checks or audits of search engines.)

At the same time, the universal service/public good aspects of search engines need to be borne in mind.[23] Their users are entitled to minimum guarantees in respect of the way their operators work: they need to be properly informed, and misleading them has to be prevented.

References

Elkin-Koren N (Winter 2001) Let the Crawlers Crawl: On Virtual Gatekeepers and the Right to Exclude Indexing. University of Dayton Law Review, 26 U Dayton L Rev. 179, Winter 2001.

Grimmelmann J (2007) "The Structure of Search Engine Law" in: 93 Iowa L Rev. (2007, forthcoming). http://works.bepress.com/cgi/viewcontent.cgi?article=1012&context=james_grimmelmann.

Introna L, Nissenbaum H (2000) Shaping the Web: Why the Politics of Search Engines Matters. *The Information Society*, vol. 16, no. 3, pp. 169–185.

Lawrence S, Giles CL (1999) Accessibility of Information on the Web. *Nature*, vol. 400, pp. 107–109.

Liddy (2002) E.D. Liddy, 'How a Search Engine Works', in: A.P. Mintz (ed.), Web of deception: misinformation on the Internet, Medford: Cyber Age books 2002, p. 197–208.

Louderback J (2007) Google's Gambit. PC Magazine, 17/7/2007, p. 7.

Nicholson S (2005) How Much of It Is Real? Analysis of Paid Placement in Web Search Engine Results. Journal of the American Society for Information Science and Technology 57(4), 448–461.

Rainie L, Shermak J (2005) Search Engine Use November 2005. Memo Pew Internet & American Life Project/Comscore Communications, 2005 http://www.pewinternet.org/pdfs/PIP_SearchData_1105.pdf.

[22]As laid down in the following directives: Framework Directive, OJ L 108/33 (24.04.2002); Access Directive, OJ L 108/7 (24.04.2002); Authorisation Directive, OJ L 108/21 (24.04.2002); Directive on privacy and electronic communications, OJ L 201/37 (31.07.2002); Universal Service Directive, OJ L 108/51 (24.04.2002).

[23]Introna and Nissenbaum (2000).

Schulz W, Held T, Laudien A (2005) Search Engines as Gatekeepers of Public Communication: Analysis of the German Framework Applicable to Internet Search Engines Including Media Law and Anti Trust Law. *German Law Journal*, vol. 6, no. 10, pp. 1419–1433.

Van Eijk NANM (2004) Regulating Old Values in the Digital Age. In: Möller C, Amouroux A (eds.), The Media Freedom Internet Cookbook. Vienna: OSCE, pp. 31–38.

Van Eijk NANM (2006) Search engines: Seek and ye shall find? The position of search engines in law. IRIS plus (supplement to IRIS – Legal observations of the European Audiovisual Observatory), http://www.obs.coe.int/oea_publ/iris/iris_plus/iplus2_2006.pdf.en) 2006 (2), pp. 2–8.

Zittrain J, Edelman B (2003) Documentation of Internet Filtering Worldwide. In: Hardy C, Möller C (eds.), Spreading the Word on the Internet. OSCE: Vienna, pp. 137–148.

E-Commerce Use in Spain*

Leonel Cerno and Teodosio Pérez Amaral

Abstract This paper analyzes the factors that influence private e-commerce from the demand side in Spain. We use econometric models and a survey of 18,948 individuals for 2003, of which 5,273 are internet users.

First, we analyze the determinants of the decision to purchase or not to purchase in the Web taking into account the link between e-commerce and the access and use of the Internet service. Then we characterize the e-consumer profile. The model suggests that the main factors that influence the decision to use e-commerce are accessibility to the Net, income and gender.

Second we use models specific to the users of e-commerce to measure the effects of the determinants on the number of purchases and the expenditure in the Web. For the expenditure equation we use the algorithm RETINA, which improves the forecast ability.

These models can be used to assess the adoption, use and expenditure of new users. This may help the operators to guide investment decisions and the administrations to reduce e-exclusion.

Introduction

An important difference between virtual and conventional markets is that there are fewer frictions in the virtual markets. In turn, one way of creating market power by virtual traders is to reduce the search costs. Internet has substantially increased the

*The authors acknowledge the financial support of the Cicyt Project SEJ2004-06948 of the Spanish Ministry of Education. They also thank the participants of the different seminars and conferences, especially Cristina Mazón, Lourdes Moreno, Lola Robles, Brigitte Preissl and two anonymous referees.

L.Cerno and T. Pérez Amaral
Universidad Carlos III de Madrid and Universidad Complutense de Madrid
e-mail: lcerno@uc3m.es, teodosio@ccee.ucm.es

B. Preissl et al. (eds.), *Telecommunication Markets*, Contributions to Economics,
DOI: 10.1007/978-3-7908-2082-9_9, © Springer Physica-Verlag HD 2009

readiness of information on prices and products, allowing consumers to identify the best option and to improve their position with respect to the on-line suppliers. The e-consumers have strengthened their position as e-commerce has developed. However, there is an important literature on consumer information overload that casts doubt on this optimistic view, see Lee and Lee (2004).

Consumer behavior, starting from the onset of e-commerce, is analyzed by Salop and Stiglitz (1982), Clemons et al. (1998), and Brynjolfsson and Smith (1999).

The on-line markets seem more efficient in terms of prices and costs. However, some studies find substantial price dispersion (Bakos 2001). This dispersion can be explained in part by the heterogeneity of specific factors such as the trust in or knowledge of the website or the brand name.

The analysis of the friction level of the Internet markets can be made from two points of view: comparing the characteristics between the two types of markets or analyzing the behavior within the electronic market. This paper adopts the second approach.

In the electronic markets, efficiency is measured in four dimensions: the price levels, the price dispersion, the elasticities, and the costs of distribution and other inputs (Smith et al. 1999). Regarding price creation, there are factors that have similar effects on price formation as in conventional markets. For example search costs put a downward pressure on the prices and intensify the differentiation by suppliers to try to keep prices above marginal costs (Peddibhotla 2000).

In turn, the after-sale service for some types of goods, or even the effect of the Web size in equilibrium, could be treated with the traditional microeconomic tools. Using the traditional assumptions, but going beyond the common belief that the prices in Internet are low because the consumers can find them easily and cheaply, Shapiro and Varian (1999) analyze under what conditions this happens.

One of the motivations of this research is the concern of some governments with avoiding e-exclusion. In this paper we study the subset of e-commerce buying referred to individuals. We use an empirical approach for a sample of 5,273 Internet users and potential e-commerce buyers in Spain in 2003. The social and demographic impacts of different factors are measured and their implications for different formulations of e-demand are estimated.

In section "E-Commerce and Internet Use", we study what variables influence the decision to **use or not to use** e-commerce. Our first model analyzes the factors that influence the decision of each individual to buy or not to buy via Internet.

Then in section "Descriptive Analysis and Definition of Variables", we analyze which factors affect the **number of times that an individual uses** e-commerce. We measure how much the characteristics of each individual influence the quantity of transactions.

In section "Specification of the Demand Models" we analyze the e-demand from two perspectives: number of purchases and expenditure. We study the effects of the determinants of the number of purchases and **how much money is spent** in e-commerce by each consumer.

Section "Conclusions" contains some conclusions and suggestions for further research.

E-Commerce and Internet Use

To understand the profile of e-buyers, we consider the determinants of the decision to buy or not to buy through Internet. We use a binary choice model to measure the impact on the probability of using e-commerce of its determinants (Train 2002). For explaining the behavior of the demand for e-commerce, it is useful to consider Internet access and use as explanatory variables (Cerno and Perez-Amaral 2005).

Descriptive Analysis and Definition of Variables

The data are part of survey TIC-H (2003) of the National Institute of Statistics (INE). It contains 5,273 individuals who are Internet users out of a total of 18,948.

In our dataset, only 3.7% out of the total sample of 18,948 individuals have bought through Internet in the last 3 months. In Fig. 1 we show the percentages of the number of e-commerce goods and services in each of the categories.

We observe that, in our sample, the more demanded products are leisure activities such as travel (41.1%) and show tickets (26.5%), while stock trades and bets are only demanded by 4.9% and 0.9% of all e-buyers, respectively.

Characteristics of the Data

Starting with this first analysis, we estimate a model for the e-buy determinants including economic and demographic attributes like income, gender, age and habitat

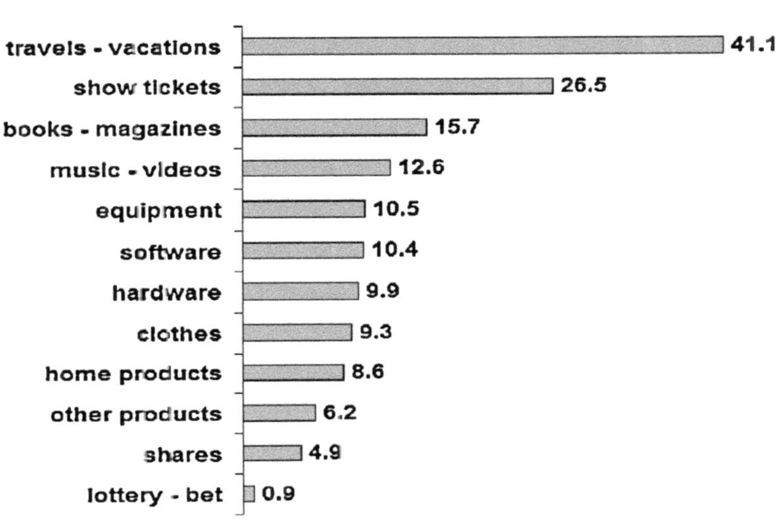

Fig. 1 Percentage of individuals who purchase from each category

size. We also consider the individual's characteristics regarding the Internet service such as access from the home, use from other places besides the home, and the intensity of use measured in hours per week (Tables 1 and 2). The binary logit model is:

$$\ln(\frac{P_i}{1-P_i}) = \beta_0 + \beta_1 IS\_Q_i + \beta_2 HOMEINT_i + \beta_3 USE_i$$
$$+ \beta_4 USAGEINT_i + \beta_5 USAGECOMP_i + \beta_6 SEXIM_i$$
$$+ \beta_7 AGE_i + \beta_9 AGE_i^2 + \beta_{10} POPUL_i + \beta_{11} MEMB_i + u_i$$

Where P_i=Pr $(PURCH_i)$is the probability that respondent i has bought through Internet in the last 3 months. Next we present the estimation results and the odd ratios (Table 3).

The global significance of the model is high since the log-likelihood of 397.09 is highly significant for a chi-square with ten degrees of freedom. The goodness of fit is also good according to the in-sample predictions of Table 4.

Table 1 Definition of variables

Variable	Definition
PURCH	Dummy = 1 if the individual has bought through internet
IS_Q	Income quintile, sequential variable 1–5
HOMINT	Dummy = 1 if the individual has Internet access at home
USE	Number of different access modes used (1 to 4)[1]
USAGEINT	Internet intensity of use (hours per week)
USAGECOMP	Computer intensity of use (hours per week)
SEXM	Dummy =1 if the individual is male
AGE	Age of the individual
POPUL	Relative population size (provincial level)[2].
MEMB	Number of residents in the household

Table 2 Summary statistics of internet users

Variable	Average	Standard deviation	Minimum	Maximum
PURCH[3]	0.126	0.336	0	1
IS_Q	4.030	0.850	1	5
HOMINT	0.615	0.487	0	1
USE	1.490	0.653	1	4
USAGEINT	38.001	29.805	0	70
USAGECOM	46.642	29.257	0	70
SEXM	0.519	0.499	0	1
AGE	33.862	12.465	15	88
POPUL	4.237	4.429	0.163	13.277
MEMB	3.333	1.192	1	6

[1]The four places where there is individual access to Internet are home, workplace, center of study and other places such as hotels, libraries, cybercoffees, etc.
[2]Referred to the individual's provincial population size divided by the total population in Spain.
[3]Note: PURCH is the proportion of Internet users that purchase via e-commerce.

Table 3 Logit results for the use or non-use of e-commerce equation

Dependent variable: *PURCH*	Coefficient	Odd ratio	*Z*
Constant	−6.19	–	−13.67
IS_Q	0.23	1.26	3.75
HOMEINT	0.48	1.62	4.39
USE	0.39	1.48	5.84
USAGEINT	0.01	1.01	6.31
USAGECOMP	0.01	1.01	2.61
SEXM	0.38	1.46	4.24
AGE	0.09	1.09	4.01
AGE^2	−0.001	0.99	−4.22
POPUL	0.04	1.04	4.77
MEMB	−0.08	0.92	−2.16

Sample: 5,223, Log-likelihood: −1773.45, $\chi_{(10)}^2$: 397.09 (Prob. = 0.000), Pseudo R^2: 0.1007.

Table 4 Goodness of fit. In-sample predictions

		Actual value		
		PURCH= 0	*PURCH*= 1	Total
Predicted value	*PURCH*= 0	3,063	208	3,271
	PURCH= 1	1,521	451	1,972
	Total	4,584	659	5,243

The data presents little collinearity as shown by the large values of the individual significance tests, z's. Here the model is nonlinear on the parameters, so the coefficients do not represent the partial effect. The odd ratios give a better idea of the marginal effect of each explanatory variable on the dependent variable.

The model correctly predicts 3,514 of the 5,243 observations (67.02%). Due to the large number of zeros in the sample, we use as threshold the proportion of zeros in the endogenous variable (0.126). Thus the specificity (percentage of correct "zeros" predicted inside the sample) is 66.81%, and the sensibility (percentage of correct "ones" predicted inside the sample) is 68.43%.

Since the sample is composed of individual private Internet users, the results only apply to this group. We conclude that:

- All the explanatory variables have positive effects on the probability of e-buying, except family size.
- Internet access at home (*HOMEINT*) and the different access modes used (*USE*) refer to the individual Internet habits and have high odd ratios. Internet access at home has the highest odd ratio of the two (1.62), followed by the different access modes used with 1.48. In other words, Internet access at home contributes 62% to the e-buy probability and the number of different access modes used contributes 48%.
- The male gender has a positive influence with an odd ratio of 1.46.
- The influence of income is positive with an odd ratio of 1.26.
- Population habitat (*POPUL*), Internet intensity of use (*USAGEINT*) and computer intensity of use (*USAGECOMP*) will also have positive impacts on the e-buy

probability, but less so than those above. The odd ratios are close to 1 (1.04 for the *POPUL*, and 1.01 for both variables *USAGEINT* and *USAGECOMP*).

- The positive effect of POPUL is picking the effects of low computer literacy, and scarcity of supply of internet in small populations. Therefore the positive effect that we have measured is consistent with our a priori experience.
- The case of *AGE* is special since it is included in the model both in levels and in squares. This is a common specification in applied econometrics. It is a quadratic function in *AGE*. Its first derivative is a linear function in *AGE*: $\partial \widehat{PURCH} / \partial AGE = 0{,}09 - (2).(0{,}001) AGE$. That means that, at early ages, the probability increases with age but at a diminishing rate. For older ages, when age increases the probability decreases. The maximum effect is at 45 years.
- The only variable that diminishes the probability is the household size, *MEMB*. Its odd ratio is 0.92 and is the only one below 1.

Finally we evaluate the predicted probability variation by levels of income, age, connection at home and number of places of Internet use.

Of these results we can say:

- The highest increment in the probability is when the individuals use the Internet in several places, with 7.86% when used from three to four places.
- Changing income quintile doesn't always result in the same increment in the e-buy probability. The probability differences increase with income.
- Having Internet at home increases the probability of purchases in the Web, with a higher increment than the differences estimated for the different income categories.

Therefore, as a conclusion, we can say that, of all the determinants for purchasing through the Internet, the most important will be the characteristics of the individual such as Internet access at home and the places of use besides the home. Surprisingly the Internet use intensity (weekly hours connected to the Web) has a positive impact but less than Internet access at home and places of Internet use. The other two variables that have positive impacts on probability are the male gender and income level, both with important but smaller impacts.

Another factor that may be relevant is proficiency in the use of the English language, which is very limited in Spain. However this should have a diminishing influence due to the increasing availability of contents in Spanish. We will analyze this issue using the more recent and richer information of the latest surveys.

Once we have analyzed the decision to buy or not to buy in Internet, we study the determinants of the next decisions: first how many times to purchase, and second how much to spend.

Specification of the Demand Models

Modeling demand or expenditure in telecommunications is not easy. Frequently the available data are binary and/or incomplete. Sometimes we do not have information on the prices paid by the consumers, or lack information on income or other variables. The functional form is usually unknown. Heterogeneity of the sample is frequent.

Before estimating e-commerce demand and expenditure functions, we describe the variables that will be used. Then the heterogeneity will be treated by using a methodology called k-means (Mac Queen 1967; Hartigan and Wong 1979) to identify groups with similar characteristics in the sample.

The additional variables in this section are shown in Table 8. Most of them are dummies. Notice that here the proxy for individual income is continuous, as originally constructed.

Next we make descriptive and graphic analyses of the variables by comparison with the expenditures in the Web. Observing the histogram of the variable *Gi* both in levels and logarithms, the evidence of the outliers is clear (Fig. 2).

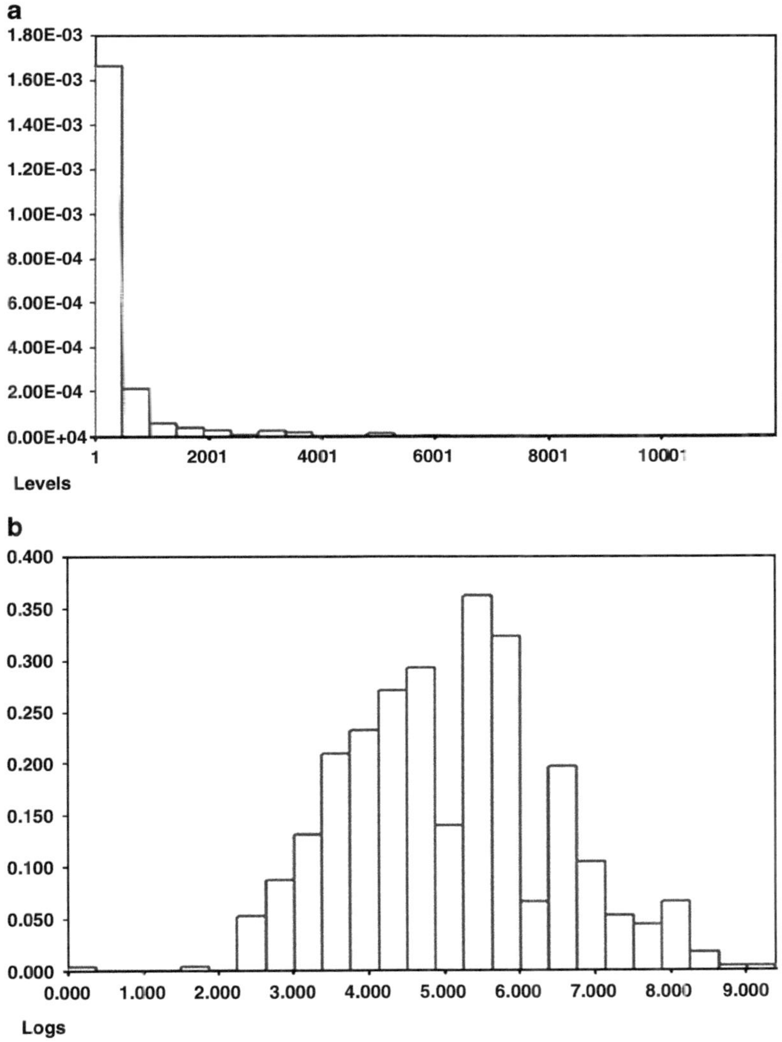

Fig. 2 Histogram of expenditures in Internet G, in levels and logs

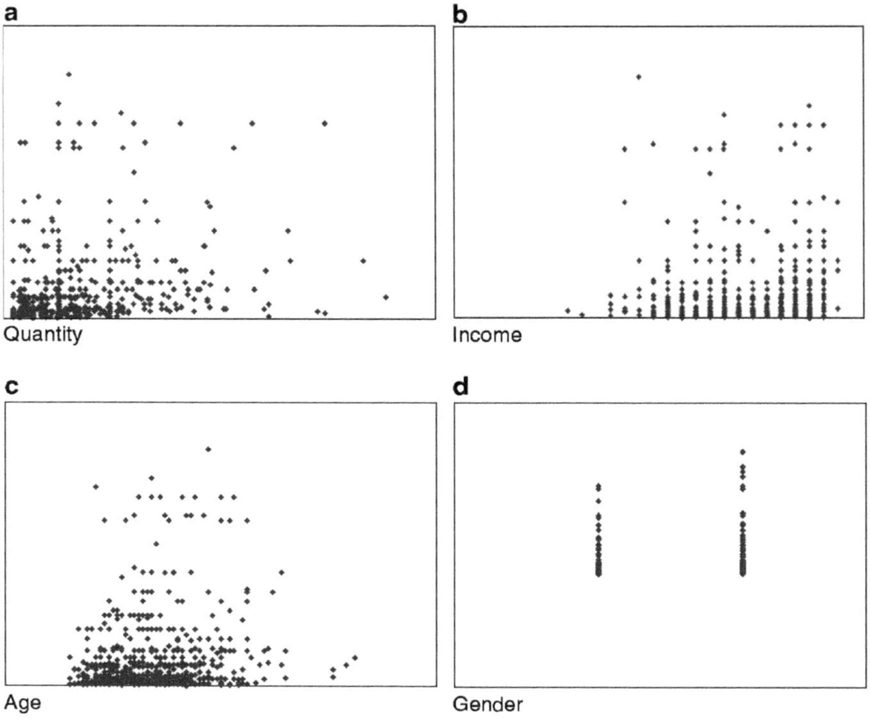

Fig. 3 Expenditure versus selected variables. The vertical axis is always expenditure, G

We use the algorithm of Peña and Yohai (1999) for detection and exclusion of extreme values. When applied to the logarithm of expense, it excludes 22 extreme values. Next we plot bivariate graphs with the expenditures in e-commerce versus selected candidate explanatory variables (Fig. 3).

The information was collected by INE using a combination of CATI[4] and PAPI[5] The data was subsequently filtered, so it is hard to imagine that any outlier is due to incorrectly registered data. However heterogeneity in the expenditure behavior is likely to be present.

Next we characterize groups of individuals that have homogeneous profiles. For that we use the probability of an individual buying in Internet, estimated in the previous section, and the k-means algorithm of Mac Queen (1967), a classic method to detect homogeneous subgroups. The homogeneity criterion is to minimize the sum of squares within each group for all the variables.

We have classified the sample in five groups, based on the probability intervals calculated from the model and seen in Tables 5 through **7**. This could change

[4]CATI: Acronym of Computer Assisted Telephonic Interview.

[5]PAPI: Paper and Pencil Interview. procedures.

Table 5 Probability of internet purchase by levels of income

Income quintile	1°	2°	3°	4°	5°
Probability	0.0815	0.1004	0.1232	0.1503	0.1821
Difference		0.0189	0.0228	0.0271	0.0318

Table 6 Probability of internet purchase by places of internet use

Table	1	2	3	4
Probability	0.1283	0.1786	0.2431	0.3217
Difference		0.0503	0.0645	0.0786

Table 7 Probability of internet purchase by internet access at home

Internet at home	No	Yes
Probability	0.1171	0.1765
Difference		0.0479

Table 8 Definition of additional variables

Variable	Definition
$QUANT_i$	Quantity of purchase transactions in internet
G_i[6]	Expenditure in internet in the last 3 months (in euros)
IS_i	Individual income index[7].
$STUDYLEVEL_i$	Years of schooling
P_1 to P_{12}	Dummies = 1 if individual buys, respectively, home products, music, books, clothes, etc. (see Fig. 1)
$FP1_i$ to $FP4_i$	Dummies =1 if the payment is through credit card, bank transfer, pay on delivery or subscription

Table 9 Internet users by probability of purchase in the internet

Group	Probability of purchase in the Internet	Group size
Less prone (Group 1)	0.010–1.101	1,757
Low propensity (Group 2)	0.101–0.191	1,341
Average propensity (Group 3)	0.191–0.298	1,037
High propensity (Group 4)	0.299–0.429	684
Very high propensity (Group 5)	0.430–0.701	411

depending on the level of aggregation and homogeneity. We have the following groups (Table 8).

The five groups in Table 9 can be treated separately, or alternatively together incorporating restrictions in the parameters. Here we have chosen an specification that allows for heterogeneity using different in the intercepts while maintaining

[6] G_i are euros spent on products through Internet in the last 3 months.

[7] Constructed at the individual level as weighted average of non-human and human capital. See Cerno and Pérez Amaral (2005)

equal slopes. This allows us to learn more about common characteristics of the sample, the slopes, while allowing heterogeneity.

We treat these groups by using group-specific constants in the models of demand for e-commerce.

Model of Demand Through the Internet

Here we estimate the parameters of demand for the number of purchases of goods and services in Internet. The endogenous variable $QUANT_i$, is the number of purchases made via Internet by an individual. Since the data are natural numbers, we use a conditional Poisson model. Here $QUANT_i$ is a function of income, the four forms of payment, gender, age, education and population size. We treat the heterogeneity by using specific dummies for each of the groups of Table 9. Now the model is

$$\text{QUANT} = \alpha_0 + \alpha_1 IS_i + \alpha_2 STUDYLEVEL_i + \alpha_3 SEXM_i + \alpha_4 AGE_i + \alpha_5 POPUL_i + \\ + \sum_{k=1}^{4} \phi_{ik} FP_k + \sum_{m=1}^{5} \xi_{im} GR_m + u_i$$

where $^{i=1,\dots,5,218}$. These are Internet users. The variables are as described above. FP_k refers to the four modalities of payment considered, while GR_m refers to the groups by probability of purchase.

Results and Discussion

From Table 10 we see that:

- The expected number of purchases in Internet is mostly related to access to a credit card, (*FP*1).
- The incidence of the rest of means of payment is smaller. The influence of payments via bank transfer (*FP*2) and pay on delivery (*FP*3) is 2.27 and 3.75, respectively.
- Payment by subscription is also important, but its incidence is the smallest.
- The individual income *(IS)* has a positive influence on *QUANT*. Its incidence is 1.51. A unit increase in *IS* is associated with a 51% increase in *QUANT*.
- Other factors such as age, education, and population size seem to have little influence on the expected quantity of transactions in the Internet.

Prediction of the Expenditure in E-Commerce

The forecast of expenditures is important for several reasons. It helps to evaluate the costs and benefits of different policy measures, and to compare different scenarios of evolution of consumer habits. In this section we seek to formulate a forecast model for the expenditures in e-commerce. For that we use a general linear model

Table 10 Frequency of use of e-commerce. Poisson model

Dependent variable	Coefficient	Incidence ratio	z
QUANT			
Constant	−3.63	–	−17.08
IS	0.42	1.51	4.67
STUDYLEVEL	−2.2	0.11	−4.74
SEXM	0.12	1.13	1.93
AGE	0.002	1	0.77
POPUL	0.1	1.01	1.69
*FP*1	2.42	11.23	36
*FP*2	0.82	2.27	10.93
*FP*3	1.32	3.75	19.79
*FP*4	0.36	1.42	2.72
*GR*1	−0.79	0.45	−7.69
*GR*2	−0.22	0.81	−2.84
*GR*3	−1.14	0.32	−5.14
*GR*4	0.11	1.01	0.13

Sample size: 5,218, Log-likelihood: −1,914.34, $\chi_{(13)}^{2}$: 3,635.19 (Prob. = 0.000), Pseudo R^2: 0.4870.

The high values of z's suggest there is very little multicollinearity present, since most of them are individually very significant.

in which we treat the different groups with constant specific dummies. The dependent variable is the log of the expenditure in e-commerce.

$$\log(G_i) = \gamma_0 + \gamma_1 \log(IS_i) + \gamma_3 \log(AGE_i) +$$

$$\gamma_4 SEXM_i + \sum_{g=1}^{4} \delta_{ig} GR_i + \sum_{p=1}^{12} \xi_{ip} P_i + u_i$$

The specific constants GRi and Pi correspond to the five groups of individuals of the previous section and to the 12 types of products, respectively.

Next we use the algorithm RETINA of Pérez Amaral et al. (2003), which chooses the model with the best out-of-sample predictive ability in terms of its AIC (Akaike Information Criterion).

The models suggested by RETINA are:

1. **Linear in the parameters** for quick and precise estimation.
2. **Nonlinear in the inputs** (explanatory variables), to enhance its approximation capabilities. The regressors used by RETINA are squares, cross products and ratios of the original inputs.
3. **Parsimonious** in the use of parameters for improved out-of-sample prediction capabilities. and
4. Can be thought of as a **generalization of ARIMA for cross section** data and other types of data.

RETINA is used as a model building and selection device. It is used only to build a model with good out of sample prediction performance. The transformations used by RETINA may not have any economic interpretation.

Table 11 Variables used by RETINA

Endogenous variable	ln (*G*)
Original continuous variables	ln (*IS*), ln (AGE)
Specific constants	SEXM, *Gg* where g=1,...,5
Specific slopes	SEXM, *Gg*, *Pp* where *p*=1, ...,12

Table 12 The basic linear model and the RETINA models for the expenditure function

	GLM (BLM)	RETINA
Parameters	8	33
AIC	0.356	−0.813
$\bar{R}^2$	0.26	0.779
RCMSPE	1.195	0.689

Table 11 shows the transformations used by RETINA. More results are provided in the Appendix.

GRg is referred to the five groups of individuals previously detected and *Pp* to the 12 groups of goods and services that are traded in Internet.

Table 12 shows the main statistics for comparing the basic linear model that uses no interactions and the model chosen by RETINA. We notice a significant decrease of the robust mean square prediction error (RCMSPE) from 1.195 to 0.689, which is indicative of better out-of-sample prediction ability.

GLM means General Linear Model, and BLM is Benchmark Linear Model. The in-sample goodness of fit $\bar{R}^2$ increases from 0.260 to a high value of 0.779. From the results in the Appendix, we also conclude that:

- The improvement in goodness of fit and RCMSPE is very important, but has been achieved by using more than four times more parameters in the RETINA model.
- In the RETINA model the original inputs enter basically as transformations, i.e. interactions, squares and ratios, so the transformations seem to improve the prediction capabilities.
- Of all the groups of consumers, the only one that shows up as significantly different in the RETINA model is the second one.
- Summarizing, RETINA suggests a model for predicting expenditure in e-commerce that is substantially better than the baseline linear model. Moreover it is easy to estimate and to use for prediction. This is obtained at the cost of using more than four times more parameters.

Conclusions

In this paper we analyze the demand for electronic commerce in Spain by individuals, using models that help us respond to three questions.

The results of the first model tend to confirm that the main determinants of the buyer are his/her income level, age and study level. However, gender and population size also have an influence.

The typical individual demanding goods and services through Internet in Spain in 2003 is a young person with higher education, living in a medium-sized population, and with a median level of income that allows him/her access to credit and allows the use of credit cards. The generalization of credit cards could be one of the main public policy measures to impulse e-commerce.

For that we use a Poisson model and conclude that access to the different forms of payment through the net (credit card, bank transfer, postage due and subscription) will influence the number of transactions. Individual income and several socioeconomic attributes are also relevant. Again, access to a credit card seems to be an important driver of e-commerce. Public policy could promote the use of credit cards for speeding the adoption of e-commerce and the advancement of the information society.

We use the Yohai and Peña method for the outliers and the k-means algorithm to detect five groups of consumers. Next we use a basic linear model together with models suggested by the new methodology of RETINA. We obtain new models that are linear in the parameters but non-linear in the inputs with enhanced out-of-sample prediction capabilities and improved in-sample fit. The types of models used here for cross section data with heterogeneity can be used in other studies for different years or regions in Spain or elsewhere.

These models can be used as tools for guiding decisions on investment and to enhance the development of the Information Society in Spain. They help to assess the costs and benefits of policy measures for promoting e-commerce and to avoid the e-exclusion of some geographical areas.

Appendix

Linear expenditure function		
Sample size	614	
Adjusted R^2	0.26	
Standard error of estimation	1.199	
RCMSPE	1.195	
AIC	0.356	
Variable	Coefficient	t statistic
Constant	2.38	3.96
Continuous original variables		
log (*IS*)	0.17	0.65
log (*AGE*)	0.51	3.17
Specific constants		
SEXM	0.29	2.75
P_1	0.69	4.81
P	−0.03	0.23
P_2	−0.09	0.82
P_3	−0.14	0.88
P_4	−0.1	0.66
P_5	−0.68	4.75
P_6	−0.55	3.7
P_7	0.32	1.52
P_8	1.12	10.59
P_9	−0.12	1.11
P_{10}	−0.96	2.34
P_{11}	–	–
GR_I	0.22	1.19
GR_{II}	−0.004	0.01
GR_{III}	0.11	0.68
GR_{IV}	–	–
GR_V	0.24	1.69

Linear expenditure function recommended by RETINA		
Sample size	614	
Adjusted R^2	0.779	
Standard error of estimation	0.647	
RCMSPE	0.689	
AIC	−0.813	
Variable	Coefficient	*Tstatistic*
Constant	7.87	5.65
Interactions		
$1/(\log(AGE))^2$	−29.28	3.7
$(\log(AGE))^2$	−0.19	3.24
$1/(\log(IS))$	−0.12	1.59
Specific slopes		
$GR_{II}/\log(AGE)$	−3.73	5.53
$P_1/\log(AGE)$	0.72	2.71

(continued)

$P_1/[\log(AGE)]^2$	−0.12	1.58
$P_2/[\log(AGE)*\log(IS)]$	4.4	2.77
$P_2/[\log(AGE)]^2$	−0.2	2.37
$P_2*[\log(AGE)/\log(IS)]$	−0.43	2.46
$P_2*\log(IS)$	−9.54	4.35
$P_2*[\log(IS)]^2$	−6.94	4.33
$P_3/\log(AGE)$	3.04	13.63
$P_3*\log(AGE)*\log(IS)$	−0.45	2.01
$P_4*[\log(AGE)/\log(IS)]$	−0.11	1.46
$P_4/[\log(AGE)]^2$	−0.07	0.9
$P_5*\log(AGE)*\log(IS)$	−0.67	2.91
$P_5[\log(AGE)]^2$	7.85	2.66
$P_5*[\log(IS)/\log(AGE)]$	6.04	1.87
$P_6*\log(AGE)*\log(IS)$	−0.21	3.36
$P_6[\log(AGE)*\log(IS)]$	−0.87	3.98
$P_7/\log(AGE)$	2.59	2.57
$P_7*\log(AGE)*\log(IS)$	−0.14	1.84
$P_7/[\log(AGE)*\log(IS)]$	1.47	1.23
$P_8*[\log(AGE)]^2$	0.25	1.97
$P_8/[\log(AGE)]^2$	100.78	1.92
$P_8/\log(AGE)$	−37.06	1.79
$P_9*[\log(AGE)/\log(IS)]$	−0.07	5.46
$P_9*\log(IS)$	−0.95	6.21
$P_{10}/[\log(AGE)*\log(IS)]$	−3.15	2.75
$P_{10}/\log(IS)$	0.9	2.49
$P_{10}*[\log(AGE)]^2$	0.08	4.75
$P_{11}*[\log(AGE)]^2$	0.05	2.63

References

Bakos Y (2001) The Emerging Landscape for Retail E-Commerce. Journal of Economic Perspectives 15 (1): 69–80.

Brynjolfsson E, Smith M (1999) Frictionless Commerce? A Comparison of Internet and Conventional Retailers. Working Paper, MIT Sloan School of Management.

Cerno L, Pérez-Amaral T (2005) Demand of Internet Access and Use in Spain. Documento de Trabajo 0506. Instituto Complutense de Análisis Económico (ICAE).

Clemons E, Hann I, Hitt L (June 1998) The Nature of Competition in Electronic Markets: An Empirical Investigation of Online Travel Agent Offerings. Working Paper, The Wharton School of the University of Pennsylvania.

Hartigan J, Wong M (1979) A k-means Clustering Algorithm. Applied Statistics 28: 100–108.

INE (2003) Encuesta Sobre Equipamiento y Uso de Tecnologías de Información y Comunicación de los Hogares del Segundo Trimestre de 2003. In www.ine.es/prensa/np388.pdf.

Lee BK, Lee WN (2004) The Effect of Information Overload on Consumer Choice Quality in an On-Line Environment. Psicology and Marketing 21 (3): 159–184.

Mac Queen J (1967) Some Methods for Classification and Analysis of Multivariate Observations. Proc. Symposium of Mathematics, Statistics and Probability, 5th, pp. 281–297, Berkeley, CA.

Peddibhotla N (2000) Are Economic Theories Good Enough to Understand E-Commerce? In Wiseman A (ed.), The Internet Economy: Access, Taxes, and Market Structure. Brookings Institution Press, Washington, DC.

Peña D, Yohai V (1999) A Fast Procedure for Outlier Diagnostic in Large Regression Problems. Journal of the American Economic Association 94: 434–445.
Pérez-Amaral T, Gallo G, White H (2003) A Flexible Tool for Model Building: The Relevant Transformation of the Inputs Network Approach (RETINA). Oxford Bulletin of Economics and Statistics 65: 821–838.
Shapiro C, Varian H (1999) Information Rules: A Strategic Guide to the Network Economy. Harvard Business School Press, Cambridge, MA.
Salop S, Stiglitz J (December 1982) The Theory of Sales: A Simple Model of Equilibrium Price Dispersion with Identical Agents. The American Economic Review 72 (5): 1121–1130.
Smith M, Bailey J, Brynjolfsson E (1999) Understanding the Digital Markets: Review and Assessment. In: Brynjolfsson E, Kahin B (eds.), Understanding the Digital Economy. MIT Press.
Train K (2002) Discrete Choice Methods with Simulation. Cambridge University Press, Cambridge.

Part III
Broadband Issues

The Diffusion of Broadband-Based Applications Among Italian Small and Medium Enterprises

Massimo G. Colombo and Luca Grilli

Abstract This paper develops an empirical model which aims at analyzing the determinants of broadband-based applications adoption among small and medium enterprises (SMEs). Focusing on a large and representative sample of Italian SMEs the econometric analysis reveals that, among other characteristics, *intra-firm capabilities* (i.e. firm labour productivity and IT skills level of employees) and *extra-firm capabilities* (i.e. the presence within the local labour market of young and skilled workforce) play a major role in explaining a firm's willingness to adopt applications. These findings are in line with those highlighted by the *skill-biased technological change* (SBTC) literature and have important implications for policy makers.

Introduction

Nowadays, the adoption of broadband technology (access and applications) is an important breakthrough innovation into the production function of most economic activities of modern economies. In particular, in so far as small and medium enterprises (SMEs) are concerned, it allows this typology of firms to achieve permanent connectivity to the global market at affordable prices (OECD 2003) and to obtain sizable productivity gains by the adoption of specific software applications. Nevertheless, many obstacles may hinder its diffusion among SMEs. In fact, the extant empirical literature on the determinants of firms' adoption of Internet has highlighted the presence of several firm-, geographic- and industry-specific factors that may influence the decision of a firm to adopt the new technology or not. In this paper, we start from the evidence provided by these previous studies in order to study the determinants of use of broadband-based applications by Italian SMEs. In particular, following a well-established model of new technology diffusion (see Karshenas and

M.G. Colombo (✉)
Politecnico di Milano, Department of Management, Economics and Industrial Engineering
e-mail: massimo.colombo@polimi.it

B. Preissl et al. (eds.), *Telecommunication Markets*, Contributions to Economics,
DOI: 10.1007/978-3-7908-2082-9_10, © Springer Physica-Verlag HD 2009

Stoneman 1993; Geroski 2000), we adopt a *rank* (or *probit*) view which, oppositely to the *epidemic* approach, assume that probabilities of adoption are inherently different across firms since these latter are heterogeneous and have different potential returns from adoption. Accordingly, the main reasons for different diffusion patterns of broadband-based applications across SMEs are mainly due to specific characteristics possessed by the firms. Relying on the *skill-biased technological change* (SBTC) literature (see Bound and Johnson 1992; Berman et al. 1994), we focus on two different but related factors that we define *intra-firm* and *extra-firm capabilities*. In particular, as long as the introduction of ICT capital into an economic organization requires an up-skilling level of employees in order to be effective (as stated by the SBTC literature), the probability of adoption will sensibly vary across firms according to the different levels of both IT preparedness and productivity of their employees (*intra-firm capabilities*) and IT familiarity and skills level of the workforce of the local area on which they are located (*extra-firm capabilities*). In fact, everything else being equal, a SME will be more willing to adopt broadband-based applications the more it is convinced to be able to exploit them. Oppositely, a low level of capabilities characterizing both firm's employees and the local workforce will presumably depress a SME's readiness to jump into the broadband wagon.

This represents the main hypothesis investigated in this paper. In particular, a sample selection framework is applied to a new longitudinal dataset of Italian SMEs in order to investigate whether, among other factors, the level of *intra-firm* and *extra-firm capabilities* affects their extent of adoption of broadband-based applications. The dataset collects information on 904 Italian firms with a number of employees comprised between 10 and 249, that operate in both manufacturing and service sectors (excluding public administration, finance and insurance). The sample is stratified by industry, size class, and geographical area so as to be representative of the Italian population of small and medium enterprises, and it contains detailed survey-based information on firm-specific characteristics and about firm adoption of broadband connection and broadband-based applications over the period from 1998 to 2005. The econometric analysis shows that both *intra-firm* and *extra-firm capabilities* figure prominently as drivers of deployment of broadband applications. As to policy implications, the finding strengthens the view that pure monetary subsidies towards firms in order to incentive them to adopt ICT capital so to increase their efficiency may have a limited impact if they are not inserted in a wide-breadth set of policy measures which aim at sustaining both types of capabilities.

The paper is organized as follows. In the next section we briefly describe the dataset. Section "The Empirical Model" is devoted to the specification of the empirical model and to illustrate the econometric methodology. Then, we highlight the main results and the implications of the econometric estimates. Some summarizing remarks and delineation of future research opportunities in section "Conclusions and Future Research Directions" conclude the paper.

The Dataset

In this paper we consider a sample composed of 904 Italian firms. The firms included in the sample are *small* and *medium enterprises* (i.e. number of employees comprised between 10 and 249) operating in both manufacturing and service sectors (excluding public administration, finance and insurance). The sample has been developed by ThinkTel in 2005 and it is stratified by industry, size class and geographical area so as to be representative of the Italian population of SMEs. Firms are observed from 1998 to 2005.

The dataset contains detailed survey-based information on firm-specific characteristics (e.g. single- or multi-plant structure, whether firms belong to groups or not) and about eventual firm adoption of broadband connection and broadband-based applications. This dataset has been complemented with firms' economic and financial data (source: AIDA, Bureau van Dijk 2007), information on the socio-economic characteristics of the area on which firms are located (Tagliacarne Institute 2006; ISTAT 2001) and longitudinal information on price levels of Internet broadband technologies (European Commission 2000, 2001/2002/2004).[1]

In this paper broadband access is defined as an Internet wired connection via ADSL or other dedicated lines with an upstream speed higher of equal to 256 Kbps (for analogous definitions see among others OECD 2002; Arbore and Ordanini 2006). We identify 15 broadband-based applications ranging from very basic (e.g. E-mail) to advanced (e.g. Supply Chain Management system) Internet use: VPN (Virtual Private Network); Data and disaster recovery system; Local protection system; VoIP (Voice over Internet Protocol) system; Video-communication, streaming, or video-conference system; E-mail; File-sharing or file distribution system; E-learning system; CRM (Customer Relationship Management) system; SCM (Supply Chain Management) system; Co-design system with suppliers and customers; E-banking system; Internet Access system; Web Site; Human Resource and Administration Management systems.

The penetration rate of broadband connection among Italian SMEs has constantly increased over time: starting from 4.8% in 1999 it has reached 66.5% in 2005. Along with access, broadband-based applications use has increased from an average number of 0.3 applications per SME in 1999 up to 5.8 applications per firm in 2005. Allegedly, if access and use by firms of broadband technologies are sensibly increased since the initial period, these figures also suggest that market saturation is still far from materializing.

[1] A complete description of collected information will be presented in Table 1.

The Empirical Model

The Econometric Methodology

The determinants of the extent of adoption of broadband-based applications by Italian SMEs are investigated through the implementation of a sample selection framework. First, a selection equation is defined:

$$\begin{aligned} z^*_{it} &= \beta' w_{it} + u_{it} \text{ with} \\ z_{it} &= 1 \text{ if } z^*_{it} > 0 \text{ and} \\ z_{it} &= 0 \text{ if } z^*_{it} \leq 0; \end{aligned} \tag{1}$$

where Z_{it} is the latent variable that captures firms' willingness to connect to Internet via broadband, z_{it} is the actual realization (equal to unity for connection, zero for non-connection) and the vector w_{it} is a set of explanatory variables including firm-specific, location-specific and time-specific variables plus other controls. Then the equation of primary interest will be given by:

$$y_{it} = \beta' x_{it} + \varepsilon_{it}; \tag{2}$$

where y_{it} is a measure of the extent of adoption of broadband-based applications based on the number of applications adopted by firms up to time *t* and the vector x_{it} includes firm-specific and location-specific factors. The sample rule is that y_{it} is observed only when z_{it}^* is greater than zero. In order to deal with and exploit the longitudinal nature of the dataset we proceeded as follows.

Equation (1) is estimated through a survival data model. As is frequent in this type of literature, we model the hazard function (i.e. the instantaneous probability of adopting, provided that this has not occurred by *t*) by the semi-parametric approach proposed by Cox (1972):

$$h_i(t) = h_0(t) \exp(\beta' w_{it}); \tag{3}$$

where $h_0(t)$ is the baseline hazard rate at time *t* (i.e. the hazard rate when all explanatory variables equal zero), w_{it} is the vector of firm-specific, location-specific and time-specific explanatory variables plus other controls and β is the vector of parameters to be estimated. This semi-parametric estimation method has the advantage of being insensitive to the specification of the baseline hazard, alleviating any possible miss-specification problem related to the hazard rate. Results obtained by the estimation of Equation (3) permit us to generate a selection correction variable (lambda) to be inserted in Equation (2) in order to properly estimate the impact of the explanatory variables x_{it} on y_{it} (i.e. the number of broadband-based applications at time *t*). In fact, failure to control for selectivity may lead to biased estimates of Equation (2) in so far as variables included in the vector w_{it} (i.e. affecting firm's decision to connect or not) are also present in the vector x_{it} (i.e. have an impact on the extent of adoption of broadband-based applications). To correct for this problem, we have used Lee's (1983) generalization of the Heckman selection model (1979) to create a selection correction variable.[2] Accordingly Equation (2) is transformed into:

$$y_{it} = \beta' x_{it} + \rho\lambda_{it} + \varepsilon_{it}; \tag{4}$$

where the selection correction variable (lambda) is given by:

$$\lambda_{it} = \frac{\varphi[\Phi^{-1}(F_i(t))]}{1 - F_i(t)}$$

where $F_i(t)$ is the cumulative hazard function computed from Equation (3), φ is the standard normal density function, and Φ^{-1} the inverse of the standard normal distribution function (see Lee 1983). Equation (4) is estimated through a random effects panel data model.

The Independent Variables

A complete list of the explanatory variables used in the estimation of the econometric models is reported in Table 1. Covariates include all those firm-, location- and time-specific variables which previous studies on this and related topics (see in particular

Table 1 Definition of explanatory variables.

Variable	Description
Intra-firm capabilities	
Value Added/ Employees$_{t-1}$	Ratio between the value added generated by the firm at time t-1 and the number of firm employee at time t-1 (source: AIDA).
Age$_t$	Number of years since firms foundation at time t.
Extra-firm capabilities	
Employee Age	Weighted average of employees' age by province (Average is weighted on the number of employees). Employees' age is measured on a scale from 1 (15-19 years) to 13 (more than 75 years) (source: ISTAT Italian census, 2001).
Employee Education	Weighted average of employees' level education by province (Average is weighted on the number of employees). Employees' level of education is measured on a scale from 1 (low level of education) to 6 (high level of education) (source: ISTAT Italian census, 2001).
Firm-specific control variables	
Employees$_{t-1}$	Number of firm employees at time t-1 (source: AIDA)
Group	One for firms belonging to business groups (source: ThinkTel).
Multi-plant	One for firms with a multi-plant structure (source: ThinkTel).
Employees Growth$_{t-1}$	Percentage growth of firm employees between time t-2 and time t-1: $(Employees_{t-1} - Employees_{t-2})/ Employees_{t-2}$ (source: AIDA).

(continued)

[2] For a similar technique used in a different context see Delmar and Shane (2006).

Table 1 (continued)

Salaries/ Employees$_{t-1}$	Ratio between the total salaries paid by the firm at time t-1 and the number of firm employee at time t-1 (source: AIDA).
Cash Flow/ Total Assets$_{t-1}$	Ratio between the cash flow generated by the firm at time t-1 and the total assets value of the firm at time t-1 (source: AIDA).
Location-specific control variables	
South	One for firms located in the South of Italy.
Average Income$_t$	Ratio between the provincial income per inhabitant and the national income per inhabitant at time t. Data are available over the period 1991-2001. Missing data have been estimated (source: Tagliacarne Institute database).
Telecommunication Network	Value of the index measuring provincial infrastructural development of telecommunication network in 2000 (source: Tagliacarne Institute database).
Time-specific control variables *(only for broadband connection)*	
Price$_t$	Broadband Internet connection normalized price. The price is the monthly rental charges for 1Mbit/s bitrate. Upload and download bitrates are added to get the total bitrate. Non-recurring charges are discounted over 3 years and added to the price. Data for years 1998-2000 refer to leased-lines rental price (source: European Commission, Directorate General for Information Society, Telecommunication Tariffs Report, 2000). Data over the period 2001-2004 refer to ADSL connection (source: European Commission, Directorate General for Information Society, Annual Report on Internet Access Costs, 2001, 2002, 2004). Missing data have been calculated basing on the same methodology (source: www.tariffe.it).
Expected Price Change$_t$	$Price_{t+1} - Price_t$, where $Price_t$ is defined as above.
Industry Adoption$_t$	$N_{adopt,t} / N_{firm,t}$, where $N_{adopt,t}$ is the expected within industry cumulated number of adopters and $N_{firm,t}$ is the within industry number of firms.
Other control variables	
Year	Value of the index measuring the year: 1=1998, 2=1999, ... 8=2005.
Industry Dummies	7 Industries Dummies: Sector1: One for Science Based manufacturing firms; Sector2: One for Scale Intensive manufacturing firms; Sector3: One for Specialized Supplier manufacturing firms; Sector4: One for Traditional manufacturing firms; Sector5: One for Utilities and Construction firms; Sector6: One for Trade firms; Sector7: One for Other Services firms.

Legend. Monetary values adjusted for inflation.

Forman 2005; Arbore and Ordanini 2006; and the companion paper Colombo et al. 2008 for a brief survey) have individuated as possible determinants of firms' adoption of broadband technology.

As to the primary interest of the paper, intra-firm capabilities are first proxied by a variable capturing the level of labour productivity reached by a SME: Value Added/Employees$_{t-1}$ is the ratio between the yearly firm value added and the number of employees. Secondly, as long as younger firms tend to select younger employees which are more likely to possess good IT knowledge, firm's age (Age_t) should capture "IT-familiarity" of firm workforce.

Extra-firm capabilities are proxied by two different covariates: *Employee Age* and *Employee Education*. The former is the weighted average of employees' age in the province where the focal firm is located (weights are based on the number of employees of the province). Age is measured on a scale from 1 (15–19 years) to 13 (more than 75 years). The latter is the weighted average of employees' level of education in the province of the focal firm. The level of education is measured on a scale from 1 (low level of education) to 6 (high level of education). Data source for both variables is the national Italian census.

Many controls are added to the models. As to firm-specific variables we include firm size (*Employees*$_{t-1}$), affiliation to a business group (*Group*), the presence of more than a plant (*Multi-plant*). This group also includes the average employee salary (*Salaries/Employees*$_{t-1}$) which aims at capturing the overall quality level of firm workforce, since SMEs characterized by a higher average employee salary are likely to have employed higher qualified personnel. Finally, the variable *Cash Flow/Total Assets*$_{t-1}$ is a measure of the availability of financial funds, while *Employees Growth*$_{t-1}$ is another indicator of firm performance and inversely proxies the degree of competitive pressure faced by SMEs.

As to location-specific factors, *Average Income*$_t$ and *Telecommunication Network* catch the overall socio-economic conditions and the quality level of the telecommunications infrastructure, respectively, of the area in which SMEs are located. A geographical dummy (*South*) is included in order to control for the firm decision to locate in the South of Italy, which represents the most economic disadvantaged area in the country.

Three additional time-specific variables are used only in the survival data analysis model. They are *Price*$_t$, *Expected Price Change*$_t$ and *Industry Adoption*$_t$ which represent the hedonic broadband price, the expected price variation over time and the industry rate of broadband adoption, respectively.

Finally, we always control for industry (*Industry Dummies*)[3] and for time of adoption of broadband connection (*Year*) in the broadband-based applications model. Note also that all time-varying variables which may generate reverse causality concerns are one period lagged so as to mitigate possible biases in the estimates.

The Results

Table 2 reports the estimates of the random effects panel data model on the determinants of adoption of broadband-based applications. *Intra-firm capabilities* exert a positive and significant impact on the extent of adoption of broadband-based applications by Italian SMEs. Everything else being equal, the number of applications is found to decrease with firm age and to increase with the labour productivity level (proxied by

[3] The sample covers a broad range of industries, with different characteristics as to processes of production, input structures and producer–customer interfaces. In turn, these factors are likely to play an important role in adoption of broadband applications, as they influence the benefits firms can derive from them. Nonetheless, the analysis of this issue goes beyond the scope of the paper and we leave it for future research.

Table 2 Determinants of SMEs adoption of broadband-based applications

	Variable	Coefficient
α_0	*Constant*	-0.070 (3.611)
α_1	*Year*	0.294 (0.058)***
α_2	*Employees*$_{t-1}$	0.501 (0.091)***
α_3	*Group*	1.006 (0.219)***
α_4	*Multi-plant*	0.779 (0.201)***
α_5	*Employees Growth*$_{t-1}$	-0.011 (0.004)**
α_6	*Age*$_t$	-0.014 (0.006)**
α_7	*Value Added/ Employees*$_{t-1}$	0.207 (0.102)**
α_8	Salaries/ Employees$_{t-1}$	0.006 (0.006)
α_9	*Cash Flow/ Total Assets*$_{t-1}$	0.471 (0.488)
α_{10}	*Employee Age*	-1.865 (0.690)***
α_{11}	*Employee Education*	2.738 (1.234)**
α_{12}	*South*	1.029 (0.455)**
α_{13}	*Average Income*$_t$	2.222 (0.875)***
α_{14}	*Telecommunication Network*	-0.553 (0.170)***
α_{15}	Lambda	0.715 (0.202)***
	$R^2 =$	0.24

Legend. *Significance level greater than 90%;
**Significance level greater than 95%;
***Significance level greater than 99%. Robust standard errors in parentheses. Random effects panel data model. Number of firms: 547; number of observations: 1759. Control variables coefficients are omitted for sake of synthesis.

the ratio between value added and number of employees) reached by the firm. The negative association between firm age and broadband-based applications adoption reveals that younger firms, often hiring younger people, are more likely to possess in-house valuable IT knowledge that leads them to more extensively use broadband-based applications. The importance of IT competencies is confirmed by the results concerning *extra-firm capabilities*. In particular, use of broadband applications by SMEs is positively influenced by location in geographic areas characterized by a labour market with a predominance of young and highly educated workforce: the coefficients of the variables *Employee Age* and *Employee Education* are negative and positive, respectively, and both significant at 95% significance level.

As to control variables, most of the results are consistent with *a priori* expectations. The coefficient of the selection correction variable (*Lambda*) is positive and significant, suggesting quite reasonably that broadband connection and use of complementary applications are positively interrelated events. Firm's demand for communication is found to positively and significantly impact the extent of adoption of broadband-based applications. In particular, large SMEs with a multi-plant structure and belonging to a business group are more likely to use broadband-based applications: the estimated coefficients of *Employees*$_{t-1}$, *Multi-plant* and *Group* are positive and significant at 99%. A high degree of competitive pressure (inversely proxied by the variable

Employees Growth$_{t-1}$) is found to exert a positive effect (at 95% confidence level) on firms' decision to adopt broadband softwares, as to the *Average Income*$_t$ variable which has a positive and significant coefficient (at 95% confidence level). Quite interestingly, the *Telecommunication network* variable and the dummy variable *South* have a statistically significant impact on the number of applications used by SMEs (99% and 95%, respectively); the former is negative and the latter is positive. On the one hand, firms located in economic disadvantaged and less-infrastructured areas may find serious obstacles in accessing broadband connection due to scarcity of suppliers and a possible low quality level of the delivered service. Clearly, this may negatively influence their broadband connection behavior.[4] On the other hand, the firm's opportunity cost to face and overcome these difficulties decreases along with the number of applications the firm decides to adopt. This means that once a SME located in less-equipped areas acquires broadband access, it will do it for using a large number of specific applications. Finally the analysis highlights quite unsurprisingly, that diffusion of broadband-based applications has been increasing over time: the coefficient of *Year* is positive and significant at 99%.

Results of the estimation of the survival data model related to the selection equation are presented in Table 3. The only key determinants of adoption appear to be vari-

Table 3 Selection equation: determinants of SMEs adoption of broadband connection

	Variable	Coefficient
α_1	*Employees*$_{t-1}$	0.060 (0.015)***
α_2	*Group*	0.257 (0.12)**
α_3	*Multi-plant*	0.412 (0.111)***
α_4	*Employees Growth*$_{t-1}$	0.001 (0.001)
α_5	*Age*$_t$	0.001 (0.001)
α_6	*Value Added/ Employees*$_{t-1}$	0.001 (0.001)
α_7	*Salaries/ Employees*$_{t-1}$	-0.001 (0.001)
α_8	*Cash Flow/ Total Assets*$_{t-1}$	-0.060 (0.115)
α_9	*Employee Age*	-0.082 (0.347)
α_{10}	*Employee Education*	0.151 (0.594)
α_{11}	South	-0.001 (0.222)
α_{12}	*Average Income*$_t$	-0.042 (0.082)
α_{13}	*Telecommunication Network*	0.002 (0.001)*
α_{14}	*Price*$_t$	-0.08 (0.039)**
α_{15}	*Expected Price Change*$_t$	-0.002 (0.001)
α_{16}	*Industry Adoption*$_t$	-0.508 (0.206)***
	Log-likelihood	-3075.31

Legend. *Significance level greater than 90%;
**Significance level greater than 95%;
***Significance level greater than 99%. Robust standard errors and number of restrictions in parentheses. Cox proportional hazards model. Breslow method for ties. Number of firms: 904; number of observations: 3678. Control variables coefficients of six industries dummies are omitted for sake of synthesis.

[4] If we look at the results of the selection equation (Table 3) the coefficient of the variable *South* is positive but statistically insignificant, while the variable *Telecommunication Network* is found to negatively affect (albeit at 90%) the probability of SMEs to adopt broadband connection.

ables that reflect the demand of communication by SMEs: both $Employees_{t-1}$ and *Multi-plant* have a positive and significant at 99% impact on the hazard rate; the coefficient of *Group* is positive and significant at 95%. Therefore, large SMEs with a multi-plant structure and belonging to a business group are those firms most likely to be early adopters. Conversely, the decision to adopt broadband connection does not seem to be affected neither by *intra-firm* nor by *extra-firm capabilities*. Quite unsurprisingly, diffusion of broadband connection is found to be driven by the decline over time of the (hedonic) price of broadband connection and to increase with the industry rate of adoption.

Conclusions and Future Research Directions

This study is an econometric investigation of the determinants of the extent of adoption of broadband-based applications by Italian SMEs. Particular attention is given to those variables capturing *intra-firm* and *extra-firm capabilities*. According to the *skill-biased technological change* (SBTC) literature, the introduction of a new technology into the production system often requires an up-grade of the existing skills of employees in order to properly exploit the process innovation. If this line of reasoning also applies to broadband technology, firms would adopt more broadband-based applications, the greater is their ability to use them. In turn, this will depend on the productivity and IT familiarity of their employees (*intra-firm capabilities*) and on the possibility to hire highly educated and IT skilled workers from the local labour market (*extra-firm capabilities*).

The empirical analysis is based on a new longitudinal dataset composed of 904 Italian SMEs (i.e. number of employees comprised between 10 and 249), which operate in both manufacturing and service sectors (excluding public administration, finance and insurance). The sample has been developed by ThinkTel in 2005, it is stratified by industry, size class, and geographical area so as to be representative of the Italian population of SMEs, and it contains detailed survey-based information about adoption of broadband connection and broadband-based applications over the period from 1998 to 2005. Data provided by the Thinktel survey has been supplemented with firms' financial data and location-specific data collected from other public and private sources.

The results of the econometric analysis may be summarised as follows. Among other important factors, *intra-firm capabilities* (i.e. firm labour productivity and IT skills level of employees) and *extra-firm capabilities* (i.e. the presence within the local labour market of young and skilled workforce) play a major role in explaining a firm's willingness to adopt applications. Moreover, these factors affect the use of applications but do not exert any influence on the decision of firms to connect via broadband. Overall, these findings are in line with those highlighted by the SBTC literature and with its vision of ICT and human capital as strictly complementary production inputs.

From a policy perspective, a clear implication derives. First of all, subsidies to the adoption of broadband connection should be accompanied by more structural and medium-long time horizon policy interventions. Policy makers should target and try to fill the IT skills and competencies gap potentially suffered from SMEs, which in turn prevents adoption of (advanced) broadband applications. In the medium term there is the need for policy schemes favouring employees training activities, the purchase of other supporting services, and the recruitment of skilled personnel. Taking a longer term view, the results of this research confirm that investments in human capital play a crucial role for economic development. While making it easier for Italian SMEs to adopt broadband-based applications, they may enable them to increase their efficiency, innovativeness and competitiveness in international markets.

Finally, we are conscious that this empirical analysis is only a first step towards a full understanding of the decision of firms to use broadband-based applications and other investigations are needed on the issue.

This study on the determinants of broadband connection and adoption of complementary applications among SMEs raises many new questions for future research. First of all, we (as most of the extant empirical literature on the topic) have not considered possible determinants that may hinder a firm's willingness to use broadband technologies, such as security issues and management's concerns about a possible increase through the use of the new technology of unproductive activities by employees. Absence of this information may help justify the relatively low amount of total variance explained by our model. Then, it would also be interesting to investigate not only the number of applications adopted by the organizations but also the "intensity" of adoption of these applications within firms. Moreover, broadband-based applications range from basic to very advanced softwares and determinants of adoption may differ accordingly.

Finally, this is a study on firms' access to broadband technology. The analysis on the effects of the use of broadband-based applications on firm productivity is high in our research agenda. In this respect, the adoption of broadband-based applications is likely to have a positive effect on productivity only as long as a firm is able to efficiently use them. According to the *skill-bias organizational change* literature (Brynjolfsson and Hitt 2000; Bresnahan et al. 2002), we would expect a rise in productivity caused by the adoption of (advanced) applications only if complementary managerial and organizational innovations are introduced into the adopting firm.

Acknowledgments Financial support from Thinktel (International Think Tank on Telecommunications) is gratefully acknowledged. The authors are also grateful for useful comments and suggestions to Thinktel committee components and to participants in the 2007 ITS Conference held in Istanbul (Turkey).

References

Arbore A, Ordanini A (2006) Broadband divide among SMEs: the role of size, location and outsourcing strategies. International Small Business Journal 24: 83–99.

Berman E, Bound J, Griliches Z (1994) Changes in the demand for skilled labour within US manufacturing industries. Quarterly Journal of Economics, 109 (2): 367–398.

Bound J, Johnson G (1992) Changes in the structure of wages in the 1980s: an evaluation of alternative explanations. American Economic Review 82 (3): 371–392.
Bresnahan T, Brynjolfsson E, Hitt LM (2002) Information technology, workplace organization and the demand for skilled labor: firm-level evidence. Quarterly Journal of Economics 117 (1): 339–376.
Brynjolfsson E, Hitt LM (2000) Beyond computation: information technology, organizational transformation and business performance. Journal of Economic Perspective 14 (4): 23–48.
Bureau Van Dijk (2007) AIDA Dataset.
Colombo M G, Grilli L, Verga C (2008) Broadband access and broadband-based applications: an empirical study of the determinants of adoption among Italian SMEs. In: Dwivedi Y K, Papazafeiropoulou A, Choudrie J (eds.) Handbook of Research on Global Diffusion of Broadband Data Transmission. IGI Global: 466–480.
Cox RD (1972) Regression models and life tables. Journal of the Royal Statistical Society 34: 187–220.
Delmar F, Shane S (2006) Does experience matter? The effect of founding team experience on the survival and sales of newly founded ventures. Strategic Organization 4 (3): 215–247.
European Commission, Directorate General for Information Society (2000) Telecommunication Tariffs Report.
European Commission, Directorate General for Information Society (2001/2002/2004) Annual Reports on Internet Access Costs.
Forman C (2005) The corporate digital divide: determinants of Internet adoption. Management Science 51 (4): 641–654.
Geroski P (2000) Models of technology diffusion. Research Policy 29 (4/5): 603–625.
Heckman J. (1979) Sample selection bias as a specification error. Econometrica 47 (1): 153–162.
Istituto Nazionale di Statistica ISTAT (2001) National Census Database.
Karshenas M, Stoneman P (1993) Rank, stock, order and epidemic effects in the diffusion of new process technologies: an empirical model. The RAND Journal of Economics 24 (4): 503–528.
Lee L (1983) Generalized econometric models with selectivity. Econometrica 51 (2): 507–512.
OECD (2002) Broadband access for business. Working Party on Telecommunication and Information Services Policies, Organisation for Economic Co-operation and Development (OECD).
OECD (2003) Broadband driving growth: policy responses. Working Party on Telecommunication and Information Services Policies, Organisation for Economic Co-operation and Development (OECD).
Tagliacarne Institute (2006) Geo Web Starter Database.

Drivers and Inhibitors of Countries' Broadband Performance – A European Snapshot

Nejc M. Jakopin

Abstract There is large variance regarding diffusion of broadband Internet among European countries. Some nations are criticised for a presumably underdeveloped broadband market. This study analyses broadband Internet access take-up using a wide range of variables to explain European countries' broadband position and to anticipate future market developments. Correlation results for the EU-25 countries are presented using a time lag design of introducing the influencing variables for the year 2003 and the outcome criteria for 2005/2006. Findings show that (1) economic prosperity and (2) computer literacy initiate broadband penetration differences. Further, strong effects are identified for (3) English language proficiency, which affects the attractiveness of global Web content for Internet subscribers, (4) teleworking, which increases the base of potential early broadband adopters, (5) service sector employment that positively correlates with the need for information access, and (6) unemployment, which reduces the spending power of consumers. Privatisation, independent regulator, and LLU lead times have a significant positive impact on broadband development, while intra-technology and general market concentration are negatively associated with broadband uptake. Inter-technology (e.g., Cable vs. DSL) competition is not significant for broadband take-up in the EU-25 sample.

Introduction

The development of broadband Internet markets has received significant attention in public policy discussions. The European Commission, United States Congress, as well as other governments and development agencies set high broadband availability as one of their key information society goals (e.g., European Commission 2002, 2005, 2006a). Broadband Internet – usually defined as connections with download speed equal to or greater than 128 kbit/s (European Commission 2006b)

N.M. Jakopin(✉)
Arthur D. Little GmbH, Düsseldorf
e-mail: jakopin.nejc@adlittle.com

B. Preissl et al. (eds.), *Telecommunication Markets*, Contributions to Economics, 187
DOI: 10.1007/978-3-7908-2082-9_11, © Springer Physica-Verlag HD 2009

or 256 kbit/s (cf. OECD 2007)[1]– is propagated as driver of economic growth and other public interest benefits (e.g., Lehr et al. 2006, pp. 4–6).[2]

Overall, the number of broadband connections has reached approximately 75 million in Europe compared with an estimated 245 million Internet users at the end of 2006 (EITO 2007, pp. 246–247). The average penetration per head in 19 European countries was 17.5%, indicating that, overall, the broadband Internet market was in an early growth stage of the life cycle at year-end 2006 (OECD 2007). Still, markets are not uniform in their broadband development: Broadband penetration ranged from 31.9% (Denmark) to 4.6% (Greece) in the EU 25 countries alone at year-end 2006 (OECD 2007). Such differences or also below-expectation levels of broadband penetration may be attributed to market characteristics. Therefore, a wider set of factors such as English literacy and proficiency in computer usage, teleworker density, service sector employment, narrowband-price levels and other idiosyncrasies that potentially affect demand for broadband subscriptions at an aggregate country-level should be analysed.

A growing number of studies tackles broadband diffusion across countries. Bauer et al. (2003), Cava-Ferreruela and Alabau-Muñoz (2004), Distaso et al. (2006), Garcia-Murillo (2005), Turner (2006), and Wallsten (2006) analyse broadband take-up as a function of various economic, societal and country specific conditions for OECD countries. Others focus on demand drivers within the United States or in one country (see for many Bauer et al. 2002; Chaudhuri and Flamm 2005; Gerpott 2007).[3] Further, for mobile telecoms, the diffusion across countries was found to be related to characteristics of the competitive environment and to general country parameters in several articles (see overview in Jakopin 2006, pp. 65–69), e.g., showing that larger and more economically prosperous countries and mobile markets with higher competitive rivalry achieve faster service diffusion and higher penetration rates (cf. Dekimpe et al. 2000). Nevertheless, European markets remain under-researched and larger indicator sets and frameworks, as well as causality considerations are still missing.

Therefore, in this paper broadband Internet access take-up is analysed for a comprehensive European (EU-25) dataset using a wide range of explanatory variables to enhance the understanding of each individual nation's current broadband position

[1] "Broadband" is used for any Internet access technology that provides faster data transfer than traditional dial-up connections by using higher bandwidth and some form of packet switching and "always on" capability. Cf. Bauer et al. (2003, p. 4), Gerpott (2007, p. 799), Maldoom et al. (2005, p. 3). Commonly, DSL (Digital Subscriber Line), Cable, Fixed Wireless, Satellite, third generation mobile, and fiber-to-the-home are technologies discussed as broadband Internet access platforms. Cf. Distaso et al. (2006, pp. 89–90), Maldoom et al. (2005, pp. 11–24), Papacharissi and Zaks (2006, pp. 65–67). DSL and (in some countries) cable are by far the dominant broadband technologies today.

[2] Benefits of broadband are usually debated qualitatively (cf. Bauer et al. 2003, p. 3; Frieden 2005, pp. 596–599; Maldoom et al. 2005, pp. 8–11) but are seldomly quantified (cf. Lehr et al. 2006).

[3] A comprehensive summary of the literature is provided in Gerpott (2007, pp. 801–805); see also Bauer et al. (2003, pp. 9–10) for additional references.

and anticipate future market developments. First, a framework is presented to reflect relevant dimensions that explain a country's broadband development. Second, samples and the operationalization of variables are described. Third, results of the empirical analyses are presented. Finally, limitations of the analyses and future research opportunities are discussed.

Broadband Development Drivers

To identify broadband development drivers, we first distinguish general factors that affect Internet demand overall, irrespective of the speed or bandwidth available (see Fig. 1). For example, service sector employment, teleworking and English literacy should increase Internet penetration and usage in general. Until quite recently, a substantial share of customers had narrowband access via traditional dial-up connections. Despite a shift towards broadband, which also means that new subscriptions are predominantly broadband services, the actual Internet user base still shows an average of 40% narrowband subscribers in European countries (ranging from 11% in Estonia up to 85% in Greece; European Commission 2006c, p. 48) – making causal explanations (incl. time-lags) of this metric important.

Second, factors determining how Internet users are distributed among the narrowband versus broadband category of access modes are taken into consideration (see Fig. 1). The coverage of a particular technology and the price level of various narrowband services (e.g., dial-up Internet access but also local or long-distance/international calls) compared to broadband may affect which services Internet users will choose for their subscription.

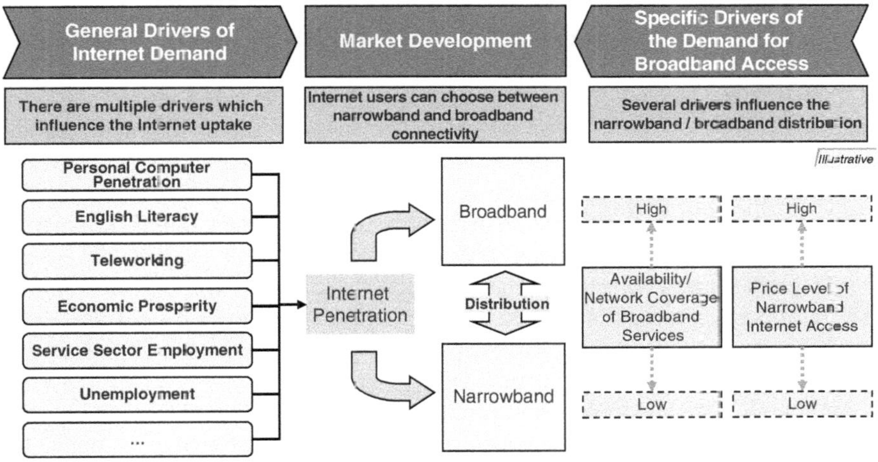

Fig. 1 Antecedents of markets' broadband Internet "performance" (From Arthur D. Little 2007; Jakopin and Von den Hoff 2007)

General Drivers of Internet Demand

The diffusion potential of Internet and broadband services is first of all shaped by general economic conditions. With increasing development levels, countries move from "low-tech" production and service modes to activities that require more information and long-distance data communication both in private and professional areas. At the same time, more financial resources become available to purchase broadband subscriptions or Internet access. A common measure to reflect the development level of countries is the gross national income per inhabitant in purchasing power equivalents (cf. Garcia-Murillo 2005, p. 89). Of course, several macro and micro factors influence economic prosperity. Unemployment, the share of service sector activities, and urbanisation are exemplary drivers of economic development. At the same time they are important indicators for broadband demand.[4] Web access is more and more relevant in sectors that rely on information and knowledge to create their products and services. Financial services, logistics, research and other services today draw on communication and information access via the Internet or closed Intranets to increase efficiency. In line with these expectations, previous studies in general found some positive effect of economic prosperity on Internet and/or broadband development (Bauer et al. 2003, pp. 15–16; Chinn and Fairlie 2007, pp. 30–35; Garcia-Murillo 2005, pp. 96–102; Turner 2006, pp. 9–10).

Broadband markets may advance faster in more densely populated areas (Cava-Ferreruela and Alabau-Muñoz 2004, p. 3; Frieden 2005, p. 598; Garcia-Murillo 2005, p. 89), which is due to lower cost and higher speed of covering the relevant market with broadband services, i.e., the roll-out of broadband access to the population ("coverage"), better leverage of communication and distribution instruments, higher interaction levels within the population, and generally speaking, in the presence of strong innovation drivers.

Previous studies confirm the spatial agglomeration-performance relation for broadband services (Bauer et al. 2003; Cava-Ferreruela and Alabau-Muñoz 2004; Turner 2006) but not for Internet penetration in general (Chinn and Fairlie 2007).

Similarly, from an operator's perspective more attractive or prosperous countries promise the highest and quickest return on investment – thereby inducing earlier market entries. Thus such markets are further advanced on the diffusion curve and exhibit higher broadband penetration.

Further, "soft" country characteristics, such as societal and cultural factors may affect the value of the Internet and broadband services to potential users. Prerequisites such as computer usage and English skills, teleworking, and cultural mindset affect the penetration level that can be reached relative to the overall population of a country and the push for earlier and/or quicker acceptance of Internet and broadband offers at the macro level.

First of all, individuals (or a household as relevant unit) require basic knowledge about working with a personal computer, a factor sometimes referred to as "digital lit-

[4] In turn, broadband is also hoped to positively impact economic development (cf. the discussion of broadband benefits by Bauer et al. (2002), Chinn and Fairlie (2007), Lehr et al. (2006, pp. 4–6), European Commission (2006a)).

eracy" (Frieden 2005, p. 599). If people are not confident with using the technology required as user interface for Internet- or broadband-enabled services, they will be reluctant to subscribe at all. Related to these capabilities, the penetration with personal computers (PCs) in a country may be taken as indicative of digital literacy, since this indicator may be hard to measure or compare across countries (Chinn and Fairlie 2007, pp. 25–30). Hence, higher levels of computer penetration are perceived to be driving broadband take-up too (Cava-Ferreruela and Alabau-Muñoz 2004, p. 3; Chinn and Fairlie 2007, pp. 25–30; Garcia-Murillo 2005, p. 90; JP Morgan Chase 2006, p. 44). In this context Bauer et al. (2003, pp. 13–14) mention "preparedness", measured as an index of factors such as openness for technical innovation and availability of complementary products such as personal computers as positively associated with broadband saturation. Similarly, Cava-Ferreruela and Alabau-Muñoz (2004, p. 7) found PC penetration to positively affect broadband penetration. However, PC penetration may also experience a positive feedback from broadband development: Attractive broadband packages and broadband-enabled services may induce new purchases of computers by previous non-users (that may still have basic PC skills due to their work environment).

Users' English language capabilities are important due to the significantly more extensive base of services and information that is available in this particular language compared to other local languages. Estimations indicate that – despite a relative decline over time – more than 50% of the indexable Web pages are available to potential users in English. The utility derived from a broadband connection is larger, if a subscriber understands and can appreciate the diverse English language content in supply worldwide. The common argument that content drives broadband makes the circumstance that information and entertainment is predominantly in English all the more relevant.

Teleworking, home office or offsite presence are additional society specific circumstances shaping demand for broadband services. More broadly, cultural factors potentially impact the openness of a target market regarding new technologies. A cultural dimension used in earlier analyses of new product adoption or technology diffusion is avoidance of uncertainty, which concerns the level to which a country's residents strive for reliability, rituals, rules, and institutionalized structures. A target audience that is reluctant to accept technical uncertainty or novelty tends to wait until an offer is "mainstream" and properly tested and tried by a sufficient number of customers. At the same time, network operators may avoid entry in markets that are known for slower innovation adoption. As a result, introduction as well as growth of Internet and broadband penetration may be significantly delayed and/or slowed down. While not studied for Internet and broadband offers, an indication for such effects can be found for mobile voice services, where avoiding uncertainty increases imitation and later adoption across countries (Sundqvist et al. 2005, p. 109).

Regulatory conditions are a further indirect contributor to Internet and broadband performance. General and telecoms-specific regulatory settings shape evolution of competition and market dynamics at the macro level. Restrictive regulation of the consolidation attempts in the German cable market is a prime example of public policy that affected broadband uptake (cf. Maldoom et al. 2005, p. 54).

Telecoms market liberalisation changed the industry in the 1990s in most countries with only a few earlier exceptions (e.g., US and UK) and a larger number of less developed markets that followed later. Lead time in terms of earlier market liberalisation may create an environment with more advanced market-oriented competitive landscapes, which can in turn provide better preconditions for Internet services in general and new broadband markets in particular. While broadband diffusion is seldom viewed in relation to lead time in liberalisation, grouping of countries by market openness did not yield significant results in one study (Bauer et al. 2003, pp. 15–16).

To sum, the expectation is to find a higher level of broadband development in countries with higher income per inhabitant, of lower unemployment, more service sector activity, and more urbanisation. Further, culturally innovation-minded countries that take a less restrictive regulatory approach and open markets earlier are expected to achieve earlier introduction and stronger broadband Internet take-up especially when their computer and English skill bases and teleworking environment are well developed.[5]

Specific Drivers of the Demand for Broadband in Contrast to Narrowband Access

Distinguishing further, factors that shape the distribution between broadband and narrowband subscriptions within the overall Internet subscriber base are taken into account (see Fig. 1). Such factors are broadly attributable to public telecoms policy and countries' broadband market settings.

Public infrastructure funding and government initiatives are potential drivers cited for broadband development (cf. Frieden 2005; Miralles 2006; Papacharissi and Zaks 2006). While qualitatively discussed in various commentaries and articles, a common pattern of initiatives was not identified. Rather, public institutions and governments use a broad range of strategies in order to promote broadband take-up within a country or region. Activities span from tax reductions for operators or also consumers, sponsorship, administrative online offerings, to public–private partnerships and public infrastructure investments (Miralles 2006, pp. 20–22). Limited comparability of these activities explains the lack of quantitative tests of their impact on market performance. However, some case studies and evidence from markets like South Korea or Japan support the notion of supportive government policy effects (cf. Frieden 2005; Papacharissi and Zaks 2006).

Both intra- and inter-modal platform competition (between providers vs. between technologies) are assumed to positively influence broadband development (cf. Distaso et al. 2006; Maldoom et al. 2005, p. 33; Polykalas and Vlachos 2006) and are characterized by a similar and overlapping reasoning: Competition potentially

[5]Additional socio-demographic characteristics cited in micro-level studies of broadband adoption (e.g., age, gender, residence area, ethnic group, income, or education, cf. Chaudhuri and Flamm 2005; Gerpott 2007) have also been analysed at an aggregate level across countries (see partially Cava-Ferreruela and Alabau-Muñoz 2004, pp. 6–10; Chinn and Fairlie 2007, pp. 30–35) but are neglected here.

increases price pressure and service quality/diversity and induces additional promotion and distribution efforts irrespective of the underlying facilities- or access-based business model. The continuing debate within the industry, public interest, and scholars about high-speed access regulation (e.g., VDSL in Germany) and the importance of intra- and inter-modal competition is an indication that the relevance for market performance of a certain regulation approach is far from clear.

Local Loop Unbundling (LLU) is presumably "driving effective price and bandwidth competition" (JP Morgan Chase 2006, p. 44). Work by Garcia-Murillo (2005, pp. 96–102) confirms this view by reporting a significant positive LLU effect for broadband availability and a similar, but weaker association for broadband penetration. Still, many analyses did not present strong LLU-effects in data until mid-year 2004 (Cava-Ferreruela and Alabau-Muñoz 2004; Distaso et al. 2006; Wallsten 2006, 2007), but focused on inter-modal competition (measured by a concentration index) of various technological platforms such as cable and DSL for which they found significant positive relations to broadband development (cf. Distaso et al. 2006).

The competitive strategy followed by operators, levels of broadband competition, the price level of telephone calls and the price-spread between narrow- and broadband Internet access are relevant variables of broadband markets. Overall, competitive intensity affects service pricing, network coverage, and bandwidth (down- and upload speed available to subscribers), which in turn shapes the attractiveness of broadband services and eventually country penetration levels relative to other markets.

Competition between broadband providers irrespective of technology used, which would correspond with the traditional view of measuring competition or market concentration (and may be identical to the intra- and inter-modal perspective in some cases) may drive broadband take-up. While Bauer et al. (2003) did not find such an effect, Garcia-Murillo (2005) provided supporting evidence.

Higher prices for fixed voice calls may drive broadband take-up: Subscribers search for lower cost telephony alternatives, which are becoming available to the mass market of broadband Internet users via voice over IP services (e.g., Skype). Distaso et al. (2006) reported some link between fixed voice call pricing and broadband customer pull-incentives.

Dial-up cost/narrowband Internet charges and broadband subscription pricing are relevant for broadband take-up (Cava-Ferreruela and Alabau-Muñoz 2004, p. 3), although narrow- and broadband access are not seen as substitutes but vertical complements (cf. Maldoom et al. 2005, pp. 5–7). Especially a higher spread between narrow- and broadband Internet access prices potentially lowers incentives to switch or upgrade to broadband by increasing perceived "cost of upgrading" (Arthur D. Little 2007; Papacharissi and Zaks 2006, pp. 71–72). As a component of subscription charges, high LLU prices (input cost) were found to be associated with lower broadband penetration (Distaso et al. 2006, pp. 100–103). However, this finding may not be directly applicable – previous studies only presented weak indications of stronger broadband performance when broadband prices were lower and/or dial-up prices higher (Arthur D. Little 2007; Bauer et al. 2003, pp. 15–16; Cava-Ferreruela and Alabau-Muñoz 2004, pp. 6–10).

Further, technical broadband availability or coverage of households is obviously required for strong market diffusion (Cava-Ferreruela and Alabau-Muñoz 2004).

Especially, the low coverage of Internet-ready cable connections, which restrict inter-technology competition, has been cited as obstacle for broadband performance. The most prominent European example is Germany, with cable broadband growth suffering from restrictive antitrust policies and misled strategic moves in the marketplace (see Fig. 2).

Research Model and Method

To shed light on causal linkages of potential "driving" factors for broadband development indicators, the following correlation analyses by design include a time lag between independent and dependent variables (see variable descriptions in Table 1 for further details – time lags may vary depending on data availability but are in general around 2–3 years). In addition, alternative indicators and data sources were considered to achieve a coherent view on the topic. Multivariate regressions were omitted because of the low sample size of this study that focuses on the 25 European Union (EU) countries, high multicollinearity of some indicators, and the large number of variables. Variable definitions, sources and descriptive statistics are presented in Table 1. The sample covers all 25 European Union (EU) member states as of year-end 2006.

In addition to commonly used data sources such as ITU and OECD reports (cf. Cava-Ferreruela and Alabau-Muñoz 2004; Chinn and Fairlie 2007; Jakopin 2006), a variety of secondary statistics was leveraged to collect information on indicators like English literacy, service sector activity, lead times or price-spread that have not received significant research attention previously (see Table 1).

Dependent Variables

Broadband development is measured with three indicators: (1) Broadband penetration, i.e., the ratio of broadband subscriptions to total population at year-end 2006, (2) the ratio of Internet users to total population at year-end 2005 and (3) the time of initial broadband introduction (see variables I–III in Table 1). These variables are not comprehensive with regard to the quality of broadband subscriptions (i.e., bandwidth, type of services) or the overall usage/demand satisfaction. However, they include the most common broadband indicator – broadband penetration,[6] as well as a general Internet metric, the Internet user ratio, and the overall country adoption covered with the broadband launch lead time. The broadband penetration ratio was highly correlated with the Internet user ratio ($r = 0.83$, $p = 0.003$, $n = 19$). Following the diffusion curve logic, commercial broadband introduction lead time had a strong correlation both to broadband penetration rates ($r = 0.54$, $p = 0.03$, $n = 19$) but not to the Internet user ratio ($r = 0.08$, $p = 0.75$, $n = 19$).

[6] Alternative definitions of penetration, particularly with different baselines, e.g., households or people aged 14 to 65, are highly overlapping (correlations exceeding $r = 0.98$). Therefore, the inclusion of additional common broadband indicators with low differentiation is not taken up.

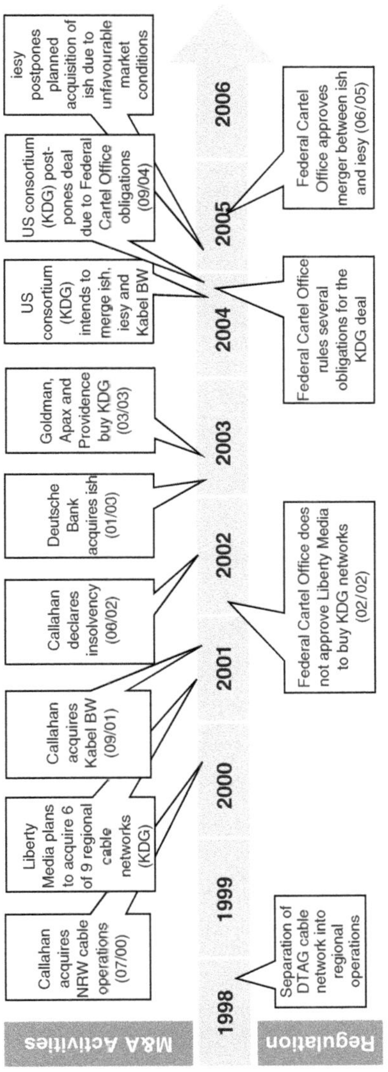

Fig. 2 Timeline of mergers and acquisitions activity in the German cable market and regulatory intervention (From Arthur D. Little 2007)

Table 1 Indicators, measurement, EU-25 descriptive statistics, and data source.

	Indicators	Measurement and Descriptive Statistics[a]
Dependent Variables	**I. Broadband Penetration 2006**[b]	• Ratio of total broadband subscriptions to total population at year-end 2006 • m = 17.48%; s.d. = 8.11; n = 19
	II. Internet User Ratio 2005[c]	• Percentage of the population that were regular Internet users at year-end 2005 • m = 43.89%; s.d. = 14.58; n = 25
	III. Broadband Launch Lead Time[d]	• Months since first commercial introduction of broadband services (DSL or Cable) in a country until year-end 2005 (alternative factual and estimated variables) • m = 68.5; s.d. = 13.5; n = 19
General Country Characteristics	**1. Economic Prosperity**[e]	• Purchasing power parity adjusted gross domestic product per capita in a country in 2003 in USD • m = 23,831.1; s.d. = 11,125.1; n = 24
	2. Prosperity Growth[e]	• Percentage growth of gross domestic product per capita in a country from 2002 to 2003 • m = 2.13%; s.d. = 2.80; n = 25
	3. Unemployment[e]	• Percentage of the total laborforce that are unemployed at year-end 2002 • m = 8.92%, s.d. = 4.23, n = 25
	4. Service Sector Activity[f]	• Employment in services as percentage of total employment in 2003 • m = 64.25%; s.d. = 8.63; n = 25
	5a. Urbanisation[e]	• Percentage of population living in urban areas in a country in 2003 • m = 73.40%; s.d. = 13.14; n = 25
	5b. Population Density[e]	• Residents per square kilometer in a country in 2003 • m = 176.7; s.d. = 246.2; n = 25
Societal Country Characteristics	**6a. Computer skills**[g]	• Percentage of the population aged 16 to 74 that have basic computer skills in 2003 • m = 61.78%; s.d. = 16.84; n = 18
	6b. Personal Computer Penetration[h]	• Personal Computers per 100 inhabitants at year-end 2003 • m = 34.01%; s.d. = 17.98; n = 25
	7. English Literacy[i]	• Percentage of the population speaking English as a foreign language in 2005 • m = 45.2%; s.d. = 26.7; n = 25
	8a. Share of Teleworkers[j]	• Percentage of the population that were teleworkers (i.e., worked at home or at a distance from their official place of work) in 1999 • m = 8.50%; s.d. = 5.44; n = 10
	8b. Home Office[k]	• Proportion of workforce that worked from home at least one day per week in 2004 • m = 1.72%; s.d. = 2.04; n = 25
	9a. Uncertainty avoidance (Hofstede)[l]	• Uncertainty avoidance index from the Hofstede survey, indicating a strive for reliability, rituals, rules, and institutionalisation • m = 66.14; s.d. = 28.55; n = 14
	9b. Uncertainty avoidance (Globe)[m]	• Uncertainty avoidance index from the Globe study. Scores for "societal cultural practices scales", i.e., current society practices • m = 4.35; s.d. = 0.72; n = 16
Telecoms Regulation	**10. Liberalisation Lead Time**[n]	• Number of months since official telecommunications market liberalization until year-end 2005 • m = 87.1; s.d. = 50.2; n = 21
	11. Independent Regulator Lead Time[n]	• Number of years since an independent regulator was established in a country until year-end 2005 • m = 9.2; s.d. = 5.1; n = 25
	12. Incumbent Privatisation Lead Time[o]	• Number of months since a significant (i.e., > 25%) share of the incumbent was sold to the public or private investors until end 2005 • m = 127.8; s.d. = 48.0; n = 10

(continued)

Table 1 (continued)

	Indicators	Measurement and Descriptive Statistics[a]
General & Telecoms Regulation	**13. Broadband Regulation Scorecard**[p]	• ECTA index of 11 variables mainly measuring local loop unbundling (LLU) conditions and market transparency at year-end 2005 • m = 0.61; s.d. = 0.22; n = 16
	14. Regulatory Scorecard Index[p]	• ECTA index of 99 variables covering e.g. speed, transparency, effectiveness, and independence of regulation as well as general and sub-sector specific market access conditions at year-end 2005 • m = 0.58; s.d. = 0.12; n = 16
	15. Intra-Technology Concentration[q]	• Hirschman-Herfindahl-Index: Sum of squared subscriber market shares of DSL providers in a country at year-end 2003 • m = 84.56; s.d. = 19.56; n = 22
	16a. LLU Lead Time[r]	• Number of months since local loop unbundling (LLU) was introduced until year-end 2005 • m = 68.9; s.d. = 25.3; n = 18
	16b. Unbundling Variants[r]	• Count of available unbundling types (Full LLU, Line Sharing, Bitstream Access, and Sub-Loop Unbundling) at mid-year 2003 • m = 2.42; s.d. = 1.43; n = 19
	17. Inter-Technology Concentration[q]	• Hirschman-Herfindahl-Index: Sum of squared subscriber market shares of different technological platforms for broadband access in a country at year-end 2003 • m = 57.67; s.d. = 17.03; n = 22
Broadband Market Environment	**18. Market Concentration**[q]	• Hirschman-Herfindahl-Index: Sum of squared subscriber market shares of all broadband providers (irrespective of technological platforms used) in a country at year-end 2003 • m = 47.35; s.d. = 18.05; n = 22
	19a. DSL Coverage[s]	• DSL coverage of the population in percent at year-end 2003 • m = 74.9%; s.d. = 21.1; n = 18
	19b. Cable Coverage[t]	• "Upgraded", i.e., Broadband-Internet-access ready cable coverage of the population in percent in the third quarter 2006 • m = 31.09; s.d. = 25.66; n = 11
	20a. Local Call Prices[u]	• Average cost of local telephone call (US$ per three minutes) in 2003 • m = 0.19; s.d. = 0.22; n = 17
	20b. International Call Prices[v]	• Residential users international telephone call basket (OECD basket methodology) at year-end 2003 • m = 1.65; s.d. = 1.35; n = 25
	21. Price Spread[v]	• Difference between 40 hour peak-time local call basket (proxy for dial-up Internet) and cheapest ADSL Internet price at year-end 2003 • m = 17.89; s.d. = 20.74; n = 15

[a]The study design is to introduce dependent variables with a time lag of 2 to 3 years into the analyses. Therefore, the independent criteria are collected for the year 2003 or the closest year for which data was available. m = Mean. s.d. = Standard deviation. n = Number of cases/countries.

[b]Variable I was taken from OECD 2007 broadband subscriber statistics, viewed June 19, 2007.

[c]Variable II was taken from the ITU World Telecommunication Indicators Database 2006.

[d]Variable III was collected from OECD Communications Outlook 2005 and OECD 2001.

[e]Variables 1–3, 5a–5b, and C1 were taken from the Worldbank World Development Indicators Database 2005. For coding of variable 3 some additional information was drawn from the International Labour Office Bureau of Statistics website at URL: http://laborsta.ilo.org, viewed July 1st, 2007.

[f]Variable 4 was collected from data provided by the International Labour Office Bureau of Statistics website at URL: http://laborsta.ilo.org, viewed July 1st, 2007. Some additional information was drawn from the Worldbank World Development Indicators Database 2005.

[g]Data for variable 6a was collected from the Eurostat Database 2007 accessible at URL: http://epp.euro- stat.ec.europa.eu, viewed January 15, 2007.

Table 1 (continued)

[h]Data for variable 6b was taken from the ITU World Development Indicators Database 2006.
[i]Variable 7 was drawn from the European Commission 2006 Report "Europeans and their Languages". For Ireland, United Kingdom, and USA English skills were assumed for 100% of the population.
[j]Variable 8a was taken from the Empirica ECaTT-Report "Benchmarking Telework in Europe 1999".
[k]Data for variable 8b was collected from the PWC et al. 2004 Report "Technical assistance in bridging the "digital divide": A cost benefit analysis for broadband connectivity in Europe".
[l]Variable 9a is based on the Hofstede-Survey as documented in Hofstede 2001, p. 151.
[m]Variable 9b is based on Globe-Study and is documented in De Luque/Javidan 2004, p. 622.
[n]Data for variable 12 was taken from URL: http://www.itu.int/ITU-D/treg/index.html (March 17, 2005).
[o]Data for variable 12 was drawn from the OECD Communications Outlook 2005.
[p]Variables 13 and 14 were aggregated from the ECTA 2005 Regulatory Scorecard Report.
[q]Data for variables 15, 17, and 18 was computed from Analysys 2006 broadband market share statistics.
[r]Variables 16a–16b stem from OECD 2003 "Development in Local Loop Unbundling", pp. 16–19.
[s]Data for variable 19a was collected from the OECD Communications Outlook 2005.
[t]Variable 19b was collected from the JP Morgan Chase 2006 Report "The Fibre Battle".
[u]Variables 20a stem from the ITU World Telecommunication Indicators Database 2006.
[v]Variables 20b–21 draw on the 2004 and 2005 Reports prepared by Teligen for the European Commission.

Independent Variables

Five variables reflect general country conditions: Economic prosperity, prosperity growth, unemployment, service sector activity, and urbanisation/population density (see variables 1–5b in Table 1). While some of these indicators are frequently used in country-level market entry studies, unemployment and service sector activity received less research attention so far. Four indicators are tested as societal country features: Computer skills/computer abundance, English literacy, teleworking, and uncertainty avoidance, with three of these criteria also tested with alternative indicators/sources due to data availability/scope (see variables 6a–9b in Table 1).

Telecommunication-specific market regulation is covered with eight indicators. Liberalisation lead time, independent regulator lead time, incumbent privatisation lead time, broadband regulation scorecard, regulatory scorecard index, intra-technology concentration, LLU lead time, unbundling variants, and inter-technology concentration cover broadband-relevant telecoms regulation circumstances in a market (see variables 10–17 in Table 1). This is especially important, since it was pointed out that measuring regulatory conditions for broadband quantitatively is difficult and that additional measures should be tested (Bauer 2003, p. 2, 19). The concentration criteria were used in one recent study (Distaso et al. 2006).

Further, four indicators for the broadband market environment are tested: Market concentration, DSL and cable coverage, average local call prices and average international call prices during the year 2003, and the narrow-to-broadband-price-spread at year-end 2003 (see variables 18–21 in Table 1).

The considerable effect-overlap between some of the independent variables should be noted. Therefore interpretations of individual bivariate effects have to be taken with caution. At the same time, such interdependencies and low size of the sample universe prevent advanced statistical analysis.

Results

To provide evidence for the expectations and arguments presented so far, bivariate correlation analyses were conducted for the set of dependent and independent variables. Results for EU-25 countries are summarized in Table 2. Interpreting associations as relevant that have both a significant Pearson and a significant Kendall correlation and acknowledging a time lag of 2 to 3 years between dependent and independent variables that was used by design (see Table 1), most of the indicators have a strong significant causal effect on broadband development indicators – and broadband penetration in particular.

General country characteristics explain both the broadband subscription and Internet user penetration reached in a country: Economic prosperity, service sector activity, and urbanisation had strong positive correlations with the broadband (Internet) market development across EU-25 markets, while prosperity growth and unemployment had a negative effect. Broadband launch lead time was not significantly higher in more prosperous countries, which is not surprising given the rather low disparity of broadband starting conditions in these markets.

Findings for "soft" society- and culture-related country characteristics mostly support our expectations. Computer penetration as indicator of computer skills as well as the computer skill assessment variable were strongly associated with broadband and Internet developments (see Table 2). However, these indicators also exhibit a strong correlation with economic prosperity, making a conclusive assessment of the effect of computer skills, independently of available income of a market's inhabitants problematic.

Further, English literacy had a strong positive impact on broadband and Internet development – lending support for the assumption that use of such services is more attractive and therefore more widespread if (freely accessible) global web content is meaningful for subscribers. Figure 3 presents plots of English literacy versus the broadband penetration and Internet user ratios that show how the European markets compare against each other. Especially for non-native English-speaking countries there is a strong trend of Web affinity associated with this criterion (i.e., higher correlation if United Kingdom and Ireland are excluded).

The share of teleworking or home office employees also correlated with broadband and Internet take-up but not with broadband launch lead time (see Table 2). The size

Table 2 Indicators explaining broadband development in European countries

	I. Broadband Penetration 2006		II. Internet User Ratio 2005		III. Broadband Launch Lead Time	
Variables[a]	*r*	*t*	*r*	*t*	*r*	*t*
1. Economic Prosperity	0,47 *	0,50 **	0,53 **	0,32 *	0,05	0,19
2. Prosperity Growth	-0,64 **	-0,44 **	-0,24	-0,19	-0,46 *	-0,24
3. Unemployment	-0,45 +	-0,28 +	-0,28	-0,26 +	-0,27	-0,05
4. Service Sector Activity	0,37	0,28 +	0,50 *	0,35 *	-0,13	-0,15
5a. Urbanisation	0,63 **	0,51 **	0,38 +	0,19	0,40 +	0,29 +
5b. Population Density	0,36	0,17	-0,04	-0,08	0,16	0,07
6a. Computer Skills	0,71 **	0,51 *	0,73 ***	0,46 **	0,08	0,09
6b. Personal Computer Penetration	0,77 ***	0,59 ***	0,81 ***	0,63 ***	0,16	0,12
7. English Literacy	0,58 **	0,52 **	0,39 +	0,31 *	0,13	0,18
8a. Share of Teleworkers	0,82 **	0,55 *	0,81 **	0,64 *	0,34	0,26
8b. Home Office	0,72 ***	0,56 ***	0,65 ***	0,44 **	0,33	0,29 +
9a. Uncertainty Avoidance (Hofstede)	-0,62 *	-0,46 *	-0,63 *	-0,52 **	0,17	-0,08
9b. Uncertainty Avoidance (Globe)	0,76 ***	0,63 ***	0,67 **	0,49 **	0,50 +	0,41 *
10. Liberalisation Lead Time	0,62 **	0,76 ***	0,49 *	0,46 **	0,49 *	0,46 **
11. Independent Regulator Lead Time	0,40 +	0,37 *	0,10	0,13	0,55 *	0,42 *
12. Incumbent Privatisation Lead Time	0,18	0,27	0,20	0,32	0,17	0,05
13. Broadband Regulation Scorecard	0,55 *	0,42 *	0,42	0,37 *	0,31	0,20
14. Regulatory Scorecard Index	0,58 *	0,46 *	0,42	0,46 *	0,25	0,12
15. Intra-Technology Concentration	-0,59 **	-0,46 **	-0,43 *	-0,36 *	-0,33	-0,15
16a. LLU Lead Time	0,59 **	0,56 ***	0,52 *	0,56 ***	0,49 *	0,35 +
16b. Unbundling Variants	0,42 +	0,33 +	0,36	0,25	0,16	0,06
17. Inter-Technology Concentration	-0,26	-0,06	-0,09	-0,03	0,05	-0,12
18. Market Concentration	-0,27	-0,10	-0,23	-0,10	-0,10	-0,05
19a. DSL Coverage	0,67 **	0,51 **	0,47 *	0,29	0,66 **	0,36 *
19b. Cable Coverage	0,74 **	0,44 +	0,47	0,15	0,25	0,30
20a. Local Call Prices	0,18	0,31	0,58 *	0,30	-0,38	-0,14
20b. International Call Prices	-0,65 **	-0,54 ***	-0,47 *	-0,36 *	-0,30	-0,04
21. Price Spread	-0,49 +	-0,31	-0,54 *	-0,33 +	-0,19	0,14

[a]For each dependent variable, Pearson (=r) and Kendall (=τ) correlations are shown in the first (second) row. The number of cases mostly ranges from 18 to 25 (only three variables have sample sizes <15, see number of cases given in Table 1).
+ $p < 0.10$; * $p < 0.05$; ** $p < 0.01$; *** $p < 0.001$ (two-tailed test).

of this user group may not be large enough in relation to the overall market size to make earlier introduction feasible for operators compared across countries. Nevertheless, as indicated by the significant correlations this target group contributes to more rapid take-up and overall stronger penetration levels at the macro level. Figure 4 plots the population share with at least 1 day per week of home office work against the (broadband) Internet penetration indicators. Particularly the countries with the most developed home office or teleworking culture, such as the Netherlands, Sweden, Finland or Denmark, have much higher broadband development levels.

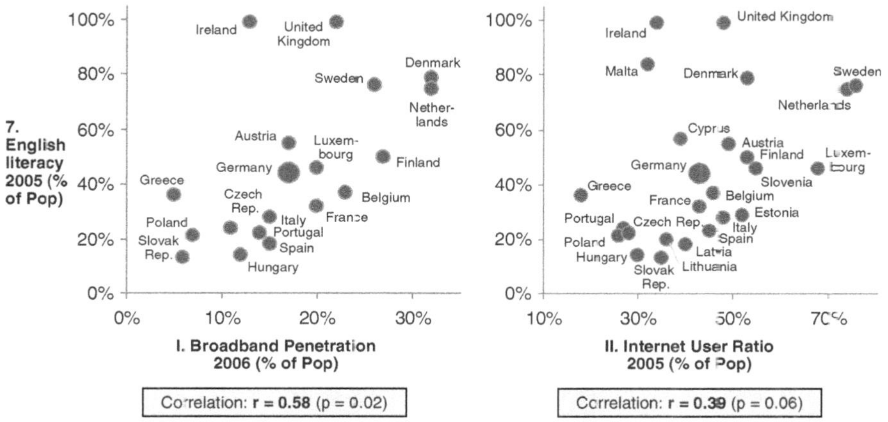

Fig. 3 Relation between English literacy and (broadband) Internet performance

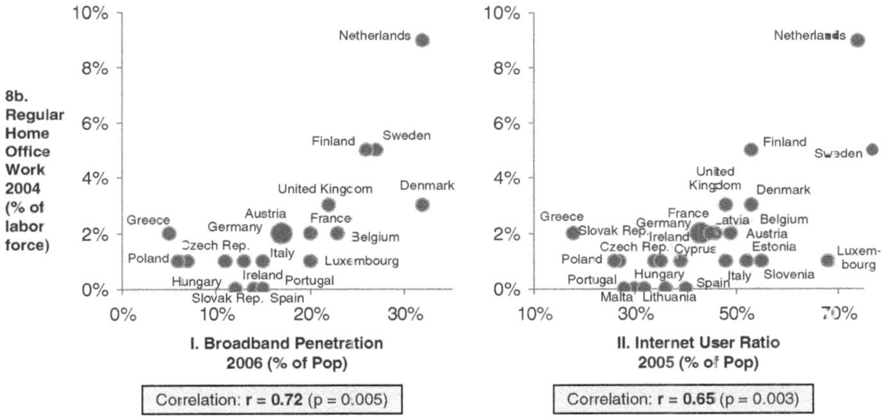

Fig. 4 Relation between home office working culture and (broadband) Internet performance

Results for uncertainty avoidance are inconclusive (see Table 2): While the Hofstede-based index correlates negatively, the Globe-study index has a positive association with broadband development (see discussion of these criteria in a different context in Jakopin 2006, pp. 224–225). Results for the Hofstede-index, despite being more than 25 years older than the Globe-study database, are more intuitively appealing: Broadband development is slower or less widespread if a society is rather uncertainty avoiding. In contrast, the Globe-study correlations suggest that countries with an uncertainty avoiding population both introduce broadband earlier and have stronger take-up.

Further, telecommunications specific market regulation affects broadband and Internet development (see Table 2): Liberalisation lead times correlates significantly positively with the three dependent variables, while lead time of independence of the telecoms regulator is not significant for the Internet user ratio and incumbent privatisation lead time does not achieve any statistically significant association (presumably due to incomplete data for this variable). Overall, the association patterns support the general diffusion curve logic and provide evidence for the positive effect especially of liberalisation and regulator independence as government actions that were carried out in the 1990s in most European markets.

The broadband regulation scorecard, as well as the aggregate regulatory scorecard index correlate positively with broadband take-up (see Table 2) – indicating that there is some face validity for these indices in capturing regulatory actions that should by design be useful for market development. Still, the construction of additive indicators as the ECTA scorecard used is subject to criticism as there is only limited conceptual foundation for inclusion and weighting of the criteria used to determine the index. Therefore, additional work with alternative operationalizations for regulatory conditions should be explored to test the assumptions even further.

Intra-technology concentration and LLU lead time are significant, and indicate that higher diversity of service offers within one technological platform and earlier LLU introduction coincide with better broadband development (see Table 2). The number of unbundling variants available to alternative operators only has marginally significant correlations with broadband penetration but not Internet development per se.

Inter-technology competition, i.e., a less uneven dispersion of subscribers across cable, DSL or other broadband platforms was not driving broadband take-up significantly (see Table 2). This result is mainly explained by the relative predominance of DSL – although some of the more advanced countries also show higher cable broadband usage, which does not seem to be a general trend.

Market outcomes and strategy related macro indicators provide mixed results (see Table 2). Market concentration, i.e., a lower number of competitors with higher shares of subscribers is not significant for broadband development at the European level. To wit, market concentration should not be referred to as one of the main drivers of (too) low/slow broadband development across and within some European markets.

DSL and cable coverage, i.e., technical availability of such services was strongly associated with broadband take-up (see Table 2 and Fig. 5). Higher local call prices had a weak positive effect on broadband and Internet development, while higher international call prices exhibited a strong significant negative correlation. The pattern of results, given the other relevant explanatory variables and relatively recent introduction of voice over IP/Internet-based telephony, is difficult to interpret. It may rather be caused by other underlying factors than a real effect of telephony price levels on Internet take-up. The correlation with international call prices may be more a result of the specific overall economic and telecoms market development of the countries characterised by high international call prices than a voice over IP substitution effect.

Finally, based on the available data set, initial analyses provide some evidence for the assumed price differential effect between narrow- and broadband Internet subscription and usage cost for an average customer that may prevent or promote switching to broadband subscriptions (see Table 2 and Fig. 6). Some low penetra-

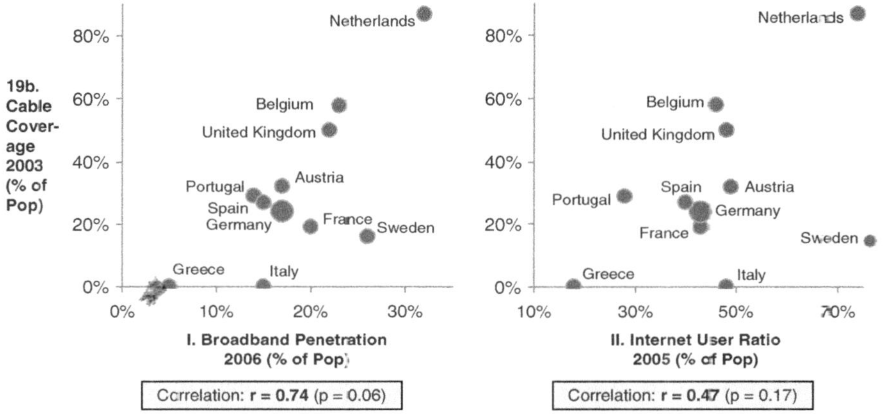

Fig. 5 Relation between cable coverage and (broadband) Internet performance

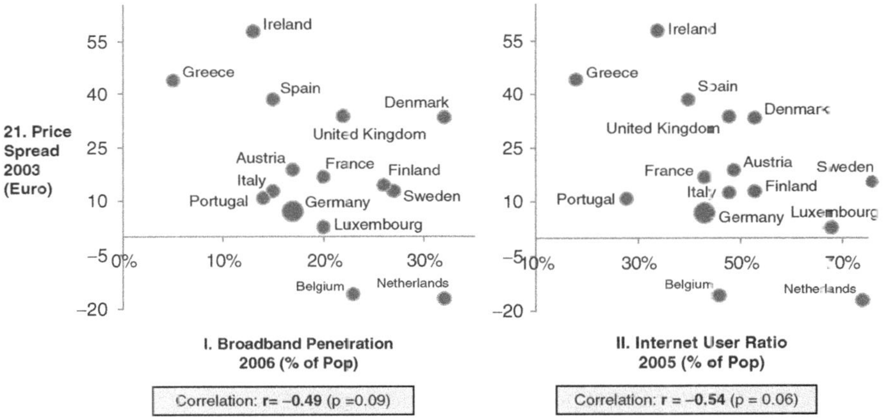

Fig. 6 Relation between narrow-to-broadband price spread and (broadband) Internet performance

tion countries had very high relative broadband prices compared with a standard dial-up package (Greece, Ireland) – in contrast, high penetration markets such as the Netherlands, and Belgium had much lower broadband prices than dial-up fees, while Luxembourg is an example of quite similar prices for standard broadband access and a usual dial-up customers' cost (see Fig. 6).

Overall, the main result of the study is that one should refrain from narrow normative categorisations of market developments without referring to the various influencing effects that contribute to the stage of market development that a country can reach at a certain point in time. Simple comparison of broadband penetration rankings is not enough. Regulators and commentators should be more aware and conscious about the specific economic and societal situations of countries being compared or benchmarked against each other – service sector focus, population agglomeration, English literacy, teleworking culture, and many other aspects shape the ground for a country's broadband performance. Operators considering entry into a foreign broadband market should take into account a broader range of indicators both to identify attractive opportunities and in endeavours attempting to predict broadband development speeds and penetration thresholds.

Conclusions

This chapter analysed indicators that have the potential to untangle differences in the international broadband development among the 25 European Union countries at year-end 2006. Explanatory variables were collected mainly for the year 2003, while the dependent variables refer to the years 2005 or 2006 to provide a time lag for the analysis, which is better suited to allow causal interpretation of results. The main areas covered by the variables are general economic conditions, societal and cultural antecedents, telecommunications-related regulatory environment, and broadband market conditions. Findings indicate that most of the 21 indicators (such as English literacy, teleworking, service sector activity, or unemployment) are significantly related to broadband development, i.e., the broadband subscriber penetration, Internet user ratio and commercial introduction timing of broadband.

Therefore, analysts and regulators should refrain from citing simple country broadband penetration rankings, but also take into account the specific situations faced by each individual country before making too single-minded comparisons. Operators reviewing foreign broadband market entry opportunities are well advised to pay attention to a broader range of variables and criteria to evaluate their stakes and options, as well as the planning outlook in a structured and wholistic manner.

The present study points towards five open research questions in areas for future advancement: (1) conditionality of effects, (2) broadband outcomes, (3) variable aggregation, (4) interacting/moderating and mediating effects, and (5) non-linearity of association patterns.

Conditional effects were not explicitly discussed in the present study. However, underlying variables could constitute an interrelated set of conditions required to achieve superior broadband development. Bauer (2003, p. 17) points out that a high price level is never associated with high broadband take-up while low prices can either coincide with high or low broadband penetration and therefore only constitute a necessary but not a sufficient condition. However, no common set of sufficient conditions to explain high broadband penetration emerged from earlier analyses (Bauer 2003, pp. 17–18). Future studies should further sharpen the understanding of necessary conditions under which follower broadband markets may be able to prosper and catch up with advanced countries.

The effects of broadband on economic development as pointed out in some commentaries, were not taken into account in this chapter. However, the main reason for the heated debate about differences in broadband development stems from the presumed positive relation between broadband access and the desired outcomes like economic growth, higher employment levels, lower inequality and others (cf. discussion in Bauer et al. 2002; European Commission 2006a; Lehr et al. 2006).

Therefore, future studies should capture effects associated with broadband Internet access and resulting outcomes in simultaneous equation models.

To enable advanced consolidation of the various indicators proposed, data should be transformed into a common scale in relation to the sample universe. The resulting common-scale variable values can then be used for aggregation according to the standard reliability testing procedure for reflective constructs that form indices that are used for regression analyses. The use of formative partial least squares constructs to aggregate overlapping variables and omit multicollinearity problems is another option to consider in this context. Formative constructs should be used, when indicators can have varying degrees of influence on a construct. Formative indicators can but must not necessarily be highly intercorrelated. Further, such multivariate modelling can be used with low sample sizes.

Further, potential interactions and moderating effects were not reviewed. However, as indicated throughout the chapter, some of the variables and constructs may not impact broadband development directly, but have a moderating effect on other potential broadband affecting criteria (e.g., regulation on market conditions). Therefore, future work should explicitly test interactions of the main study variables.

Finally, non-linearity, i.e., u-shaped or other curvilinear patterns of association between the explanatory variables and broadband development indicators were neglected in the present study. While an initial review of the variables used here suggests that they all should have monotonous rather than non-linear effects on broadband development, it may still be useful to further review arguments towards such relation patterns and conduct empirical tests on them.

References

Arthur D. Little (2007) Germany's Broadband Performance in Comparison to Other European Countries. Duesseldorf: Arthur D. Little GmbH.

Bauer J M (2003) Prospects and limits of comparative research in communications policy-making. Paper presented at the 31st Annual TPRC Communication, Information and Internet Policy Conference, Arlington, VA, September 19–21.

Bauer J M et al. (2002) Broadband: Benefits and Policy Challenges. A Quello Center Report. East Lansing, MI: Michigan State University.

Bauer J M, Kim J H, Wildman S S (2003) Broadband uptake in OECD countries: Policy lessons and unexplained patterns. Paper presented at the 14th European Regional Conference of the International Telecommunications Society, Helsinki, Finland, August 23–24.

Cava-Ferreruela I, Alabau-Muñoz A (2004) Key constraints and drivers for broadband development: A cross-national empirical analysis. Paper presented at the 15th European Regional Conference of the International Telecommunications Society, Berlin, Germany, September 4–7.

Chaudhuri A, Flamm K (2005) An analysis of the determinants of broadband access Paper presented at The Future of Broadband: Wired & Wireless? UFL-LBS Workshop, University of Florida, Gainesville, FL, February 24–25.

Chinn M D, Fairlie R W (2007) The determinants of the global digital divide: A cross-country analysis of computer and internet penetration. Oxford Economic Papers 59: 16–44.

Dekimpe M G, Parker P M, Sarvary M (2000) Global diffusion of technological innovations: A coupled-hazard approach. Journal of Marketing Research 37(2): 47–59.

De Luque M S, Javidan M (2004) Uncertainty avoidance. In: House R Culture, Leadership, and Organizations. Thousand Oaks, CA: Sage, 602–653.

Distaso W, Lupi P, Manenti F M (2006) Platform competition and broadband uptake: Theory and empirical evidence from the European union. Information Economics and Policy 18: 87–106.
ECTA (2006) Regulatory Scorecard. Brussels: European Competitive Telecommunications Association/Jones Day/SPC Networks.
EITO (2007) European Information Technology Observatory. Frankfurt a.M.: EITO.
European Commission (2002) eEurope 2005: An Information Society for All. Action plan presented in view of the Sevilla European Council, June 21–22, 2002 (COM 263). Brussels: European Commission.
European Commission (2005): i2010 – A European Information Society for Growth and Employment (COM 229). Brussels: European Commission.
European Commission (2006a): Bridging the Broadband Gap (COM 129). Brussels: European Commission.
European Commission (2006b) Broadband Access in the EU: Situation at 1 July 2006 (COCOM 06–29). Brussels: European Commission.
European Commission (2006c) Special Eurobarometer 249: E-Communications Household Survey. Brussels: European Commission.
Frieden R (2005) Lessons from broadband development in Canada, Japan, Korea and the United States. Telecommunications Policy 29, 595–613.
Garcia-Murillo M (2005): International Broadband Deployment: The Impact of Unbundling. Communications & Strategies 57, 83–105.
Gerpott T J (2007) Einflussfaktoren der Adoption eines Breitbandanschlusses durch Privatkunden. In: Hausladen I (ed.) Management am Puls der Zeit – Band 1: Unternehmensführung. Munich: TCW Transfer-Centrum: 797–820.
Hofstede G (2001) Culture's Consequences. 2nd Ed. Thousand Oaks, CA: Sage.
Jakopin N M (2006) Einflussfaktoren des Internationalisierungserfolgs von Mobilfunknetzbetreibern. Wiesbaden: DUV.
Jakopin N M, Von den Hoff K (2007) Breitbandmarktentwicklung in Deutschland – Eine indikatorpluralistische Betrachtung. Ifo Schnelldienst 60 (21): 15–19.
JP Morgan Chase (2006) The fibre battle – Changing dynamics in European wireline. European Equity Research. London: JP Morgan Chase.
Lehr W H et al. (2006) Measuring Broadband's Economic Impact. Paper presented at the 33rd Research Conference on Communication, Information, and Internet Policy (TPRC), September 23–25, 2005, Arlington, VA (revised in January 2006).
Maldoom D et al. (2005) Broadband in Europe. Berlin: Springer.
Miralles F (2006) Efficiency of Public Promotions Policies in the Diffusion of Broadband Networks. An Exploratory Analysis. Presented to the 34th Research Conference on Communication, Information, and Internet Policy, Arlington, VA. September 29, 2006.
OECD (2007) OECD Broadband Statistics to December 2006. Paris: Organisation for Economic Co-operation and Development.
Papacharissi Z, Zaks A (2006) Is broadband the future? An analysis of broadband technology potential and diffusion. Telecommunications Policy 30: 64–75.
Polykalas S E, Vlachos K G (2006) Broadband penetration and broadband competition: Evidence and analysis in the EU market. Info 8 (6): 15–30.
Sundqvist S, Frank L, Puumalainen K (2005) The effects of country characteristics, cultural similarity and adoption timing on the diffusion of wireless communications. Journal of Business Research 58: 107–110.
Turner S D (2006) Why does the U.S. lag behind? Broadband penetration in the member nations of the Organization for Economic Cooperation and Development. Discussion Paper. Washington, DC: Free Press.
Wallsten S (2006) Broadband and Unbundling Regulations in OECD Countries. Working Paper 06-16. Washington, DC: AEI-Brookings Joint Center For Regulatory Studies.
Wallsten S (2007) Towards Effective U.S. Broadband Policies. Washington, DC: The Progress & Freedom Foundation.

The Telecom Policy for Broadband Diffusion: A Case Study in Japan

Koshiro Ota

Abstract Though the penetration rate of broadband services in Japan is not outstanding, the transmission speed is much faster and the price is lower than in other advanced countries. Moreover, FTTH is now spreading much faster than DSL. There are two aims of this paper: first, to derive the principle and the substance of the recent Japanese telecom policy; and second, to examine the factors that contribute to the diffusion of broadband services. I will focus not only on the level of unbundled local loop rates but also on the rate-setting methodology for group center and zone center interconnections. I will also argue against some conventional views that "telecoms regulators in Japan have tended to take a stronger line with their incumbent than in Western countries." In my view, the Japanese telecom policy as a whole was not so hard on NTT, rather it supported NTT's and other carriers' business development by taking into consideration NTT's business situation.

Introduction

Broadband communications bring various benefits to people's daily lives and economic activities. There is a great concern regarding the telecom policy for the development of broadband services. The services that telecom carriers offer contain digital subscriber line (DSL) and fiber to the home (FTTH). Regarding DSL, mandatory unbundling that allows an applicant (e.g. a new entrant) to use some specific network elements of an incumbent carrier, is nowadays frequently used in developed countries.[1] Here, the copper loop (or subscriber line) is used as a representative of

[1] Another means is co-location that allows an applicant to co-locate its equipment in the buildings of an incumbent carrier. Co-location also needs a proper charge-setting for the space.

K. Ota (✉)
Faculty of Economic Sciences, Hiroshima Shudo University, Japan
e-mail: k-ota@rf7.so-net.ne.jp

B. Preissl et al. (eds.), *Telecommunication Markets*, Contributions to Economics,
DOI: 10.1007/978-3-7908-2082-9_12,

unbundled network elements (UNEs). It should be noted, however, that FTTH plays a limited role in broadband policies, though it benefits from financial incentives, such as low interest or interest free loans and tax concessions.

There are two aims of this paper: first, to derive the principle and the substance of the recent Japanese telecom policy[2]; and second, to examine the factors that contribute to the diffusion of broadband services. I will focus not only on the level of unbundled local loop rates but also on the rate-setting methodology for group center (GC) and zone center (ZC) interconnections. This is particularly important if one aims at understanding the recent Japanese policy. I will also argue against some conventional views that "telecoms regulators in Japan [] have tended to take a stronger line with their incumbent than in Western countries" (Fransman 2006b).[3] In my view, the Japanese telecom policy as a whole was not so hard on NTT, rather it supported NTT's and other carriers' business development by taking into consideration NTT's business situation.

Though the penetration rate of broadband services in Japan is not outstanding, the transmission speed is much faster and the price is lower than in other advanced countries (see the International Telecommunications Union [ITU] Internet Report in 2005). Moreover, FTTH is now spreading much faster than DSL, and parallel services such as IP phone (also called Voice over IP phone) become very common. Therefore, this paper might give some pieces of advice on the telecom policy to other countries.

In section "Means for Broadband Diffusion", the rationale for employing various means for broadband diffusion will be explored. In section "Unbundling and New Entrants in Japan", unbundling practices adopted in Japan and business strategies of Softbank Corp. and other new entrants in the broadband markets will be reviewed. Section "The Japanese Telecom Policy" examines NTT's policies regarding optical fibers based on the interdependence between the Japanese Government and NTT. In section "Summary of Sections 'Unbundling and New Entrants in Japan' and 'The Japanese Telecom Policy'", results of sections "Unbundling and New Entrants in Japan" and "The Japanese Telecom Policy" will be synthesized; and finally in section "The Evaluation of the Japanese Telecom Policy", Japanese telecom policies will be evaluated.

[2] Of course, this does not mean that all telecom policies are decided based on a clearly defined principle.

[3] Fransman (2006b) bases that statement on the fact that the unbundled local loop rate is low and that the share of an incumbent carrier in the DSL market is low in Japan (36%; 99% in the United Kingdom, 91% in Germany, 85% in the United States and 82% in Korea in 2003–2004). See also Wallsten (2006) and Crandall (2005b). However, it is not appropriate to evaluate the Japanese telecom policy only on the DSL market.

Means for Broadband Diffusion

DSL

DSL service is usually based on an incumbent carrier's subscriber line. The subscriber line is regarded as an essential facility for a competitive carrier to run a successful business, so it is necessary to open it up to competitive carriers in order to advance competition between DSL carriers. There are two views that lend support to the above assumption. The first is called the "stepping stone" theory on the capital investment of new entrants. According to this theory, unbundling is expected first to enable new carriers to enter into the market and establish a customer base of their own, then to have them invest in facilities in order to create infrastructure competition. However, there is no solid evidence to support this view (therefore, it is not a theory but a hypothesis). Furthermore, the conditions for inducing new carriers to enter the market could also restrain their capital investment after their entries. The second view is based on the "competitive stimulus" (or "investment deterrence") hypothesis on the capital investment of incumbent carriers.

Many empirical studies have tried to verify these theories, and there are few, if any, studies that support them.[4] However, this does not necessarily mean that unbundling is useless for advancing dynamic competition. Rather, some conditions will have to be met for unbundling to work well, and unbundling may not have been applied properly in the United State and some other countries.

FTTH

FTTH is usually characterized by much faster transmission speed than DSL. Additionally, its transmission speed is constant irrespective of the distance from a telecom carrier's central office. It has also a potential to advance some parallel services. However, FTTH has not been diffused even in developed countries. Some stress the need for means to promote the construction of FTTH that demands a great amount of money and implies considerable economies of scale. For example, Oniki (1996) proposes to put a government guarantee to a bond issued by a telecom carrier for that purpose, and to allow only one carrier (NTT) to construct and manage its network under regulation. He also points out that "the construction of analog telephone network in the U.S. was pushed ahead by (old) AT&T's monopoly power" (p. 81, note 39). As explained in detail in section "The Japanese Telecom Policy", the Japanese telecom policy that supported NTT's construction of FTTH could be said, to make good use of

[4]See, for example, Crandall (2005a), Crandall et al. (2004) and Hazlett and Bazelon (2005). Hausman and Sidak (2005) conclude that unbundling experiences in the U.K., New Zealand, Canada and Germany do not support the "stepping stone" theory. The only exception that I know is Willig et al. (2002). However, their data have some trouble as pointed out by Crandall (2005a).

monopoly power not for cost reduction as the case in the U.S., but for the increase of profits (financial resources). There are some countries where unbundling is applied to FTTH, however if the terms and conditions of business are favorable for users who rent the lines, every carrier will hold off from investing in its own infrastructure.

Unbundling and New Entrants in Japan

Interconnection Rules and the Local Loop Rate

In Japan, any telecom carrier is supposed to agree to the request for interconnection except in some specified cases. Moreover, any carrier with Category I designated telecom facilities will establish interconnection tariffs and other terms and conditions of interconnections and shall obtain authorization from the Minister for International Affairs and Communications (Telecommunications Business Law, Articles 32, 33). NTT's local loops and telecom facilities installed to be integrated with the loops are designated as Category I in every prefecture.[5]

DSL service is based on the NTT's copper loop and the rate for the usage is set at the incremental cost of NTT. In December 2000, the total amount of ¥2,062 (¥: yen) was authorized as the incremental cost for full unbundling and this amount was composed of the following elements; the local loop cost of ¥1,905, the line database management cost (1), and the charge billing and collection cost (2). The total amount of ¥187 was authorized for line sharing and this amount is composed of (1), (2) and the additional main distribution frame (MDF) cost. In 2003, not the end-user's rate but the carrier's rate (¥1,405) was applied to the loop cost. Their difference reflected the cost difference between offering service to a carrier and to an individual customer. The rate for line sharing in Japan is cheaper than that in any EU country, and the rate for full unbundling is similar to that in the cheaper EU countries (Tajiri 2007, p. 78, note 13).

However, what is important for competition policy is whether rates exceed total costs or not. Here of vital importance is the treatment of the local loop cost. In Japan it was not clear how closely the carrier's rate corresponded with cost, because the end user's rate was set low enough to secure universal service. However, the facts that it took Softbank 3 years and a half to go into the black and that not a few scholars had doubted Softbank's success until recently, suggest that the local loop rates were not so low to give preference to new entrants and further to guarantee them financial success.

Unbundling can also be applied to the optical fiber loop.[6] While its rate has been set at ¥5,074 per month, its total cost was said to be over ¥10,000 (Kanzaki 2006;

---

[5] Before unbundling and co-location rules were established, a new entrant needed longer lead-time and higher cost to launch DSL service based on NTT's copper loops. In December, 2000, the Fair Trade Commission (FTC) warned NTT East on the suspicion of its obstructive conduct toward a new entrant.

[6] A carrier who has more than 50% of total loops (metal loops plus optical fiber loops) takes the duty of unbundling (or its loops are designated as Category I). Because NTT dominates metal loops, its optical fiber loops as well as metal loops are mandatorily unbundled.

Nikkei Communications 2005). The rate was calculated by NTT based on the estimate of the cost and the demand from fiscal year (FY) 2001 to FY 2007 (such rate-setting methodology is called the "future cost methodology" in Japan), and the return was included in the calculation. Even with this significant difference between the rate and the cost, competitive carriers using NTT's FTTH services displayed no particular growth of their business.

The Pioneer of the DSL Market: Softbank

Broadband service in Japan was popularized by DSL companies and the leading part was played not by NTT but by Softbank (whose founder, Masayoshi Son, is known in Japan as "the ultimate risk taker").[7] When Softbank announced its entry into the DSL business (its entry was in September 2001), NTT and some other companies had been offering DSL services and they had less than 300,000 users in total. "Tokyo Metallic Communications Corp. that first started DSL services in Japan, was brought nearly to financial default because of capital overspending" (Nikkei Communications 2003). Softbank, on the other hand, "lacked completely management resources necessary for the []DSL business" (Miki 2006, p. 131). Furthermore, Softbank had an interest-bearing debt of about ¥400 billion as of June 2001, and still needed an investment of "allegedly over ¥100 billion" (Kodama 2005, p. 337) to set the business on its way. In such a situation, Softbank, with setting the break-even point at 2–3 million subscribers, launched much faster and cheaper service (8 Mbps/¥2,280) than its rivals (NTT: 1.5 Mbps/¥6,550) and simultaneously launched an audacious sales promotion activities such as free distribution of DSL modems. The total estimated cost to acquire a subscriber amounted to ¥37,000 (Nikkei Communications 2003). In addition, Softbank offered a new business model of vertical integration with DSL service and Internet service provider (ISP) service linked together. To "build a high-speed trunk network with cheap devices" (Hiramiya, a former engineer of Softbank; Nakamichi 2004), Softbank, based on NTT's optical fibers, built its own Gigabit Ethernet (IP network) "which nobody had believed to work" (Ikeda 2003, p. 13, footnote 18). And, for a technical reason, it adopted the ITU standard transmission method, called Annex A, not a Japanese original Annex C that the other rivals have adopted before. This helped Softbank to minimize the procurement cost of modems and DSL Access Multiplexers (DSLAMs) for DSL service. Later, Softbank purchased Tokyo Metallic Communications to make better use of its management resources. In April 2002, Softbank launched cheap IP phones service with the DSL modem connected with a standard telephone set which became an important characteristic of its DSL service. Consequently,

[7] eAccess Ltd. and Acca Networks Co. were (and are still) among several nation-wide DSL companies. According to Kanzaki (2006), "NTT had a plan to shift from the integrated services digital network (ISDN) to the optical fiber network at first. …. However, they could not overlook the great advance of Softbank and made strenuous efforts unavoidably in the[]DSL business" (p. 98).

Softbank gained 3 million users in 11/2 years and counted 5.14 million users in total at the end of September 2006. It is now enjoying a profit in this business. Son remarked that Softbank was not able to enter into the DSL market without unbundling (Nikkei Communications 2005). It should be noted that Softbank was not a telecom carrier, but a software wholesaler at that time which supports Son's remark.

NTT's Rivals in the FTTH Market: USEN, K-Opticom and KDDI

FTTH, as noted above, has been diffusing rapidly in Japan. The pioneer in this market was a landline music broadcasting company, USEN Corp. It started offering service at ¥6,000 per month (a third of NTT's price in a trial service) in March 2001. However, USEN prioritize profit incentive and restricted its FTTH network to some lucrative large cities. At present, the main rivals of NTT in this market are power companies (or their subsidiaries) that have their own optical fiber networks. Kansai Electric Power Co. is perhaps the most active participant in this market. Its subsidiary, K-Opticom Corp., covered about 70% of households in the Kinki region including Osaka and Kobe, and exceeded the share of NTT West in three prefectures in that region as of March 2005. On the other hand, Tokyo Electric Power Co., the largest power company in Japan, sold its optical fiber business with about 140,000 km optical fiber in total to KDDI Corp. in January 2007. As a result, KDDI, the second largest telecom carrier behind NTT, obtained a footing to construct its own optical fiber network in the Kanto region including the Tokyo metropolitan area, and gained an additional 340,000 subscribers. Still, regarding the FTTH market, there is no company which can favorably compare with NTT, such as Softbank in the DSL market. (Softbank still devote itself to DSL service.[8])

The Japanese Telecom Policy

The U.S.–Japan Negotiations and Interconnection Charges

The Japanese telecom policy on broadband requires examination not only on the level of the unbundled local loop rate but also on the rate-setting methodology for GC and ZC interconnections.[9] Its particular importance has been revealed in the

---

[8] Softbank (2007) describes as follows: "[our DSL service] has received broad support from customers in such area as cost and speed" (p. 22). "Based on our judgment that it has sufficient capacity to enjoy various contents for broadband transmission at present, we keep on [] acquiring new customers" (Softbank 2007, Japanese ed., p. 5).

[9] Such vision and some opinions expressed in this paper are shared with Tilton (2004). Ikeda (2003) states that "NTT admitted unbundling in exchange for a compression of reduction in the interconnection fee after it became an issue in the [U.S. – Japan Negotiations]" (p. 14).

process of the U.S. – Japan Negotiations on Telecommunications and the development following from them.

NTT's interconnection charges were regarded as "the largest pending economic problem between the U.S. and Japan since the Birmingham Summit in 1998" (Nikkei Sangyo Shimbun, July 21, 2000, p. 1). NTT's charge was ¥5.81 for GC interconnection (for 3 min) and ¥11.98 for ZC interconnection in January 1, 2001 and both charges were higher than those in the U.S. (¢4.20≈¥4.99 and ¢5.53≈¥6.56 respectively; ¢: cent).[10] The reason why the U.S. interfered in NTT's interconnection charges was to introduce more competition in the telecom market and therefore to stimulate economic activities in Japan ("Japanese strong economy is advantageous to the U.S.," Richard Armitage, the former U.S. Deputy Secretary of State; Asahi Shimbun, October 29, 2000, p. 9).[11] The other motive was to "reduce entry barriers in Japanese markets for the U.S. companies to access them more easily" (the U.S. Embassy, March 2, 2001). The U.S. demanded the reduction of the interconnection charge of 22.5% in 2 years from the end of 2000 and of 40% or more in total, and the adoption of the new rate calculation methodology based on long-run incremental costs (LRIC) in early 2001. On the first point, Japan insisted on the reduction of 22.5% in 4 years, and both countries agreed on the reduction of about 20% in 2 years, 22.5% in 3 years from FY 2000, and that Japan would examine a further reduction (of much more than 22.5%) in 2002, taking into consideration the business situation of NTT East and West (Nikkei Sangyo Shimbun, July 21, 2000, p. 1). According to this agreement, NTT's charge decreased from ¥4.95 in FY 2000 to ¥4.60 in FY 2001 and to ¥4.50 in FY 2002 for GC interconnection and from ¥7.65 to ¥5.88 and to ¥4.50 for ZC interconnection in the same period. Moreover, though the U.S. insisted that non traffic-sensitive (NTS) cost should be recovered by a fixed charge,[12] a decision whether to allocate the fixed-cost portions of the local loop to the interconnection charge as before or to allocate them to the basic charge was postponed to 2002.

This agreement did not mean the controversies between the U.S. and Japan regarding telecom charges were over. Japan built an original LRIC model based partially on the foreign models. It aimed at protecting business operations of NTT and therefore at maintaining universal service and promoting its infrastructure construction (Telecommunications Council 2000; Information and Communications Council 2004). For that reason, for example, the application of the LRIC methodology to local loops was postponed until April 2003. Moreover, to secure universal service,

[10] The interconnection charges in the U.K., France and Germany were p1.18≈¥2.27 (GC) and p1.72 ≈¥3.31 (ZC); c16.53≈¥3.26 (GC) and c32.40≈¥6.39 (ZC); and pf5.13≈¥3.39 (GC) and pf11.07≈¥7.32 (ZC) respectively (p: pence, c: centime, pf: pfennig; Telecommunications Council 2000). Higher interconnection charge brings higher usage-based call rate, and then it could increase the demands for fixed-priced DSL and FTTH for the Internet access.

[11] Richard Fisher, the former Deputy U.S. Trade Representative, had a similar remark.

[12] To recover NTS costs by the fixed basic charge and traffic sensitive (TS) costs by the usage-based call rate are efficient in the viewpoint of surplus maximization. On the other hand, because subscribing to a telephone network is premised on the surplus of telephone calls exceeding the basic charge, a low basic charge is needed to securing universal service.

some NTS costs such as feeder remote terminals were being recovered not from fixed basic monthly charges, but from usage-based call rates and interconnection charges (the share of their costs in the total NTS costs counted 49.4% in 2003).[13] To calculate interconnection charges, those NTS costs were divided by the traffic that got through NTT's switches for a certain period and this traffic volume turned to decrease in 2000 partially because of the diffusion of the IP phone. In FY 2000–FY 2002, the traffic data in FY 1998 was used, so the charges were calculated higher. In FY 2003–FY 2004, the immediate data, namely those in the second half of FY 2001 and the first half of FY 2002 was used, and the ZC interconnection charge was raised to ¥5.36, an 11.2% increase over FY 2002 (the GC interconnection charge decreased to ¥4.37). Moreover, as the actual traffic volume was lower by 15% from the estimate used in the calculation, a retroactive settlement system was applied and NTT's burden was decreased.[14] The interconnection charges also depend on the accounting method of equipment depreciation. Japan extended depreciation periods in 2002 (e.g. for transmission equipment, from 6 to 8.4 years), but these were still shorter than those adopted in the U.S. In January 2007, the universal service fund system was started and universal service cost that had been borne only by NTT was to be shared among all telecom carriers in proportion to the usage of telephone numbers (NTT East's and West's deficit of universal service amounted to ¥36.6 billion even after getting financial support from the fund).[15] The U.S., the U.K. and the EU Commission criticized the conceptual basis for interconnection charges as well as the model for its calculation and also some numerical values applied to the model. Such criticism was based on doubts regarding management efficiency of NTT. The U.S. estimated that "the current monthly charge of NTT East and West seems, in comparison with a credible cost estimate of local networks, to have a margin enough to recover the NTS costs confirmed under the current LRIC model" (March 2003). In fact, NTT has been dealing with the efficiency issue, and in 2000–2002 NTT West has reduced its costs by ¥600 billion with its income falling by ¥500 billion.

"e-Japan Strategy," "u-Japan Policy" and NTT

In January 2001, the Government proclaimed the "e-Japan Strategy," which aimed at "mak[ing] Japan the world's most advanced IT nation within five years" by "provid[ing] high-speed constant access networks to at least 30 million households"

[13] However, the recovery of NTS costs is gradually shifting to fixed monthly charges.

[14] Though the competitive carriers resisted to a series of events and filed an administrative suit against the Ministry of International Affairs and Communications (MIC), they lost the case at a district court (However, the retroactive settlement system was abolished).

[15] The cost of universal service is the total of costs of subscriber telephone services, public telephone services and emergency call services. It should be noted that NTT East and West are also required to contribute to the fund.

"at affordable rates." In September 2004, the MIC published the "u-Japan Policy" to develop the way to utilize information and communications technology (ICT) in the progressively aging society and to materialize the ICT society based on it. In November, 2004, NTT drew up the NTT Group's Medium-Term Management Strategy (NTT 2004) to support these policies, and promised to provide 30 million customers with optical fiber access and high quality and security next-generation network services by 2010 and to invest a cumulative total of ¥5 trillion for fixed communications operation (according to Nikkei Communications 2005, ¥3 trillion of the total for the next-generation network). Actually, while NTT East's and West's operating and ordinary profits decrease, their capital investments, especially in the area of FTTH grew (see Tables 1 and 2). According to NTT, "Japan has problems of a declining birthrate and a growing proportion of elderly people (among others), and to cope with these social problems it is very important to construct an optical fiber broadband network that enables interactive and highly value-added image communications" (NTT 2003). In November 2005, NTT announced the second reorganization of group companies to execute the Medium-Term Management Strategy efficiently regardless of the criticism of its rival carriers who claimed that this was just a reintegration of the NTT group (Nikkei Communications 2005).

Why does NTT continue to develop optical fibers, even though this area of business seems to be unprofitable? One reason may be to gain a first-mover advantage which was the core business strategy of Softbank (see Tajiri 2007). The advantage comes from the fact that users are hesitant to change their carrier once they sign a contract (according to an inquiry of the FTC [2004], 84.3% of FTTH users have no plan to change the carriers). While the number of FTTH users has increased from about 2.9 million (at the end of FY 2004) to about 5.46 million in FY 2005 and to about 8.8 million in FY 2006, NTT, coupled with strong brand loyalty and sales promotion, increased its market share from 57.5% to 62.6% and to 69% respectively.[16]

Table 1 The capital investments of NTT East and West (unit: ¥100 million) (From Information NTT East 2006; Data Book NTT West 2006)

	FY 1999	FY 2000	FY 2001	FY 2002	FY 2003	FY 2004	FY 2005
NTT East	5,309	5,496	3,656	3,342	3,778	3,991	4,222
NTT West	5,059	5,481	4,006	3,624	3,976	3,978	4,629

Notice: FY 1999 is defined to July 1, 1999–March 31, 2000.

Table 2 The construction of optical fibers (unit: 100 km) (From Information NTT East 2006)

	FY 1999	FY 2002	FY 2003	FY 2004	FY 2005
NTT East	338	832	1,189	1,572	2,056
NTT West	323	876	1,365	1,623	1,862

Notice: Figures are as of the end of each FY. However, figures in FY 1999 are as of July 1, 1999.

[16] NTT needs a business model to gain from its "enclosed" users. However, if the use of monopoly power is regulated strictly, making a profit is not easy even with vertical integration or cooperation.

Another reason may be the interdependence between the Japanese telecom policy and NTT. Because the Japanese Government needs NTT's co-operation for the implementation of the "e-Japan Strategy," it adopts telecom policy that takes into consideration NTT's business conditions. NTT, on the contrary, could enhance its raison d'être by cooperating with the Government and lobby for such consideration.[17] The Japanese telecom policy includes other objectives besides introducing competition by decreasing access charges. These objectives are not intended to impose a handicap on NTT. As one third or more of NTT stocks are possessed by the Government by law, NTT suffers less pressure from general stockholders. This may make it easier for NTT to commit itself to investing more in optical fiber networks (Tajiri 2007). A consistent telecom policy in recent years enabled NTT to implement medium- and long-term plans under the conditions where the progress of contents downloadable on FTTH and alternative wireless broadband techniques such as WiMAX was uncertain.[18] The fact that the MIC approved NTT's proposal for a uniform rate for FTTH from FY 2001 to FY 2007 may also contributed. The progress of competition in the FTTH market discussed in section "NTT's Rivals in the FTTH Market: USEN, K-Opticom and KDDI", and the repeal of certain unbundling rules in the U.S. might also have influenced NTT's view of the regulatory environment in the future (NTT has been urging the repeal of the unbundling obligation for FTTH).[19] This point is raised in a different context by Crandall (2005a) and Hazlett (2002). In the U.S. where the telecom policy is a complex judicial issue, the uncertainty regarding regulation is a prime motive behind general restraint toward capital investment in broadband networks not only by incumbent local exchange carriers (ILECs) but also by CATV operators who were out of telecom regulation in the early 2000s[20].Regarding the investment ability of NTT, Ida (2006) insists that not completely dividing NTT and restructuring it as a holding company and subsidiaries enables tacit cross-subsidization from NTT DoCoMo (a mobile phone company) to NTT East and West and stimulates growth of the FTTH networks by both companies. However, cross-subsidization as a capitalization vehicle is not easy under the strict guidance and control of the MIC and the FTC. In theory, the holding company, that owns about two third of DoCoMo stocks, can nominally finance NTT East and West using the dividend from DoCoMo, but

[17] Some scholars point out that NTT has exercised political power on the ruling Liberal Democratic Party (LDP) and has induced the telecom policy to its advantage. See Tilton (2004) and Nikkei Communications (2006).

[18] However, in June 2006, an advisory panel to Internal Affairs and Communications Minister proposed to review exhaustively the telecom laws in 2010 to functionally separate the NTT East's and West's bottleneck facilities. Then, the Government agreed with the LDP, that was against a hasty reorganization of NTT, to prepare fair competition rules such as the opening of NTT' networks and to discuss the NTT's reorganization in 2010 taking into account the spread of broadband services and the progress of NTT' Medium-Term Management Strategy.

[19] On unbundling in the U.S., see Bauer (2006).

[20] In Japan, there is little administrative suit around the telecom policy and therefore little involvement of a court in the policy. An exception is referred in footnote 14.

in reality, it just borrows money from banks and issues corporate bonds to finance them. Rather, one may observe that core target of the Japanese telecom policy is to directly support the investment abilities of both NTT East and West.

Summary of Sections "Unbundling and New Entrants in Japan" and "The Japanese Telecom Policy"

This section provides a short summary of sections "Unbundling and New Entrants in Japan" and "The Japanese Telecom Policy". With regard to the attitude of the Japanese telecom policy toward NTT:

- The rate for the NTT's copper loop that was rented by DSL companies was comparatively low, but it was calculated based on the principle of incremental cost. The rate for FTTH was calculated by NTT based on the estimate of the cost and the demand from FY 2001 to FY 2007. (Then, a short-term deficit does not indicate whether the rate level is too high or too low.)
- The U.S. and the U.K. repeatedly asked Japan to further open the telecom market and therefore to reduce the interconnection charge. However, Japan gave preference to NTT's business operations and the maintenance of universal service, and postponed the application of the LRIC method to interconnection charges which allowed recovering a part of fixed costs from usage-based call rates and interconnection charges.
- Softbank introduced many inventive ideas and put them into practice, and in consequence succeeded in gaining significant share in the DSL market.

Therefore, one cannot support the view that telecom policies aimed at weakening NTT's dominant position in the Japanese market.

With regard to the diffusion of high-speed and low priced broadband services, the emergence of a new player, i.e. Softbank, was a major factor in market dynamics. However, this fact alone does not explain sufficiently the rapid growth of FTTH. If these services were unprofitable, NTT could wait for a change in the regulatory environment in the same manner as ILECs did in the U.S. Therefore, the following points should be added[21]:

- The interdependence between the Japanese telecom policy and NTT, or competition policy that takes into consideration NTT's business conditions, gave NTT the financial basis to invest in FTTH networks when the traffic volume on the public switched telephone networks decreased.
- A consistent telecom policy enabled NTT to implement medium- and long-term plans even under the uncertainty of market and technology.
- The construction of FTTH, coupled with strong brand royalty and sales promotion, consequently brought NTT many FTTH users.

---

[21] In the Kinki region, the fact that K-Opticom advanced capital investment without fear of regulation on its optical fiber networks contributed to FTTH diffusion.

The Evaluation of the Japanese Telecom Policy

The facts that high-speed and low-priced broadband services diffuse and that both an incumbent carrier and new entrants contribute to it (i.e. both service-based and facilities-based competition has developed) provide a good empirical and conceptual background for discussion on the telecom policy.

In my view, the Japanese telecom policy may provide an appropriate but not an optimal solution for other countries. First, the postponement of the application of the LRIC methodology to interconnection charges and the recovery of a part of fixed costs from usage-based call rates or interconnection charges seem arbitrary compared with the U.S. views that were based on economic principles. As those policies relate closely to universal service, the system of universal service and the rate-setting methodology for interconnection should be designed and evaluated in block. The U.S., though in its criticizing the uniform interconnection charge between NTT East and West, proposed to make use of the universal service fund system. Second, NTT's FTTH activities were pursued in a growth stage of the business cycle, where an increase in users encourages capital investment and the capital investment enables the further acquisition of users. I believe that this is the "unintentional benefit of Japan's telecommunications policy" (from Ida 2006). Without this benefit, it is doubtful whether the capital investment of NTT would have been implemented according to the Medium-Term Management Strategy. Third, the addition of the NTS cost to interconnection charges raised the rates of ordinary telecom services. Moreover, the competition in those markets was restricted, because only the costs (effectively, the payments to NTT) of competitive carriers rose. Of course, it was the users who finally bore the real burden of higher rates and less competition.

The Japanese telecom policy for broadband diffusion was based on NTT's monopoly of local loops and there were some peculiar factors for its success. Therefore, a country that aims to diffuse broadband services should not copy or follow the Japanese policy, but should extract factors that are effective in achieving that goal and reflect them in its own policy, while promoting competition between incumbent carriers and new entrants.

Acknowledgments This paper is based on a presentation at the 18th European Regional ITS Conference, October 2–5, 2007, Istanbul. I would like to thank Professor Hitoshi Mitomo (Waseda University) for his encouragement and assistance. I would also like to thank Dr. Brigitte Preissl (the editor of this book) and Professor Chris Czerkawski (Hiroshima Shudo University) for their invaluable comments.

References

Bauer JM (2006) Broadband in the United States. In: Fransman M (ed.) (2006a).
Crandall RW (2005a) Competition and Chaos: The U.S. Telecommunications Sector Since 1996. Brookings Institution Press, Washington, DC.

Crandall RW (2005b) Broadband Communications. In: Majumdar SK, Vogelsang I, Cave ME (eds.), Handbook of Telecommunications Economics, Vol. 2, North-Holland, Amsterdam.
Crandall RW, Ingraham AT, Singer HJ (2004) Do Unbundling Policies Discourage CLEC Facilities-Based Investment. Topics in Economic Analysis & Policy, 4(1).
Fransman M (ed.) (2006a) Global Broadband Battles: Why the U.S. and Europe Lag While Asia Leads, Stanford Business Books, Stanford.
Fransman M (2006b) Introduction. In: Fransman M (ed.) (2006a).
Hausman JA and Sidak JG (2005) Did Mandatory Unbundling Achieve Its Purpose? Empirical Evidence from Five Countries. J. Comp. Law Econ., 1(1):173–245.
Hazlett TW (2002) Broadband Regulation and the Market: Correcting the Fatal Flaws of Glassman-Lehr. Working Paper, Manhattan Institute.
Hazlett T, Bazelon C (2005) Regulated Unbundling of Telecommunications Networks: A Stepping Stone to Facilities-Based Competition? TPRC.
Ida T (2006) Broadband, Information Society, and National System in Japan. In: Fransman M (ed.) (2006a).
Ikeda N (2003) The Unbundling of Network Elements: Japan's Experience, RIETI Discussion Paper Series 03-E-023.
NTT (2004) NTT Group's Medium-Term Management Strategy. NEWS RELEASE.
Softbank (2007) Consolidated Financial Report: For the Fiscal Year Ended March 31, 2007
Tilton M (2004) Nonliberal Capitalism in the Information Age: Japan and the Politics of Telecommunications Reform. JPRI Working Paper, No. 98.
Wallsten S (2006) "Broadband and Unbundling Regulations in OECD Countries," Working Paper 06-16, AEI-Brookings Joint Center for Regulatory Studies.
Willig RD, Lehr WH, Bigelow JP, Levinson SB (2002) Stimulating Investment and the Telecommunications Act of 1996. Report filed by AT&T in FCC Docket 01-338. (in Japanese).
Fair Trade Commission (2004) Broadband Service tō no Kyōsō Jittai ni kansuru Chōsa Hōkokusho.
Information and Communications Council (2004) Heisei 17 Nendo ikō no Setsuzokuryō Santei no Arikata ni tsuite.
Kanzaki M (2006) NTT Min-eika no Kōzai: Kyojin no 'Dokusen Kaiki wo Tou'. Nikkan Kogyo Shimbun, Tokyo.
Kodama H (2005) Gensōkyoku: Son Masayoshi to Softbank no Kako, Ima, Mirai. Nikkei BP, Tokyo.
Miki N (2006) Softbank 'Jōsikigai' no Seikō Hōsoku. Toyo Keizai, Tokyo.
Nakamichi O (2004) Yahoo! BB Network Kōchiku Ki. http://itpro.nikkeibp.co.jp/prembk/NBY/techsquare/20040921/1/?ST=print
Nikkei Communications (2003) Sirarezaru Tsūsin Sensō no Shinjitsu (Unknown Facts under the Communication War): NTT, Softbank no Antō. Nikkei BP, Tokyo.
Nikkei Communications (2005) Hikari Kaisen wo meguru NTT, KDDI, Softbank no Yabō: Sirarezaru Tsūsin Sensō no Shinjitsu. Nikkei BP, Tokyo.
Nikkei Communications (2006) 2010 Nen NTT Kaitai: Sirarezaru Tsūsin Sensō no Shinjitsu. Nikkei BP, Tokyo.
NTT (2003) Shachō Kaiken. http://www.ntt.co.jp/kaiken/2003/030730.htm.
NTT East (2006) Information NTT East 2006. NTT East.
NTT West (2006) Data Book NTT West 2006. NTT West.
Oniki H (1996) Jōhō High Way Kensetsu no Economics. Nihonhyōronsha, Tokyo.
Tajiri N (2007) Broadband no Fukyū Yōin to sono Seisaku teki Gan-i ni kansuru Kenkyō (A Study of Factors Contributing to Broadband Diffusion and their Policy Implications). Doctoral dissertation, Waseda University, Tokyo.
Telecommunications Council (2000) Setsuzokuryō Santei no Arikata ni tsuite (Policy on Calculation of Interconnection Charge).

Part IV
Mobile Drivers

Mobile Termination Carrier Selection*

Jörn Kruse

Abstract Adopting the proposal to introduce "Mobile Termination Carrier Selection" (MTCS) would turn mobile termination into a competitive market. Predominantly, four parallel GSM networks offer coverage in a specific area and, after minor software changes to the GSM-standard, would be able to deliver the termination service to any mobile device (for all F2M and M2M calls). Thus, the mobile termination service would be supplied competitively. The demand side decisions of selecting the terminating networks would be made by either the calling parties themselves (i.e. MTCS at the retail level, call-by-call or preselection) or by the originating networks (i.e. MTCS at the wholesale level). Several arguments, including transaction cost economies, suggest that MTCS at the wholesale level would be most likely to prevail in the market. Efficient prices could be expected as a result of competition. Introducing MTCS would dispense of any need for mobile termination rate regulation.

Introduction

Mobile telephony has been an ongoing success story ever since the GSM standard was introduced and competing mobile network operators were licensed beginning in the early 1990s. This success is owed largely to the fact that most European countries feature three or four mobile network operators (and additional service providers) offering their services. Most mobile markets are highly competitive. As a consequence, prices have dropped and the mobile penetration rate in most countries has proven to be very high.

It is worth noting that, as opposed to most other network industries, the mobile communications sector is characterized by several parallel physical network

*Many Thanks to Florian Lauer whose support I gratefully acknowledge.

J. Kruse
Helmut-Schmidt-University Hamburg
e-mail: joern.kruse@hsu-hh.de

B. Preissl et al. (eds.), *Telecommunication Markets*, Contributions to Economics,
DOI: 10.1007/978-3-7908-2082-9_13, © Springer Physica-Verlag HD 2009

infrastructures, including base stations, transmission lines, switching units (MSC), etc. Most European countries feature three or four parallel GSM infrastructures operating competitively. Basically, the countries' prevalent mobile market structures are the result of licensing policies by national regulatory authorities that assigned the GSM spectrum. The authorities required the licensees to operate as vertically integrated entities. The licensees were thereby instructed to establish individual cellular infrastructures in addition to offering their mobile services to customers.

Despite the sector's overall competitiveness and remarkable market performance, regulatory authorities have identified the wholesale submarket of mobile termination as being monopolistic. In an attempt to remedy the adverse implications, most European countries have applied an ex ante regulation. It is doubtful whether price regulation can serve as an appropriate response to the perceived termination monopoly. Nevertheless, this will not be discussed in this paper (see for example Gans and King 2000; Kruse 2003; Crandall and Sidak 2004). Regulation, however, is faced with severe methodological problems associated with large common costs and demand complementarities (Competition Commission 2002; Valletti and Houpis 2005; Newbery 2004).

This paper focuses on the currently regulated mobile termination markets. In particular, alternative modes of transaction will be discussed in the following sections. It is important to note that the mobile termination monopoly is basically a result of previous governmental and/or regulatory decisions. This refers to standards and regulations before and during GSM implementation and licensing. It reveals that during standardization, licensing, and regulation, the authorities failed to formulate an economic analysis that would have allowed competition to prevail in all transaction segments of mobile telephony. As a consequence, specific transaction schemes and market structures have emerged, which, in turn, have led to the current problems. These transaction relations in the mobile sector will be described in further detail in the following section.

A fundamental message of this paper is that regulatory authorities should focus on changing the specific transaction scheme, thereby introducing competition to the termination segment, rather than regulating prices. Two economic alternatives are available to deal with the mobile termination problem. Since the conventional calling-party-pays principle is often regarded as the root cause of the termination problem, the alternative receiving-party-pays principle (section "Receiving-Party-Pays and Bill-and-Keep") has been suggested as a possible remedy.

This paper then goes on to identify another element of GSM as being the most crucial factor leading to the termination problem: The exclusive relationship between any specific handset and only a single cellular network carrying out the termination service. It is suggested in section "The Principle of Mobile Termination Carrier Selection" that this should be replaced by the mobile termination carrier selection principle, whereby any handset may communicate with a variety of different GSM networks. Section "Mobile Termination Carrier Selection at the Retail Level and at the Wholesale Level" deals with mobile termination carrier selection at both the retail and wholesale level in more detail, and section "Merits and Problems" goes on to mention potential merits and problems associated with this approach.

Calling-Party-Pays

When the cellular mobile telephony standard, GSM, was standardized and introduced by European regulatory agencies, all the basic elements of the transaction scheme had already been determined and, as a consequence, the course of the retail and wholesale market structures had been set.

Almost all European countries voted for the calling-party-pays-principle (CPP) which is nowadays considered to be responsible for the termination problems addressed here. The respective transaction relations are depicted in Fig. 1. Let's assume the calling party A_O, either from a fixed network or from a mobile network (both denoted as origination network O_A), calls a mobile handset B_B (receiving party) subscribed to a mobile network B.

In technical terms, the originating network provides the first segment of the phone call (origination) from the calling party A to the interconnection point (IC) with the mobile network B which is the gateway MSC_B of the latter From IC to the mobile device B_B of the receiving party (termination segment of the call), the service is delivered by the mobile network B.

Under CPP, the calling party A_O pays O_A for the whole call (Z_{AO}). Z denotes the payment (and the transaction relation) for the whole call, whereas Y represents the payment for the originating segment (from the calling party to the interconnection point at the gateway MSC of the mobile network), and T represents the payment for the terminating segment (from the interconnection point to the receiving device B_B). Z_{AO} denotes a market transaction between the calling party A and its network O and the payment (for the whole call) from A to O.

For the termination segment of the call there is a market transaction T_{OM} in which originating network O_A pays the mobile network B. In the conventional view, B has a monopoly over the termination service under CPP, since, under the given setting, network B is solely capable of communicating with Handset B_B.

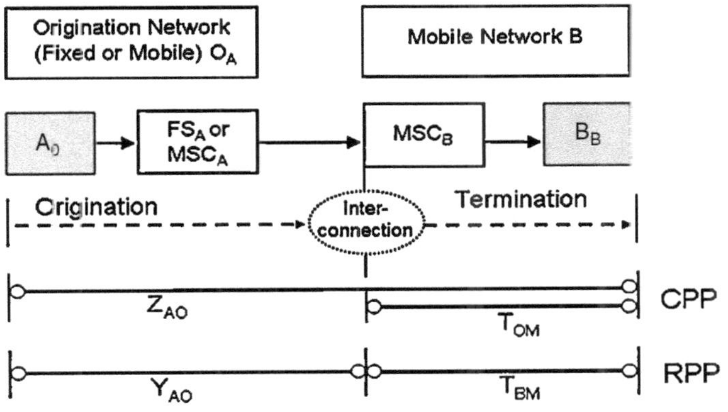

Fig. 1 Calling-party-pays and receiving-party-pays

Certain reform approaches believe the main problem to lie in the CPP principle itself (see section "Receiving-Party-Pays and Bill-and-Keep"), whilst another focuses on the exclusive communication patterns (see section "The Principle of Mobile Termination Carrier Selection").

Receiving-Party-Pays and Bill-and-Keep

The implementation of the calling-party-pays-principle (CPP) in most countries is seen as the main reason for the competitive problems related to mobile termination and the consequential governmental ex-ante regulation. The corresponding alternative which modifies the transaction relationships is known as the receiving-party-pays-principle (RPP) (Littlechild 2006).

With RPP, the (fixed or mobile) calling party pays its originating network only for the first segment of the call up to the interconnection point (gateway MSC). This is represented by Y_{AO} in Fig. 1. The termination service from the interconnection point to the handset of B_B is charged to the receiving party B_B. If this charge T_{BM} is positive ($T_{BM} > 0$), the receiving party has to pay for incoming calls.

In RPP, the termination of incoming calls is a part of the service bundle a network operator provides to its subscribers. They will take the rate for the incoming calls (among the rates for other services) into account before subscribing to a specific network. Thus, mobile termination is under competitive pressure.

The pricing decision for incoming calls is up to any individual mobile network operator. Since the marginal cost of terminating a call is low, the network operator may choose not to charge his customers for incoming calls at all in order to be attractive to potential and actual subscribers. If incoming calls are not charged to the receiving party, RPP is equivalent to bill-and-keep (B&K).

Receiving-party-pays is not a new concept. A number of countries, e.g. the USA, Canada, Hong Kong and China implemented RPP, also known as Mobile-Party-Pays (MPP), from the outset. A number of other countries, especially in Latin America, initially applied RPP, but have since switched to CPP (Zehle 2003; Dewenter and Kruse 2006). In most European countries the RPP principle is applied to calls to mobiles roaming abroad. The calling party pays only for the national segment of the call, whereas the receiving party itself pays for the mobile service in the international segment including termination abroad. The RPP, resp. B&K, has been suggested for mobile as well as fixed networks. The discussion has been quite controversial (Wright 2002; Quigley and Vogelsang 2003; Crandall and Sidak 2004; Marcus 2004; Hausman 2004; Littlechild 2006).

The main advantage of RPP as a structural alternative to CPP has already been mentioned: Since mobile termination would cease to be a monopoly, it would no longer elicit the need for regulation.

A potent argument against RPP is based on the assumption that the receiving parties might attempt to avoid payments for incoming calls by switching off their handsets altogether. This argument gains weight in the light of potentially significant numbers of commercial or other unwanted calls (junk calls).[1] Even if networks opt not to charge for incoming calls, or the regulatory agency introduces bill-and-keep, junk calls might yet pose a problem, seeing as they are based on low prices for the calling party.

Switching off devices would reduce the demand for the mobile networks' airtime minutes, which, in turn, would lead to higher average costs and thus potentially higher prices. There is a concern that the penetration rate would perhaps decrease because RPP may lead to lower attractiveness of mobile telephony. There has been evidence from countries that switched from RPP to CPP that the number of incoming calls and the number of terminated mobile minutes have increased. In general, the empirical picture with respect to CPP/RPP and penetration is not significant (Dewenter and Kruse 2006).

There is some evidence that regulatory authorities are seriously considering introducing bill-and-keep, be it to get rid of the termination regulation problem or as a reaction to lobbying pressure from fixed networks which would have to pay less for calls to mobiles.

The Principle of Mobile Termination Carrier Selection

Another structural alternative to the conventional mobile termination transaction scheme is the concept of mobile termination carrier selection. This concept was first presented in Kruse and Haucap (2004) and Kruse (2006). Its application would entirely avoid the existence of a monopoly and turn the mobile termination service into an individual market that can be expected to be highly competitive.

Mobile termination carrier selection (MTCS) basically applies the conventional calling-party-pays principle. The calling party would pay for the origination as well as for the termination segment of the call. The abovementioned problems associated with receiving-party-pays would therefore be avoided.

The most fundamental technical reason for the existence of the termination monopoly (and therefore for the prevalence of governmental ex ante rate regulation) is the fact that, under the conventional setting, the subscribed mobile network is exclusively capable of communicating with the mobile device of the receiving party.

The principle of MTCS is based on the technical feasibility that terminating a call to a specific handset could also be carried out by other GSM networks offering coverage in that specific area. If this were the case, the calling party (or the originating network, respectively) would be able to choose between alternative mobile networks to terminate the call to a specific receiving device. The GSM networks would compete for delivering that service.

[1]This may be the case despite technical countermeasures that may be implemented, such as spam filters, different ring tones depending on the origin of the call, or simply the fact that people are getting more used to checking the display revealing the number of the calling party before answering the call.

The proposed MTCS principle would work at the retail level as well as the wholesale level. This will be outlined in section "Mobile Termination Carrier Selection at the Retail Level and at the Wholesale Level". With MTCS at the retail level, the individual calling party would select the terminating network by either call-by-call or preselection. Alternatively, the origination network would select the terminating network at the wholesale level.

From an economic viewpoint, mobile telephony offers an important advantage over fixed-line telephony with respect to multiple infrastructures. In fixed networks, most parties are connected to the rest of the world by only a single physical subscriber line. Under these technical conditions, in order to call someone, this specific subscriber line has to be used. It can be seen as a monopoly and will therefore usually be regulated. In this respect, GSM mobile communication is completely different. Normally, three or four parallel GSM networks are in place which cover almost the entire country concerned. A specific mobile device always enjoys cellular coverage by mostly four different networks. With respect to already existing hardware, it would technically not be a problem to reach this handset.

There is, however, a software problem. The conventional GSM standard does not provide the capability of differing networks being able to reach a specific handset. This capability is exclusively reserved to the network the receiving party has subscribed to which therefore holds a monopoly. This shows that termination regulation is basically a consequence of former standardization decisions. These should be revised in such a way as to enable multiple access.

The technical setting is demonstrated in Fig. 2. The calling party A_O in the fixed or mobile originating network O_A wishes to call the handset B_B which is subscribed to the mobile network B. Under the conventional GSM standard, only cellular network B is able to terminate the call. Switching on handset B_B initiates the signaling traffic

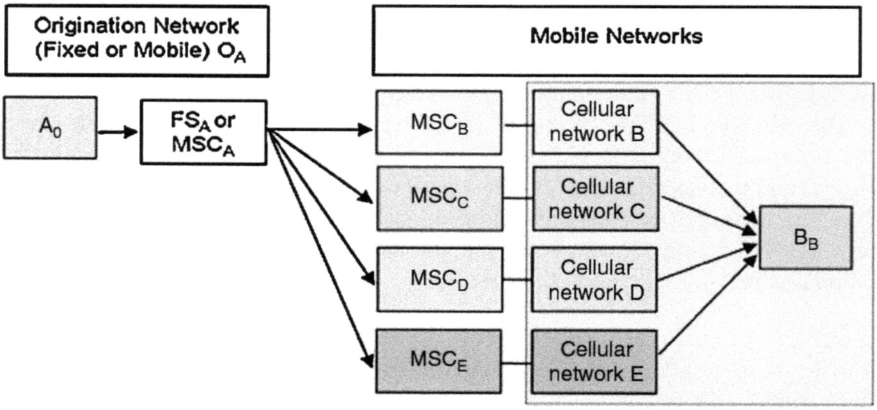

Fig. 2 Mobile networks terminating a call to B_B

exclusively with network B and allows outgoing calls to be placed and incoming calls to be received solely via network B.

If the GSM standard were to be revised such that networks C, D, and E were able to gain access to handset B_B, MTCS could be introduced and all four networks would be able to compete for the service to terminate the call. This presupposes the condition that the other operators are able to receive signaling traffic from B_B in order to locate B_B in their own cellular networks at any given time and store the information in their registers.

The fact that other networks' communication with a specific third party mobile device would not meet significant problems is illustrated by considering the service of international roaming. In this case, a specific GSM handset from one country roaming abroad is able to place calls in other countries where operators also use the GSM standard. Most other countries have not just one but mostly three or four GSM networks that are each capable of providing the international roaming service including termination. A technical requirement for international roaming (and for MTCS) is that both the mobile device (handset) as well as the foreign network (third network) is operating the GSM standard in the corresponding spectrum, 900 or 1,800 MHz.

In order to introduce MTCS, the regulatory agency would have to rule that the GSM software used by mobile operators needs to be adapted. This basically implies changing the GSM standard in such a manner as to allow different networks to conduct signaling traffic with receiving handsets.

After implementation, each individual mobile operator would then decide whether and how it wishes to supply terminating calls to handsets subscribed to other networks. Essentially, this decision would depend on the relationship between incremental costs and incremental revenues.

Incremental revenues would basically consist of the fees collected for terminating services to handsets of other cellular networks. The operators would also have incentives to prevent competing GSM networks from terminating traffic to their own subscribers by setting attractive prices. The demand side will be discussed in section "Mobile Termination Carrier Selection at the Retail Level and at the Wholesale Level".

The incremental costs of MTCS for a mobile operator would include the (modest) outlays for larger capacities of registers, etc. as well as operating costs which are associated with an increase in signaling traffic. Each operator would have to provide the signaling traffic of all the handsets he wishes to supply with the termination service. Whether or not capacities for payload traffic (calls to mobiles), especially base transceiver stations, mobile switching centers, transmission lines, etc. have to be scaled up basically depends on success in the mobile termination market.

Under these incremental cost and revenue conditions one can reasonably assume that every GSM operator would actively supply the termination service in the MTCS market. Thus, effective competition would be on the way. Because of competition, any price regulation of the terminating service would become completely obsolete. Under mobile termination carrier selection an individual market for the termination service would emerge.

Mobile Termination Carrier Selection at the Retail Level and at the Wholesale Level

The markets for mobile termination carrier selection can be established at the retail level as well as at the wholesale level. The latter is based on market transactions between the terminating networks and the originating networks. Retail MTCS, on the other hand, characterizes market transactions between the individual calling parties and the terminating networks. Let's consider this variant first.

With mobile termination carrier selection at the retail level (Fig. 3), the individual caller (either from a fixed or from a mobile network) would select the mobile network he wishes to terminate his calls to a mobile number. In principle, this could work on a call-by-call as well as on a preselection basis. Regarding retail MTCS on a call-by-call basis, the customer would select the terminating service for each individual call. To do so, he would have to append a specific carrier code to the mobile number of the desired receiving party. In the case of retail MTCS and preselection, a calling customer would subscribe to a contract with a specific mobile network to terminate all future calls to mobiles.

In retail MTCS (call-by-call or preselection), the individual calling party would pay for the complete call to a mobile, thus covering both segments ($Y_{AO} + T_{AMi}$). A_O would therefore face two different transaction partners for both particular segments of his call. The originating network O_A would be the transaction partner for the originating segment of the call up to the interconnection point, whilst the selected mobile network would be the transaction partner in the terminating segment (from the interconnection point up to B_B). The entire billing process would be managed by the origination network which would charge the customer for both segments of the call and transfer the termination fee T_{AMi} to the selected mobile network i.

Since the calling party would pay for the mobile segment of his call (as always under CPP), he would be incentivised to select the most favorable offer, either by call-by-call or preselection. The calling party would also have an incentive to remain informed on different termination rates, thus incurring information cost.

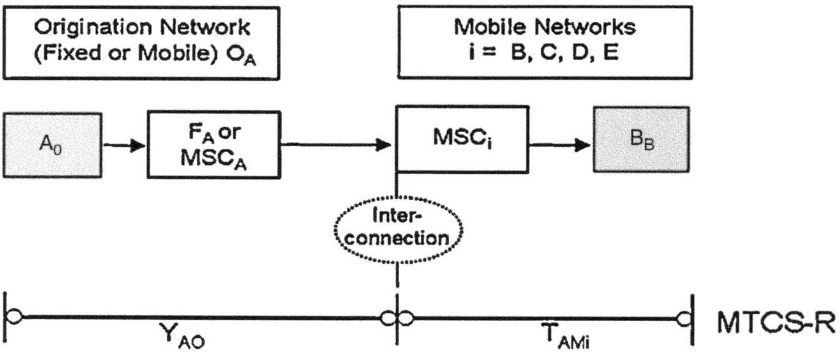

Fig. 3 Mobile termination carrier selection at the retail level

The termination service is nearly homogenous, the only relevant quality parameter being regional coverage. If the selected termination network should lack coverage in the relevant location of the receiving device, the terminating service would have to be carried out by the subscribed network or by any other carrier offering coverage.

Under retail MTCS a carrier might offer third parties terminating services (as preselection and/or call-by-call) for all fixed-to-mobile and/or mobile-to-mobile calls or only for specific market segments. In particular, mobile originating networks would have incentives to offer particularly favorable conditions to their own customers for off-net-calls (calls to mobiles subscribed to other networks). Thereby, the original off-net-calls would turn into on-net-calls.

Another transaction scheme is MTCS at the wholesale level (network level). The fixed and mobile originating networks would constitute the demand side and engage in market transactions with mobile terminating networks. Each origination network would negotiate favorable termination rates for calls to mobiles, since termination rates represent major input costs. These rates would influence their competitiveness on their own retail markets where the price for calls to mobiles is a major criterion for potential subscribers.

Figure 4 shows that the transaction scheme of MTCS at the wholesale level is similar to the conventional setting of calling-party-pays (Fig. 1) used in European countries today. The only, yet decisive, difference is T_{OMi} instead of T_{OM}. This represents the central element of MTCS: The originating networks would be able to choose between competing mobile termination networks.

Under wholesale MTCS, each mobile network would have strong incentives to offer competitive termination rates, since each originating network would buy a considerable number of terminating minutes per month. On the termination cost side, not only short run but also long run incremental cost would be low since termination uses the same network elements that are also necessary for outgoing calls.

Mobile originating networks generally have cost incentives to terminate calls to mobiles on their own network, and this is also (in economic terms) technically efficient.

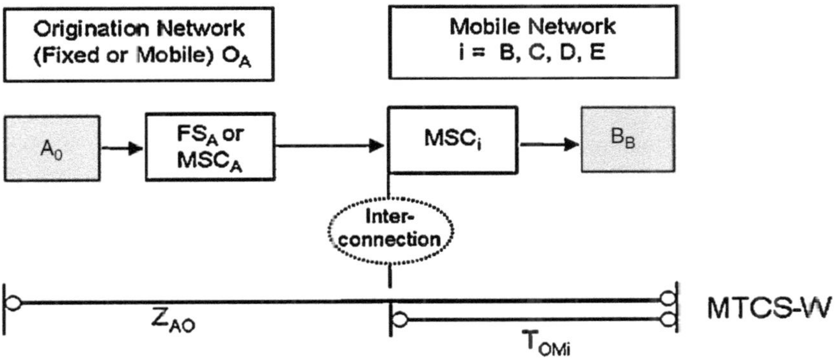

Fig. 4 Mobile termination carrier selection at the wholesale level

The regulatory authority would not have to opt for retail or wholesale MTCS. In general, this could be left to the market. Retail MTCS and wholesale MTCS might coexist. The originating networks would (on the basis of their wholesale agreements with terminating carriers) offer their customers a tariff for all fixed-to-mobile or mobile-to-mobile calls. At the same time, mobile networks might offer preselection and/or call-by-call options to calling parties from other networks to terminate their calls.

The calling party would compare its network's prices for complete calls to mobiles Z_{AO} with the sum of the prices for the originating (Y_{AO}) and the terminating (T_{AMi}) segment. If $Y_{AO} + T_{AMi} < Z_{AO}$, the individual caller would opt for the retail option. Vice versa, he would take advantage of his network's comprehensive offer for complete calls to mobiles which would entail additional significant information and transaction cost advantages for the caller.

Presumably, in most cases $Y_{AO} + T_{AMi} > Z_{AO}$ holds because the originating networks have informational and bargaining advantages. They would probably be able to negotiate more favorable termination rates with mobile networks than their customers would receive on the retail termination market for either call-by-call or preselection ($T_{OMi} < T_{AMi}$). The originating networks would set their prices for Y_{AO} in the retail scheme as well as Z_{AO} in the wholesale scheme. Therefore, the originating networks would be able to design their price structures such that their customers would prefer the wholesale option. They would have incentives to do so due to cost reasons (scale economies in transmission lines to MSCs) as well as for reasons of termination input prices T_{OMi} which might tend to decrease with larger volume.

Thus, the actual scale of retail and wholesale MTCS, respectively, could be left to market forces and consumer preferences. For most (if not almost all) transactions it can be expected that wholesale MTCS would prevail as has been outlined above. It would not only be technically cost efficient for the carriers but also transaction cost efficient from an economic point of view.

From a consumer perspective, the predicted outcome that the wholesale MTCS would prevail possesses the important advantage that the callers would not be required to constantly remain informed on retail termination rates, since they could rely on favorable terms of origination networks based on wholesale MTCS. Retail MTCS would then mostly function as an element securing contestability.

It might then be advisable for regulatory authorities to rule that all originating networks have to allow retail MTCS which would include offering the originating service separately and to announce the respective rate for Y_{AO}. This would hamper collusion, if it should be a problem. Generally speaking, the fact that only three or four networks exist that are capable of providing the service, one may be concerned whether mobile termination markets would actually be competitive or in fact subject to collusion. The market structure for termination services would be equivalent to that of other mobile services (subscription, outgoing calls, etc.) in which collusion is not likely to occur (Kruse 2004) and actually does not occur. Among the reasons are high fixed and very low marginal costs, market homogeneity, vertical market transparency and high elasticity of demand, excess capacity in UMTS, etc. Additionally, the mobile operators have quite different incentives. This is especially true with respect to the larger GSM firms that were licensed early on the one hand, and the respective third and fourth operators aggressively vying for market shares on the one other.

Merits and Problems

The main advantage of MTCS is the avoidance of any mobile termination monopoly such that regulation of terminating rates would be rendered completely obsolete and could be abandoned altogether. In MTCS wholesale as well as in MTCS retail markets the mobile termination rates would be determined by network operators' decisions in competitive markets, so efficient prices could be expected to prevail.

Under MTCS the termination service would be supplied in a separate competitive market. It would not be a part of larger package as is the case under RPP. The specific problems regarding RPP which were discussed in section "Receiving-Party-Pays and Bill-and-Keep" would not appear.

If MTCS were to be implemented initially at both the retail and the wholesale level, it could be expected to develop predominantly into wholesale MTCS which is more efficient from an economic point of view. Essentially, the introduction of MTCS would maintain the conventional CPP principle, avoiding a reversal of transactional relationships between the calling and receiving parties. In this respect, the regulatory authority could therefore rest assured that consumers would not oppose the new scheme. Most of them would not even realize the change, other than perhaps paying less for calls to mobiles, depending on the pricing policy of the carriers.

Certain requirements and potential problems would be associated with the introduction of MTCS. These are discussed in the following four points.

1. The introduction of mobile termination carrier selection necessitates an explicit decision by the regulatory agencies. The authorities might hesitate for two reasons. Firstly, they prefer, whenever possible, to avoid any economic and political risk that is necessarily associated with any regulatory change. Secondly, they might not espouse the idea of abandoning termination regulation which is associated with budgets and jobs in regulatory agencies.
   The introduction of MTCS could be carried out at the national level as well as at the European or global level. If international agreements could not be reached, it would not at all be a problem for a single country to implement this system on its own. Because the CPP principle would essentially be maintained, this solitary move would not result in any problems for international telecommunications traffic, being most obvious in the event that wholesale MTCS prevails.
2. As mentioned above, the GSM standard needs to be revised in order to allow other GSM networks to communicate with a specific handset. Some technical modifications in the network elements as well as in the end users' devices would also be necessary, depending on the specific technical solution that would be implemented. It would determine whether or not the technical functionalities of MTCS in the handsets can be implemented by software updates and/or simply by replacing the conventional sim cards by new ones. The network operators would have to implement some new features in the next software update, in order to support MTCS and to enable communication with every GSM handset in a specific region.
3. With MTCS, the volume of data on the active handsets, their location, the billing information, etc. that would need to be stored would be higher. More signaling traffic would be generated. The mobile networks would have to expand the

capacity of specific registers and network elements. This would mostly depend on their market strategy and revenue policy.

4. A more general aspect relates to the changing of regulatory rules ex post, i.e. after licensing and after mobile operators' investments. If we interpret a license agreement as a contract between the regulatory agency and the licensed firm, a change of rules raises the question of institutional stability and regulatory credibility. From an economic viewpoint, any new regulatory intervention after major sunk investments gives rise to problems.

Generally, this would also apply in the case of regulatory introduction of MTCS, since it would represent an intervention in market and revenue structures. But this was also the case regarding ex post introduction of an ex-ante-regulation of terminating rates, just as it would be with respect to any other regulatory change such as the introduction of RPP or bill-and-keep.

Mobile termination carrier selection needs to be judged in the light of the prevalent alternative, governmental ex ante price regulation. Since the concept of MTCS transforms the regulated monopoly into a competitive market, the proposed changes would seem to be highly justified, especially considering the calling and receiving parties not having to adapt to a noticeably new framework.

Conclusion

Any form of governmental monopoly regulation is highly unsatisfactory for a variety of reasons. This also holds for the mobile termination market. However, contrary to the case of "real monopolistic bottlenecks", institutional alternatives are available here that would place the mobile termination service under competitive pressure. One is the concept of receiving-party-pays or bill-and-keep in which the terminating service is only one element of a larger bundle of services offered to mobile customers. The application of this principle would imply significant changes for network operators and for users, aside from additional problems (junk calls, etc.).

The other alternative would be to apply mobile termination carrier selection, whereby the mobile termination service is transformed into an individual competitive market. Here, mobile termination carrier selection at the wholesale level would represent the most efficient form, with neither the calling nor receiving party having to adapt to new transactional schemes, due to the conventional calling-party-pays principle remaining unchanged.

Since mobile termination would represent a competitive market resemblant of other mobile markets characterized by large common costs, the pricing decisions would be left to mobile operators and would depend on demand elasticities as well as firms' market and revenue strategies. It can be assumed that efficient price structures would prevail.

From an economic perspective, mobile termination carrier selection has no significant disadvantages and can be regarded as the first choice solution for the termination problem.

References

Competition Commission (2002) Vodafone, O2, Orange and T-Mobile: Reports on References under section 13 of the telecommunications act 1984 on the charges made by Vodafone, O2, Orange and T-Mobile for terminating calls from fixed and mobile networks. Presented to the Director of telecommunications (December 2002)

Crandall RW, Sidak JG (2004) Should Regulators Set Rates to Terminate Calls on Mobile Networks. Yale Journal on Regulation 21: 1–46

Dewenter R, Kruse J (2006) Calling Party Pays or Receiving Party Pays? The Diffusion of Mobile Telephony with Endogenous Regulation. Discussion Paper

Gans JS, King SP (2000) Mobile Network Competition, Customer Ignorance and Fixed-to-Mobile Call Prices. Information Economics and Policy 12: 301–327

Hausman JA (2004) Economic analysis of regulation of CPP. Paper (19th November 2004)

Kruse J (2003) Regulierung der Terminierungsentgelte der deutschen Mobilfunknetze. Wirtschaftsdienst 83 (3): 203–209

Kruse J (2004) Competition in Mobile Communications and the Allocation of Scarce Resources: The Case of UMTS. In: Buigues, P, Rey, P (eds.): The Economics of Antitrust and Regulation in Telecommunications: Perspectives for the New European Regulatory Framework. Cheltenham (Edward Elgar): 185–212

Kruse J (2006) Mobilterminierungswettbewerb: Eine neue Lösung für ein aktuelles Problem. Multimedia und Recht 9 (12) MMR aktuell: VI–IX

Kruse J, Haucap J (2004) Remedies bei der Terminierung im Mobilfunk. Unpublished Economic Report

Littlechild SC (2006) Mobile Termination Charges: Calling-Party-Pays vs. Receiving-Party-Pays. Telecommunications Policy 30: 242–277

Marcus JS (2004) Call termination fees: The US in global perspective. Paper, ZEW-conference, Mannheim

Newbery D (2004) Application of Ramsey Pricing for Regulating Mobile Call Termination Charges. In: Vodafone (eds.) Regulating Mobile Call Termination. Vodafone, London, p. 12

Quigley N, Vogelsang I (2003) Interconnection Pricing: Bill and Keep Compared to TSLRIC. Final Report for Telekom NZ (April 2003)

Valletti TM, Houpis G (2005) Mobile Termination: What Is the "Right" Charge. Journal of Regulatory Economics 28 (3): 235–258

Wright J (2002) Bill and Keep as the Efficient Interconnection Regime. Review of Network Economics 1 (1): 54–60

Zehle S (2003) CPP Benchmark Report. Coleago Consulting (February 2003)

Countervailing Buyer Power and Mobile* Termination

Jeffrey H. Rohlfs

Introduction

Different countries have different practices with regard to charging for calls to mobile.

- In some countries, including: Canada, China, Hong Kong, Russia, Singapore and the United States, mobile network operators (MNOs) charge their subscribers airtime on calls that they receive. This regime is known as mobile party pays or MPP.
- In most of the rest of the world, however, mobile subscribers are not charged for incoming calls. Instead, the MNO levies a mobile-termination charge on other network operators for terminating calls. The originating network operator generally passes the mobile termination charge on to its subscriber who made the call. The regime is therefore known as calling party pays (CPP).

This paper discusses the analysis required to support public policies regarding mobile termination.[1] The analytical issues include:

- Specification of the relevant product market
- Determination of market power
- Assessment of countervailing buyer power
- Regulatory intervention
- Relaxing of regulation

*The author thanks Justus Haucap for helpful comments.

[1]The economic issues associated with setting mobile termination rates are discussed from a somewhat different perspective by Rohlfs (2006). See also Thomson et al. (2006).

J.H. Rohlfs
Analysys Mason, Washington,
e-mail: jhr@analysys.com

B. Preissl et al. (eds.), *Telecommunication Markets*, Contributions to Economics,
DOI: 10.1007/978-3-7908-2082-9_14, © Springer Physica-Verlag HD 2009

Policies of the European Union

We focus largely on the regulatory policies of the European Union (EU). Those policies are important in their own right, since they are applied throughout all of Europe. In addition, many other countries have adopted the EU policy framework.

The general policy of the EU is that telecommunications charges should be regulated in a particular relevant market if, but only if, the following three criteria are satisfied:

- The market is subject to high and non-transitory entry barriers.
- The market has characteristics such that it will not tend over time towards effective competition.
- Competition law does not suffice by itself to deal with the market failure (absent ex ante regulation) (Commission of the European Communities 2006, SEC[2006]837, p. 10).

The first two criteria, if satisfied, would indicate the presence of non-transitory significant market power (SMP) in the market. The third criterion, if satisfied, would indicate that enforcement of competition law would not suffice to obviate regulation.

This general policy of the EU has been applied to mobile-termination charges (as well as to other telecommunications charges).

In most European countries, several competitive operators supply mobile services. The market for retail mobile services is therefore usually found not to satisfy the three criteria, and retail mobile prices are usually not regulated.

In contrast, mobile termination rates are generally subject to regulation (though the precise scope of regulation varies somewhat from country to country). This policy generally follows a finding that the three criteria are satisfied *in the market for mobile termination*. In that market, MNOs are found to have non-transitory SMP. The logic underlying this finding is described in the following sections.

Specification of the Relevant Product Market

Relevant product markets are specified solely with respect to conditions on the demand side of the market.[2] A relevant product market includes the product itself (in this case, mobile termination) and other products that are sufficiently close substitutes (Office of Fair Trading 2004).

---

[2]This point is stated explicitly in the "Horizontal Merger Guidelines" of the U.S. Department of Justice and Federal Trade Commission, (issued April 2, 1992, revised April 8, 1997) in the initial overview section. *See*http://www.usdoj.gov/atr/public/guidelines/horiz_book/10.html. Of course, the associated SSNIP tests for market power, as discussed below, depend on supply considerations, as well.

Mobile termination is purchased by a telecommunications operator that wishes to complete a call from one of its subscribers to a mobile subscriber (on another network). Regulators within the EU have generally reasoned that there is no good substitute to purchasing mobile termination from the MNO that serves the called party. It follows that the relevant product market consists solely of mobile termination supplied by a particular MNO. It further follows that each MNO has a monopoly in the market for its mobile termination; i.e., the "terminating-access monopoly."[3]

This logic is valid only if the market includes all types of calls that terminate on a particular mobile network. In particular, the market must include mobile-to-mobile (MTM), as well as fixed-to-mobile (FTM) calls. Those calls are substitutes for each other and must all be considered to be in the same relevant market.[4]

The finding of monopoly in the relevant product market must be qualified to some extent. Suppose that Mobile Network A had extremely high charges for mobile termination. Those charges would be passed on from Mobile Network A to other network operators, then on to callers in other networks in the form of higher call prices. Those callers would likely respond by declining to make calls to subscribers of Mobile Network A, except in emergencies. The subscribers to Mobile Network A might well find this outcome unsatisfactory and switch to another mobile network. In this example, the demand for termination on Mobile Network A does depend on demand conditions on other mobile networks.

Relevant product markets are not, however, usually specified with regard to extremely large price increases. The usual practice is to include substitutes that would be used if there were a small, but significant, non-transitory increase in price (SSNIP).[5]

Empirical evidence has demonstrated that the cross-elasticities of demand for retail mobile services with respect to the price of mobile termination are quite small – both in absolute terms and relative to the magnitudes of other mobile cross-elasticities.[6] A simple (perhaps simplistic) explanation for this finding is that subscribers care more about charges that they pay themselves than about charges that subscribers to other networks pay. It follows from this finding that, although a MNO faces some competitive discipline with respect to setting mobile-termination rates, that discipline is quite weak. In the absence of regulatory constraints, mobile termination rates could be unacceptably high. Indeed, in the past, where mobile termination rates in many countries were not regulated, the rates were often quite high.

[3]For further discussion of the terminating-access monopoly, Laffont and Tirole (2001).

[4]Hausman and Wright (2006) emphasise the point that substitution between MTM and FTM calls is significant.

[5]For example see OFT403 (2004, Paragraphs 2.7, 2.10).

[6]For example, Oftel found that the cross-elasticity was so small that it could reasonably be disregarded. See Competition Commission (2003), Appendix 9.1, Table 3.

In summary, the finding that the relevant product market is mobile termination of a particular MNO is reasonable – notwithstanding the weak cross-elasticities with other mobile markets.

Determination of Market Power

The determination of market power might seem to be a trivial exercise. Each MNO has a monopoly in the market for its mobile termination. Regardless of the precise criteria for determining how much market power is significant, a monopoly would seem to qualify as having SMP.

The official guidelines of the EU are, however, more sophisticated. They recognise that market power of sellers is diminished to the extent that buyers have countervailing buyer power (CBP).[7]

Historically, CBP was not a major consideration in regulatory decisions whether to regulate mobile termination. Recently, however, the issue has become more prominent, as discussed below.

In the broader economy (apart from telecommunications), CBP usually arises where there is concentration on the buyers' side of the market; i.e., monopsony or oligopsony. In that case, buyers may be able to obtain outcomes that are satisfactory to them, notwithstanding SMP on the sellers' side of the market. Indeed, the buyers may have the upper hand.

Significant CBP is an inherent aspect of telecommunications markets. To be sure, each MNO has a terminating-access monopoly. At the same time, however, the originating operator, whether fixed or mobile, has an "originating-access monopoly". That is, if a MNO wants its subscribers to be able to receive calls from another operator, it must use the originating services of that operator.

The originating-access monopoly always limits the market power of firms that have a terminating-access monopoly. It must be considered for a proper determination of whether the terminating operator has SMP.

Binmore and Harbord made this point in arguing for relaxing regulation with regard to the setting of mobile-termination rates (Binmore and Harboard 2005). Hutchinson, a small MNO, used the Binmore-Harbord article to appeal SMP findings in the UK and Ireland. It argued that the network operators to whom it sells mobile termination possess relatively great CBP.[8] Vodafone has made a related point in numerous filings that describe mobile termination as a "two-sided market."

---

[7]Directive (2002/19/EC) on Access and interconnection, Annex II.

[8]See Competition Arbitration Tribunal (CAT) Case No: 1047/3/3/04 November 2005 judgement paragraph 35 and "Assessment of whether H3G holds a position of SMP in the market for wholesale mobile voice call termination on its network" – Ofcom statement March 2007.

Exercise of CBP

In this section we address the issue of whether CBP can be an adequate substitute for regulation. In order to do so, we consider the various ways that CBP might be exercised *in the absence of regulation*. In a later section we then consider how these market dynamics are or can be affected by regulation.

CBP can be exercised by either fixed or mobile operators who buy mobile termination. The incumbent fixed operator is often subject to more regulatory constraints than are mobile operators. These constraints apply, *inter alia*, to the exercise of CBP. This distinction is not, however, relevant to the exercise of CBP *in the absence of regulation*.

Some important possibilities for exercising CBP in the absence of regulation are described in the following sections.

Failing to Reach Agreement on Terms of Interconnection

In the absence of regulation, the buyer and seller may not reach an agreement on the terms and conditions of interconnection. With no regulatory or legal recourse, the inevitable outcome would be that calls from one network to the other would be blocked. That is, subscribers of one network would not be able to call subscribers of the other network. That outcome is harmful to the subscribers of both networks.

The outcome may additionally be harmful to both network operators, because the value that they can deliver to their subscribers is much reduced. They cannot complete calls from one network to the other.

Another possibility, however, is that a large operator may reap competitive gains from non-interconnection at the expense of a small operator. The subscribers of the large operator will be able to complete most of their calls, even if there is no interconnection to the small operator. In contrast, the small operator's service may be wholly unsatisfactory if its subscribers cannot complete calls to the large operator. The large operator in this case has much greater CBP than the small operator.

In reality, regulatory intervention is quite likely to occur if the operators fail to interconnect. Such intervention can ameliorate the harms to subscribers deriving from lack of interconnection. It can also prevent the lessening of competition where a large operator does not interconnect with a small one. In Europe and in many other countries, operators are required to interconnect. Regulatory intervention can therefore be expected in the event that networks do not interconnect.

Regulatory intervention is discussed in a later section of this paper. The prospect of regulatory intervention obviously affects all aspects of the bargaining between the two operators.

Raising Termination Rates on the Buyer's Network

Another way for a buyer of mobile termination to exercise CBP is to threaten to raise its own mobile termination rates. One possibility is for MNOs simply to insist on symmetrical rates with each other. Much more onerous threats, however, are also possible. For example, a buyer could threaten to charge extremely high rates for mobile termination on its own network unless the (other) MNO agrees to a sufficiently low rate for mobile termination.

In order for this exercise of CBP to be really effective, the buyer must be able to threaten to charge different termination rates to different network operators. Such discrimination is, however, limited by the ability of the buyer to route its calls through a third network. Rerouting does, however, involve additional costs, including transactions costs. Thus, the CBP from threatening to raise termination rates for a particular operator could be significant, albeit not devastating.

Raising Retail Prices

Another way for a buyer to exercise CBP is to raise retail prices for calls to another operator's network. It would, of course, not be unexpected for an operator to pass on mobile-termination charges to its subscribers. Much more onerous threats are, however, also possible. An operator could threaten to charge extremely high retail prices for calls to a particular network, unless that network's charge for mobile termination is sufficiently low.

In order for this exercise of CBP to be really effective, the buyer must be able to charge different retail prices for calls to different networks. The price would depend on the identity of the terminating operator – not simply on whether the call is off-net. Prices would differ, even though unit costs are virtually the same.

Withholding Payment for Mobile Termination

Finally, the buyer may exercise CBP by withholding payment for mobile termination. Withholding payment does not fit into the neat confines of economic theory, where one generally assumes that buyers pay for the goods and services that they consume. Nevertheless, this tactic has frequently been used by buyers of interconnection (and access) services.

In the broader economy, apart from telecommunications, a seller generally (eventually) responds to non-payment by discontinuing supply of the product (in addition to pursuing legal remedies). In telecommunications, discontinuing supply is the same as not interconnecting, as discussed above. A small operator may find it impractical to decline to interconnect with a large network, because it would then be offering wholly unsatisfactory service to its subscribers.

In addition, sellers of mobile termination can and do seek legal and/or regulatory redress. Doing so, however, generally involves substantial legal costs and long delays before the money is actually collected.

Regulatory Intervention

It will not have escaped the reader that the tactics described in the previous section for exercising CBP, especially the most severe tactics, are rarely seen in the real world. The reason is not, of course, that the operators are too stupid to think of the tactics or too kind and gentle to exercise them. Rather, the reason is that regulators substantially restrict the exercise of CBP, especially by the incumbent fixed operator.

We have already mentioned that regulatory mandates to interconnect prevent any exercise of CBP that leads to non-interconnection. At the same time, anti-discrimination rules prevent the buyer's exercise of CBP through:

- *Raising its own mobile termination rate* for calls from a particular network whose mobile termination rate is too high or
- *Raising retail prices* for calls to a particular network whose mobile termination rate is too high

Regulators also typically exert pressure on operators to make payments for mobile termination (though the payments may nevertheless be received after long delays).

For these reasons, regulators in many countries, especially those in Europe, have argued that MNOs satisfy the two criteria for SMP in the market for mobile termination on their network – notwithstanding CBP. The reason, they argue, is that buyers are not permitted to exercise their CBP. CBP cannot act as a check on the terminating-access monopoly unless the buyer is permitted to exercise it. It follows that regulation is needed to prevent abuse of the terminating-access monopoly.

Nevertheless, that argument does raise the issue of whether relaxing regulation with respect to setting mobile termination rates would be efficacious, if combined with relaxing regulation with respect to the exercise of CBP. This issue is addressed in the next section.

Relaxing of Regulation

Let us suppose, for purposes of argument, that the buyers of mobile termination do, indeed, have substantial CBP. Let us further suppose that regulators (counterfactually) give the buyers full scope to exercise their CBP. What would be the market outcomes?

The outcomes depend on the ways in which CBP is exercised, as previously discussed.

Failing to Reach Agreement on Terms of Interconnection

We have previously noted that subscribers of both networks are harmed if network operators do not reach agreement on the terms of interconnection and decline to interconnect. Those harms can be quite serious, because telecommunications is such an important part of the modern economy. Telecommunications is additionally relied upon for use in emergencies. Furthermore, as previously discussed, anti-competitive consequences can ensue if a large network operator declines to interconnect with a small one.

For these reasons, it is almost surely *not* in the public interest to allow buyers of interconnection services (including mobile termination) to exercise their CBP in any way that leads to networks not being interconnected.

Raising Termination Rates on the Buyer's Network

As previously noted, anti-discrimination rules generally prevent a network operator from charging different termination rates to different operators. But what if those rules were relaxed?

Truly excessive termination rates charged to particular operators have much the same consequences as not interconnecting at all. The inevitable outcome would be very little communication between the two networks.

Another discomfiting possibility is that two MNOs will reach an agreement whereby both charge high prices for call termination. Such an agreement can benefit both MNOs, allowing them to earn supra-competitive profits as shown below, even though retail markets are quite competitive.

The high termination rates will presumably be flowed through to subscribers. Each network will then have high charges, well above cost, for off-net calls. Under this regime, a subscriber will tend to choose a network on which he/she has a large community of interest in order to have a high percentage of on-net calls. The communities of interest give each MNO some degree of market power. That is, a subscriber will be reluctant to change operators even if another operator has somewhat lower prices. The lower price schedule will be balanced to a significant extent by the higher percentage of off-net calls, because the subscriber has less community of interest on the other network.

One might have presumed that buyers would use their CBP to lower the prices that they pay. In the context of mobile telecommunications, the presumed result would then be lower call prices for subscribers, as the reduction in mobile-termination rates is flowed through.

In the above example, however, the outcome of the bilateral monopoly may be that prices are *raised*, to the detriment of consumers.

Raising Retail Prices

Raising termination rates on the buyer's network affects consumers, as the higher rates are flowed through in the form of higher call charges.

The effects on consumers are precisely the same if the operators agree jointly to raise their retail prices; i.e., price fixing. The effects, if this practice is permitted, are the same as discussed in the preceding sub-section.

Withholding Payment for Mobile Termination

Markets work properly only when buyers pay for the goods and services that they purchase. Regulators cannot reasonably encourage or enable buyers to withhold payment for mobile termination.

Summary

The combination of not regulating the price of mobile termination and allowing buyers full scope to exercise their CBP does not work well. Existing policies, under which mobile-termination rates are regulated and the exercise of CBP is restricted, lead to better outcomes.

The foregoing discussion applies to a CPP regime. Under the alternative regime of MPP, mobile termination charges are small, possibly zero. Given that the market for retail mobile services is effectively competitive, which is the case in most countries, a MNO's charges its retail subscribers for incoming calls need not be regulated. For these reasons, the scope of regulation can be much narrower under MPP than under CPP without untoward consequences. Littlechild has suggested that a narrower scope for regulation may be an important advantage of MPP (Littlechild 2006).

Conclusions

The EU policy is to regulate telecommunications charges if the network operator that sets the charge satisfies two criteria for non-transitory SMP that cannot be ameliorated through enforcement of competition law. The relevant product market for mobile termination is specified to be that of a particular mobile network operator (MNO). That operator has a monopoly in the relevant market – the terminating access monopoly. Consequently, mobile-termination rates are generally regulated, in Europe and elsewhere, to be cost-oriented.

That policy has recently been challenged on the basis that buyers of mobile termination may have significant countervailing buyer power (CBP). In principle, CBP could be exercised in any or all of the following ways:

- Failing to reach agreement on terms of interconnection
- Raising termination rates on the buyer's network
- Raising retail prices
- Withholding payment for mobile termination

In practice, the exercise of CBP is restricted by regulatory mandates to interconnect, non-discrimination rules, and enforcement of the obligation to make payments for mobile termination. Since buyers are not permitted to exercise their CBP, it cannot ameliorate the terminating access monopoly.

Our analysis shows that unrestrained exercise of CBP would likely harm consumers. More generally, the bilateral monopoly between buyers and sellers of mobile termination would, if unregulated, lead to perverse results for consumers. We conclude that existing policies of regulating mobile-termination rates and restraining the exercise of CBP are beneficial. The alternative of deregulating mobile-termination rates and allowing full scope for exercise of CBP would be much worse for consumers.

Our finding is that in a CPP regime, mobile-termination rates should be regulated, regardless of CBP. This finding applies generally to interconnection prices wherever at least one operator has a terminating access monopoly. (It does not, however, necessarily apply to the pricing of unbundled network elements.)

References

Binmore K, Harboard D (September 2005) Bargaining over Fixed-to-Mobile Termination Rates: Countervailing Buyer Power as a Constraint on Monopoly Power. Journal of Competition Law and Economics 1 (3), 449–472.

Commission of the European Communities (2006) "Commission Staff Working Document, Public Consultation on a Draft Commission Recommendation, on Relevant Product and Service Markets within the electronic communications sector susceptible to ex ante regulation in accordance with Directive 2002/21/ED of the European Parliament and the Council on a common regulatory framework for electronic communication networks and services," second edition, Brussels, 28 June 2006, SEC(2006)837.

Competition Commission (2003) "Vodafone, O2, Orange and T-Mobile: Reports on references under section 13 of the Telecommunication Act 1984 on the charges made by Vodafone, O2, Orange and T-Mobile for termination calls from fixed and mobile networks".

Hausman J, Wright J (June 2006) Two Sided Markets with Substitution: Mobile Termination Revisited.

Laffont J J, Tirole J (2001) Competition in Telecommunications. Cambridge, MA: MIT Press.

Littlechild S C (2006) Mobile Termination Charges: Calling Party Pays vs Receiving Party Pays. Telecommunications Policy 30(5), 242–277.

Office of Fair Trading [of the U.K.] document OFT403 (December 2004) Market Definition, Understanding competition law.

Rohlfs J H (2006) Bandwagon Effects in Telecommunications (pp. 79–115). In: Sumit Kumar Majumdar, Martin Cave, Ingo Vogelsang (eds.), Handbook of Telecommunications Economics, Volume 2. Elsevier.

Thomson H, Renard O, Wright J (2006) Mobile Termination (pp. 277–302). In: Dewenter R, Haucap J (eds.), Access Pricing: Theory and Practice. Elsevier.

U.S. Department of Justice and Federal Trade Commission, Horizontal Merger Guidelines (issued April 2, 1992, revised April 8, 1997).

National Roaming Pricing in Mobile Networks*

Jonathan Sandbach

Abstract This paper develops a practical model of optimal and competitive neutral national roaming access prices. This method takes account of the geographical cost structure of networks, and thus allows for the "cream-skimming" effect whereby a new entrant will concentrate its own network build in low cost (higher traffic density) urban areas, especially when its uses a technology that has a cost advantage in these areas. Both incumbent and new entrant networks will invest in more geographic coverage when the national roaming access price is set higher – the incumbent will do so because of the extra revenue it will get from roaming charges, and the new entrant will do so in order to avoid paying roaming charges to the incumbent. The paper provides an illustration of how the method could be applied to a situation where the host incumbent is restricted to GSM 900 against a new entrant deploying WCDMA 2.1 GHz. Under realistic assumptions we have calculated that a competitively neutral national roaming access price will be about 38% above the average cost on the host incumbent's network, although this result will depend on the specific distributions of traffic against geography in the country concerned. An access price set at this level will ensure competitive neutrality between networks, and provide efficient investment signal for the new entrant network.

Introduction

National roaming enables a new mobile entrant to compete against existing incumbent mobile networks in the absence of full national coverage by its own infrastructure. A common scenario is where the new entrant provides its own network infrastructure

J. Sandbach
Head of Regulatory Economics, Vodafone Group, Vodafone House,
The Connection, Newbury, Berkshire, RG14 2FN, UK
e-mail: jonathan.sandbach@vodafone.com

*The views expressed in this paper are those of the author, and should not necessarily be attributed to Vodafone.

B. Preissl et al. (eds.), *Telecommunication Markets*, Contributions to Economics,
DOI: 10.1007/978-3-7908-2082-9_15,

only in urban areas, and relies on roaming onto an incumbent's network in the rest of the country. New 3G network operators invariably require national roaming to provide full national coverage, because of the high cost of building a WCDMA network at 2.1 GHz in rural areas.

What wholesale price should the new entrant pay in any commercially negotiated national roaming deal, or what rate should be set under regulatory oversight?[1] There are a number of possible answers to this question.

Firstly, a national roaming access price could equal the long run incremental cost of the new entrant's traffic on the host incumbent's network. An efficient economic price needs to cover the long run incremental cost imposed on the host network (including the full economic cost of additional capacity). A price below this will mean that the new entrant could provide services at a price to the end user below the true economic cost of the resources it uses – a level against which the host incumbent network could not compete. However, even long run incremental cost of traffic (that includes the costs of replacing retiring capacity) will exclude costs that are incurred purely for network coverage (i.e. are not incremental to traffic in either the short or long term).[2] These fixed costs of building and operating the network in the geographic regions where the new entrant used national roaming would fall on the host incumbent. If we reasonably presume that the new entrant would only require national roaming in rural areas, where coverage costs are high compared to capacity costs for traffic, these fixed costs would be large, and the new entrant would be at a cost advantage over the incumbent.

Secondly, a national roaming access price could equal the average cost of traffic on the host incumbent's network. However, where the new entrant focuses its own network build on urban areas, the new entrant would still be at a cost advantage, since it could combine the low costs of self-building urban areas with a national roaming access price based on national average costs in rural areas. Effectively the host incumbent would be subsidising the new entrant's traffic in rural areas. The incumbent network could fall victim to "cream skimming" by the new entrant, with the result that competition will be distorted and ultimately the dynamic efficiency of the competitive market will be damaged.

Thirdly, a national roaming access price could leave the incumbent's profits unaltered. This is the so-called Efficient Component Pricing Rule (ECPR) originally proposed by Baumol (Baumol 1983; Baumol and Sidak 1994). The rationale for this rule is that the new entrant should only be successful in the market to the extent that it is equally or more efficient as the incumbent. A less efficient new entrant (supported by a low national roaming access price) would be detrimental to the overall economic efficiency of the industry and, ultimately, would be detrimental to

---

[1] National roaming access prices are normally commercially negotiated, since the new entrant will have a competitive choice of host networks (Ofcom (2004), Paragraphs 3.12 and A.4).

[2] These costs will include those of establishing a minimum number of base station sites to provide coverage, but exclude equipment costs that depend on the quantity of traffic (e.g. transceivers) or additional sites built purely for traffic capacity reasons.

consumer interests. The ECPR leads to the conclusion that the national roaming access price should reflect the incremental cost of the roaming traffic on the host incumbent's network, plus the lost profit margin from the incumbent not supplying this traffic itself at a retail level.[3] This leads to a national roaming rate set equal to the incumbent's retail price less avoided retail costs ("retail-minus"). The principal objection to the ECPR is that it takes the existing retail prices as already efficient or competitive, and denies the possibility that there is scope for an efficient new entrant (albeit one that required national roaming) to provide additional competitive pressure to lower prices further (Economides and White 1995). Thus, national roaming prices set on this basis would be incompatible with the objective of enhancing competition through a new entrant.

Lastly, a national roaming access price could allow both networks to make the same level of expected profit given equal retail market shares – the competitive equality criterion. We concentrate on this criterion since it is the only one that is consistent with the objective of enhancing efficient competition through new entry. Notice that we are not proposing to equate actual profits, or even expected profits under unequal market shares; but rather we equate expected profit under equal market shares, calculated from an efficient cost base subject to the different technology available to each network. It is this criterion that provides incentive for economic efficiency and market growth on the part of both networks.

Achieving the correct rate for the national roaming access price becomes particularly important if the new entrant operates a WCDMA 2.1 GHz network, and has no technological flexibility (perhaps because of its license).[4] Other things being equal, spectral efficiencies should provide 3G operators with greater traffic capacity, and so in areas where the network is dimensioned for capacity (rather than coverage), as will often be the case in urban areas, unit costs can be expected to be lower than for a GSM 900 MHz network carrying the same amount of traffic and with the same absolute amount of allocated spectrum. However, the situation reverses in rural areas, where the network needs to be dimensioned for coverage rather than capacity. Here the cost advantage will lie with GSM 900, rather than WCDMA at 2.1 GHz.

Therefore, the national roaming access price that seeks to preserve a competitive neutral market (to maximise the dynamic efficiency between competitors) will need to allow for:

- The higher costs that the incumbent faces in rural areas, where the demand for national roaming will be greatest
- The cost disadvantage it may face in urban areas, if it is restricted to GSM 900 technology

---

[3]This is similar to a "retail minus" rule as proposed by Ofcom (2004, Paragraphs A.9 and A.10).

[4]Throughout this paper we assume that neither network has flexibility in the technology it deploys (or at least the spectrum it has been assigned). For example, if the new entrant has flexibility it should be able to achieve a cost base at least as low as the incumbent (for the same volume of traffic). The exposure of the incumbent to cream skimming under a national roaming agreement would then be even higher. Likewise, if the incumbent has access to higher frequencies it will be advantaged.

It follows, therefore, that in determining a national roaming access price (by either commercial negotiation or regulatory over-sight) the geographical cost structure of both the incumbent and the new entrant network will be relevant. The time profile of national roaming prices is also relevant. As the new entrant network expands its network coverage, a greater proportion of the roaming traffic is in more remote areas, with progressively higher unit cost. Setting a time profile for roaming according to an ex-ante anticipated efficient network build will provide the correct incentives for the new entrant to complete a geographical network build consistent with efficient investment.

There is a very large body of on-going research dealing with "horizontal" two-way interconnection between competing (mobile) networks, starting with Armstrong (1998) and Laffont et al. (1998). This literature looks at mobile networks that have the same geographical coverage and so only rely on each other to terminate calls to subscribers on the competing network (through interconnection). This paper, however, deals with "vertical" one-way interconnection – a smaller network needing to access a larger established network to originate calls in areas where it has no coverage of its own. This situation is analogous to competition in the fixed telecommunications sector, where a new entrant needs to purchase network access from an incumbent network because it has no local customer access network of its own. Such situations are fully discussed in Dewenter and Haucap (2007).

The model in this paper leads to a positive relationship between the national roaming access price and the level of investment by both the host incumbent and new entrant network: the incumbent because a higher access price will lead to greater wholesale revenue, and the new entrant because of the greater incentives to build its own network.[5] This is an interesting conclusion that should be considered alongside other contributions on this subject in the telecommunications sector. For example, Cave and Vogelsang (2003) argue for initially low access prices to encourage new entrants, but possibly increasing over time to incentivise the new entrant to build its own infrastructure.

The situation analysed in this paper has similarity to that in Foros et al. (2002). Foros, Hansen and Sand model first two equally placed facilities providers who are able, to some extent, to roam onto each other networks, and second introduce a virtual network with no infrastructure of its own. Our case is somewhere between the two, where a new entrant has some network, but not complete coverage. Foros, Hansen and Sand also concentrate on the consequences of voluntary and mandated roaming under cooperative or non-cooperative investment decisions by the network operators.

Section "Competitive Neutral National Roaming Rate" of this paper develops a formal model which is used to specify a competitive neutral national roaming access price. Section "The National Roaming Access Prices and Incentives to Invest" develops this model further to investigate the impact of the access price on incentives for

[5] Note, of course, that this is not to say that higher access prices are always a good thing. Although higher access prices may increase aggregate industry investment, it may also lead to an increase in retail prices and a reduction in consumer welfare. Although a full analysis of the consumer welfare implications are beyond the scope of this paper, we do refer to this point at a later stage.

network investment, and how the access price should be set to give the correct signal for efficient investment by the new entrant. Section "Conclusion" presents conclusions.

Competitive Neutral National Roaming Rate

In this section we introduce the basic model in which we determine a competitive neutral national roaming access price, and illustrate this model through a calibration.

Model

Assume that the costs of building a mobile network,*i*, can be expressed as the sum of coverage and capacity costs:

$$C_i = v_i V_i + m_i (q + r) \tag{2.1}$$

Where:

V_i is geographical coverage (in km²) of network *i*;

v_i is the coverage cost (per km²) of network *i* (see footnote 4), dependent on the network technology (e.g. GSM 900 or WCDMA 2.1 GHz);

q is the volume of airtime minutes originated or terminated by subscribers to a network;

r is the volume of national roaming airtime minutes originated or terminated by subscribers;

m_i is the long run marginal cost of a minute of airtime on network *i*, dependent on the network technology (e.g. GSM 900 or WCDMA 2.1 GHz).

We will assume that there are two networks: an incumbent operating a GSM 900 network which we take to be network *i*, and a new entrant operating a WCDMA 2.1 GHz network which we take to be network *j*. The characteristics of the respective technologies are such that:

$$v_i < v_j \tag{2.2a}$$

$$m_i > m_j \tag{2.2b}$$

It will evidently be the case that network *i* will have the greater own-network geographical coverage of the two networks. We suppose, however, that there will be a national roaming agreement that will provide network *j* with the same geographical coverage. We suppose, therefore, that there will be no quality or consumer preference differences between the two networks,[6] and that in competitive equilibrium both networks offer the same market price, *p*, and have potential to win the same

[6] We do not model any advantage that the WCDMA network may have in offering 3G services.

volume of own subscriber traffic, q.[7] The main question of this paper is to determine the national roaming access price, a, paid by network j to network i that ensures that this competitive equilibrium will be achieved.

Reflecting the national roaming relationship between the two networks, we can write the profit functions as:

$$\pi_i = pq - v_i V_i - m_i\,(q + r) + ar \tag{2.3a}$$

$$\pi_j = pq - v_j V_j - m_j\,(q - r) - ar \tag{2.3b}$$

where p is the retail price, applying to both the incumbent and new entrant (since they compete). Initially we will treat p as exogenous. This allows us to solve for the optimum level of investment (which depends on p). However, we will later also consider the case where investment by the host incumbent is fixed (to provide full-coverage), and here we will calculate p at a level that results in zero economic profit for both networks – assuming that excess profit is competed away, and the national roaming access price is set at a level that ensures competitive neutrality.

We now need to specify the relationship between V_i and q:

$$q = Q\varphi(V_i) \tag{2.4}$$

where Q is the traffic volume once full national coverage has been achieved, thus $0 \le \varphi\,(V_i) \le 1$ with $\varphi\,(0) = 0$. We can further restrict $\varphi'\,(V_i) \ge 0$ and $\varphi''(V_i) \le 0$, since we would expect the network to spread out from the most traffic rich areas. Figure 1 shows this relationship for one of Vodafone's developed country networks by the thick dark line (the thin line marked "gamma = 0.3" is a fitted line for a particular functional form that will be introduced in the next section).

We can now determine the volume of airtime on network j as:

$$q - r = Q\varphi(V_j) \tag{2.5a}$$

And the volume of national roaming traffic as:

$$r = Q\,[\varphi\,(Vi) - \varphi\,(V_j)] \tag{2.5b}$$

Substituting into the profit functions of Equations (2.3) gives:

$$\pi_i = \{(p - m_i)\,\varphi\,(V_i) + (a - m_i)\,[\varphi\,(V_i) - \varphi\,(V_j)]\}Q - v_i V_i \tag{2.6a}$$

$$\pi_i = \{(p - m_j)\,\varphi\,(V_j) + (p - a)\,[\varphi\,(V_i) - \varphi\,(V_j)]\}Q - v_j V_j \tag{2.6b}$$

We see that the profit of the host incumbent is the sum of (1) the margin it makes on its own retail calls: $(p - m_i)\varphi(V_i)Q$, (2) the margin it makes on roaming traffic from the competitor: $(a - m_i)[\varphi\,(V_i) - \varphi\,(V_j)]Q$, less (3) its fixed cost of network coverage: $v_i V_i$. Similarly, the profit made by the new entrant is the sum of (1) the margin it makes from calls on its own network: $(p - m_j)\varphi(V_j)Q$, (2) the margin it

[7] We do not deny that in the short term the new entrant will have a lower traffic share, but it is the long term competitive equilibrium for efficiently operated networks that needs to be considered for the competitive neutral national roaming access price.

makes on roaming traffic on the host network: $(p - a)[\varphi(V_i) - \varphi(V_j)]Q$, less (3) its fixed cost of network coverage: $v_i V_i$.

We now consider three variants of a model:

- *Variant 0*: Coverage of both networks is exogenous. The host incumbent provides full national coverage and the new entrant provides only limited coverage.
- *Variant 1*: Coverage of both networks is endogenous, i.e. optimised to maximise profits given respective cost functions. We would expect the incumbent to have a near national coverage network, and the new entrant a limited coverage network, depending on costs and the national roaming access price.
- *Variant 2*: Coverage of the incumbent network is exogenous (full national coverage), and coverage of the new entrant network is endogenous, i.e. optimised to maximise profits given its cost functions and the national roaming access price.

Variant 0: Exogenous Coverage of Both Networks

This variant proceeds on the basis that network i has full national coverage of the land mass (so that $\varphi(V_i) = \varphi(A) = 1$) and network j has some lesser coverage, but nevertheless taken as being fixed.

In this case it is simple to calculate the competitive neutral national roaming access price by setting $\pi_i = \pi_j$ and solving:

$$a = m_i + \frac{v_i A - v_j V_j + (m_i - m_j)\,\varphi(V_j)Q}{2\left[1 - \varphi(V_j)\right]Q} \tag{2.7}$$

This result simply says that the national roaming access price should be set equal to the marginal cost of traffic on the incumbent host network plus a term equal to half of:

- The cost difference between the two networks in respect to the overall coverage costs (taking account of the larger coverage of the host incumbent network)
- The capacity costs of the two networks in respect to that proportion of the traffic carried on the new entrant's own network
- With both the above spread over the volume of roaming traffic

This makes intuitive sense. The host incumbent network is compensated for the incremental costs caused by the new entrant's roaming traffic, and in addition there is an adjustment for the intrinsic cost advantage or disadvantages between the networks over the geographical area in which the networks overlap, spread over the volume of roaming traffic. Thus the competitive neutral national roaming access price is:

- Increased (reduced) if the incumbent host network has larger (smaller) marginal costs of traffic
- Increased if it provides greater network coverage (which is made available to the new entrant)
- Increased if the new entrant has access to a technology with lower capacity costs within the coverage of its own network

- Increased by the fact that the new entrant will only require roaming over a portion of the host incumbent's network where traffic is lower relative to coverage costs

In the extreme case, where the new entrant has no own-network, the competitively neutral national roaming access price is simply the marginal cost on the host incumbent's network, plus a half share of its coverage costs spread over all the new entrant's (roaming) traffic.

Note that, unlike the ECPR, the competitive neutral national roaming access price does not depend on the retail price, and thus avoids the criticisms of the ECPR (e.g. perpetuating pre-entry retail prices that may not be at competitive levels).

Calibration

We now illustrate how this calculation may look in practise. We first need to quantify the relationship between traffic and coverage. This can readily be done by the incumbent network, as shown by the example in Fig. 1. We parameterise this by a particular functional form:

$$\varphi(V) = \frac{(V / A)[1+\gamma-(V / A)^{\gamma}]}{\gamma} \tag{2.8}$$

where A is the total land mass of the country in km². This parameterisation gives a very close fit to the actual data from Fig. 1 when $\gamma = 0.3$. This is typical of many developed networks, whereby 50% land mass coverage allows 81% of the traffic to be captured, and 90% land mass coverage allows over 99% of traffic to be captured.

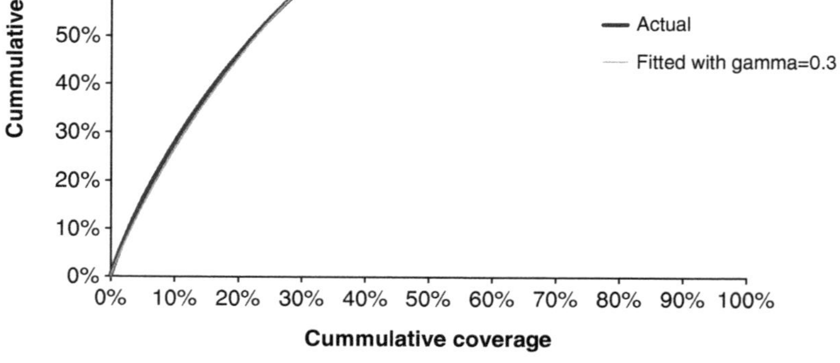

Fig. 1 $\varphi(V)$ in developed country

We next need cost functions for the two networks. Table 1 shows an analysis of the costs of an omni-sector base station in the UK, which can be taken as being indicative of the non-traffic costs of coverage.[8]

We also require estimates of the marginal cost of traffic. We assume a cost of 4 ppm (price per minute) for a GSM 900 network (noting that the OFCOM model estimates 4.9 ppm for a 2G network, including coverage costs). Costs will be lower for a WCDMA 2.1 GHz network due to spectral efficiencies. We can estimate the WCDMA 2.1 GHz costs to be 2.8 ppm.[9]

Figure 2 shows the resulting competitive neutral national roaming access price. At low levels of new entrant coverage this reflects the geographically averaged costs on the host incumbent network (a marginal cost of 4 ppm, plus coverage cost, giving 4.25 ppm). As the new entrant expands its own network into initially high traffic density urban areas, and restricts its national roaming requirements to lower traffic density rural areas, the national roaming access price rises accordingly. It continues to rise until coverage of about 75% is achieved. Beyond this point WCDMA 2.1 GHz coverage becomes uncompetitive and, if this level of coverage

Table 1 Illustrative coverage costs (From OFCOM model and Vodafone analysis)

	Investment	Asset life (years)	WACC	Annuitised investment cost	Opex	Total cost
Site acquisition, preparation and civil works	£80,526	20	15%	£12,865	£9,018	£21,883
Equipment (omni-sector)	£61,369	10	15%	£12,228	£6,692	£18,920
Total	£141,895			£25,093	£15,710	£40,803
	Cell coverage		Cost/km^2			
GSM 900	51.0 km^2		£800			
WCDMA 2.1 GHz	13.8 km^2		£3,200			

[8] Although multi-sector base stations (usually three-sector) are more common, an omni-sector base station gives a better indication of the underlying coverage costs, excluding any costs of traffic capacity.

[9] WCDMA provides more airtime capacity that GSM on each cell site. WCDMA transceivers use 5 MHz of spectrum, but allow around 60 voice channels on each transceiver, compared to only 8 channels on 200 kHz of spectrum for a GSM network (assuming full rate voice codec). More importantly, WCDMA allows significantly more efficient use of the spectrum, effectively providing re-use of spectrum in neighbouring sectors, compared to an average spectrum re-use factor of around 12 under GSM networks. Therefore, in rough terms, WCDMA allows for approximately 12 channels/MHz, compared to only 3.3 for GSM. Therefore, the incremental cost of capacity at a WCDMA cell site is lower by a factor of about 3.6. In practise the difference is not so pronounced if half rate voice codec is used within the GSM network in order to better utilise capacity. In some situations half rate voice codec can be used for up to about 40% of call volumes without seriously compromising voice quality, thus increasing capacity within the GSM network by a factor of 1.4. In conclusion, therefore, the capacity difference between a GSM and WCDMA network is reduced from 3.6 to 2.6, which can be considered as translating into a marginal cost reduction of about 62% at the air interface. However, the air interface related costs account for only 50% of the total marginal costs (others being backhaul and core network), and so the actual WCDMA cost saving is more likely to equal about 31%.

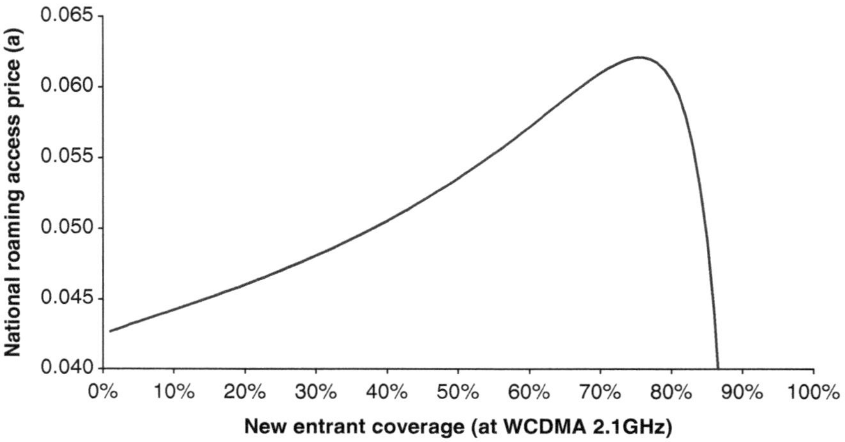

Fig. 2 Competitive neutral national roaming access price

were to be provided the new entrant would require a subsidy which, in the model, is reflected in a reduced national roaming rate for the remaining area it does not cover. Since the amount of roaming traffic becomes very small at these high levels of coverage, the WCDMA 2.1 GHz subsidy becomes very large when expressed in terms of a national roaming access price.[10]

The National Roaming Access Prices and Incentives to Invest

The methodology of the previous section points to the possibility of applying a national roaming access price conditional on the network coverage achieved by the new entrant. As the new entrant expands network coverage in low cost urban areas, and restricts its roaming requirements to high cost rural areas, the roaming rate will rise to preserve the competitive neutrality of the market, effectively neutralising any "cream skimming" by the new entrant.

A drawback with this approach is that a national roaming access price that rises with the new entrant's network coverage may disincentivise the new entrant to invest. The way to avoid this disincentive is to set the access price, not conditional on actual network coverage, but on the new entrant's "optimal" coverage, or a time path leading to an optimal network coverage. In the event that the new entrant fails to achieve this level of network build, it will still be required to pay the roaming rate that would apply if it did. This will provide incentive for the new entrant to achieve the optimal level of network build in order to achieve a competitive neutral

[10] This is not a realistic scheme for subsidising WCDMA 2.1 GHz coverage.

national roaming access price, and will not reward an inefficient new entrant (that under-builds) with a lower access price.

We develop two Variants of the model. In the first we endogenise the network coverage of both the host incumbent and the new entrant. In the second we assume that the host incumbent has full national coverage and endogenise only the network coverage of the new entrant.

Variant 1: Endogenous Coverage in Both Networks

The first variation will proceed on the basis that neither network has complete national coverage (although network *i* will have greater coverage than network *j*). We need to consider the optimal (or profit maximising) build of both networks, which will depend on *p* and *a*.

Both networks will seek a level of network coverage (V_i and V_j) that will maximise profits. First order conditions with respect to V_i and V_j, valid whenever $a > 2m_i - p$ and $a > m_j$, give[11]:

$$\varphi'(V_i) = \frac{v_i}{(p + a - 2m_i)Q} \tag{3.1a}$$

$$\varphi'(V_j) = \frac{v_j}{(a - m_j)Q} \tag{3.1b}$$

Taking the specific functional form in Equation (2.8) we have:

$$\varphi'(V) = \frac{(1 + 1/\gamma)[1 - (V/A)^{\gamma}]}{A} \tag{3.2}$$

And so:

$$V_i = A\left[1 - \frac{\gamma}{1+\gamma}\frac{v_i}{(p + a - 2m_i)Q}\frac{A}{}\right]^{\frac{1}{\gamma}} \tag{3.3a}$$

[11] Second order conditions are easily checked:

$$\frac{\partial^2 \pi_i}{\partial V_i^2} = (p - 2m_i + a)\varphi''(V_i)Q < 0 \text{ if } a > 2m_i - p$$

$$\frac{\partial^2 \pi_j}{\partial V_j^2} = (a - m_j)\varphi''(V_i)Q < 0 \text{ if } a > m_j$$

Both these conditions will be fulfilled if *a* and *p* at least exceed marginal cost on both networks (which we would expect).

$$V_j = A\left[1 - \frac{\gamma}{1+\gamma}\frac{v_j}{(a-m_j)Q}\frac{A}{}\right]^{\frac{1}{\gamma}} \tag{3.3b}$$

The interesting conclusion of Equations (3.3) is that both networks will invest in more geographic coverage when the national roaming access price is set higher – the incumbent will do so because of the extra revenue it will get from roaming charges, and the new entrant will do so in order to avoid paying roaming charges to the incumbent.

It is also interesting that whilst the incumbent decides the extent of investment in geographical coverage on the basis of retail prices (since a higher retail price makes investment in marginal areas more profitable), the new entrant is not directly concerned with the price. This is because the new entrant's geographic coverage at a retail level is determined by that of the incumbent (through national roaming), not its own investment. Rather, the new entrant's network investment objective will be solely cost minimisation – a "make-or-buy" decision.

Figure 3 shows the optimal network coverage of both networks as a function of the national roaming access price.

We are interested in the level of a that will ensure the competitive neutrality between the two network, i.e. $\pi_i = \pi_j$, assuming that both networks invest optimally (efficiently) in geographic network roll-out. Solving this problem with the assumptions stated in Table 1 gives $a = 5.3$ ppm with coverage of $V_i = 89\%$ and $V_j = 52\%$ by the host incumbent and new entrant networks respectively.

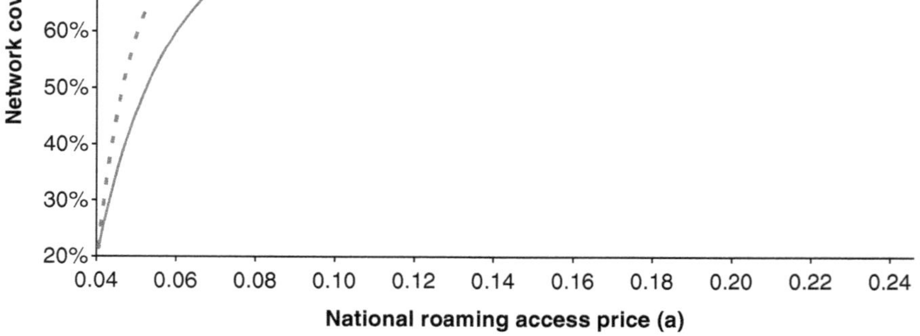

Fig. 3 Network coverage and national roaming rate

Variant 2: Complete Geographical Coverage

We now look at the variant where the incumbent has built a complete national coverage network, i.e. $\varphi(A) = 1$. This allows us to relax our assumption of a fixed value of p (which in the previous case determined the incumbent's geographic network roll-out). We can now assume that, in a mature market, super-normal profit will be eliminated by competition, and so that p will be set such that $\pi_i = 0$. In this way we recognise that, rather than being fixed, p will depend on a. In particular, a higher value of a will directly generate more national roaming revenue for the incumbent, but also cause the new entrant to extend its own geographic network coverage, thus off-setting the incumbent's roaming revenues. Both these factors will affect p.

The incumbent's profit equation from Equation (2.6a) becomes:

$$0 = [p - 2m_i + a - (a - m_i)\varphi(V_j)]Q - v_i A \tag{3.4}$$

So that we solve:

$$p = \frac{v_i A}{Q} + 2m_i - a + (a - m_i)\varphi(V_j) \tag{3.5}$$

Substituting this into Equation (3.4), the new entrant's profit function (as a price taker) becomes:

$$\pi_j = v_i A - v_j V_j - 2aQ + (2a - m_i - m_j)\varphi(V_j)Q \tag{3.6}$$

The first order condition with respect to V_j, valid whenever $2a > m_i + m_j$, is[12]:

$$\varphi'(V_j) = \frac{v_j}{(2a - m_i - m_j)Q} \tag{3.7}$$

We see that the new entrant's optimal coverage under the Variant 2 model is positively dependent on two margins: the margin between the national roaming access price and the marginal cost of traffic on both the new entrant and the host incumbent network. The first determines the national roaming outpayment savings that the new entrant will receive by expanding its network coverage, whilst the second (which does not occur under the Variant 1 model) is the loss in national roaming profit that the incumbent passes through to an increase in the market retail price. Thus, by expanding its network, the new entrant has a detrimental effect on the incumbent, and so causes a rise in the retail price.

For this reason, new entrant network coverage under the Variant 2 model will always exceed that under the Variant 1 model, whenever the national roaming

[12] Second order conditions are easily checked:

$$\frac{\partial^2 \pi_j}{\partial V_j^2} = (2a - m_i - m_j)\,\varphi''(V_j)Q < 0 \ \text{if } 2a > m_i + m_j.$$

access price exceeds the marginal cost of traffic on the host incumbent network. Mathematically, this is seen from comparing Equations (3.1b) and (3.7) and noting that $\varphi''(V_j) < 0$ and $2a - m_i - m_j > a - m_j$ whenever $a > m_i$.

Taking the specific functional form in Equation (2.8), we have:

$$V_j = A\left[1 - \frac{\gamma}{1+\gamma}\frac{v_j}{\left(2a - m_i - m_j\right)Q}\frac{A}{}\right]^{\frac{1}{\gamma}} \tag{3.8}$$

Figure 3 shows the new entrant's optimal coverage as a function of the national roaming access price. As expected, the coverage is always higher under the Variant 2 model, since the new entrant finds that it can increase the retail price by enlarging its own network and so reducing the amount of roaming profit that the host incumbent receives.

We are primarily interested in the level of a that will ensure the competitive neutrality between the two network, i.e. $\pi_i = \pi_j = 0$, assuming that both networks invest optimally (efficiently) in geographic network roll-out. Solving this problem under the numerical assumptions of Table 1 gives a value $a = 6.2$ ppm with new entrant coverage of $V_j = 75\%$. As anticipated the Variant 2 model calculates a larger network roll-out, and a corresponding higher national roaming access price compared to the Variant 1 model, due to roaming only being required in lower traffic density areas. It should also be noted that this optimal and competitive neutral national roaming charge (at 6.2 ppm) is significantly higher than the average cost of traffic on the host incumbent network, which can be calculated under assumptions of Table 1 as 4.5 ppm excluding roaming traffic, or 4.25 ppm including roaming traffic. This difference is a result of the restricted geographical coverage of the new entrant network, which means that national roaming is required only in lower traffic density areas of the host incumbent's network.

Finally, note that the solution to the Variant 2 model ($a = 6.2$ ppm with new entrant coverage of $V_j = 75\%$) is, as we would expect, consistent with the more general solutions to the Variant 0 model (shown in Fig. 2) where new entrant coverage is taken to be exogenous. What is particularly interesting is that the optimal national roaming access price from the Variant 2 model is equal to the "maximum" price under the Variant 0 model. It can easily be verified mathematically that this must be the case for any $\varphi(V_j)$,[13] but there is also an intuition for this as follows. The new entrant builds a network to a size that maximises profits for any given national roaming access price. However, because of the competitive neutrality condition, a greater profit means a higher access price. Thus the network size that maximises the competitor's profit will coincide with the network size for which it can afford

[13] This can be done by differentiating Equation (2.7) with respect to V_j to find that the turning point of a occurs when $\varphi'(V_j^*) = \dfrac{v_j[1-\varphi(V_j^*)]}{v_i A - v_j V_j^* + (m_i - m_j)Q} = \dfrac{v_j}{(2a - m_i - m_j)Q}$ on substitution from Equation (2.7).

to make the greatest access charge outpayment and still remain in a competitive neutral position relative to the incumbent.

Consumer Welfare

A formal analysis of consumer welfare is beyond the scope of this short paper. In further work it would be interesting to look at a national roaming access price set to maximise consumer surplus or economic welfare (which will equate to the same in this model since produce surplus is a fixed margin). The issue will be whether the level of network investment resulting from the national roaming access price maximises consumer surplus. At the moment it seems there is no reason why it should, leading to a potential cost in terms of static consumer surplus, in order to achieve a dynamic consumer surplus benefit through competition from a competitively neutral access price.

Conclusion

This paper has proposed a practical model that can be used to calculate optimal and competitive neutral national roaming access prices. As with any model, we need to be cautious that results may be depend on how the model has been set-up. For example, in this paper we have made specific assumptions concerning the retail price, treating it either as exogenous, or set to reduce economic profit to zero (to capture the long term outcome of effective competition). It is always possible, that introducing an endogenous retail price, alongside an explicit consumer demand function, might modify the results.

Nevertheless, the model described in this paper should provide insight into a national roaming access price set achieve competitive neutrality between networks, whilst taking into account the interaction with the industry cost structure of geographic network roll-out by both the incumbent and new entrant. By taking account of the geographical cost structure of the networks, the model allows for the "cream-skimming" effect whereby a new entrant will concentrate its own network build in low cost (higher traffic density) urban areas, especially when its uses a technology that has a cost advantage in these areas (e.g. WCDMA 2.1 GHz).

Both networks will invest in more geographic coverage when the national roaming access price is set higher – the incumbent will do so because of the extra revenue it will get from roaming charges, and the new entrant will do so in order to avoid paying roaming charges to the incumbent.

The paper provides an illustration of how the method could be applied to a situation where the host incumbent is restricted to GSM 900 against a new entrant deploying WCDMA 2.1 GHz. Under realistic assumptions we have calculated that a competitively neutral national roaming access price will be about 38% above the average cost on the host incumbent's network, although this result will depend on the specific distribution of traffic against geography in the country concerned. An access price set at this level will ensure competitive neutrality between networks, and provide an efficient investment signal for the new entrant network.

The model could be adapted to the situation where the host incumbent also has access to WCDMA 2.1 GHz. This would require a more complex composite cost function, but with essentially the same model. Although the new entrant would no longer have a cost advantage in urban areas, it would nevertheless benefit from using the host incumbent's network in higher cost rural areas.

References

Armstrong M (1998) Network Interconnection in Telecommunications. Economic Journal, 108(448): 545–564

Baumol WJ (1983) Some Subtle Issues in Railroad Deregulation. International Journal of Transport Economics, 10(1–2): 341–355.

Baumol WJ, Sidak G (1994) Towards Competition in Local Telephony. Cambridge, MA: MIT Press

Cave M, Vogelsang I (2003) How access pricing and entry interact. Telecommunications Policy 27: 717–727

Dewenter R, Haucap J (2007) Access Pricing: An Introduction. In: Dewenter R, Haucap J (eds.) Access Pricing: Theory and Practice, Amsterdam: Elsevier

Economides N, White LJ (Fall 1995) Access and Interconnection Pricing: How Efficient Is the 'Efficient Component Pricing Rule'? Antitrust Bulletin XL(3): 557–579

Foros O, Hansen B, Sand JY (2002) Demand-Side Spillovers and Semi-collusion in the Mobile Communications Market. Journal of Industry, Competition and Trade, 2(3): 259–278

Laffont JJ, Rey P, Tirole J (1998) Network Competition: I. Overview and Nondiscriminatory Pricing. Rand Journal of Economics, 29(1): 1–37

Ofcom (2004) National roaming, a further consultation. July 2004

Can Competition Be Introduced Via the Issue of New Mobile Telephony Licences: The Experience of 3G Licensing in Europe

Peter Curwen and Jason Whalley

Abstract This chapter focuses on the aftermath of 3G licensing within Europe. More specifically, the chapter examines whether 3G licensing has brought about the enhanced competition of mobile markets that was sought at the time of licensing. After analysing 3G licensing across Europe, the chapter identifies four operators – Sonera (which is now part of TeliaSonera), Telefónica, France Télécom/Orange and Hutchison Whampoa – that used the opportunities presented by the issuing of licences to expand into new markets. The analysis draws a distinction between Hutchison Whampoa and the other operators, noting how its strategy differs and its mixed success to date. The chapter concludes by questioning the long-term viability of Hutchison Whampoa.

Introduction

Several years have now passed since the licensing of third-generation (3G) mobile telecommunication licences commenced in Europe.[1] The heady days of the German and British auctions were soon replaced by a much more sober assessment of the prospects for 3G. Subsequent auctions and beauty contests alike raised substantially less money than the German and UK auctions, and the share price of many telecommunication companies collapsed due partly to the debts that they had accumulated while acquiring the licences. Some 3G licensees responded to mounting debts and uncertainty by either returning their licences or putting their proposed networks into abeyance, while others scaled back their 3G ambitions.

Nevertheless, an increasing number of operators have succeeded in launching some kind of service, using data cards in laptops and/or handsets. Indeed, during

[1]For an overview of the 3G licensing process see, for example, Curwen (2002) or Gruber (2005).

P. Curwen (✉)
University of Strathclyde, Glasgow, Scotland, UK
e-mail: pjcurwen@hotmail.com

B. Preissl et al. (eds.), *Telecommunication Markets*, Contributions to Economics,
DOI: 10.1007/978-3-7908-2082-9_16, © Springer Physica-Verlag HD 2009

2007Q2 8 million of the 14.5 million new connections in Europe were 3G-enabled – the first time that the proportion had exceeded 50% – although this did not necessarily imply that all 3G handset owners would access high-speed services (Cellular-news 2007). It is accordingly a suitable point in time to ask whether, as one might have expected, all of the initial 3G launches have been by 2G (GSM) incumbents, and, if not, how the new entrants have fared. New entrants are interesting because the problems that they face in the mobile markets are compounded by their initial lack of a dedicated 2G network which would provide them with revenue-generating customers and, in turn, some of the substantial cash-flow needed to pay for the licences and roll out their 3G networks.

With this in mind, the remainder of this paper is structured as follows. In the first section, an overview of licensing in Europe is provided that enables 3G new entrants to be identified. In the second section, the focus shifts to describing what has happened to the 3G new entrants since they were licensed. A distinction is made here between the relatively successful Hutchison Whampoa (which trades as '3') and other generally less successful 3G new entrants. The following section discusses the impact that 3G new entrants have made on the EU mobile telecommunications landscape. Conclusions are drawn in the final section.

3G Licensing in Europe

The starting point for our analysis is Table 1. This table depicts the current (as of end-December 2007) state of 3G licensing across Europe, which is defined here as encompassing the 27 Member States of the European Union (EU), the European Economic Free Trade Area, prospective accession countries to the EU and all other countries or territories having some form of independent government within the post-Communist understanding of Europe. In total, there are nearly 50 independent entries – referred to as countries for convenience in what follows – in the table.

Drawing on Table 1, the first observation that can be made is that not every country within Europe has as yet issued 3G licences. As of end-August 2007, 41 countries had awarded 3G licences. The first 3G licence in Europe was awarded in March 1999 by Finland, which was joined at the forefront of 3G in April 1999 by the Isle of Man. A significant proportion of the listed countries awarded their 3G licenses during 2000 and 2001, in good part because of the timetable for the launch of 3G laid down by the European Union. Since 2001, 3G licensing has steadily permeated the rest of Europe and the few remaining unlicensed countries are located on the peripheries of Europe and possess small populations.

A second observation that can be made is that 23 countries opted to increase the number of companies in the mobile market by issuing more 3G licences than there were 2G incumbents at the time. Most countries increased the number of 2G mobile licences by one when choosing how many 3G licences to issue, although five countries – Austria, Germany, Italy, Luxembourg and Norway – opted to issue two additional 3G licences.

Table 1 3G licensing across Europe, 31 December 2007(From Annual reports, company websites, regulators' websites, other websites, media reports.)

Country/territory	2G licences	3G licences available	3G licences awarded	Method (BC = beauty contest)	Date	3G license winners[a]
Albania[b]	3	–	–	–	–	–
Andorra[b]	1	1	1	Allocated	Jan 2005	STA
Austria	4	4–6	6	BC + auction	Nov 2000	Hutchison 3G, max.Mobil, Mobilkom Austria, ONE, tele.ring[a], 3G Mobile[m]
Belarus[b]	4	–	–	–	–	–
Belgium	3	4	3	Auction	Feb 2001	KPN Mobile 3G, Mobistar, Proximus
Bosnia-Herz.[b]	3	–	–	–	–	–
Bulgaria	3	3	3	Tender (1) Allocated (2)	Mar 2005[e]	MobilTel, Viva Ventures, GloBul
Croatia[h]	2	3	2	Tender – allocated[d]	Oct 2004	T-Mobile, VIPnet
Croatia[b]	3	1	1	Tender - allocated	Dec 2004	Treca Sreca
Cyprus (S)	2	2	2	Allocated	Dec 2003	Investcom, CyTA
Czech Repub.	3	3	2	Auction – allocated[d]	Dec 2001[f]	EuroTel Praha, RadioMobil
Czech Repub.	3	1	1	Allocated	Feb 2005	Oskar
Denmark	4	4	4	Auction	Oct 2001	Hi3G Denmark, Orange[n], TDC, Telia Denmark
Denmark	3	1	1	Auction - allocated	Dec 2005	Sonofon[n]
Estonia	3	3	3	BC – allocated[d]	July 2003[g]	Eesti Telecom, Radiolinja, Tele2
Estonia	3	1	1	Auction	Dec 2006	Grosson/Renberg/RealGroup/ProGroup[o]
Faroe Isles[b]	2	–	–	–	–	–
Finland	4	4	4	BC + annual fee	Mar 1999	Radiolinja, Sonera, Suomen 3G[p], Telia Finland
France	3	4	2	BC + fee	July 2001	Orange, SFR
France	3	2	1	BC – allocated[d]	Sept 2002	Bouygues Télécom
France	3	1	1	Tender	Oct 2007	–
Germany	4	4–6	6	BC + auction	July 2000	E-Plus Hutchison, Group 3G, Mannesmann, MobilCom Multimedia[q], T-Mobile, Viag Interkom
Gibraltar[b]	1	–	–	–	–	–

(continued)

Table 1 (continued)

Country/territory	2G licences	3G licences available	3G licences awarded	Method (BC = beauty contest)	Date	3G license winners[a]
Greece	3	4	3	Auction - allocated	July 2001	CosmOTE, Panafon, Stet Hellas
Guernsey[b]	1	2	1	BC – allocated[d]	Mar 2003	Wave Telecom
Guernsey[b]	2	1	1	BC	Sept 2006	Guernsey Telenet
Hungary	3	4	3	Auction	Dec 2004	Pannon, T-Mobile, Vodafone
Iceland[b]	3	4	3	BC - allocated	Mar 2007	Og fjarskipti, Novator, Síminn[r]
Ireland[b]	3	4	3	BC + fee	June 2002[h]	Hutchison 3G Ireland, mmO_2, Vodafone
Ireland	3	1	1	BC + fee	Nov 2005	Smart Telecom[s]
Ireland	3	1	1	Allocated	Mar 2007	eircom (Meteor)[s]
Isle of Man[b]	1	1	1	Allocated	Apr 1999	Manx Telecom
Isle of Man[b]	1	2	2	Allocated	May 2006	Cable & Wireless, Wire9 Telecom
Italy	4	5	5	BC + auction	Nov 2000	H3G, IPSE 2000, TIM, Wind, Omnitel
Jersey[b]	1	2	1	BC – allocated[d]	Sept 2005	Jersey Telecom
Jersey[b]	2	2	2	Allocated	May 2006	Cable & Wireless, Jersey AirTel,
Latvia	2	3	2	Auction - allocated	Sept 2002	LMT, Tele2
Latvia	3	1	1	Auction	May 2005	Bité
Liechtenstein[b]	4	4	3	Allocated	July 2001[i]	mobilkom, Tele2/Tango, Viag Europlattform
Liechtenstein[b]	4	1	1	Allocated	Oct 2003[j]	Liechtenstein TeleNet
Lithuania	3	3	3	BC - allocated	Feb 2006	Bité, Omnitel, Tele2
Luxembourg	2	4	3	BC - allocated	May 2002	EPT, Orange[t], Tele2
Luxembourg	2	1	1	Tender	July 2003	LuXcommunications
Macedonia[b]	3	–	–	–	–	–
Malta	2	3	3	BC – allocated[d]	Aug 2005	Go Mobile, Vodafone, 3G Telecoms
Moldova[b]	3	1	1	Allocated	July 2006	Moldtelecom[u]
Monaco[b]	1	1	1	Allocated	June 2000	Monaco Telecom
Montenegro[b]	2	1	1	Auction	Mar 2007	m:tel
Montenegro[b]	3	2	2	Allocated	May 2007	ProMonte, T-Mobile
Netherlands	5	5	5	Auction	July 2000	3G-Blue, Dutchtone, KPN Mobile, Libertel-Vodafone, Telfort
Norway[b]	2	4	4	BC + fee	Dec 2000	Broadband Mobile, NetCom GSM, Telenor, Tele2
Norway[b]	2	2[c]	1	Auction - allocated	Sept 2003	Hi3G Access

Norway[b]	3	1[c]	1	Allocated	Dec 2007	Mobile Norway
Poland	3	4	3	Tender – allocated[d]	Dec 2000	PKT Centertel, Polkomtel, Polska Telefonica Cyfrowa
Poland	3	1	1	Tender + annual fee	May 2005	Netia (P4)
Portugal	3	4	4	BC + fee + annual fee	Dec 2000	ONI-Way[v], Optimus, Telecel, TMN
Romania	4	4	2	Tender	Nov 2004	MobiFon, Orange
Romania	4	2	2	Tender	Oct 2006	RCS&RDS, TeleMobil[w]
Serbia[b]	2	2	2	Allocated	n/a	Telekom Srbija, Telenor
Serbia[b]	2	1	1	Allocated	Nov 2006	mobilkom Austria
Slovakia	2	3	2	Auction - allocated	July 2002	EuroTel, Orange, Profinet.sk[x]
Slovakia	2	1	1	Tender	Aug 2006	Telefónica
Slovenia	3	3	1	Auction – allocated[d]	Nov 2001[k]	Mobitel
Slovenia	3	3	2	Auction	Sept 2006	Si.mobil, T-2
Spain	3	4	4	BC + fee + annual fee	Mar 2000	Airtel, Amena, Telefónica, Xfera
Sweden	4	4	4	BC + annual fee	Dec 2000	Europolitan, Hi3G Access, Orange Sverige[y], Tele2
Switzerland[b]	3	4	4	Auction	Dec 2000	Dspeed, Orange, Swisscom, Team 3G[z]
Turkey[b]	3	–	–	–	–	–
UK	4	5	5	Auction	May 2000	BT3G, Hutchison 3G, One-2-One, Orange, Vodafone
Ukraine[b]	5	1	1	Allocated	Nov 2005	Ukrtelecom

[a]Licensees are cited under the names used when the licence was first issued. The number of 2G licences is applicable to the time of the event.

[b]Not an EU Member State.

[c]Two of the 2003 licences had been returned by the original licensees – Broadband Mobile in August 2001 and Tele2 in November 2002 (which became the licence acquired by Hi3G Access) with the other returned licence being bought by Mobile Norway (50% owned by Tele2) which had recently acquired a GSM licence.

[d]The initial intended licensing method was abandoned in favour of an allocation as the number of applicants did not equal/exceed the number of licences.

[e]The three licences were not awarded at the same time or through the same method. MobilTel was awarded its licence in March 2005 after a tendering process was completed, while Viva Ventures and GloBul were allocated their licences in April 2005.

[f]The award of two licences in December 2001 was actually the third occasion on which the Czech Republic had attempted to award 3G licences. The previous two attempts at a tender, in September and October 2001, both failed to attract bidders.

[g]The three licences were not awarded at the same time. Eesti Telecom and Radiolinja received their licences in July 2003, and Tele2 in August 2003

(continued)

Table 1 (continued)

[h]The three licences were not awarded at the same time. Hutchison 3G Ireland received its licence in June 2002, mmO_2 in August 2002 and Vodafone Ireland in September 2002.

[i]The three licences were not awarded at the same time. Viag Europlattform finally accepted its licence in March 2001 while Tele2/Tango and Mobilkom received their licences in July 2001

[j]Telecom FL initially refused the offer of a licence. Its owner, Swisscom, then sold the company to the government in July 2003 and when the transfer was completed in October, the now re-named Liechtenstein TeleNet accepted the licence.

[k]At the second attempt. An auction planned for May 2001 attracted no bidders.

[l]As a condition for its acquisition of tele.ring in April 2006, T-Mobile was obliged to dispose of tele.ring's two sets of 5 MHz paired 3G spectrum, of which at least one had to go to existing licensee Hutchison 3G Austria.

[m]3G Mobile sold its Austrian licence to mobilkom Austria in December 2003. Half the spectrum (5 MHz paired) was sold on to T-Mobile.

[n]When TeliaSonera acquired Orange Denmark it was obliged to return one of its two 3G licences which was re-auctioned.

[o]Originally awarded to Grosson Capital in November which failed to pay so it was subsequently offered to Renberg Investments (which declined), RealGroup (which failed to pay) and ProGroup.

[p]The licence held by Tele2 was revoked in July 2005.

[q]MobilCom returned its licence in December 2003.

[r]One licence was awarded in April.

[s]The licence acquired by Smart was finally revoked in November 2006 and offered to eircom which had bought 2G incumbent Meteor.

[t]Orange (which had yet to roll out its 2G network) returned its licence in December 2004.

[u]The licence was for cdma2000.

[v]The licence was revoked in January 2003 and the spectrum divided up among the other licensees.

[w]The first installment was not paid until January 2007 when the award became official.

[x]Although it technically won the licence, Profinet.sk did not make the required down-payment and the licence was revoked in August 2002.

[y]Orange Sverige sold its licence to Svenska UMTS-nät in December 2003, but this was not sanctioned by the regulator. In November 2004, the regulator recalled the licence.

[z]The licence was revoked in April 2006 and handed back in April 2007.

Thirdly, not all 3G licences on offer have as yet been awarded. Across the countries that have licensed 3G networks to date, 136 licences have been offered counting only the first occasion that licences were available in each country but 165 including those re-offered and those newly created for a second round of licensing. However, only 143 of these licences have been taken up of which one (Profinet.sk in Slovakia) was immediately revoked. A variety of licensing mechanisms can be identified that is more complex than the normal distinction that is made between auctions on the one hand and beauty contests on the other. A closer examination of the licensing process column in Table 1 identifies six different mechanisms, namely: an allocation without a competition; an allocation due to insufficient competing bidders; an auction; a minimum-price tender; a beauty contest followed by an auction; and a beauty contest together with either a lump-sum fee or an annual fee or both. It is worth noting that although the British and German auctions grabbed the headlines after raising $43.2 and $46.1 billion respectively, only a small minority of countries have opted in practice to use an auction with most preferring instead some form of a beauty contest together with a fee and, more recently, the allocation of licences (although this has increasingly reflected a shortage of bidders).

Related to this is a fourth observation, namely that in 19 instances so far – Croatia, Czech Republic, Denmark, Estonia, France, Guernsey, Ireland, Isle of Man, Jersey, Latvia, Liechtenstein, Luxembourg, Montenegro, Norway and Poland, Romania, Serbia, Slovakia and Slovenia – 3G licences were awarded over two or more rounds. In the case of France, for example, the method adopted was a beauty contest together with a substantial lump-sum fee. Although a wide range of bidders were initially attracted, their numbers rapidly declined as they acknowledged either that their priorities lay elsewhere or that they could not afford to participate in the licensing process. KPN, for example, stated that its priority was Belgium where it already had a 2G licence, while T-Mobile withdrew citing its lack of a 2G licence. In contrast, Bouygues Télécom withdrew due to its poor financial position, arguing that GPRS was sufficient to provide most of the services that it intended to provide.

In July 2001, two 3G licences, costing $4.5 billion apiece, were awarded to Orange and SFR. The following year, the government sought to re-issue the remaining two 3G licences. The license period was extended to 20 years, and the fee reduced from $4.5 billion to $550 million plus 1% of non-handset revenues, and the same conditions were offered retrospectively to the existing licensees. Despite this – although it must be said that France was seen as providing a particularly unfriendly environment for new entrants – Bouygues Télécom was the only bidder in this second round, and was awarded its licence in September 2002.

In October 2006, the regulator once again began to seek a buyer for the fourth licence. It did this by the unusual expedient of fixing a date – 17 November – beyond which the 5 MHz of 2G spectrum – available to the 3G licensee from end-2009 – would not be awarded along with the 3G licence. Failing the appearance of a new applicant, the 2G spectrum would be reallocated to the existing licensees. On the final day, ISP Iliad expressed its interest in a licence, thereby ensuring it would be offered, but did not guarantee to bid for it. Cable operator Noos-Numericable was also said to have declared an interest, but Neuf Cégétel opted out.

The price was expected to be 620 million, the equivalent price to the previous licence. The timetable announced in November 2006 was for applications in the form of a tender to be lodged by spring 2007 with the licence to be awarded later in the year. However, the tender in October 2007 attracted no applications because the up-front licence fee was deemed to be too high.

In the Czech Republic, in contrast, four attempts have been made to award 3G licences since the process began in September 2001. The process was protracted primarily due to the government valuing the licences at considerably more than the operators, whether incumbents or potential new entrants, were prepared to pay. The May 2001 valuation of $167 million per licence was double what most commentators thought the licences were worth. This high price was not attractive to bidders, and although the price was subsequently reduced and the terms improved, the incumbent operators continued to argue vocally that the price was too high. At the third attempt, the November 2001 auction still attracted only two bids – from incumbents EuroTel Praha and RadioMobil – and they were accordingly awarded licences. In December 2004, the regulator initiated the process of awarding another licence by stating that sufficient spectrum was available, and that the licence would be awarded through a tender. The government wanted the third incumbent, Oskar, a Czech company, to be awarded the licence for $88 million. As this was less than the sum paid previously by the existing licensees, they quite naturally complained. Although the government refused to reduce the licence fee retrospectively, arguing that market conditions had changed, it did agree to an extension of their launch date until 2007. Oskar duly accepted its licence, agreeing to pay the fee over 3 years.

In Croatia, the two rounds were necessitated by the failure in October 2004 of the Tele2-led consortium Treca Sreca, to apply for a 3G licence at the same time that it made its application for a 2G licence. The two incumbents were duly awarded 3G licences for 17.6 million apiece. Two months later, Treca Sreca successfully applied for the outstanding 3G licence. In Latvia, the third licence was won by Bité, an incumbent in neighbouring Lithuania which as of the end of 2005 had yet to offer licences. In the case of Poland, new entrants were put off from bidding in the initial beauty contest in December 2000 by the absence of clear-cut rights to roam on to incumbents' 2G networks, and controversy dogged the re-offering of the outstanding licence until May 2005 when it was finally awarded to a consortium led by a subsidiary of a Polish fixed-wire operator, Netia. Liechtenstein, for its part, is an unusual case because the government of this tiny state offered four licences, of which three were eventually accepted some considerable time after the offer had been made. The fourth licence was refused, but the owner of the company rejecting the licence sold it to the government and the licence was eventually awarded to the now re-named company. The multiple Irish and Norwegian rounds will be dealt with below, as new entrants were integral to both.

Fifthly, it is possible to identify 35 3G new entrants across Europe. A new entrant is defined here either as a 3G licensee that does not have already a 2G licence or as a bidding consortium not majority owned by a 2G licensee in the market where it is bidding for a 3G licence. On this basis, listed by country, the 3G new entrants are:

- Austria: Hutchison 3G, 3G Mobile
- Croatia: Treca Sreca
- Cyprus (South): Investcom
- Denmark: Hi3G Denmark
- Estonia: Grosson Capital/Renberg Investments/RealGroup/ProGroup
- Finland: Telia Finland
- Germany: Group 3G, MobilCom Multimedia
- Guernsey: Wave Telecom, Guernsey Telenet
- Iceland: Nova
- Ireland: Hutchison 3G Ireland, Smart Telecom
- Italy: H3G, IPSE 2000
- Jersey: Jersey AirTel
- Latvia: Bité
- Luxembourg: LuXcommunications, Orange
- Malta: 3G Telecoms
- Montenegro: m:tel
- Norway: Broadband Mobile, Hi3G Access
- Poland: Netia (P4)
- Portugal: ONI-Way
- Romania: RCS&RDS
- Serbia: mobilkom Austria
- Slovakia: Profinet.sk
- Slovenia: T-2
- Spain: Xfera
- Sweden: Hi3G Access, Orange Sverige
- Switzerland: 3G Mobile (Suiza)
- UK: Hutchison 3G

In addition, it should be borne in mind that one 2G/3G licensee in Liechtenstein, Telekom FL, was sold to the government and, re-launched as Liechtenstein Telenet, took over the 3G licence. Further, that when Smart Telecom in Ireland forfeit its 3G licence, it was acquired by eircom which had recently acquired a 2G licensee Meteor. Neither are accordingly treated here as new entrants.

As these 35 new entrants can be found in countries both small and large as well as in EU Member States and non-EU countries, it cannot be argued that 3G new entrants are associated with any particular geographical characteristic of the countries in which they can be found. It is, perhaps, unsurprising that new entrants are to be found in four of the five largest telecommunications markets in Europe,[2] with the exception, France, proving unattractive initially even to one of the three 2G incumbents. The UK, uniquely, reserved for new entrants the largest of the five 3G licenses that it offered, but the likelihood of new entry was obviously enhanced in general by the offer of more 3G licences than there were 2G incumbents as in, for

[2]The largest five telecommunication markets in Europe, as defined by population size, are France, Germany, Italy, Spain and the UK.

example, Germany. Perhaps the most interesting development was incumbent Telia's failure to win a 3G licence in Sweden where all four successful licensees scored maximum points in the beauty contest, although Telia has subsequently made arrangements to operate as a virtual network operator (MVNO).

It is worth noting that whereas 2G MVNOs are fairly widespread in Europe, there have so far only been a small number of instances where – Telia excepted because it was a 2G incumbent – new entrants have launched as MVNOs over 3G networks. We have respectively, Bravocom in Estonia in July 2006 (which subsequently secured a licence via ProGroup); Saunalahti in Finland in January 2005 (which was subsequently acquired by licensee Elisa); M6 in France in March 2007; debitel in Germany in August 2006 (being acquired by SFR of France); and TDC Song Norge in Norway in April 2006. Whether it is economic to resell bulk capacity acquired from a 3G network owner is a moot point, and it is noteworthy that the above companies have tended not to remain independent, but the MVNO model may prove to be more popular as data downloads increase in volume and value.

3G New Entrants

A closer inspection of the 35 new entrants allows a clear distinction to be made between those that have been able to launch their 3G services and those that have, for whatever reason, failed (so far) to do so. As shown in Table 2, just 17 of the 35 new entrants had launched their 3G services by the end of 2007 and as the footnotes to Table 1 demonstrate, a significant proportion of the sample no longer have licences and hence never will launch. However, this table also highlights the fact that four companies – Sonera (now part of TeliaSonera), Telefónica, France Télécom/Orange and Hutchison Whampoa – originally set out to use the 3G licensing process to enter new markets.

Sonera, a 3G licensee in its home market of Finland, sought to exploit 3G licensing to expand its geographical footprint aggressively across Europe. In partnership with Telefónica, Sonera successfully bid for 3G licences in Germany and Italy, while with Enitel the company won a licence in Norway. Sonera was also an original shareholder in Xfera in Spain, acting mainly with Vivendi Universal.

The cost of these licences varied considerably. The German licence cost $7,680 million, while the Italian licence cost $2,713 million. In contrast, the two Scandinavian licences were considerably cheaper; the Norwegian licence cost roughly $23 million while only a nominal fee was paid for the Finnish licence.[3] Nevertheless, the cumulative impact on Sonera was to undermine its financial stability – the company's share price collapsed, its credit rating fell and eventually the chief executive resigned.

Sonera endeavoured to stabilise its financial position by withdrawing from its 3G investments in Germany and Italy. In the second quarter of 2002, Sonera wrote

[3] A nominal administration fee of 1,000 per 25 kHz was charged.

Table 2 European 3G new entrants, 31 December 2007 (Annual reports and regulator, company and other websites)

Country	Company	Date service launched[a]	Number of subscribers[b]	Main shareholders[c]
Austria	Hutchison 3G	Apr 2003	–	Hutchison Whampoa
Austria	3G Mobile	–	–	Telefónica
Croatia	Treca Sreca	–	–	Tele2
Cyprus (S)	Investcom	Dec 2004	–	Investcom
Denmark	Hi3G Denmark	Nov 2003	–	Hutchison Whampoa, Investor
Estonia	ProGroup	–	–	Bravocom Mobil
Finland	Telia Finland	Oct 2004	–	Telia
Germany	Group 3G	–	–	Sonera, Telefónica
Germany	MobilCom Multimedia	–	–	France Télécom, MobilCom
Guernsey	Wave Telecom	July 2004		Jersey Telecom Group
Guernsey	Guernsey Telenet	–	–	Bharti Group
Iceland	Nova	–	–	Novator
Ireland	Hutchison 3G Ireland	July 2005	–	Hutchison Whampoa
Ireland	Smart Telecom	–	–	Private investors
Italy	H3G	Mar 2003	–	Hutchison Whampoa
Italy	IPSE 2000	–	–	Sonera, Telefónica
Jersey	Jersey AirTel	June 2007	–	Bharti Group
Latvia	Bitė	June 2006	–	TDC
Luxembourg	LuXcommunications	May 2005	–	Mobistar (Orange)
Luxembourg	Orange	–	–	Orange
Malta	3G Telecoms	–	–	n/a
Montenegro	m:tel	July 2007	–	Telekom Srbija, Ogalar
Norway	Broadband Mobile	–	–	Sonera, Enitel
Norway	Hi3G Access Norway	–	–	Hutchison Whampoa, Investor
Poland	Netia (P4)	Mar 2007	–	Netia
Portugal	ONI-Way	–	–	ONI, Telenor
Romania	RCS&RDS	Dec 2007	–	RCS&RDS
Serbia	mobilkom Austria	July 2007	–	Telekom Austria
Slovakia	Profinet.sk	–	–	Profinet
Slovenia	T-2	–	–	Zvon Ena Holding
Spain	Xfera	Dec 2006	–	Vivendi Universal, Sonera, ACS
Sweden	Hi3G Access Sweden	Apr 2003	–	Hutchison Whampoa, Investor
Sweden	Orange Sverige	–	–	Orange
Switzerland	3G Mobile (Suiza)	–	–	Telefónica
UK	Hutchison 3G	Mar 2003	–	Hutchison Whampoa, NTT DoCoMo

[a] 'Launch' is taken here to be the date when the service is first made available, usually via laptops to corporate customers.
[b] No data as yet available for 31 December 2007.
[c] When licensed.

down the value of its investments in Group 3G and IPSE 2000 to zero at a combined cost of SEK39.2 billion (TeliaSonera 2003, p. 53). In addition, Sonera was released from any future obligations, and in the case of IPSE 2000 the company said that it would not be making any further investments in Italy. Telefónica also wrote down the value of its investments in these two markets in July 2002 by 4.9 billion (Telefónica 2002, p. 6) as well as exiting the Austrian market in late 2003 when it transferred its stake in 3G Mobile, including frequencies, to mobilkom Austria.

Sonera also tried to turn its back on its Spanish and Norwegian 3G investments. In late 2001, Broadband Mobile handed back its 3G licence in Norway after Enitel concluded that it could not afford to roll out the network. In Spain, Sonera wrote down the value of its investment in Xfera by SEK660 million in December 2002, but, in its new guise as TeliaSonera, it remains a major stakeholder in Xfera in conjunction with ACS. The Spanish government made various concessions over the ensuing period to help ensure the launch of Xfera, and with some reluctance it finally agreed to launch in June 2006. When even this date appeared likely to be missed, the regulator became more aggressive and the launch eventually took place in December (Telegeography 2006a, b).

Although these write-downs and exits did provide Sonera with a degree of financial stability, the company's forays into 3G had weakened the company to such an extent that it was eventually forced into a 18 billion merger with Telia (Brown-Humes 2002). In order to gain EU approval for the merger, Telia was required to sell Telia Finland (Guerrera and Brown-Humes 2002, p. 27). Telia sold the company, which was its sole 3G new entrant, to Suomen 2G in June 2003.

France Télécom expanded its geographical footprint by entering three new markets via the 3G licensing process, once (in Germany) in its own right and twice (in Luxembourg and Sweden) via majority owned subsidiary Orange (now once again wholly owned by its parent). In early 2000, France Télécom invested 3.7 billion in exchange for 28.5% of MobilCom in Germany. This investment would be used to fund the purchase of a 3G network, giving MobilCom, previously a reseller, a network of its own for the first time. MobilCom would fund the rest of the purchase price of the 3G licence through bank loans and vendor loans that were tacitly guaranteed by France Télécom (Johnson 2002, p. 22).

Relations between France Télécom and Gerhard Schmidt, the founder and controlling shareholder of MobilCom, deteriorated in the aftermath of winning the 3G licence. In essence, France Télécom wanted a rapid rollout of 3G services so that the 3G operator, MobilCom Multimedia, would begin to generate revenues as soon as possible, while Schmidt wanted to delay the rollout of 3G services due to lower than expected demand. France Télécom accordingly decided to walk away from MobilCom. In the process, it would absorb 7 billion of MobilCom's debt in exchange for 90% of any proceeds from the sale of the 3G licence (Benoit 2002). MobilCom sold its 3G assets in May 2003 to E-Plus for 20 million (Spiller 2003), a tiny fraction of the $7,650 million purchase price. The actual licence was returned to the regulator in December 2003.

Orange was awarded a 3G licence in Sweden where a beauty contest was held in December 2000. The four licensees paid $42,800 between them in fees and agreed to pay 0.15% of their annual revenues to the government. Although Orange Sverige started life as a joint venture between Orange and four other companies, by the end of 2002 Orange had bought out its partners.[4] In order to reduce the costs of their 3G licences nationwide, the four licensees and Telia formed two infrastructure-sharing companies that would build infrastructure outside of the main population centres of Stockholm/Uppsala, Gothenburg and Malmö/Lund. Along with Vodafone and Hi3G Access, Orange Sverige formed 3GIS.

Orange Sverige also sought to reduce the percentage of the population to be covered by its network from 99.9% to 93%. Although this apparently modest reduction would affect only a relatively small number of people, it would significantly reduce the cost of rolling out its network across Sweden. However, the Swedish regulator, PTS, refused to agree to such a reduction with the consequence that Orange announced in December 2002 that it would be closing Orange Sverige at a cost of 252 million.

Hutchison Whampoa

Given the tribulations of Sonera, Telefónica and France Télécom/Orange as outlined above, it becomes evident from Table 2 that only Hutchison Whampoa remains as a major active 3G new entrant. Hutchison Whampoa is present as a new entrant in seven European markets and has launched networks in all cases bar Norway.

Hutchison Whampoa re-entered the UK mobile communications market – having sold off Orange, its original network – by acquiring the fifth and largest of the 3G licences which, as noted, had been reserved for new entrants. Hutchison 3G UK was formed as a joint venture between NTT DoCoMo (20%), KPN (15%) and Hutchison Whampoa (65%). However, the relationship between the three partners was fraught. In early 2003, KPN refused to contribute to a £1 billion call for additional funds by Hutchison 3G UK, claiming that it broke the shareholder agreement that was in place (Leahy and Nutall 2003, p. 31).

While the legal validity of the cash call was being resolved in court, Hutchison Whampoa made up the shortfall with the consequence that some voiced their concerns that the company had over-exposed itself to 3G. Shareholder disharmony was resolved in November 2003 by KPN agreeing to sell its shares back to Hutchison Whampoa for £90 million. This was followed shortly afterwards, in May 2004, by NTT DoCoMo's decision also to sell back its shares for £120 million. Although in both cases a 2007 deadline was agreed, Hutchison Whampoa completed the purchase of these shares during 2005 so that an IPO of Hutchison 3G UK could go ahead in 2006 (Lau 2005, p. 28). Freed of its partners, Hutchison has proved to be a fierce competitor in a market dominated by four roughly equal 2G incumbents, and at the time of writing had 4 million 3G subscribers.

---

[4] The other shareholders were Bredbandsbolaget, Skanska, NTL and Schibsted (Curwen 2002, p. 174).

Hutchison Whampoa is also present in a second large European market, namely, Italy. Hutchison Whampoa acquired a majority stake in Andala, a consortium led by Tiscali, which successfully bid for a 3G licence in November 2000. In February 2001, Hutchison Whampoa acquired the bulk of the stake held by Tiscali as well as part of the stakes held by other shareholders. As a consequence, Hutchison Whampoa increased its stake to 78.3% and changed the consortium's name to H3G. This stake was further increased to 88.2% when one of the other shareholders failed to acquire its full stake and is currently 90%. Italy is the most successful market by far, with over 7 million 3G subscribers at the time of writing.

Outside of these two large markets, Hutchison Whampoa has also launched its 3G services in four smaller markets: Austria, Denmark, Ireland and Sweden. In Austria Hutchison 3G is wholly owned by Hutchison Whampoa which paid $120 million for the licence. Interestingly this sum was only just above the reserve price set by the government. 3G services were launched, albeit initially covering only 35% of the population, in April 2003.

In Sweden the operator, Hi3G Access, is a 60–40 joint venture with Investor, the quoted investment arm of the Wallenberg family. Although the licence was acquired for a nominal sum, the rollout deadlines and coverage requirements of the licence were exacting and hence necessitated heavy investment. To mitigate the financial burden, Hi3G Access formed 3GIS, initially with Vodafone Sweden and then with Orange Sverige. Hi3G Access opened its network in December 2001, and launched its 3G services commercially in April 2003.

Unusually among 3G new entrants, Hi3G Access has acquired 3G licences in other markets. Hi3G Access entered the Danish mobile communications market by establishing Hi3G Denmark to bid for a 3G licence. The company successfully bid for a licence, paying $118 million in September 2001, and launched its service almost 2 years later in November 2003. More recently, Hi3G Access has also acquired a licence in Norway. Norway has issued its 3G licences over two rounds, with the second round being necessitated by the decision of both Broadband Mobile and Tele2 to return their licences. In September 2003, the Norwegian government sought to re-issue the two returned licences although in practice it only succeeded in awarding one of them. Hi3G Access Norway paid NOK62 million ($8.2 million) and agreed to provide services to at least 30% of the population within 6 years (3G Newsroom 2003). Unusually for a Hutchison operator, this has not rushed to launch its network.

The final European market where Hutchison Whampoa has a licence is Ireland. Hutchison 3G Ireland, wholly owned by Hutchison Whampoa, paid $50 million in June 2002 to acquire a 3G licence. This licence was larger than the other licences offered, in part because the licence holder was required to make spectrum available to MVNOs. Services in Ireland were launched in July 2005 (Hutchison Whampoa 2005, p. 56), and since then the licence left unallocated during the first offer has been taken up. A national 2G roaming agreement with Vodafone was arranged and Pure Telecom, a fixed operator, announced that it wished to become a MVNO using Hutchison 3G Ireland's network.

Discussion

The licensing of 3G across Europe has provided an opportunity for companies to enter new markets. In all, 34 new entrants have emerged from the licensing process although only four companies – Sonera (now part of TeliaSonera), Telefónica, France Télécom/Orange and Hutchison Whampoa – have been particularly active at any point in time. Significantly, only the latter company, Hutchison Whampoa, remains as a truly committed 3G new entrant in Europe.

All things considered, it is clear that Hutchison Whampoa is something of an oddity within the European mobile communications industry. First of all, it is axiomatic that, since Hutchison Whampoa has no installed 2G customer base to fund its expansion into 3G, it must rely on other sources for funds. The foray into 3G has been financed by a very deep-pocketed parent company that seems willing to suffer huge short-term losses and to persevere against the odds that have led incumbents like Sonera and Telefónica to abandon most or all of their 3G new entrant investments. Notwithstanding the minority stakes taken by, for example, NTT DoCoMo in the UK or Investor in Sweden, the combined licence costs and subsequent network rollout represent a substantial investment by Hutchison Whampoa in 3G. Furthermore, Hutchison has now repurchased (albeit at a considerable discount) the stakes held by DoCoMo and KPN in its UK operation.

Secondly, the rollout strategy adopted by Hutchison Whampoa is almost diametrically opposite to that of every incumbent. By and large, incumbents decided early on that they were earning massive revenues from their 2G networks, the technology was immature and that handsets were either unavailable and/or clunky and that if one of them was holding back there was every incentive for the others to follow suit. In contrast, Hutchison Whampoa needed to obtain a revenue flow as early as possible and hence chose to be the first to launch in every market if humanly possible.

Such a strategy was, and remains, very risky, not least because it assumes that the initial outlays can be recouped once a subscriber base has been accumulated. Although Hutchison Whampoa has launched in six European countries and had accumulated over 10 million subscribers by the end of 2005,[5] its subscriber base remains relatively small in global terms at 12½ million of which 11 million are accounted for by just two networks. As a consequence, Hutchison Whampoa cannot spread its costs over as large a subscriber base as it 2G rivals, nor for that matter can it use a large installed subscriber base to achieve significant scale economies. The result is to place Hutchison Whampoa at a disadvantage against its much larger 2G incumbent rivals like Orange or Vodafone.

[5]In terms of equity-adjusted subscribers, it was still only the 25th largest international mobile operator in December 2004 with 9.5 million subscribers across all of its international operations.

Hutchison Whampoa is also disadvantaged by the inherent tension of growing its subscriber base through handset subsidies and competing largely on price.[6] Handset subsidies are costly and only pay for themselves after several years of subscriber revenue. However, this payback period is extended when prices are lower than those of competitors. The challenge for Hutchison Whampoa is accordingly to attract and then retain subscribers while moving away from price competition.

To date, the evidence as to whether Hutchison Whampoa is capable of managing such a transition is mixed. In Italy, for example, the company initially targeted high-end business users before broadening its appeal by introducing lower tariff packages as network coverage improved. As its rivals have launched their 3G services, the company's response appears to be one of ongoing handset subsidies while introducing new services. It can be argued that this strategy did eventually show some signs of success given that, for example, Hutchison Whampoa claimed that its average revenue per user (ARPU) in Italy had risen above that of its rivals, with 23% coming from non-voice services in 2005 (Hutchison Whampoa 2005, p. 53). Although this could be taken to imply that the company was able to manage the transition, it is worth noting that 90% of Italian subscribers are prepaid and thus the ones most likely to switch providers to take advantage of lower prices elsewhere.

A similar mixed picture can be found in the UK. The initial service-focused launch strategy was soon dropped in favour of reduced tariffs that bundled voice with large numbers of text messages. This change was brought about by a combination of unsatisfactory handset quality and the company's lack of network coverage. Although handset quality has improved and the network has been expanded nationwide and been enhanced with HSDPA, the company has continued to emphasise price rather than services. However, from mid-2004 onwards the company has drawn attention to the successful mass download of content – if only because depicting itself as a media rather than a telecommunications company theoretically implies a higher stock market valuation. The opening day of the English 2004/2005 football season saw 400,000 downloads, while the 6 months to February 2005 saw 10 million music video downloads. These successes can be interpreted as suggesting that ARPU was becoming less reliant on voice and more on new services such as downloads than was previously the case.

Given the huge cost of rolling out networks, it was Hutchinson's intention to float minority stakes in the more promising networks, but this plan was effectively abandoned in early 2006 when it was finally acknowledged that with little product differentiation, investors would not find the strategy of buying market share particularly attractive. True, a 10% stake in the Italian business was sold to an investment bank for 420 million in February 2006, but only after a 7 billion ($8.3 billion) float was cancelled – the original valuation was 12 billion (Guerrera and Lau 2006; Michaels 2006). By the year-end the outlook appeared to be so poor that commentators were

[6]In October 2006, the '3' UK CEO admitted that voice services accounted for 75% of annual turnover, and that '3' UK had not pushed its Internet services because they were not any good. However, the latter were henceforth expected to grow massively and to that end '3' UK would be acquiring further retail premises, starting with 95 outlets bought from The Link and O2 (Morris 2006).

touting the prospect that Hutchison might either exit the European market or merge with incumbents (Cellular-news 2006; TelecomDirectNews 2006).

Not surprisingly, Hutchison Whampoa has remained defiant about its 3G prospects in Europe. In August 2007, the company announced a modest improvement in its 3G unit – which has assets outside Europe although Italy and the UK are its biggest markets – with losses before interest and tax narrowing to HK$11.3 billion ($1.45 billion), a 6% improvement over the same period 1 year previously. The CEO, as ever citing the favoured metric of traditional earnings minus customer acquisition costs (CACs), claimed that an important internal benchmark had been achieved with the arrival of 'positive monthly ebitda after all CACs' (Mitchell 2006, 2007). A less optimistic observer would probably have noted that Hutchison had so far invested roughly HK$200 billion (roughly $25 billion) in its 3G operations and that the losses in 2006, albeit halved compared to 2005, still amounted to $1.5 billion worldwide.

In September 2007, rumours began to do the rounds that '3' Italia was up for sale to a trade buyer, with T-Mobile and Vodafone cited as interested parties. Further developments are expected during 2008.

Conclusions

Although 35 new entrants were initially formed to take advantage of the licensing of 3G, Hutchison Whampoa has emerged as the only significant new entrant that remains active in the European mobile communications market. Telefónica, TeliaSonera and France Télécom/Orange have all withdrawn either wholly or in part for a variety of reasons. In contrast, Hutchison Whampoa has so far launched services in six markets. However, while a substantial subscriber base has indeed been accumulated, it has been achieved at considerable financial cost, raising doubts as to the long-term viability of the company.

The attitude of the otherwise highly profitable parent company remains central to determining whether Hutchison Whampoa will continue to be active in Europe and thus bring competition to mobile markets. Recent indications suggest that the parent company will continue to support its European investments, thereby providing some of the additional competition that was sought by licensing more 3G players than 2G incumbents. If the parent company is not supportive, or Hutchison Whampoa remains unsuccessful in the markets that it has chosen to enter, then the competition enhancing role that it brings to these markets will be lost. In other words, prices will not decline as fast as they would have done and the pace of innovation will be less than would have otherwise been the case.

The experience of Hutchison Whampoa and the other new entrants suggests a number of areas for further research. Only a limited number of new entrants have so far emerged from the 3G licensing process in Europe, and considerably fewer have actually launched a service. This begs the question as to whether the advantages of incumbency are such that only the most determined and well-resourced of new entrants – either because they have a wealthy parent or because they are themselves incumbents elsewhere – can enter the market in the first place and then launch 3G services. Not only do the advantages of incumbency need to be clarified, but if competition is to be enhanced through new entrants then the means by which these advantages can be offset by, for example, regulatory initiatives also needs to be investigated. Further research is also required into what lessons can be learnt from 3G licensing across Europe. This is particularly important if some countries opt to use the licensing of new technologies such as 4G (whatever form that may take) to increase the number of players as they did with the licensing of 3G. The experience of 3G would suggest that replicating existing licensing methods in the future is unlikely to achieve this objective. Consequently, other strategies need to be identified that allow for competition to be enhanced.

References

3G Newsroom (2003) Hi3G awarded 3G licence in Norway. www.3gnewsroom.com. Cited 15 May 2005

Benoit B (2002) Mobilcom and France Telecom strike a deal. Financial Times, 22 November

Brown-Humes C (2002) Telia and Sonera seek partners after 18bn merger. www.ft.com. Cited 11 April 2002

Cellular-news (2006) Is a Hutchison 3G withdrawal from Europe imminent? www.cellular-news.com. Cited 16 October

Cellular-news (2007) European 3G subscriber base passes 60 million. www.cellular-news.com. Cited 26 September

Curwen P (2002) The Future of Mobile Communications: Awaiting the Third Generation. Palgrave, Basingstoke

Gruber H (2005) The Economics of Mobile Telecommunications, Cambridge University Press, Cambridge

Guerrera F, Brown-Humes C (2002) Commission to clear Sonera link with Telia. Financial Times, 9 July, p. 27

Guerrera F, Lau J (2006) Rare disappointment for Hutchison Whampoa. www.ft.com. Cited 15 February

Hutchison Whampoa (2005) Annual Report 2004. Hong Kong

Johnson J (2002) On the line. Financial Times, 9 September, p. 22

Lau J (2005) Hutchison Whampoa buys out UK 3G partners. Financial Times, 11 May, p. 28

Leahy J, Nutall C (2003) Li Ka-Sing's 3G plan hits a snag, Financial Times, 13 June, p. 31

Michaels, A (2006) Market woes derail 3 Italia IPO plans. www.ft.com. Cited 12 June

Mitchell T (2006) 3 Group points the way. www.ft.com. Cited 27 March

Mitchell T (2007) Losses at Hutchison 3G unit narrow. www.ft.com. Cited 23 August 2007

Morris A (2006) 3 UK steps up focus on non-voice business. www.totaltele.com. Cited 25 October

Spiller K (2003) E-Plus to buy Mobilcom's 3G network. Financial Times, 4 May

TelecomDirectNews (2006) Speculation deepens on Hutchison Whampoa's possible sale of the 3 Group. www.telecomdirectnews.com. Cited 22 November

Telefónica (2002) Annual Report 2002, Madrid, Spain

Telegeography (2006a) Government begins process to remove Xfera's 3G licence, reports say. www.telegeography.com. Cited 3 April

Telegeography (2006b) Xfera adds 15,000 users in first fortnight, papers say. www.telegeography.com. Cited 14 December

TeliaSonera (2003) Annual Report 2002, Stockholm, Sweden

Does Regulation Impact the Entry in a Mature Regulated Industry? An Econometric Analysis of MVNOs

Delphine Riccardi, Stéphane Ciriani, and Bertrand Quélin

Abstract Since 1998, the European telecommunications industry is entered into a liberalization phase. In mobile markets, the liberalization policy induces the introduction of competition between a larger number of competitors and a decrease in retail prices. However, the assessment of national markets reveals insufficient competition between network operators and a new regulation was proposed to facilitate private investments into this mature industry.This paper investigates the determinants of the fringe entry into European mobile telecommunications markets between 1998 and 2005. More precisely, we intend to answer the following question: how do cross-national differences in the market structure and the regulatory design (regulatory incentives and governance) affect the number of Mobile Virtual Network Operators' (MVNOs) in mobile markets? We test a set of hypothesis using internationally comparable variables of economic and regulatory determinants and allowing for ten European Member States and temporal fixed effects on 8 years. We infer the hypotheses to predict cross-national variations in the number of MVNOs entries. We then control for the potential effects of the contractual governance of the MVNOs' access to the incumbents' mobile networks. We demonstrate that the amount of fringe entry into a mature industry is the result of both the strategic behavior of the incumbents towards hosting MVNOs on their networks and the adoption of credible regulations to prevent the exercise of strategic entry-deterring activities. Our findings are salient for policy-makers and practitioners alike.

Introduction

In economics, an important question about regulation is appropriate incentives and efficiency. For a regulated industry, a key point relies on the capability to attract investors and new entrants which are able to compete with historical monopolies or incumbents.

D. Riccardi(✉), S. Ciriani, and B. Quélin
HEC, Paris

B. Preissl et al. (eds.), *Telecommunication Markets*, Contributions to Economics,
DOI: 10.1007/978-3-7908-2082-9_17, © Springer Physica-Verlag HD 2009

Two contradictory forces interact. First, the regulation must protect private investment in infrastructure through limiting the number of competitors. Second, the level of competition must be high enough for obtaining a sustainable market growth and a price decrease for customers.

So, since the enlargement of market is obtained and maturity stage reached, the key question is to know whether market does need a further step in regulation to attract extra new entrants or private contracts between players are enough. Theoretically, the key question is about the best way to implement new offers: new regulation or private ordering?

In this article, we aim to explain why new competitors entered a mature industry, such as the European mobile telecommunications industry, when other competitors did not?

Until 1998, the European telecommunications regulatory policy focused mainly on the liberalization of fixed-line telephony. After 1998, the regulators concentrated on the liberalization of the mobile industry as well. Granted with strengthened powers, they adopted three main pro-competitive economic incentives with expected impacts on prices and mobile market growth: the introduction of the dial number portability, the regulation of interconnection charges and the presence of airtime resellers. (Grzybowski 2005) Number portability should reduce consumer-switching costs and decrease prices (Klemperer 1995; Buehler et al. 2006); the regulation of interconnection charges should decrease marginal costs of providing mobile services in the industry and lower prices. (de Bijl and Peitz 2002, 2004) Finally, the presence of airtime reseller should increase the number of competitors, lower prices and promote innovative service offerings that benefit mobile users. (Valletti 2003, p. 9) At the very end, the so-called "ladder investment theory" suggests that resellers may invest into their own infrastructures in order to be less dependent on the incumbent and offer a wider range of services (Cave and Vogelsang 2003).

Due to the scarcity of the available radio spectrum capacity, allocating traditional mobile licenses was impossible so that new mobile competitors would have to negotiate access to the networks of the incumbents and eventually to ask for relevant regulatory assistance. Despite the absence of a common definition,[1] Mobile Virtual Network Operators (MVNO) are characterised by being operators who provide mobile communications services to users without its own airtime and government-issued licenses. Following Dewenter and Haucap (2006), we must acknowledge that this broad definition of an MVNO does not completely cover all MVNO business models deployed in diverse European States with different regulations. Adopting a differentiation proposed by the business literature we distinguish between three types of MVNOs models (IDATE 2006):[2]

[1] For a survey on MVNOs definitions, see Dewenter and Haucap (2006).

[2] Some business studies have argued that the definition by the ownership of certain key assets was flawed because it assumes that the use of these assets can only be achieved by acquiring them entirely. MVNOs are ranged according to the degree of control over some aspects of service design (Analysys paper 2005, "The future of MVNOs").

- Full MVNOs, which provide their own network core including a mobile switching center (MSC)[3]
- Intermediate MVNOs, which acquire a switched service, but either provide their own home location register (HLR) or share a jointly owned HLR with and Mobile Network Operator[4]
- And Thin MVNOs, which only provide additional applications and content and which are little different from pure resellers or service providers (they are also called "*enhanced service providers*", Kiesewetter 2002)[5]

To date, the number of entry has been very different from one European Member State to another. This lack of convergence in European markets development is related to many factors, such as the legal and contractual difficulty of negotiating and then implementing an access agreement with an incumbent or the wrong choice of business strategies. IDATE (2006) identified two main threats related to MVNOs strategies: first, the destruction of value if the pricing pressure is too high (switching to the low-cost model), and second, the prohibitive cost of building whole or parts of mobile networks.

The purpose of this research is to provide an empirical investigation on the impact that market structure and regulations may have on the amount of fringe entry into a mature industry. For this purpose, we use a panel of entry data for ten European Member States spanning the period of 1998–2005. We use both the individual and time dimensions of our dataset, given the time series of consistent data available. We demonstrate that the amount of fringe entry into a mature industry is the result of both the strategic behavior of the incumbents towards hosting MVNOs on their networks and the adoption of credible regulations to prevent the exercise of strategic entry-deterring activities. We organize our paper in the following manner. The first section discusses the previous literature. The second section describes our data and variables and the third one describes the empirical models we test in this paper. The last section presents the results and discusses conditions that promote competition in mobile markets.

Regulation, New Institutional Economics and Fringe Entry into a Mature Industry

The scholarly literature that underpins our hypotheses falls into two categories: the literature on fringe entry into a mature industry and the new institutional economics. We begin by summarizing the primary insights of each body of literature as they

[3]Tele2 (Denmark, Finland or Sweden), or BT (UK) or Saunalahti (Finland) belong to the "full MVNO" type.

[4]Virgin Mobile (UK) belongs to the "intermediate MVNO" type.

[5]Telmore, Tesco Mobile or Sun Telecom is belonging to the "thin MVNOs" type.

relate to the expected determinants of the number of entrants into newly liberalized European mobile markets.

During the 1990s European Member States gradually started to liberalize their telecommunications markets by promoting service competition and access to network infrastructure. For the mobile industry, this has led to a decrease of retail prices, an increase in the number of competitors and a mobile diffusion reaching a near saturation point. To date, the European mobile market can be qualified as a highly mature voice market with a regulated framework, a competitive dynamics operated by mobile network incumbents and a fringe entry of airtime resellers (some retailers but mainly MVNOs).

This competitive dynamics is explained by the literature related to fringe entry into a mature industry. Industry maturity is often synonymous with a few dominant firms, high barriers to entry and a low rate of entry. "*However, mature industries often show a dramatic increase on the number of the firms. Typically, this occurs as a result of the founding of new kinds of organizations that are different from incumbent firms.*" (Swaminathan 1998, p. 389) Two alternative explanations for firm entry into new market segments in a mature industry have been proposed: the resource-partitioning model and the niche formation process. (Swaminathan 1998).

Some authors refer the resource-partitioning model (Carroll 1985). As industries mature they come to be dominated by a few generalist firms. These generalist firms attempt to maximize their performance by drawing on the largest possible resource space, the centre of the market, opening up resources on the periphery of the market segmentation to specialist new entrants (Beesley and Hamilton 1984).

Other authors argue that new market niches may emerge as a result of discontinuities in an industry's environment: for example changes in government policy or in regulatory regimes may open oligopolies and create competitive opportunities for new entrants (Abernathy and Clark 1985, p. 18).

We argue that MVNOs' entry in the European mobile markets may be driven by both a resource-partitioning process as a result of non-targeted mobile consumers (2.1) and a niche formation process as a result of the adoption of entry regulations (2.2) when new entrants perceive those regulations as credible commitments (2.3).

Competitive Entry, Resource-Partitioning and Market Concentration

Following the resource partitioning model, MVNOs' entry into mature mobile markets may be due to the degree of market concentration, mainly oligopolies.

Resource partitioning model describes a marketplace as being made up of multi-dimensional service space, with each dimension representing a distinctive customer attribute. Organizations align themselves within such market topography by targeting their services at the various resource spaces or market segments (Mainkar et al. 2006, p. 1068).

At the early stage of market's evolution, the market is characterized by a low degree of concentration.[6] More specifically, the market is composed of a large number of generalist firms, each of which cannot individually affect prevalent price levels. The market coverage overlaps near the centre but a large proportion of the total market is covered by differentiated services so that the resource space available for specialists is smaller (Swaminathan 1998, p. 393).

Once a few generalists are concentrated in the centre space, predictions from industrial organization economics and resource-partitioning differ. Industrial organization economics (Schmalensee 1978; Judd 1985) predicts that the benefit continue to accrue to the incumbents due to the scale economies, collusion and credible commitment. In contrast, resource partitioning model predicts that with a higher level of concentration, the generalist firms are fewer in number and larger in size so that the total resource space covered is smaller than in the case of a concentrated market with differentiated services. Specialists have access to greater resource that they exploit at the fringe market segments without entering into direct competition with the larger generalists. (Swaminathan 1998, p. 393) High concentration in the market implies that specialists can draw upon fringe resources without entering into direct competition with generalists (Freeman and Lomi 1994; Lomi 1995; Swaminathan 1995, 1998; Carroll et al. 2002).

At the advanced stage of market's evolution, Dobrev (2000) applies the resource partitioning model in periods of a decrease of market concentration following market liberalization and deregulation. In such context, it is showed that declining market concentration has a negative effect on the founding of generalists and a positive effect on the entry rate of specialists. (Dobrev 2000, p. 401) This results from the fact that the observed overall disintegration in industry consolidation actually conceals increasing local concentration. Swaminathan (2001) offers an explanation related to the fact that the generalist can appropriate a portion of the resource space by developing an ability to operate in both the generalist and specialist segments. Generalists will operate either by copying specialist's routines or by extending their product lines into specialists' space so that some resemble the service features offered by specialists, albeit at a lower cost.

Recent MVNOs' studies confirm that the incumbents' incentives to voluntarily provide network access critically depend on the degree of service differentiation: "Generally, MNOs will voluntarily provide network access if the services offered by the candidate MVNOs are sufficiently differentiated, as with a high degree of product differentiation the revenue effects outweigh the competition (cannibalization) effects." (Dewenter and Haucap 2006, p. 2; Greenstein and Mazzeo 2006; Ordover and Shaffer 2006) Following MVNOs' access to mobile networks, market's evolution can be translated into a two-stage process:

---

[6] Concentration refers to the aspect of the competitive process that is driven by the size distribution of the dominant incumbent firms within a given resource space.

- A first stage characterized by a high degree of market concentration and MVNOs' entry with differentiated services (full and intermediate types) that do not enter into direct competition with generalists
- A second stage characterized by a lower degree of market concentration and MVNOs' entry with differentiated services (full and intermediate types) imitated by generalists

So, the hypothesis 1 follows:

> *Hypothesis 1: The decrease of market concentration induces an increase in the amount of fringe entry (MVNO) if new entrants deliver differentiated services into European mobile telecommunications markets.*

Competitive Entry, Entry Regulation and Asymmetric Regulation

MVNOs' entry into mature mobile markets may be also due to the emergence of a niche as a result of changes in access price and entry regulations.

Initial competition took the form of competitive access providers, companies (like Sense in Sweden) that allowed customers to bypass the incumbents' mobile network and the associated expenses. Fearing the cannibalization of their own customers, many incumbents then refused to grant access to their network or negotiated lengthy contracts with restrictive terms and conditions.

Some EU Member States decided to assist MVNOs' entry by adopting favorable access price regulation. Furthermore, as of July 2003, European Union asked for a national assessment of the competition level on the market for the wholesale mobile access and call origination. Depending on national mobile markets, regulators adopted or threatened to adopt formal decisions that designated incumbents with significant market power and that proposed regulatory remedies to correct it. One of the proposed remedy was a regulation of entry by mandating MVNOs' access to incumbents' mobile networks. As a result, national regulatory policies vary between mandatory and non-mandatory access and the mobile telecommunications industry has become one industry consisting of incumbent facing competition from a competitive fringe, the MVNOs.

When competitive fringe firms enter the mobile telecommunications industry, national regulators must decide whether to impose access price, mandatory access on the network operators, maintain or remove them. "*Asymmetric regulation occurs when a single firm, or group of firms, is subject to differential regulatory oversight*". (Abel and Clements 2001, p. 229) We classify access price and entry regulation as asymmetric when those policies apply only to network operators and not to MVNOs. Access price and mandatory access are asymmetric regulations as they concern only incumbents.

In contrast, the mobile number portability regulation cannot be qualified as an asymmetric regulation as it concerns all the mobile operators (the network and the virtual operators). However, the portability regulation is usually considered as a regulation aiming at lowering the entry barrier related to customer inertia (Armstrong and Sappington 2006). Therefore, the mobile number portability corresponds to an entry regulation.

The impact of an asymmetric regulation on the amount of entry can be illustrated as a two-stage game. (Schankerman 1996) In the first stage, potential competitors make independent decisions on whether to enter a market. In the second stage, firms engage in price competition or services differentiation. By altering competition and expected profitability in the second stage, asymmetric regulation can impact competitive entry in the first stage. The existing literature addresses asymmetric regulation in the telecommunications industry, especially that pertaining to asymmetric price regulation (de Bijl and Peitz 2002, 2004; Kim and Park 2004; Armstrong and Sappington 2006, p. 33), carrier of last resort obligations (Weisman 1994; Schankerman 1996), and quality of service regulation (Abel and Clements 2001). The general consensus is that asymmetric regulation is associated with a significant higher amount of entry.

Consistent with other forms of asymmetric regulation identified in the literature, asymmetric access price and entry regulations should induce high amount of fringe entry into mobile markets.

Regulatory Design: The Assessment of the Regulation Credibility as a Factor of Private Investments

One of the main insights of the NIE literature relates to the assessment of the state's ability to commit to a utilities-specific regulatory schema. To apprehend this ability, the NIE literature analyzes regulation via a "design" construct whose two components are "regulatory incentives" and "regulatory governance" (Levy and Spiller 1994, 1996). The former refers to the rules governing pricing, subsidies, competition and market entry, networks infrastructure interconnections, etc., and therefore applies to the utilities themselves. The latter refers to the mechanisms by means of which the state restrains a political organization's ability to reform the regulatory framework that applies to the utilities sector and the mechanisms for settling any subsequent conflicts. Following this literature, a regulatory design should be qualified as a credible regulation depending on the extent to which institutional safeguards increase the costs of reneging on previous commitments in place.

From the perspective of private investors, the same literature emphasizes the extent to which the political institutions in place support political actors' commitments not to expropriate the property or rent-streams of investing firm increases the incentives of telecommunications firms to invest. (Henisz and Zelner 2001, 2004) In other worlds, private investors in regulated markets will only believe government pledges regarding future economic incentives to the extent that they are credible.

We argue that the assessment of credibility is a question of primary importance in the case of MVNOs' entry. In order to facilitate new entries, the introduction of asymmetric regulations reformed the current regulatory framework based on the allocation of spectrum. Some incumbents have indeed complained that the introduction of further competitors would be a violation of their license conditions and that it should be regarded as a hold-up on their specific investment in network infrastructure (Dewenter and Haucap 2006).

Accordingly, the MVNOs' incentives to invest in mobile markets will depend on the credibility of those asymmetric regulations: access price and entry regulations. More particularly, entry decisions will depend on the institutional safeguards that increase the costs of reforming those asymmetric regulations. A theoretical result arising from both the literature on niche formation process and the NIE literature would be that:

Hypothesis 2: The existence of a credible regulation (access price and entry regulations) induces a high amount of fringe entry (MVNOs) into European mobile telecommunications markets.

Model, Data Collection and Variables

In order to unravel the determinants of MVNOs' entry, we propose a model relying on the previous literatures. We settle a distinction between economic and technological factors that may give way to regulatory intervention. Economic factors of MVNOs' entry are related to market structure whereas technological factors depend on the level of dependence between the MVNO and the incumbent, which is translated into contractual governance. Regulation may impact both economic and technological factors in order to assist new entries; the nature and strength of the impact depend on the credible commitment of a regulator.

The Dependent Variable

In the empirical analysis described below, we examine a dependent variable which is the cumulative number of entrants (MVNO) at the end of each year. We define an entry as the launch of a mobile communications services, excluding the mere announcement of commercial relationships in press releases. The dependent variable is constructed from the European Commission's report on implementation of the regulatory framework dated of 2005[7] and from the Telecom paper Mobile MVNO/SP List[8] with verification on each MVNO website.[9]

---

[7] European Commission, 2005, 11th report on implementation of the regulatory framework, Annex 1, p.22.

[8] The Telecompaper Mobile MVNO/SP List was previously called the "*Takashi Mobile MVNO/SP list*" and is accessible on: http://www.telecompaper.com/research/mvnos/

[9] Following external comments, the authors fully acknowledge the limits of the paper due to the availability of data on the dependent variable. However, it results from the data selection process that the selected list of MVNOs was the most detailed and updated source that distinguishes between MVNOs and other mobile service providers over the time period. It is also acknowledged that the subsequent construction of the dependent variable may have some impact on the conclusions to be drawn from the econometric model.

The Explanatory Variables

We group the explanatory variables used in our empirical analysis into the following three categories for the purpose of description and discussion: (1) variables related to market structure; (2) variables related to contractual governance; (3) variables related to regulation.

Variables Related to Mobile Market Structure

This first category includes several measures that we use to control for variations in market structures across European Member States and the level of market competition. The first variable (COMP) is the degree of competition by new entrants which is defined by the annual level of market shares of all MVNOs. (OECD 2000, 2003) We expect a high degree of competition by new entrants to reflect high level mobile market competitiveness so that we anticipate a positive correlation between this variable and MVNOs' entry (Gruber and Verboven 2001).

The second variable (CONC) controls for the degree of concentration in the mobile telecommunications markets, as measured by the HHI (Herfindhal-Hirschman Index) that is the sum of the squares of the market shares of all mobile networks operators (MNOs). We must acknowledge that our data only take into account the market shares of the incumbents excluding those of new entrants. Following Hazlett and Mu.oz (2004), we endorse the fact that the number of mobile operators is most often fixed externally, by spectrum licensing so that we expect the magnitude of concentration (HHI) to be largely the result of regulatory design. We expect that the evolution of the HHI level is linked to the credible commitment of a European Member State to improve competition because a high degree of concentration may lead to a regulatory decision on individual or collective dominance with compulsory obligations for MNO(s) which may be favorable to MVNOs. We then anticipate that the degree of concentration is positively linked to MVNOs entry.

The third variable (MNO) represents the number of mobile incumbents and most often corresponds to the number of mobile networks accessible for MVNOs. We recognized that we do not distinguish the incumbents' technologies that are 2G or 3G and that some national 3G licenses included obligations to give access to MVNOs. Nonetheless, we argue that 3G was very partially rolled-out during the studied period. Moreover, this paper is related to the potential impact of access regulation that is supposed to be technologically neutral.

The fourth variable (P) relates to the penetration rate, which is the number of connections to a service divided by the population.

The fifth variable (ARPU) represents the mobile network operator's average revenue per user. ARPU is usually related to the level of prices and/or the level of minutes of use. National regulators often relied on the ARPU level to conclude on the market competitiveness. Higher ARPU would be linked to the existence of MNOs' market power eventually due to the market concentration. However, McCloughan and Lyons showed that no evidence was found that European mobile markets concentration had any influence on ARPU. (McCloughan and Lyons 2006) Depending on country and operator, we expect ARPU to be weakly related to MVNOs entry.

The sixth variable (CHURN) measures how many customers are disconnecting from the network. Churn is calculated by the formula: gross connections minus net new connections, divided by customer base at the end of the previous year.

Variable Related to Contractual Governance

This variable is (CONTRACT) and control for the different types of MVNOs business models according to their contractual integration with the network operators: full, intermediate and thin MVNOs. This distinction between MVNOs is based on the degree of control of some network elements and it results in a differentiation between mobile services. The classification into the "thin" type does not allow any distinction between MVNOs based on trademark, pricing policy or distribution networks.

Variables Related to Regulation

The third category includes several variables that we use to control for the nature of the regulatory incentives and the credibility of the regulatory commitment, which is translated into the variations of regulatory governance.

The first regulatory variable (PRG) is a dummy variable related to the access price regulation (Sappington and Weisman 1996) that equals 1 if the regulator uses either traditional rate-of-return regulation,[10] or a price-cap regulation, a policy that allows for limited price flexibility. PRG equals 0 if the regulator adopts complete deregulation so that the access price is the result of commercial negotiation between MNO and MVNOs without any regulatory intervention.

The second regulatory variable (PORT) is related to the mobile network portability (MNP) and aims at allowing customers to retain their assigned mobile telephone numbers when changing the subscription from one mobile network operator to another. While reducing switching costs of mobile users and facilitating new entries, MNP may lead to high costs of implementation, reduction in tariff transparency. (Buehler and Haucap 2004) Depending on the commitment by regulators to implement MNP regulation, it is possible to assume that this variable may have a significant impact on MVNOs' entry.

The third regulatory variable (FD) is a dummy variable related to the formal decision that a national regulator may have adopted when assessing market competition for "wholesale mobile access and call origination" (also classified as "market 15"

[10]This means that prices are set close to costs and the incumbent only earns at some competitive return.

by the European commission). Following the revision of the European framework in 2002, the national regulators have had to conduct market analysis in order to control for the competition degree that is designating undertakings with significant market power and eventually proposing ex post regulatory remedies. FD equals 1 if the national regulator has adopted a formal decision with or without regulatory remedies. FD equals 0 either if the national regulator has not adopted any decision or if a formal decision was withdrawn or cancelled by national courts. We anticipate that a formal decision is a signal for new entrants of the commitment for a credible regulatory environment. Therefore, it is possible to assume that this variable is correlated with the dependent variable. However, the impact of this variable remains uncertain as the adoption of a formal decision can be very different from its implementation in time and constraints on incumbents so that the impact may also depend on the extent of the national judiciary control.

The fourth regulatory variable (INDEP) is an aggregate governance indicator of the quality of public services, the quality of the civil service and the degree of its independence from political pressures, the quality of policy formulation and implementation, and the credibility of the government's commitment to such policies. It measured the policy consistency and forward planning (Kaufmann et al. 2006).

The fifth regulatory variable (REGQ) is an aggregate governance indicator of the ability of the government to formulate and implement sound policies and regulations that permit and promote private sector development (Kaufmann et al. 2006).

The regulatory governance variable are both extracted from the same source, the World Bank's research project on governance indicators, that provides for data between 1996 and 2005. We rely on this source because, to our knowledge, it is the only one providing for indicators over time.

Empirical Test

In this section, we provide an empirical analysis of economic and regulatory factors that may have important and/or significant effect on the cumulative number of entrants in the mobile market. We estimate the impact of mobile market structure and regulatory framework policy on the total cumulative number of Mobile Virtual Network Operators entry at the end of each year of the sample period. Our econometric framework is closed to Alexander and Feinberg (2004). The number of entrants is of count data type. This implies a Poisson or negative binomial distribution for the dependent variable of our model. As the Huber-White estimator is used to correct for heteroscedasticic structure of residuals, all estimations are supposed to provide robust standard errors for coefficient estimates.

To take account of potential over-dispersion arising in the context of a Poisson distribution and in order to obtain consistent and valid estimates, we also provide a negative binomial framework estimation. In order to fully account for the panel structure of cur database, individual and time fixed effects are then considered.

In order to capture the lagged effect of the determinants of entry, some explanatory variables are lagged one period. Furthermore, all regulatory variables are lagged one period, while on the market structure side, competition degree and market concentration indicator (HHI) are kept in levels.[11] Given the relatively small sample size and panel data structure, we adopt a generalized linear model and maximum likelihood estimators which allow controlling for specific cross-sectional countries fixed effects[12] in the modeling of entry. Huber and White estimator is used to provide robust estimation in the context of Poisson regression.

We run a first set of regressions that offset the impact of regulatory design in the dependent variable. In this case, the cumulative number of MVNOs at the end of each period. The number of entries is exclusively explained by the market structure.[13] The results are presented in the Tables 2 and 4. It is shown that the mobile market structure has important and significant effects on the total cumulative number of MVNOs' entries, both in the Poisson and negative binomial models. The models only differ in the impact of ARPU, which is only significant in the negative binomial framework, while the impact is negative in both models and coefficient estimates are the same. The competition index COMP measured by the level of market shares of all MVNOs have a positive impact of relatively high magnitude and high significance on the dynamics of entry. The CHURN rate has also expected positive sign with high significance level, as a high disconnecting rate may foster entry of new virtual mobile competitors. We observe that a high number of incumbent operators (MNO) may hinder the incentives to enter. The impact is negative with a lower significant level but still, a relatively higher magnitude. Another variable of interest is the degree of market concentration, i.e. the Herfindahl Index. Its impact on MVNO's entry appears to be negative, which is counter-intuitive as a high number of existing operators in the mobile market may discourage entry of new potential competitors. The results, however, clearly show that the magnitude of market concentration effect is very low, while it still differs from zero. Besides, potential over-dispersion in the Poisson model does not seem to induce bias in the estimation of the market structure impact, as the results derived from the negative binomial model are quite similar in significance, magnitude and sign.

---

[11] The dynamic panel approach is ruled out, as we do not suppose correlation between explanatory variables and residuals. This allows ruling out Arellano-Bond estimator as the lagged endogenous variable is not included in the set of regressors. In this context, there is no persistence effect arising from the total number of MVNOs.

[12] The unobserved international differences across countries are captured by the fixed effects. A random structure for individual and time effects is also tested, but the results are not reported. Besides, Poisson fixed effect model can also be properly estimated by adding individual or time dummies to the set of explanatory variables. It is not the case, however, for a negative binomial distribution.

[13] Even in the absence of explicit regulatory framework variable in the first test, regulation is not neutral as the related effect may be embodied in the market structure itself. Adding these variables to the test leads to the estimation of the combined effects of regulation and market structure on entry.

Table 1 Variables descriptions and sources

Variable	Variable description	Sources
MVNO	The cumulative number of entrants (MVNO) in each national mobile market (i) at the end of each year (t)	The European Commission's report on implementation of the regulatory framework dated of 2005 the Telecompaper Mobile MVNO/SP List the verification on each MVNO website
COMP	The annual level of market shares of all MVNOs	– National regulators' websites – OMSYC, "MVNO in Europe, benefits and risks of co-opetition" (2004 and 2006)
CONC	HHI that is the sum of the market squares of all competitors	– OVUM, 2005. "Regulatory status of mobile access in Europe".
MNO	The number of Mobile network operators in each national mobile market (i) at the end of last year (t-1)	– The European Commission's report on implementation of the regulatory framework dated of 2005 – OVUM, 2005. "Regulatory status of mobile access in Europe".
P	The number of connections to a service divided by the population.	– Analysys, Mobile Networks and Services-Country reports
ARPU	Average revenue per unit (ARPU): A mobile network operator's average revenue per connection.	– Analysys, Mobile Networks and Services-Country reports – ABN-AMRO, 2005. Pan European Telecoms, Wireless Model Builder.
CHURN	A measure of how many customers are disconnecting from the network. Churn is calculated by the formula: gross connections minus net new connections, divided by customer base at the end of the previous year.	– Analysys, Mobile Networks and Services-Country reports – ABN-AMRO, 2005. Pan European Telecoms, Wireless Model Builder.
CONTRACT	The types of MVNOs business models according to their contractual integration with the network operators.	– OMSYC, "MVNO in Europe, benefits and risks of co-opetition" (2004 and 2006) – the verification on each MVNOs' website
SECTOR	The MVNOs' core business or industrial background	– OMSYC, "MVNO in Europe, benefits and risks of co-opetition" (2004 and 2006) – the verification on each MVNOs' website

(continued)

Table 1 (continued)

Variable	Variable description	Sources
PORT	The existence of a regulation on mobile network portability (MNP)	– OVUM, Country Regulation Review – IDC, European Wireless and Mobile Communications: Country and operators profiles – Analysys, Mobile Networks and Services-Country reports
PRG	The existence of any access price regulation	– OVUM, Country Regulation Review – IDC, European Wireless and Mobile Communications: Country and operators profiles – NCB, 2005, MVNOs: virtual barbarians at the gates of the mobile arena
FD	The adoption of a formal decision on market 15 analysis by a national regulator	– The European Commission's report on market analysis under article 7 – Communication from the Commission on Market reviews under the EU regulatory framework, COM (2006) 28 final; annexes I and II – OVUM 2005. "Regulatory status of mobile access in Europe".
INDEP	The quality of public services, the quality of the civil service and the degree of its independence from political pressures, the quality of policy formulation and implementation, and the credibility of the government's commitment to such policies	Kaufman et al. 2006. "Governance Matters V", the World Bank
REGQ	The ability of the government to formulate and implement sound policies and regulations that permit and promote private sector development	Kaufmann et al. 2006. "Governance Matters V", the World Bank

The regulatory variables are then included in the regressions, to provide an estimation of the combined effect of mobile market structure and regulation on entry. Interestingly, we observe little differences between both Poisson and negative binomial models, which tend to prove that over-dispersion may not induce important overestimation of covariates' impact in terms of statistical significance within our test. Both the results show evidence of ambiguous effects of regulatory decisions on MVNOs entry. The ability of government to foster private sector incentives, captured by the regulation quality variable (REGQ), appears to have a negative impact on entry, which is statistically significant at 1% confidence level in both Poisson and negative binomial models, while the regulatory independence variable (INDEP) has a positive effect on entry, which is only significant in the negative binomial model estimation, at 1% confidence interval. The adoption of a mobile network portability regulation, (PORT), has a negative impact on entry but is only significant in the Poisson model, at a relatively low confidence level (10%).

The quality of public services, the quality of public policy and the degree of government's commitment as well as regulatory independence are important determinants of the total number of entries; while the respective impact of independence slightly differ across models. The adoption of a formal market decision by a national regulator, (FD), may foster entry while the existence of access price regulation (PRG) is a barrier to entry. Both variables have respectively strong and significant positive and negative impact on entry at high confidence level (Pr (>|z|) = 0). Regulatory policies that set access price and have direct impact on the mobile market structure are relevant factors that determine the number of entrants in each country. Further research may lead to the estimation of the probability of MVNOs entry according to both the type of their business model and their specific core business or industrial background. Using a mixed effect probit model, as well as a mixed effects negative binomial equation, we estimate the impact of our set of explanatory variables on the proportion of min, medium or full sized potential entrants. In particular, a general non linear framework accounting for within group heterogeneity and unobserved correlated random effects will allow capturing the marginal effects of regulatory decisions and market structure components on the chosen entrants' type of business model. The results of estimation for both the Poisson and negative binomial regressions are reported in the following tables (Tables 3 & 5).

Table 2 Poisson fixed effects model with robust standard errors

Variable	Estimate	Std. error	z value	Pr(>|z|)
(Intercept)	7.33832470	1.70817271	4.2960	1.739e–05***
lag(ARPU)	−0.01298053	0.01524310	−0.851	0.3944542
lag(CHURN)	0.03326780	0.00561420	5.9256	3.111e–09***
lag(P)	−0.03566585	0.01406217	−2.536	0.0112032*
lag(MNO)	−0.26998101	0.08292783	−3.255	0.0011315**
CONC	−0.00076775	0.00017663	−4.346	1.382e–05***
COMP	0.09956811	0.01406355	7.0799	1.443e–12***

Significance codes: 0 '***' 0.001 '**' 0.01 '*' 0.05 '.' 0.1 ' ' 1; Chi-Square = 157.0006

Table 3 Poisson fixed effects model with robust standards errors

Variable	Estimate	Std. error	z value	Pr(>|z|)
(Intercept)	6.3077e+00	1.9635e+00	3.2125	0.0013156**
lag(MNO)	−2.8643e−01	9.0227e−02	−3.1745	0.0015007**
lag(CHURN)	2.5360e−02	5.8803e−03	4.3127	1.613e−05***
CONC	−7.6252e−04	1.7832e−04	−4.2761	1.902e05***
COMP	1.1739e−01	2.0201e−02	5.8111	6.207e−09***
lag(P)	−1.0405e−02	1.3799e−02	−0.7540	0.4508294
lag(INDEP)	6.8873e−01	6.7682e−01	1.0176	0.3088671
lag(REGQ)	−1.4363e+00	6.1362e−01	−2.3407	0.0192481 *
lag(M)	1.5934e+01	1.1275e+00	14.1329	<2.2e−16 ***
lag(PORT)	−4.6337e−01	2.4203e−01	−1.9145	0.0555549.
lag(PRG)	−1.6021e+01	1.1382e+00	−14.076	0 < 2.2e−16 ***
lag(ARPU)	−1.1848e−02	1.7031e−02	−0.6957	0.4866402

Significance codes: 0 '***' 0.001 '**' 0.01 '*' 0.05 '.' 0.1 ' ' 1; Chi-Square = 119.4003

Table 4 Negative binomial fixed effects model

Variable	Estimate	Std.Error	z value	Pr(>|z|)
(Intercept)	9.09348631	2.10121036	4.3277	1.506e−05 ***
lag(MNO)	−0.27146245	0.09801229	−2.7697	0.0056112 **
CONC	−0.00097029	0.00017735	−5.4711	4.473e−08 ***
COMP	0.11784117	0.01416493	8.3192	<2.2e−16 ***
lag(CHURN)	0.03811287	0.00791921	4.8127	1.489e−06 ***
lag(P)	−0.04362990	0.01850027	−2.3583	0.0183569 *
lag(ARPU)	−0.02718362	0.01554952	−1.7482	0. 0804299.

Significance codes: 0 '***' 0.001 '**' 0.01 '*' 0.05 '.' 0.1 ' ' 1; Log Likelihood = −283;AIC: 319.00

Table 5 Negative fixed effects model

Variable	Estimate	Std. error	z value	Pr(>|z|)
(Intercept)	7.3960e+00	2.0396e+00	3.6262	0.0002876 ***
lag(MNO)	−2.7776e−01	9.8573e−02	−2.8178	0.0048354 **
CONC	−8.3144e−04	1.7568e−04	−4.7328	2.215e−06 ***
COMP	1.1465e−01	2.0686e−02	5.5424	2.984e−08 ***
lag(PORT)	−1.8382e−01	2.6879e−01	−0.6839	0.4940528
lag(PRG)	−3.7190e+01	1.3198e+00	−28.1787	<2.2e−16 ***
lag(CHURN)	3.2066e−02	7.4824e−03	4.2855	1.823e−05 ***
lag(M)	3.6900e+01	3.3785e+00	10.9221	<2.2e−16 ***
lag(P)	−2.8792e−02	1.6009e−02	−1.7984	0.0721065.
lag(REGQ)	−1.2746e+00	6.0665e−01	−2.1010	0.0356371 *
lag(INDEP)	1.1377e+00	6.3955e−01	1.7790	0.0752422.
lag(ARPU)	−2.1010e−02	1.7239e−02	−1.2187	0.2229477

Significance codes: 0 '***' 0.001 '**' 0.01 '*' 0.05 '.' 0.1 ' ' 1; Log Likelihood = −300;AIC: 324.74

Results and Discussion

In this section, we discuss our results by comparison with our hypotheses and then we distinguish new entry without and with regulatory assistance.

New Entry into a Mature Regulated Industry Without Any Regulatory Assistance

In this first part, we focus on the impact of the market structure factors on MVNO's entry while excluding an impact from specific regulatory policies.

Part of those market structure factors, our results show that the degree of concentration has a negative impact but very small effect on the number of MVNOs' entries. This impact is reinforced by the negative but strong impact of the number of mobile network operators (MNOs).

In contrast, the churn rate (CHURN) between the incumbents, and the degree of competition between MVNOs (COMP) have both a positive and strong impact on MVNOs' entries. These results are aligned with the resource-partitioning model's predictions.

At the early stage of market evolution, the market structure is characterized by a large number of incumbents (MNOs) and a high degree of concentration (CONC) for those generalists. As a large proportion of the total market is covered by differentiated services, the resource space available for MVNOs is small. At this stage, the market structure has a negative impact on MVNOs entry.

Following the predictions of the resource-partitioning model, a high degree of market concentration should be followed by a decrease in the number of generalists and an increase in their size.[14]The total resource space covered by incumbents is smaller so that specialists have access to resource located at the fringe of the market. Our results clearly show that the evolution of the market structure depends on the strategic behavior of incumbents towards hosting (or not) MVNOs on their network. More specifically, this market evolution depends on the consequences of the competitive dynamics between incumbents characterized by the churn rate (CHURN).[15]

The migration of customers from one incumbent to another initiates a strategic reaction from the incumbents. The initial difficulty about hosting MVNOs evolved

---

[14]We must acknowledge that our data do not allow us to control for this prediction. Due to the specificity of spectrum licensing process, the market concentration is fixed externally so that the number of MNOs has not really decreased as their licenses have been renewed between 1998 and 2005. However, European Commission regularly stresses the distinction between active and non-active licenses. We can not exclude the decrease in the number of active MNOs.

[15]We do not take into account other factors of churn variation as independent variables.

and wholesale revenues from new entrants were considered significant enough to compensate the decrease of retail revenues resulting from the increase in the churn rate. MNOs started to host MVNOs.

Following the resource partitioning model, the resource space depends on the fringe entry by specialists rather than generalists firms. MVNOs should exploit resource at the fringe of the market segments without entering into direct competition with the large specialist incumbents.

Dobrev (2000) shows that the decrease of the market concentration promotes fringe entries by specialists. Regarding the mobile telecommunications services market, the question is then related to the qualification of MVNOs as specialists or generalists.

Our data do not provide us with an empirical measure of the differentiation degree between the mobile services. However, in our context, the differentiation degree is analysed as the way of delivering services adapted to specific customers and different from services delivered by incumbents.

Drawing on our distinction between MVNOs, we can limit the ability to differentiate as a function of the control of specific network elements such as Home Location Register, Commutation Centres... Controlling or not those network elements falls into the scope of the access contract between MVNOs and mobile carriers. That is why empirical reports usually link the degree of network integration to the ability to differentiate the mobile service. (OMSYC 2005; IDATE 2006) Drawing on this relationship, we use contractual integration between MVNOs and network incumbents as a proxy of the service differentiation. Access contracts between mobile carriers and MVNO have been classified into three categories ranging from the minimum integration (contract 1) to a medium integration (contract 2) and a maximum integration (contract 3).

Appendix on the entry differentiation by contract presents that result of a second test related to the impact of contractual governance type onto the causal relationship between market concentration and fringe entry. It results from the second test that the impact of market concentration is negatively related to MVNOs' entry and the significance of the relationship is stronger in the case of contract 1 (low differentiation degree) than in the case of contract 3 (high differentiation degree) and stronger in the case of contract 3 than in the case of contract 2 (medium differentiation degree).

In a context of a slight market de-consolidation, mobile carriers have decided to grant network access to new entrants whose services are characterized by either a very low or a high differentiation degree. In order to explain why incumbents have opened their networks to MVNOs with non-differentiated services, we might refer to the existence of network over-capacity that had to be exploited without any fear of cannibalization. In contrast, MVNOs entry with differentiated services can be explained by the resource-partitioning model as specialists exploit the fringe market segments without entering into competition with incumbents. This explanation of fringe entry is supported by the impact of the concentration degree, the impact of the penetration rate and the impact of the MVNOs market shares level.

In conclusion, our results show a two-sided evolution of the mobile market concentration:

- On one hand, two market structure variables (CONC and MNO) show resistance to market de-concentration resulting in barriers to MVNOs' entries.
- On other hand, two market structure variables (COMP and CHURN) show positive strengths of market de-concentration facilitating MVNOs entries.

New Entry into a Mature Regulated Industry Assisted by Entry Regulations

In this second part of the discussion, we examine the impact of the regulatory determinants on the rate of fringe entry into a mature industry. The dependent variable remains the rate of MVNOs' entry at the end of each year. The set of explanatory variables includes the market structure factors and the variables related to regulation. Those regulatory variables are to be ranged between economic incentives (PORT, PRG and FD) and regulatory governance (INDEP and REGQ).

Our results show that regulation has had some impacts on the MVNOs' entry. While signs and amplitude of the market structure variable remains the same than in the first test, the impact of the regulation is balanced between variables:

- The impact is positive in the case of the adoption of formal decision and the independence of the regulator.
- And the impact is negative in the case of the number portability regulation, the access price regulation and the regulation quality.

Globally considered, the signs and amplitudes of the regulatory impacts are neutralized for both the economic incentives and the regulatory governance. From this result, we can infer that entry regulations did not significantly impact the MVNOs' entries.

From the previous part of the discussion we know that the market structure mainly depends on the MNOs' strategic decision to host MVNOs. Therefore, we must conclude that the regulation did not impact MNOs' ability to grant access to MVNOs. Depending on national cases, MNOs' bargaining power was only limited by the level of regulated access price One of the reasons for this result may be the inability of the regulation to promote the entry of MVNOs specialists. The regulation would have eased the market de-concentration by promoting commitments to invest in mobile networks. Such commitments would have induced the adoption of the full MVNOs contractual governance type and limited the adoption of the thin type. This would have increased the differentiation between mobile services. We have empirical confirmation of such an impact. In Germany, the formal decision promotes MVNOs' investments in mobile networks with a result of entry by full MVNOs.

Besides this general discussion, we can further analyze the impact of each regulatory variable. Considering firstly the impact of economic incentives, our results show that the number portability regulation has had a negative but minor impact on MVNOs entry. This conclusion is not very surprising as the efficiency of such a regulation is strictly conditioned by the timing of the switching process. Mc Cloughan and Lyons (2006) showed that the mobile number regulation has an impact on prices and churn when the switching process is less than 5 days but it has no impact when it is slower. During our studied period, mobile number portability was barely quicker than 5 days so that this economic incentive has no significant impact on the churn between MNOs and has therefore a limited impact on the competitive dynamics of the market structure and MVNOs entry.

Our results also show that the access price regulation has had a negative impact on MVNOs entry. Following Valletti (2003) and Sarmento and Brandao (2007), the choice between retail minus and cost-based methods has an impact on competition in the downstream market; retail-minus being more efficient during a transitional period towards the full liberalization of the market. Depending on the choice made by national regulators, our data show that the access price regulation did not succeed in impacting competition in the downstream market (see Appendix).

Our results finally show that the adoption of a formal decision has had a positive impact on MVNOs entry. Formal decisions have taken different features: either excluding designation of any significant market power within the analyzed market (five Member States belong to this category) or designating significant market power and imposing remedies (two Member States are in this case). A final case is related to Member States still investigating market analysis (three Member States are in this last case). None of the national regulator which has adopted remedies has imposed access obligations to any MNO. In such circumstances, we can not conclude on the regulatory commitment towards mandatory access. Nevertheless, it seems that the process of adoption of a formal decision has induced MNOs strategic analysis about the regulatory commitments. Threatening an unfavorable decision, MNOs have been invited to reconsider MVNOs' entry.

Considering secondly the impact of the regulatory governance on MVNOs entry, we can observe different signs related to the impact of independence and regulation quality on the MVNOs entry. Our results clearly show that the independence has a positive impact on MVNOs entry, whereas the regulation quality has a negative impact on MVNOs entry. Regarding independence, our variable is related to the degree of independence from political pressures, the quality of policy formulation and implementation, and the credibility of the government's commitment to such policies. The positive impact of such a variable on MVNOs entry suggests that private investors gave credit to a regulatory independence and quality. Those potential entrants positively assessed the ability of the regulator to commit itself with credibility.

However, the negative impact of the variable related to regulatory quality neutralized the previous impact when it deals with the promotion of private sector development. MVNOs negatively assessed the soundness of economic incentives adopting to give assistance to new entries. More precisely, national regulators have been considered as being unable to formulate and implement MVNOs' entry regulations. Behind those results, we must insist on some factors of regulatory efficiency: clarity of assignment of functions, regulatory autonomy, accountability and transparency (Smith 1997; Stern and Holder 1999; Noll 2000; Kessides 2004).

Appendix: Entry differentiation by contract

Dependent variable: contract 1

Variable	Estimate	Std. error	z value	Pr(>|z|)
(Intercept)	2.8004e+00	1.8349e+00	1.5262	0.1800730
Lag(g$mno)	−2.6267e−01	1.1547e−01	1.1547e−01	−2.2748
lag(g$arpu)	−3.9553e−03	1.7947e−02	−0.2204	0.8255632
g$conc	−7.3495e−04	2.0244e−04	−3.6305	0.0002829***
g$comp	1.2475e−01	2.3871e−02	5.2259	1.733e−07***
lag(g$port)	−3.9800e−01	2.8248e−01	−1.4089	0.1588557
lag(g$prg)	−4.1698e+01	3.8686e+00	−10.7787	<2.2e−16***
lag(g$churn)	1.1294e−02	5.4149e−03	2.0857	0.0370098*
lag(g$m)	4.1225e+01	2.8654e+00	14.3872	<2.2e−16***
lag(g$p)	3.3345e−02	1.0883e−02	3.0640	0.0021842**
lag(g$regq)	−6.3112e−01	6.1389e−01	−1.0281	0.3039190
lag(g$indep)	−2.0538e−01	6.4603e−01	−0.3179	0.7505532

Significance codes: "***": 0%; "**": 1%; "*": 10%; ".": 5%

Dependent variable: contract 2

Variable	Estimate	Std. error	z value	Pr(>|z|)
(Intercept)	−2.4487e+01	9.9488e+00	−2.4613	0.013843*
g$mno	1.9946e−01	3.6226e−01	0.5506	0.581909
lag(g$arpu)	1.9060e−01	7.1532e−02	2.6645	0.007710**
g$conc	1.3731e−04	9.0653e−04	0.1515	0.879611
g$comp	−1.9355e−01	8.7358e−02	−2.2156	0.026716*
lag(g$port)	2.3840e+00	7.4138e−01	3.2157	0.001301**
lag(g$prg)	−1.5729e+01	1.0688e+00	−14.7165	<2.2e−16***
g$churn	−1.6467e−01	6.5570e−02	−2.5114	0.012027*
lag(g$m)	1.4911e+01	9.2201e−01	16.1724	<2.2e−16***
lag(g$p)	1.6923e−01	5.4478e−02	3.1064	0.001894**
lag(g$regq)	−5.3133e+00	1.9112e+00	−2.7800	0.005435**
lag(g$indep)	7.4523e+00	2.6473e+00	2.8151	0.004876**

Significance codes: "***": 0%; "**": 1%; "*": 10%; ".": 5%

Dependent variable: contract 3

Variable	Estimate	Std. error	z value	Pr(>|z|)
(Intercept)	2.6489e+00	1.7753e+00	1.4921	0.1356719
G$mno	−2.9988e−01	1.5537e−01	−1.9302	0.0535880
lag(g$arpu)	−4.6447e−02	2.4192e−02	−1.9199	0.0548647
G$conc	−5.4189e−04	2.5470e−04	−2.1275	0.0333766*
G$comp	1.8425e−01	3.1656e−02	5.8204	5.869e−09***
lag(g$port)	−1.3617e+00	3.7298e−01	−3.6508	0.0002614***
lag(g$prg)	−1.6057e+01	1.3489e+00	−11.9041	<2.2e−16***
G$churn	1.4152e−02	9.2983e−03	1.5220	0.1280012
lag(g$m)	1.5266e+01	1.2753e+00	11.9701	<2.2e−16***
lag(g$p)	4.0667e−03	1.7324e−02	0.2347	0.8144084
lag(g$regq)	3.8315e−01	1.0557e+00	0.3629	0.7166542
lag(g$indep)	−7.7939e−01	8.3168e−01	−0.9371	0.3486938

Significance codes: "***": 0%; "**": 1%; "*": 10%; ".": 5%

References

Abel, J. and M. Clements. 2001. "Entry under asymmetric regulation". *Review of Industrial Organization* 19 (2): 227–243.

Abernathy, W. and K. Clark. 1985. "Innovation: Mapping the winds of creative destruction". *Research Policy* 14: 3–22.

Alexander, D. and R. Feinberg. 2004. "Entry in local telecommunications markets". *Review of Industrial Organization* 25: 107–127.

Armstrong, M. and D. Sappington. 2006. "Regulation, competition and liberalization". *Journal of Economic Literature* XLIV: 325–366.

Beesley, M. and R. Hamilton. 1984. "Small firms' seedbed role and the concept of turbulence". *Journal of Industrial Economics* 33: 217–232.

Buehler, S. and J. Haucap. 2004. "Mobile number portability". *Journal of Industry, Competition and Trade* 4(3): 223–238.

Buehler, S., R. Dewenter and J. Haucap. 2006. "Mobile number portability in Europe". *Telecommunications Policy*. 30(7): 385–399.

Cave, M. and Vogelsang, I. 2003. "How access pricing and entry interact". *Telecommunications Policy*, 27(10/11): 717–728.

Carroll, G. 1985. "Concentration and specialization: Dynamics of niche width in populations of organizations". *American Journal of Sociology* 90: 1261–1283.

Carroll, G., S. Dobrev and A. Swaminathan. 2002. "The evolution of organizational niches: U.S. automobile manufacturers 1885–1981." *Administrative Science Quarterly* 47: 233–264.

de Bijl, P. and M. Peitz. 2002. "New Competition in Telecommunications Markets: Regulatory Pricing Principles," CESifo Working Paper Series CESifo Working Paper No., CESifo GmbH.

de Bijl, P. and M. Peitz. 2004."Dynamic regulation and entry in telecommunications markets: A policy framework." *Information Economics & Policy* 1 (16–3): 411–437.

Dewenter, R. and J. Haucap. 2006. "Incentives to license virtual mobile network operators." In R. Dewenter and J. Haucap (eds.), Access Pricing: Theory and Practice. Elsevier, Amsterdam.

Dobrev, S. 2000. "Decreasing concentration and reversibility of the resource partitioning process: Supply shortages and deregulation in the Bulgarian newpaper industry, 1987–1992". *Organization Studies* 21(2): 383–404.

Freeman, J. and A. Lomi. 1994. "Resource partitioning and foundings of banking cooperative in Italy." In J. Baum and J. Singh (eds.), *The Evolutionary Dynamics of Organizations*. Oxford University Press, New York, pp. 269–293.

Greenstein, S. and M. Mazzeo. 2006. "The role of differentiation strategy in local telecommunication entry and market evolution: 1999–2002". *Journal of Industrial Economics* 54(3): 323–350.

Gruber, H. and F. Verboven. 2001. "The evolution of markets under entry and standards regulation – the case of global mobile telecommunications". *International Journal of Industrial Organization* 19(7): 1189–1204.

Grzybowski, L. 2005. "Regulation of mobile telephony across the European union: An empirical analysis." *Journal of Regulatory Economics* 28(1): 47–67.

Hazlett, T. and R. Muñoz. 2004. "A welfare analysis of spectrum allocation policies". AEI-Brookings Joint Centre, related publication 04–18, August 2004.

Henisz, W. and B. Zelner. 2001. "The institutional environment for telecommunications investment". *Journal of Economics & Management Strategy*, 10(1): 123–147.

Henisz, W. and B. Zelner. 2004. "Explicating political hazards and safeguards: A transaction cost politics approach". *Industrial & Corporate Change* 13(6): 901–915.

IDATE, 2006. "MVNO The new deal", *Digiworld focus*, November 2006.

Judd, K. 1985."Credible spatial pre-emption". *RAND Journal of Economics* 16: 153–166.

Kaufmann, D., A. Kraay and M. Mastruzzi. 2006. "Governance Matters V". The World Bank.

Kessides, I. 2004. "Reforming Infrastructure: Privatization, regulation and competition". *World Bank Policy Research Report.*

Kiesewetter, W. 2002. "Mobile virtual network operator: Ökonomische Perspektiven und regulatorische Probleme". *WIK Diskussionsbeitrag* 233, WIK: Bad Honnef.

Kim, B. and S. Park. 2004. "Determination of the optimal network charge for the mobile virtual network operator system". *ETRI Journal* 26: 255–265.

Klemperer, P. 1995. "Competition when consumers have switching costs: An overview with applications to industrial organization, macroeconomics, and international trade." *The Review of Economic Studies* 62: 515–539.

Levy, B. and P. Spiller. 1994. "The institutional foundations of regulatory commitment: A comparative analysis of telecommunications regulation". *Journal of Law, Economics & Organization* 10(2): 201–247.

Levy, P. and P. Spiller. 1996. "Regulations, institutions and commitment: comparative studies of telecommunications." Cambridge: Cambridge University Press.

Lomi, A. 1995. "The population ecology and organization founding: Location dependence and unobserved heterogeneity". *Administrative Science Quarterly* 40: 111–144.

Mainkar, M., M. Lubatkin and W. Schulze. 2006. "Toward a product-proliferation theory of entry barriers". *Academy of Management Review* 31(4): 1062–1075.

McCloughan, P. and S. Lyons. 2006. "Accounting for ARPU: New evidence from international panel data." *Telecommunications Policy* 30: 521–532.

Noll, R.G. 2000 "Telecommunications reform in developing countries." In A.O. Krueger (ed.), *Economic Policy Reform: The Second Age*. University of Chicago, Chicago, IL.

OECD. 2000. "Regulation, market structure and performance". ECO/WKP (2000)10.

OECD. 2003. "Indicators for the assessment of telecommunications competition". DSTI/ICCP/TISP(2001)6/FINAL.

OMSYC. (2005). MVNO in Europe, benefits and risks of co-opetition.

Ordover, J. and G. Schaffer. 2006. "Wholesale Access in Multi-firm Markets: When Is It Profitable to Supply a Competitor?" Simon School Working Paper No. FR 06-08.

Sappington, D. and D. Weisman. 1996. "Designing Incentive Regulation for the Telecommunications Industry". Cambridge, MA: MIT Press.

Sarmento, P. and A. Brandão. 2007. "Access pricing: A comparison between full deregulation and two alternative instruments of access price regulation, cost-based and retail-minus". *Telecommunications Policy* 31(5): 236–250.

Schankerman, M. 1996. "Symmetric regulation for competitive telecommunications". *Information Economics & Policy* 8(1): 3–24.

Schmalensee, R. 1978. "Entry deterrence in the ready-to-eat breakfast cereal industry". *Bell Journal of Economics* 9: 305–327.

Smith, W. 1997. "Utility regulators – the independence debate", *the World Bank Public Policy for the Private Sector*, note n°127.

Stern, J. and S. Holder. 1999. "Regulatory governance: Criteria for assessing the performance of regulatory systems. An application to infrastructure industries in the developing countries of Asia". *Utilities Policy* 8: 33–50.

Swaminathan, A. 1995. "The proliferation of specialist organizations in the American wine industry, 1941–1990". *Administrative Science Quarterly* 40: 653–680.

Swaminathan, A. 1998. "Entry into new market segments in mature industries: Endogenous and exogenous segmentation in the US brewing industry". *Strategic Management Journal* 19: 389–404.

Swaminathan, A. 2001. "Resource partitioning and the evolution of specialist organizations: The role of location and identity in the U.S. wine industry". *Academy of Management Journal* 44: 1169–1186.

Valletti, T. 2003. "Obligations that can be imposed on operators with significant market power under the new regulatory framework for electronic communications Access services to public mobile networks". *Report for the European Commission.*

Weisman, D. 1994. "Why Less May Be More Under Price-Cap Regulation". *Journal of Regulatory Economics* 6(4): 339–361.

Part V
Business Strategy

Exploring Technology Design Issues for Mobile Web Services

Mark de Reuver, Harry Bouwman, and Guadalupe Flores Hernández

Abstract As IP-based 3G+ mobile networks increasingly become available, attention in research is shifting towards middleware and service platform-related issues (Killström et al. 2006; Popescu-Zeletina et al. 2003). At the same time, web services technology and underlying service-oriented architecture have enhanced flexibility and interoperability in service development in the fixed Internet and traditional IT world. A logical next step in making mobile services deployment easier, faster and more efficient would be to apply web services technology to the mobile domain in the form of mobile web services (MWS) (Pashtan 2005). MWS offers important opportunities with regard to providing generic service elements like charging, authentication, authorization, accounting, context information and billing. As such, the technology would compete with IP Multimedia Subsystem (IMS), a standard originating from the telecommunications domain. While IMS strengthens the operators' current strategic position, MWS can be offered by any business actor and may therefore be used by content and service providers to disrupt the industry's current structure. As yet, it is unclear whether or not MWS offers a feasible alternative, as it is still in the early stages of its development. In this paper we explore critical design issues in the technology domain, as part of wider business model research. We build on earlier work regarding business models for mobile data services (De Reuver et al. 2006; Haaker et al. 2006). Based on a literature survey, we suggest that critical design issues in MWS business models include overcoming the constraints and limitations of mobile devices and networks; using the SIM card model or web services-based mechanisms for user authentication; user profile management; security; billing and charging; securing an open technical architecture; the server–client model; interoperability mechanisms between IMS and MWS; and the role division regarding authentication, security, user profile management and billing/charging provisioning. A small-scale survey among 29 academic and industrial experts reflects a general concern for most of these design issues.

M. de Reuver (✉), H. Bouwman, and G. Flores Hernández
Delft University of Technology, The Netherlands
e-mail: G.A.deReuver@tudelft.nl

B. Preissl et al. (eds.), *Telecommunication Markets*, Contributions to Economics,
DOI: 10.1007/978-3-7908-2082-9_18, © Springer Physica-Verlag HD 2009

Introduction

As the mobile voice market is becoming saturated (Forrester 2006), mobile data services represent an essential next step towards increasing revenue per user, reducing churn and allowing companies to distinguish themselves from their competitors (Jenkins 2004). In the development of such services, the bottleneck is no longer infrastructural in nature, with IP-based 3G+ mobile networks becoming increasingly available. Nevertheless, content and service providers have a hard time developing new services quickly and efficiently. As a result, attention is now shifting towards middleware and service platform-related issues (Killström et al. 2006; Popescu-Zeletina et al. 2003).

At the same time, web services technology and underlying service-oriented architecture have enhanced the flexibility and interoperability with regard to service development in the fixed Internet world. Combined efforts on the part of companies (e.g., IBM, Sun, and Microsoft) and organizations (e.g., W3C, OASIS, and OMA) in the IT world have resulted in a stack of XML-based Web services standards (e.g., SOAP, UDDI, WSDL, WSFL), based on principles of interoperability and standardization. As a result, web services can combine a variety of services from different providers and present them as a single integrated business function.

Given the current challenges in the mobile domain, a logical next step towards making mobile data services deployment easier, faster and more efficient would be to apply web services technology, i.e. *mobile web services* (MWS) (Pashtan 2005). This is made possible by the increased intelligence in mobile devices, through technologies like Java 2 Platform Micro Edition. While MWS may be used to provide content-related services to end users, there are also important opportunities in bundling new and existing services (Farley and Capp 2005), and in delivering generic service elements. Among the possible generic service elements based on MWS are charging, authentication, authorization, accounting, security, context information and billing. In this paper, we focus on these generic service elements because we expect they will have an impact on business modeling issues in the mobile domain.

It is interesting to note that the same type of functionalities is also provided by another technology: IP Multimedia Subsystem (IMS), a standard that originated within the telecommunications industry (3GPP, IETF) (Camarillo and Garcia-Martin 2006; Cuevas et al. 2006; UMTS Forum 2003). The choice between IMS and MWS is not merely a technical one, but may have a considerable influence on the interdependencies between actors in the mobile industry. Because IMS is implemented in the mobile operator's core network, content providers will have to negotiate and adhere to the operator's requirements with regard to the use of the functionalities. As a result, operators retain control over end-users, which reinforces their privileged position. By contrast, MWS in principal can be hosted by any party, enabling third parties to assume roles that normally fall within the domain of operators, such as billing, authentication and providing context information. Theoretically, this could reduce operators to basic connectivity providers. From a broader perspective, the differences between IMS and MWS could be seen as the traditional clash between the more closed telecommunications sector and the open Internet world.

Because MWS may disrupt the way business is currently being conducted in the mobile domain, the question arises what the long-term impact of this technology is on business models. With MWS still in the early stages of its development, many of the issues involved are still related to technology. It is the aim of this paper to explore which technology-related design issues are currently most critical in MWS-based business models. Our focus is in particular on generic service elements and middleware applications like authentication, billing and security. We derive these design issues from relevant literature on MWS.

We begin by explaining how we define the terms business model and critical design issues. Then, we explore the design issues in business models for MWS, based on a literature survey. While discussing these issues, we will present feedback we received from a small-scale web-based survey we conducted among 29 experts in industry, consultancies and the academic community.

Business Models

As we discussed in the previous section, MWS may have the potential to disrupt the way business is currently being conducted in the mobile data services industry. However, the value-creating logic enabled by MWS remains obscure. A useful tool in exploring this area is the business model construct, which helps explain the logic involved in creating and capturing value from technological innovations (Chesbrough and Rosenbloom 2002).

Although we are fully aware that there is no commonly agreed upon definition of business models (e.g., Alt and Zimmermann 2001), we have decided to adopt the definition provided by Chesbrough and Rosenbloom (2002): "a blueprint that describes how a network of organizations cooperates in creating and capturing value from new, innovative services or products". With regard to the components that make up a business model, we distinguish four domains (Faber et al. 2003):

- Service: description of the value proposition and market segment at which the offering is aimed.
- Technology: description of the technical functionality required to realize the service offering.
- Organization: description of the multi-actor value network required to create and distribute the service offering, and the focal firm's position within this value network.
- Finance: description of the way a value network intends to capture monetary value from a particular service offering and how risks, investments and revenues are divided over the different actors of a value network.

Thus far, MWS-based business models are in the initial stages of their life cycle (Bouwman and MacInnes 2006), and there are still many unresolved questions regarding performance. Because the business models involved are still being changed and improved, they are highly unlikely to represent a definitive version.

At this stage, especially technology-related design choices are vitally important as they impact the viability of the business model in the long run. Therefore, this paper focuses on the technology domain for MWS-based business modes.

Exploring Technology Design Issues in MWS

Adopting web services technology in the mobile services domain is not a trivial matter. In addition to the generic constraints and limitations of mobile devices and networks, there are other issues on the crossroads of business and technology that also play a role. Based on an exploratory literature study regarding MWS and on our previous findings regarding design issues for mobile services (Haaker et al. 2006), we have identified eight potential technology design issues for MWS. In the remainder of this paper, we present the design issues as generic trade-offs between different design options. Which option is preferable may depend on the perspective and strategic interests of the type of stakeholder involved, e.g., content providers versus operators, and new entrants versus incumbents.

To find out how relevant the design issues we identified may be, we asked 29 experts to provide us their insights in a web-based survey. The experts involved include specialists from operators, content providers and application developers, as well as academics and consultants. Our aim was not to compile a representative, random sample, but we used convenience sampling to obtain some initial feedback with regard to the ideas we developed in the paper.

We began by asking them whether they expected the technologies to take off within their company (or within the market studied by the academic respondents), see Fig. 1. Not all the respondents were enthusiastic about IMS, with 24% expecting it would never be used. With regard to MWS, reactions were mixed: 24% said that there were probably companies that already use the technology, while 41% expected it to be used later than next year. A majority of the respondents expect solutions combining MWS and IMS technologies to emerge in the long term. Importantly, some respondents indicated they were not certain, because they found the question too technical or were unaware of the exact timelines.

Design Issue 1: Adaptation of Web Services Protocols

An important question will be whether handhelds can support MWS. Compared to desktop computers, mobile devices provide limited CPU processing power, smaller memory, limited battery capacity and smaller storage capacities (Berger et al. 2003). In addition, because mobile devices generally have a small screen and keyboard, filling in forms is more complicated, which influences the usability of MWS (Farley and Capp 2005). Also, as bandwidth is still smaller and more expensive in mobile networks compared to the fixed Internet, another issue that needs to be addressed

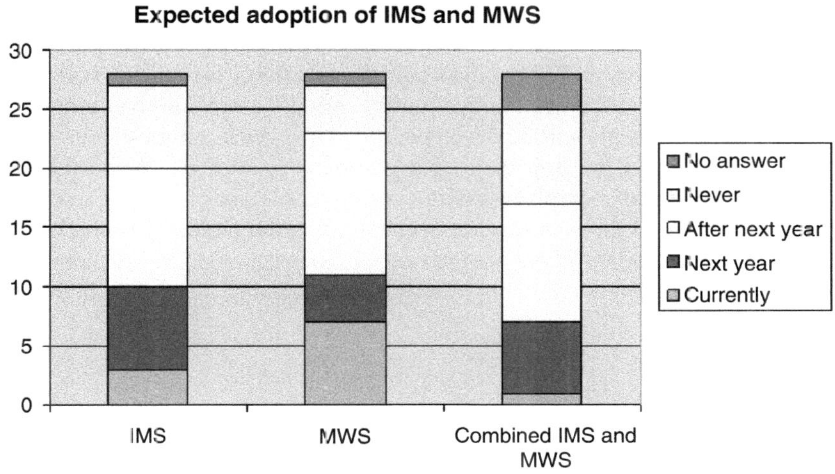

Fig. 1 Expected adoption of IMS and MWS, $n = 28$

has to do with limited data rates. Keeping in mind that web services technology was designed for powerful desktops and servers using wired networks (Limbu et al. 2004), this raises performance issues with regard to MWS, for example the processing of XML and SOAP messages, which require more processing power than HTML messages (Limbu et al. 2004). In addition, XML messages generate a larger overhead than HTML pages, as they can amount up to five times the size of a content message (Tian et al. 2004). In case bandwidth is indeed limited, or if the user is charged per packet of transmitted data, this may well become a constraining issue. In short, although XML provides flexibility and interoperability, its performance effects may become problematic.

Although some of these issues may become less constraining as technology advances, for now MWS have to be designed within the limits of the available computation, memory and energy resources (Berger et al. 2003). Several solutions to the problem regarding XML processing and transmission in MWS have been proposed in literature, including faster parsers, data compression, protocol optimizations, a reduction in the number of SOAP messages exchange, a reduction in the size of the messages by removing all unnecessary information from the messages, and a translation of the plain text XML messages to binary coding (Limbu et al. 2004). For example, Wireless SOAP may reduce the size of SOAP messages by 3–12 times, although this would require even more processing power from the user device, increase server response times and pose a threat to interoperability (Srirama et al. 2006). More advanced compression approaches have been proposed, for example the dynamic compression approach by Tian et al. (2004), who recommend including the need for compression as a quality of service requirement in MWS standards.

Generally speaking, although we expect that technical solutions will be developed to solve the challenge involved in running MWS on mobile handsets, the question is whether these solutions will be open or proprietary in nature. When web services were first developed, they were based on industry-wide, open protocols, in working groups in which industry and other organizations could freely participate. With regard to MWS, the question becomes whether mobile-specific extensions and adaptations to web services protocols will be developed in similar open initiatives, or whether companies will try to develop their own proprietary solutions. While the latter option provides established players with a strategic advantage, it may reduce the wider applicability of the web services technology in the mobile domain. In other words, the generic trade-off here is to have open, industry-wide MWS protocols or proprietary solutions in overcoming technical drawbacks of the mobile domain.

Design Issue 2: Authentication

Generally speaking, because handhelds can be carried around, authentication is an important issue in the mobile services domain. Devices may get out of range of a network access point, and temporarily disrupt network connectivity. Moreover, when a person moves to another network (e.g., from UMTS to WiFi), the IP address of his or her device may change. In addition, disrupted communication between the mobile device and the remote server is also more likely in the mobile context, because mobile networks are more vulnerable to link outages (Pilioura et al. 2003). Messages in web services technology are exchanged asynchronously, which can lead to errors in the communication. When communication is disrupted for any of the reasons described above, it is vitally important to ensure that services are resumed as soon as communication is restored. Failure to do so may have serious unwanted consequences, in particular with sensitive applications like mobile payment (Berger et al. 2003).

Traditionally, authentication of moving devices has been supported by operators using the SIM card model. SIM cards provide information about the user to the network and a secure link between the subscriber account and the operator bill, and they services to be personalized and localized. However, the SIM card model may not be the ideal solution to the problems outlined above. Moreover, there are serious questions concerning the security it provides (e.g. Ahmad et al. 2003). And finally, the information provided by SIM cards can only be accessed by operators, creating a potential conflict with content providers who are interested in delivering context-aware or personalized applications (Farley and Capp 2005). As a result, one of the design choices involved is whether to keep using the traditional SIM card authentication method, or to look for MWS-based alternatives that provide a more reliable mobility management. It is interesting to note that most of the experts in our survey preferred web services for authentication (mean = 4.93, sd = 1.12, n = 28) to the SIM card model that is currently being used (mean = 3.43, sd = 1.37, n = 28), see Fig. 2. However, most of them also agreed that single sign-on technologies should be implemented (mean = 5.36, sd = 1.52, n = 28). This indicates that MWS may have a potential to provide authentication services, provided they include a single sign-on functionality.

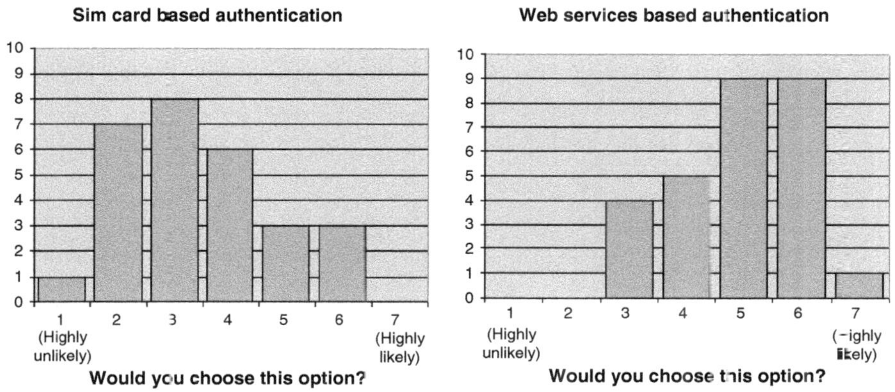

Fig. 2 Web services versus SIM card based authentication

While operators have traditionally played the role of authentication provider through their GSM network techniques, MWS enable user authentication via a generic web service. This means that, as far as MWS based business models are concerned, this role could also be played by third parties, for instance ISPs, service providers or content providers. The design choice involved here is whether to opt in favor of a network operator or of a third party to provide authentication-related services. The experts were divided on this issue (mean = 4.37, sd = 1.668, n = 27), which indicates that a consensus has yet to be reached.

Design Issue 3: User Profile Management

User profiles containing information such as preferences, personal data, interest and context need to be collected, stored, and maintained. There are several alternatives when it comes to storing user information, with variations in the level of user involvement (Pashtan 2005). The experts in our survey gave a mixed response with regard to the desirability of automatic user profile generation (mean = 4.5, sd = 1.45, n = 28).

Sometimes, user information is distributed among various companies, and the law does not always permit them to share the information (cross-domain profiling) (Ali Eldin 2005). Most experts agreed that including a policy in MWS based business model with regard to the way customer information is shared is important (mean = 5.39, sd = 1.52, n = 28).

User information could be stored on the mobile device, and then a web service would be invoked by service providers to request that information. This could provide an alternative to centralized storage. There was a high level of disagreement among the experts as to whether user profiles should be stored centrally at the service provider (mean = 4.43, sd = 1.38, n = 28) or locally on the user device

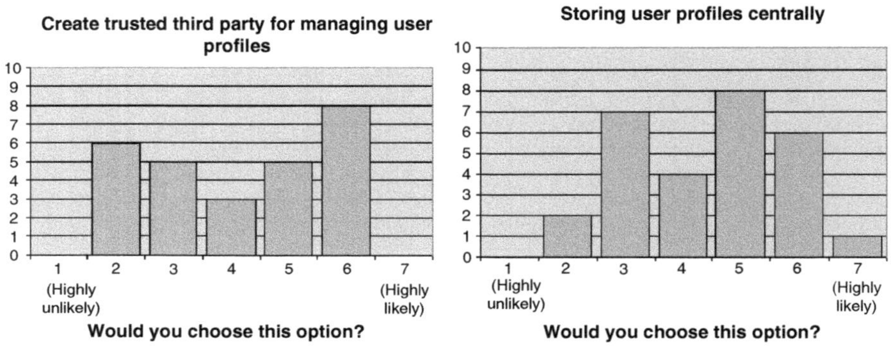

Fig. 3 Management of user profiles

(mean = 4.36, sd = 1.56, n = 28), see Fig. 3. The fact that there was no consensus among the experts would indicate that this is a particularly critical design issue.

Currently, operators manage user information and they can decide to share that information with service providers. However, web services technology enables third parties to exchange user information directly with the mobile device. Our experts provided a mixed response with regard to using a third party to store the user profiles (mean = 4.15, sd = 1.59, n = 27), see Fig. 3.

Design Issue 4: Billing and Charging

Traditionally, billing has been one of the most critical components in the operator's infrastructure. With the emergence of MWS and other advanced 3G-services, new billing and charging schemes have to be developed to match their special characteristics, for instance in terms of pre/post-paid convergence (AtosOrigin 2006). The experts in our survey gave a mixed response with regard to the likelihood of opting in favor of advanced billing schemes, such as generic billing schemes (mean = 4.86, sd = 1.51, n = 28) and personalized billing (mean = 4.21, sd = 1.45, n = 28).

Controlling the billing relationship with the end user is important, as it provides additional revenues and a potentialstrategic advantage through the collection of user data (Weill and Vitale 2001). While many operators take their billing role for granted, third parties are highly interested in deploying web-based services that allow them to take over this role. In this sense, this is another area where the emergence of MWS has a potential impact on the way the various roles are divided within the telecom industry. Third parties may be very interested in maintaining the billing relationship with their customers when offering mobile services, and excluding mobile network operators. Operators, on the other hand, have a strategic interest in maintaining the status quo in this area. Content providers and other third parties will be looking into the possibilities offered by MWS to take over the role of billing provider, reducing

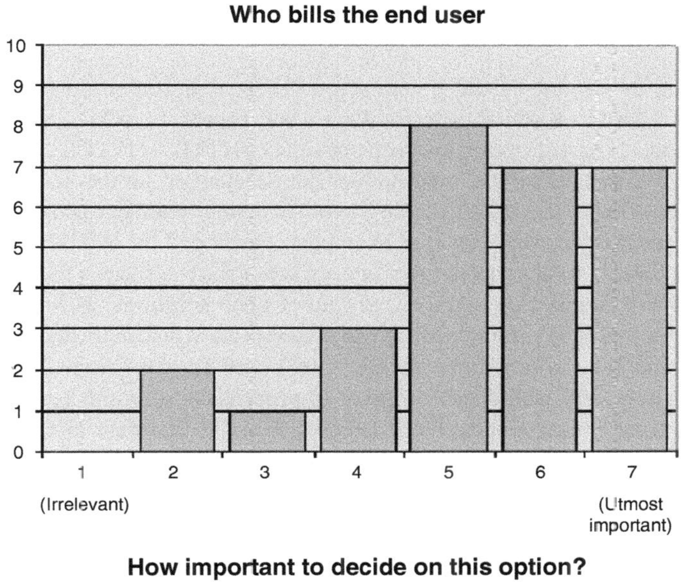

Fig. 4 Billing provider role

the revenue shares of operators. Most of the experts in our service indicated that they considered this a critical issue (mean = 5.36, sd = 1.44, n = 28), see Fig. 4.

Design Issue 5: Security

Success in the demand of a service offering depends to a large extent on the level of security perceived by consumers. Wireless networks have security issues with respect to over-the-air transmissions and additional gateways between wireless and wired domains. Standard organizations such as 3GPP and IETF deal with those security concerns.

However, the original W3C specification for web services standards provides no security. This means external mechanisms or standard modifications have to be put in place, and the issue involved has to do with what the best way is to implement the security requirements involved into MWS technology (Pashtan 2005). Association between mobile device and user ID has traditionally been implemented by the SIM card model, which is not a very sophisticated system. The design choice is thus whether to opt in favor of security in web services technology or for more generic network based security. Security based on web services technology was not a solution most experts supported (mean = 4.75, sd = 1.50, n = 28).

Design Issue 6: Technical Architecture

To take full advantage of the potential of MWS, it is important to develop an architecture based on standard technologies and protocols (XML, SOAP, UDDI, WSDL) and using Internet transport protocols (HTML, HTTPS, SMTP). This architecture would define the building blocks and standard protocols used in MWS transmission. One of the design choices involved here is whether to build proprietary solutions, for instance the ones that have been developed by Nokia and Vodafone (Nokia and Sun 2004) or to opt in favor of open solutions, for instance the Open Mobile Alliance MWS architecture. Actors may favor open solutions, as interoperability may generate value all around. Moreover, consumers would not feel attached to a company and they would have access to a wider service offer. However, big companies may very well decide in favor of proprietary solutions (especially with regard to the technical architecture), keeping in mind that they already control a large portion of the market that they are reluctant to share.

Design Issue 7: Interoperability of MWS and IMS

Many of the MWS functionalities that have been identified, such as authentication, billing, and security, are also offered by IMS. There are two industry giants who are leading in either of the trends: Nokia leads MWS (Nokia and Sun 2004) and Ericsson leads IMS (Levenshteyn and Fikouras 2006). According to North (1990), competition amongst technologies is sometimes based more on the traits of the firms representing the technologies than on the technologies themselves. Therefore, we think that MWS and IMS may be coexisting in the near future, although most operators are betting on IMS instead of MWS. However, content and service providers are implementing web services in their operations and processes, and are therefore more likely to provide mobile web services. Hence, interoperability is required to pull down technical barriers that limit user access to services and to expand the number of services available.

An important discrepancy is that MWS is based on asynchronous communication, while the IMS domain requires continuous SIP sessions that are not yet supported in most WS applications (Levenshteyn and Fikouras 2006). While suggestions have been made to tackle this issue, for instance gateway mechanisms (Levenshteyn and Fikouras 2006), whether or not to implement and develop these types of interoperability mechanisms between IMS and MWS remains a strategic choice. If such mechanisms are not developed, MWS may prove to have a very limited long-term viability. It is interesting to note that our experts were moderate with regard to the need for interoperability mechanisms between web services and IMS (mean = 4.82, sd = 1.2, n = 28), see Fig. 5.

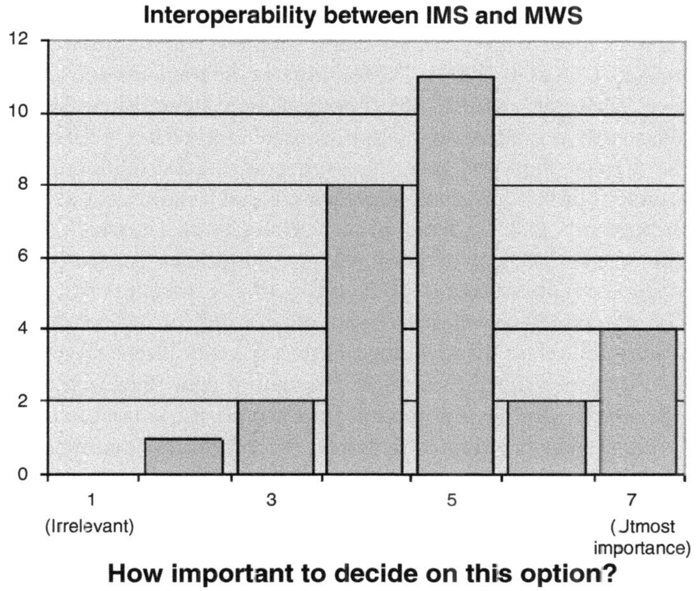

Fig. 5 Interoperability between IMS and MWS

Design Issue 8: Client–Server Architecture

According to the service-oriented architecture, there are three main roles involved in providing web services: requestor, provider and broker. In case the mobile device is the service requestor, one can choose between thick or thin client solutions. In thick client solution, the mobile device acts as the service requestor, and a remote web server as the service provider. The mobile client requests the service from the provider via SOAP messages, and processes the XML messages by itself. The benefits of this approach are that the protocols used to implement on the mobile device are readily available to be installed by handset manufacturers, and that the applications are free to request any service from any service provider based on the user's context and preferences (Pilioura et al. 2003). However, there are many challenges with regard to the thick client model, such as limited processor power to process the XML messages, limited bandwidth to transmit the large-overhead SOAP messages, limited memory in the mobile device, a limited display, finite battery power, and unavailability of the network necessitating iterations in service discovery (Pilioura et al. 2003).

The *thin client model* eliminates XML communication with the mobile device by installing a proxy or 'personal agent' that represents the mobile device (Cheng et al. 2002; Chu et al. 2004; Pilioura et al. 2003) This proxy server takes care of the

XML communication with the service provider and communicates with the mobile device via a traditional WAP gateway, using simple HTTP requests and HTML or WML responses (Chu et al. 2004). This eliminates the transmission and processing of the 'heavy' XML messages by the mobile device. In addition, the proxy could carry out tasks such as content adaptation, content conversion, bookmark management, cache management and other 'housekeeping tasks' (Pilioura et al. 2003). The disadvantages of the thin client approach are that it introduces a single point of failure in the network, making denial of service attacks easier, which raises security issues (Pilioura et al. 2003). In addition, the proxy becomes a central element in the system, putting the party in charge of the proxy in a dominant position. As Pilioura et al. (2003) note, this may well lead to a new situation of walled garden business models. A more technical disadvantage is that the thin client model can lead to unpredictable response times in case of fluctuating data rates (Chu et al. 2004). Hybrid models have also been proposed, for example the smart client model from Chu et al. (2004), which provides dynamic, reconfigurable and adaptive choices between local and external execution of applications.

In addition to the traditional client–server models discussed so far, the role of service provider can also be played by the mobile device, in which case the mobile device offers the web service to the outside world instead of consuming it. There are roughly two ways to implement an architecture in which the mobile device acts as a service provider. One is to have the mobile device acting as the party providing the web service, and the central web server as the receiver, i.e. the provider and requestor roles in the thick and thin client model are reversed (Berger et al. 2003). The advantage of this approach is that it could solve firewall problems, removing the problems of managing a proxy in the thin client model, and users gain greater control over the services they use and the information they disclose, because they have the option to turn off their device (Berger et al. 2003). Berger et al. (2003) mention a large number of applications of this approach. For example, the provider of a content service to an end-user could request information stored on the mobile device, such as context information, location information or personal information. Another possible application is mobile payment, for example a store can request the credit card number from the mobile device when a transaction takes place.

The other approach to mobile devices providing services involves mobile ad hoc networks or peer-to-peer networking (Gehlen and Pham 2005; Niemegeers and Heemstra de Groot 2003; Srirama et al. 2006). In these networks, there is no central server that can provide web services to clients, which means that the devices themselves have to be both the requestors and the providers of the web services. Applications include mobile gaming (Gehlen and Pham 2005), renting content such as songs to other mobile users, or exchanging personal information about the capabilities of the person carrying the mobile device which could be handy in case of an emergency situation (Berger et al. 2003).

There are several issues involved when a mobile device provides rather than consumes a service. For one thing, the service has to be tolerant to longer response times and to temporary unavailability as mobile devices may be switched off due to limited power supply and they may be disconnected as a result of the nature of

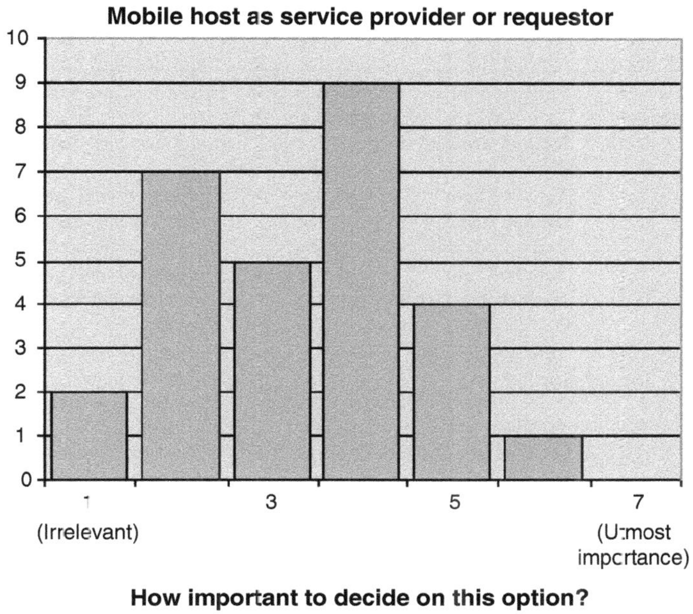

Fig. 6 Having the mobile device as web service provider or requestor

the mobile network (Berger et al. 2003). To take care of changing IP addresses, WSDL files have to be regenerated and UDDI registry has to be updated. However, these updates pose a severe security risk, as location updates in the UDDI may be forged (Pilioura et al. 2003). A related issue is that, because mobile devices may shut down, the garbage in UDDI registries can be expected to be even higher than it is with traditional web services (Berger et al. 2003). In addition, trust between devices, installation of new web services on the mobile device and firewalls could also provide a threshold with regard to the development of this architecture (Berger et al. 2003).

To summarize, according to literature, choosing between the mobile device or the service provider to host MWS may be a critical issue. However, this was not what our experts told us. Most of them did not consider it not such a critical issue after all (mean = 3.3, sd = 1.3, n = 28), see Fig. 6.

Conclusions

MWS have the potential to change the existing value networks in the mobile services industry, as they allow service and content providers rather than operators to offer generic service elements such as authentication, security and billing. However, in order to move toward a more mature stage of the technology and unlock its value, there are a number of technology-related design issues that need to be solved, which include overcoming constraints and limitations of mobile devices and wireless

Table 1 Technology design issues for mobile web services

Critical design issue	Main questions	Main trade-offs
Adaptation of web services protocols	How to improve web services protocols in order to overcome constraints and limitations of mobile devices and wireless networks?	Developing proprietary solutions versus adding new standard protocols to the web services stack
Authentication	Which technology should be used to identify the user's device in a secure and single-sign one approach	SIM card model versus web services-based authentication Operator as authentication provider versus third party
User profile management	How to generate user profiles?	User involvement versus automatic generation
	How to store user profiles?	Centralized versus decentralized storing
	Who owns and manages the user profiles?	Operator versus third party owning the information
Security	How to comply with basic security requirements? How to adapt WS protocols to support security?	Web services-based security versus generic mobile network-based security
Billing and charging	How to implement a reliable real time billing system?	Prepaid-post-paid convergence versus separated schemas
	Who should bill end users?	Operator billing entity versus third party
Technical architecture	Which technical architecture should be chosen?	Proprietary versus open solutions
Interoperability of IMS and MWS	How to assure interoperability between IMS and MWS?	Web Services technologies versus IMS
Client–server architecture	Should the mobile device host or provide web services?	Mobile devices hosting web services versus traditional client–server architecture
		Intelligence in the network versus intelligence in the device
		Thick client versus thin client solutions

networks; role division regarding authentication, billing, and user profile management; choosing between SIM cards or web services-based authentication; interoperability mechanisms between IMS and MWS; and the client–server architecture. The experts we surveyed provided a mixed response to most of these design issues, indicating that they are not yet resolved. A summary of the issues is provided in Table 1.

Limitations and Next Research Steps

We have to keep in mind that MWS is a relatively new technology and that, although expectations are high, it is a technology that has not yet been implemented on a meaningful scale. As the list of design issues is based on exploratory literature research, additional issues could turn out to be relevant. In addition, it is possible that the industry is not yet completely familiar with this technology, which means that the people involved may not be completely clear as to the issues involved. Some of the respondents did indicate that they were uncertain in particular about some of the more technical aspects we discussed.

This study should be regarded as an initial exploratory step in our research. We used the expert survey to obtain initial feedback with regard to our ideas rather than as an extensive validation. The results should thus be interpreted with care, keeping in mind the relatively small number of respondents, who were not intended to provide a representative, random sample. In addition, because we did not pre-test the survey items, interpretation may have been an issue and reliability of scales cannot be assured. Some of the next steps in our research are refining the survey questions based on the initial results and on the comments provided by the respondents, and increasing the respondent base. After these steps have been taken, our aim is to link the design issues to critical success factors, and ultimately to business model performance, taking into account the other three dimensions of the business models.

Acknowledgments The work presented in this paper was carried out within the Freeband User Experience project (www.freeband.nl). We would like to thank Timber Haaker from the Telematica Instituut and Wolter Lemstra from Delft University of Technology for their suggestions in reviewing the paper.

References

Ahmad A, Chandler R, Dharmakhikari A A, Sengupta U (2003) SIM-Based WLAN Authentication for Open Platforms. Technology @ Intel Magazine, August 2003.

Ali Eldin A M T (2006) Private Information Sharing Under Uncertainty: Dynamic Consent Decision-Making Mechanisms. Unpublished dissertation, Delft University of Technology, Delft, The Netherlands.

Alt R, Zimmermann H-D (2001) Preface: Introduction to Special Section – Business Models. Electronic Markets 11 (1): 3–9.

AtosOrigin (2006) Meeting the billing challenge for the Telecom Industry. [Electronic Version]. AtosOriging White Paper. http://www.atosorigin.com/NR/rdonlyres/C5BDE4DF-13F6-4EB3-86A0-3519F46DC958/0/wp_telco_billing.pdf.

Berger S, McFaddin S, Narayanaswami C, Raghunath M (2003) Web Services on Mobile Devices – Implementation and Experience. Paper presented at the Fifth IEEE Workshop on Mobile Computing Systems & Applications, Montery, CA.

Bouwman H, MacInnes I (2006, January 4–7) Dynamic Business Model Framework for Value Webs. Paper presented at the 39th Annual Hawaii International Conference on System Sciences, Big Island, Hawaii.

Camarillo G, Garcia-Martin M A (2006) The 3G IP Multimedia Subsystem (IMS). Merging the Internet and the Cellular Worlds. Wiley.

Cheng S-T, Liu J-P, Kao J-L, Chen C-M (2002) A New Framework for Mobile Web Services. Paper presented at the 2002 Symposium on Applications and the Internet, Nara City, Japan.

Chesbrough H, Rosenbloom R S (2002) The Role of the Business Model in Capturing Value from Innovation: Evidence from Xerox Corporation's Technology Spin-Off Companies. Industrial and Corporate Change 11(3): 529–555.

Chu H-H, You C-W, Teng C-M (2004) Challenges: Wireless Web Services. Paper presented at the Tenth International Conference on Parallel and Distributed Systems, Newport Beach, CA.

Cuevas A, Moreno J I, Vidales P, Einsiedler H (2006) The IMS Service Platform: A Solution for Next-Generation Network Operators To Be More Than Bit Pipes. IEEE Communications Magazine, 44 (8): 75–81.
De Reuver M, Bouwman H, Haaker T (2006) Testing Critical Design Issues and Critical Success Factors During the Business Model Life Cycle. Paper presented at the 17th European Regional ITS Conference, Amsterdam, The Netherlands.
Faber E, Ballon P, Bouwman H, Haaker T, Rietkerk O, Steen M (2003, June 9–11) Designing Business Models for Mobile ICT Services. Paper presented to Workshop on Concepts, Metrics and Visualization, at the 16th Bled Electronic Commerce Conference e-Transformation, Bled Slovenia.
Farley P, Capp M (2005) Mobile Web Services. BT Technology Journal, 23 (2): 202–213.
Forrester (2006) The European Mobile Landscape 2006: Forrester.
Gehlen G, Pham L (2005, January 8–10) Mobile Web Services for Peer-to-Peer Applications. Paper presented at the Second IEEE Consumer Communications and Networking Conference, Las Vegas, NV.
Haaker T, Faber E, Bouwman H (2006) Balancing Customer and Network Value in Business Models for Mobile Services. International Journal of Mobile Communications 4 (6): 645–661.
Jenkins G (2004) GSM White paper. Brilliant Past, Bright Future. GSM White paper by Deutsche Bank. http://www.3gamericas.org/PDFs/gsm_whitepaper_feb2004.pdf.
Killström U, Mrohs B, Immonen O, Pitkänen O, Galli L, De Reuver M, et al. (2006) A New Mobile Service Architecture Addresses Future Mobile Business Environments. Paper presented at the Wireless World Research Forum Meeting, Heidelberg, Germany.
Levenshteyn R, Fikouras I (2006) Mobile Services Interworking for IMS and XML WebServices. IEEE Communications Magazine, 44 (9): 80–87.
Limbu D K, Wah L E, Yushi C (2004) Wireless Web Services Clients Development – Using Web Services Standards and J2ME Technology. Synthesis Journal 5: 151–162.
Niemegeers I G, Heemstra de Groot S M (2003) Research Issues in Ad-Hoc Distributed Personal Networks. Wireless Personal Communications 26 (2–3): 149–167.
Nokia, Sun. (2004) Deploying Mobile Web Services Using Liberty Alliance's Identity Web Services Framework (ID-WSF).
North D C (1990) Institutions, Institutional Change and Economic Performance. Cambridge: Cambridge University Press.
Pashtan A (2005) Mobile Web Services. Cambridge: Cambridge University Press.
Pilioura T, Tsalgatidou A, Hadjiefthymiades S (2003) Scenarios of Using Web Services in M-Commerce. ACM SIGecom Exchanges 3 (4): 28–36.
Popescu-Zeletina R, Arbanowskib S, Fikourasc I, Gasbarroned G, Geblere M, Henninge S, et al. (2003). Service Architectures for the Wireless World. Computer Communications 26: 19–25.
Srirama S N, Jarke M, Prinz W (2006) Mobile Web Service Provisioning. Paper presented at the Advanced International Conference on Telecommunications and International Conference on Internet and Web Applications and Services, Guadeloupe, France.
Tian M, Voigt T, Naumowicz T, Ritter H, Schiller J (2004) Performance Considerations for Mobile Web Services. Computer Communications 27: 1097–1105.
UMTS Forum (2003) Strategic Considerations for IMS – the 3G Evolution.
Weill P, Vitale M R (2001) Place to Space: Migrating to E-business Models. Boston, MA: Harvard Business School Press.

Business Models for Wireless City Networks in the EU and the US: Public Inputs and Public Leverage

Pieter Ballon, Leo Van Audenhove, Martijn Poel, and Tomas Staelens

Abstract Based on a comparison between 15 cases in the EU and the US, this paper outlines and details six typical 'business model' configurations for so-called wireless cities. It attempts to identify the optimal mix between the inputs provided by public bodies to these initiatives, and the leverage and returns gained by them in order to fulfill various policy objectives.

Introduction[1]

Despite a few recent setbacks, it is estimated that a great number of cities worldwide are deploying, or have plans to deploy, wireless broadband networks over substantial parts of their territory (CDG 2005). City authorities increasingly perceive broadband Internet access as a public utility to be provided to their community at affordable prices or even free of charge. As it is often asserted that market forces alone are failing to provide inexpensive broadband services, or are relatively slow to deploy ubiquitous broadband access networks, many of these cities argue that it is their obligation to fill the void. The extents to which they engage themselves, as well as the scale of the initiatives vary significantly. Various forms of public involvement are being tested, and are often being contested by private network providers anxious to protect their investments. Also, while some public authorities,

P. Ballon (✉), L.V. Audenhove, M. Poel, and T. Staelens
IBBT-SMIT, Vrije Universiteit Brussel, Belgium
e-mail: pieter.ballon@vub.ac.be

[1] This paper is based on a research report (Van Audenhove et al., 2006) commissioned by the Centre for Informatics of the Brussels Region (CIBG) of Brussels, Belgium. The opinions expressed in this paper are those of the authors alone and do not necessarily reflect the opinion of the CIBG or the Brussels Government. The authors would like to thank the CIBG for its support. Also, we gratefully acknowledge the assistance of Tim Van Lier and Dorien Baelden (both IBBT-SMIT) in gathering case study material.

B. Preissl et al. (eds.), *Telecommunication Markets*, Contributions to Economics, 325
DOI: 10.1007/978-3-7908-2082-9_19, © Springer Physica-Verlag HD 2009

most notably in the United States, are looking into the possibility of rolling out wireless city networks over the entire city, other 'wireless cities' seem to prefer a more cautious and staged approach, building on existing infrastructure, and with specific policy objectives in mind. In these cases, full coverage is not always the end goal and many initiatives stick to providing hotspots and 'hotzones' (Shamp 2004).

These varying approaches already illustrate that there are important choices to be made regarding the rationale and design of wireless city networks, and that public bodies are coming up with widely diverging answers in this respect in different parts of the world. Various authors have described the municipal wireless movement in the US as a significant and potentially disruptive phenomenon (Shamp 2004; Bar and Park 2006; Gibbons and Ruth 2006; Gillett 2006). Developments in Europe have until now received less attention (Kramer et al. 2006). Some authors have also described the various (theoretical) 'business model options' for municipal authorities to engage in wireless city deployment (Bar and Park 2006; Lehr et al. 2006). Much less has been written about the actual business and exploitation models currently employed in wireless city networks, often in the form of public–private partnerships. Judging by the lack of documentation, the organizational, financial and architectural choices that together determine the 'business model' of a wireless city initiative are often neglected or conceived in a haphazard way. This paper aims to fill this gap by outlining and detailing the various business models behind wireless city networks in both the EU and the US, with a specific focus on the roles that were taken on by public bodies, the inputs provided by them into the initiative and the influence exercised by public authorities regarding the deployment and exploitation of these networks.

This paper features results from a comparative analysis of 15 cases in 14 cities, 8 of which European and 6 American. It focuses on large cities and/or large-scale initiatives, in which public bodies play at least a minimal role. While the main driving force behind the wireless network roll-out in the selected cases can be a public body as well as a private organization or a community of individuals, purely private roll-out of WiFi or WiMAX networks was not taken into account. As wireless city networks are a recent phenomenon, with little initiatives being fully operational at this time, the cases are in various stages of development. However, all cases studied have already made a number of basic choices regarding the 'business model' to be used, and can thus be compared in this regard. This paper distinguishes six basic business model types, each of which is illustrated with one specific case (see below).

The table below provides an overview of the cases, the phase of the initiative as of end 2006, and the key driver of the project (Table 1). The selected cases used to illustrate the basic business model types are: Bologna, Bristol, Portland, St. Cloud, Stockholm and Turku.

The paper is structured as follows. After this introduction, the second section distinguishes six typical configurations of specific 'business models' found in practice. The next sections discuss an exemplary case of each configuration. The final section of the paper offers a succinct comparison of the six models in terms of the public inputs and returns.

As a point of departure, it needs to be taken into account that heavy public involvement does not necessarily constitute the most effective way to achieve

Table 1 Description, phase and key driver of wireless cities

City	Short description	Phase	Key driver
Bologna (IT)	Iperbole wireless network: experimental WiFi network providing wireless internet access to selected groups	Pilot	Public: City of Bologna
Boston (US)	Gradual expansion of Boston Main Streets WiFi project providing wireless internet access to entire city	Request for proposal	Public: Boston Main Street
Bristol (UK)	Bristol hot zone: WiFi hotspot zone providing wireless internet access and walled garden services	Operational	Public: City of Bristol
Cardiff (UK)	BT open zone: WiFi hotspots and zones providing wireless internet access	Operational	Private: British Telecom
Leiden (NL)	Wireless Leiden: community network of wireless nodes sharing internet connections	Operational	Local Community
Paris a (FR)	Establishment of 400 WiFi access points	Information phase	Public: City of Paris
Paris b (FR)	Site provisioning to private operators with the objective of full WiFi coverage of Paris	Information phase	Public: City of Paris
Philadelphia (US)	Wireless Philadelphia: large-scale WiFi network providing wireless internet access	Roll-out	Public: City of Philadelphia
Portland (US)	WiFi/WiMAX network providing wireless internet access to citizens, companies and city workers	Tendering phase	Public: City of Portland
Sacramento (US)	Large-scale WiFi network for wireless internet access and additional services	Tendering phase	Public: City of Sacramento
San Francisco (US)	WiFi network covering the entire city for wireless internet access	Request for proposal	Public: City of San Francisco
Saint Cloud (US)	Cyber Spot: full coverage of city with WiFi/WiMAX network providing wireless internet access	Operational	Public: City of Saint Cloud
Stockholm (SW)	Stockholm mobile connect: WiMAX network providing wireless internet access	Roll-out	Public: City of Stockholm
Turku (FI)	OpenSpark: WiFi community network providing wireless internet access	Operational	Private/Local Community: Sparknet
Westminster (UK)	WiFi network for closed circuit television and other services	Operational	Public: City of Westminster

public policy goals (see also Bar and Park 2006). In a context of concerns over unfair competition and limited public funds, it can be expected (see e.g. Osborne 2000) that public authorities will opt for the optimal trade-off between minimizing their inputs for establishing and operating wireless city networks (i.e. in terms of investments and risks), and maximizing their leverage for reaching specific policy goals (i.e. narrowing the digital divide, stimulating innovation, and so on). The objective of this paper is to highlight the most promising wireless city business models in this respect. However, the case studies also illustrate that the optimal model partly depends on contextual objectives and circumstances, including national and supranational regulation related to safeguarding competition.

Business Model Configurations

Most literature on wireless city networks is rather vague on the various business and exploitation models that are conceivable and/or used in practice. Bar and Park (2006) form an exception to this rule in the sense that they propose a typology of nine business models. This typology is constituted by all potential combinations between two key roles (i.e. network ownership and network operation) that can each be taken up by three types of actors (i.e. public, one private actor, multiple others). The authors stress that the resulting models are theoretical archetypes, and admit that several of these combinations are not found in practice and even quite unlikely to ever occur.

Our paper adapts the Bar and Park typology to match more closely the real-life cases in Europe and the US. A first adaptation concerns the basic value-adding roles that are central to the business model. We distinguish network ownership and service provision (i.e. providing the service to customers and maintaining the customer relationship), instead of network ownership and network operation. This is because owning the physical assets on the one hand, and owning the customer relationship on the other hand, are the most fundamental business roles that can be distinguished in these cases, and because they are often taken up by different actors, while operating the infrastructure in itself is a less crucial role, and is usually either combined with network ownership or with service provisioning. Also, the "Multiple others" category of Bar and Park is split into "multiple private actors" and "community", as these actors are governed by a fundamentally different business logic, i.e. a wholesale logic versus a voluntary, often not-for-profit logic. Thus, at the level of network ownership we can distinguish between:

- Private player: The network is operated on the basis of a contractual arrangement in the form of a license, concession, etc. The municipality contributes by providing access to sites, existing backbones, financial support, etc.
- Public player: The municipality owns the network and operates it itself.
- Open site: The municipality provides open access to sites for the construction of networks.
- Community player: The network is operated by a community of individuals and/or organizations.

At the level of service provision we can distinguish between:

- Private player: One private player provides access to services on the network.
- Public player: A public or non-profit actor provides access to services on the network.
- Wholesale: Various private players build on a wholesale access offer and provide services to end users.
- No specific ISP: There is no specific party providing access to end user services (e.g. because only point-to-point data links are provided).

Finally, combinations that are impossible or highly unlikely have been omitted, and configurations resulting in very similar business models were joined together.

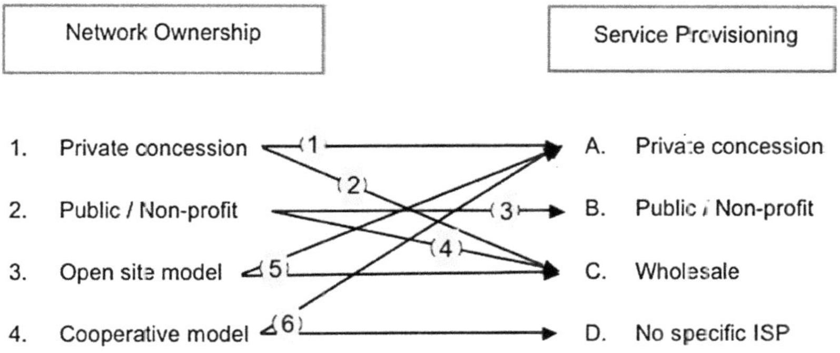

Fig. 1 Typical business model configurations for wireless cities

As a result, six basic combinations can be distinguished, which are graphically displayed in the figure above (Fig. 1).

The following sections characterize these combinations and illustrate them with an exemplary case and business model. As mentioned, specific points of attention concern the inputs provided by public bodies and their influence on the conditions and characteristics of the wireless initiatives.

Inputs provided by public bodies may range from mere facilitation (e.g. through advice or actively bringing private stakeholders together), giving access to public amenities (e.g. by providing access to public sites for putting up antennas or by providing backhaul through a municipal backbone), licensing and/or granting exclusivity, acting as launching customer, providing limited or full financial support, to public network roll-out, operation or service provision.

The resulting leverage municipal authorities obtain may refer to influencing market structure, setting rules regarding openness, promoting societal goals, stimulating or requiring specific applications to be offered, ensuring a certain geographical coverage, receiving direct or indirect financial returns, and influencing tariffs and prices (e g. by making the service free, by setting price caps, or by providing special tariffs for disadvantaged citizens or for city employees).

Private–Private Model

Bristol, Cardiff, Paris (a) and Westminster are examples of this model. Complex tendering procedures were used to select the private organization(s). An important difference exists between Bristol and Cardiff on the one hand, and Paris (a) and Westminster on the other hand. In the former instances, the network is in large part financed by the private actor. In the latter instances, the cost of the network is borne by the municipal authority, that outsources network ownership, operation and service provision to private concession holders. In all instances, the initiatives are limited in scale or scope. They either cover a small geographical area, or the service provision is only aimed at municipal employees. It is clear that these limitations are due a.o. to concerns over unfair competition. Head-on competition with completely private provision of internet services to the general market is in all instances avoided.

The overall business model for the Bristol case is depicted below (Fig. 2). It illustrates that the roles of *Network building*, *Network operation* and *Service provision* are all executed by the same private actor, i.e. CitySpace. In return for the exclusivity of site provision, together with limited initial financial support, the influence of the public authority is at a level that can be characterized as low-to-medium. In this case, this means that the city can demand that free internet access is granted to users for a limited time per day. In the longer term, the city of Bristol is striving towards the establishment of a larger wireless cloud. In this case, the city would consider opening up sites to additional private actors (and would thus be migrating towards an open site model). Also, network operator CitySpace is currently considering opening up its network to additional service providers (and thus moving towards a private–wholesale model).

On the basis of the cases considered, it can be said that the private–private model seems to be most suited for limited initiatives, given a.o. concerns over allegations of unfair competition. In attendance of clear European guidelines regarding unfair competition in this field, it largely remains an open question to what extent open tender procedures and market conform pricing can alleviate such concerns. Typical inputs by the public authority include site provision and sometimes also financial support. The extent of the public influence in the private–private model with no or limited public funding can be characterized as 'low to medium', with some influence

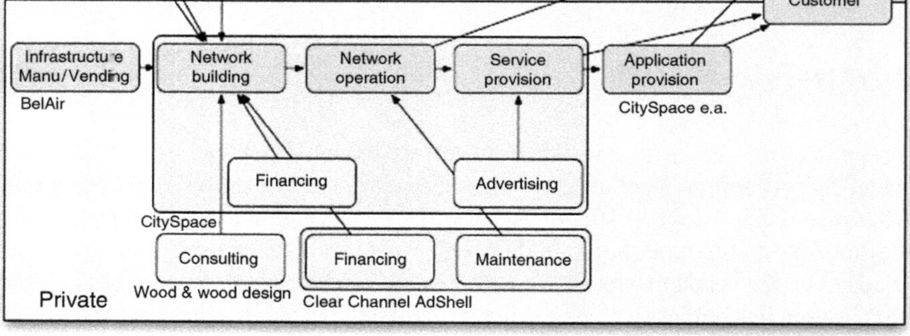

Fig. 2 Business model configuration of Bristol case

on tariffs typically being exercised by the public body. In case of an outsourcing scenario, the financial input as well as the resulting influence by the city naturally is very high, and is in that case similar to the public–public model.

Private–Wholesale Model

Most recent, large US cases in our sample follow this model, including Philadelphia, Portland, Sacramento and San Francisco. The baseline of this model is that the government grants (often exclusive) access to public sites and facilities (such as lampposts or traffic lights) on which a private actor can build its wireless network. The right to use these sites is compensated by private network owners by (1) direct financial returns to the city, (2) granting inexpensive network access to city employees, and/or (3) agreeing to price caps for (particular groups of) citizens.

The figure below depicts the business model behind Portland's wireless city (Fig. 3). Despite the fact that the city authorities were the initial drivers behind the project, the municipality plays a rather limited role in the eventual business model. The most important actor is MetroFi, a private actor who will fund the network roll-out and will operate the network.

The city of Portland grants access to lampposts, traffic lights and public buildings. In return, MetroFi pays fees for the use of these facilities. Also, the city becomes the main 'anchor tenant', or 'launching customer' of MetroFi's network. In addition to a paid, high speed data access subscription for all citizens, MetroFi will provide a free basic access service to the inhabitants of Portland. This will be financed

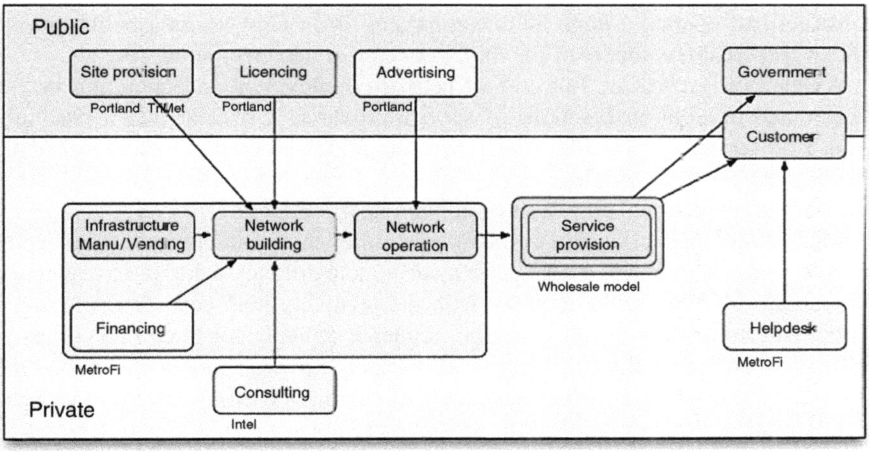

Fig. 3 Business model configuration of Portland case

by advertisements. The modalities of the advertising scheme are regulated by an agreement between the city and MetroFi.

On the whole, the private–wholesale model is regularly used for large meshed wireless clouds, whereby service is aimed at the public at large. Advantages to public authorities are that the (financial) inputs required from them remain limited, that they retain a relative high level of influence, and that there is less risk of claims of unfair or distorted competition. However, the case of Sacramento provides an example of the limits to this strategy in terms of influence exercised by public authorities. In this city, the designated operator MobilPro withdrew from negotiations because it claimed that the city's requirements in terms of caps on tariffs and free access for designated user groups rendered profitable exploitation impossible. More recently, EarthLink, the company involved in many US projects including San Francisco and Philadelphia, has scaled back and put on hold several of its activities in municipal WiFi projects. It claims that too much risk is currently being borne by the private stakeholders, implying that cities should co-finance WiFi roll-out by acting as anchor tenants, i.e. by guaranteeing paid public usage of the WiFi infrastructure.

Public–Public Model

Only one initiative in our sample falls within the public–public category. This is the US town of St. Cloud. It should be noted that this is a specific case in the sense that St. Cloud is a small city of only 23,000 inhabitants, yet the initiative still has an important scope, as it comprises complete coverage, both in- and outdoors, on the territory of the city.

The network was installed in 2005–2006. The philosophy was that internet access should be free and accessible to all. Therefore, the municipality chose to construct and operate a large all-encompassing WiFi network as a public utility. The overall business model of the St. Cloud case is depicted below (Fig. 4).

As the city carries the full cost of network deployment and operation in this model, and in addition functions as service provider, it is clear that it has 'full' impact on services and tariffs. As a result, the St. Cloud case is the only example of a completely 'free' wireless service offering. Financing is assured through taxes and through operational cost savings in municipal services.

Meanwhile, legislation and court rulings aimed at restricting municipalities to 'distort' competition in this market have made it more difficult in the US to implement this model. This evolution, together with the relatively high costs of deploying a large WiFi network with full coverage, render it unlikely for large (US) cities to follow this model. In the EU, there was no experience with this model among the cities in our sample. And while the advent of (mobile) WiMAX may diminish the cost of network roll-out considerably, the fact that (mobile) WiMAX, in order to be an efficient network technology, needs to operate in licensed spectrum, makes it unlikely that cities will build and operate their own wireless broadband network.

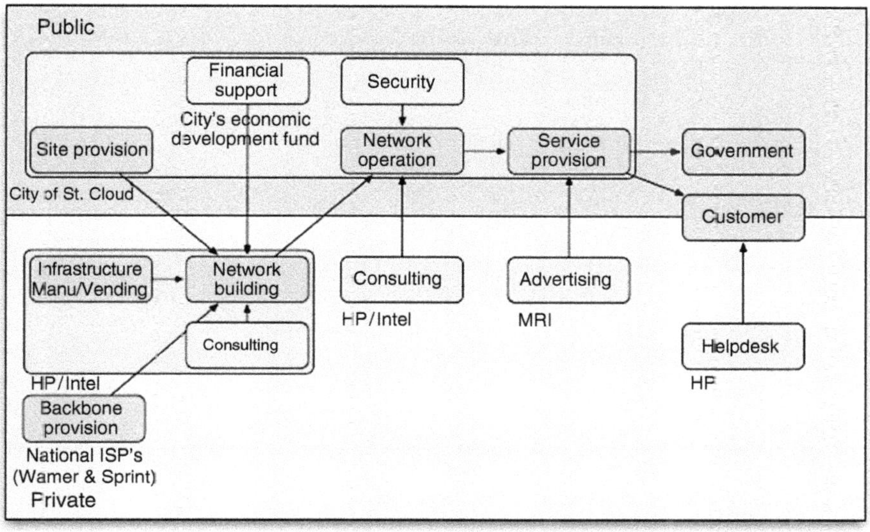

Fig. 4 Business model configuration of St. Cloud case

Public–Wholesale Model

In the public–wholesale model, the network is financed and operated by the municipality or by a non-profit organization set up by the municipality, and is subsequently opened up in a wholesale arrangement to service providers. Stockholm and Boston are examples of this model. However, both initiatives are still in the planning and information phase at the time of writing this paper, and are still considering several options. In Stockholm, Stokab is set to roll out a WiMAX network and to open it up for various service providers. Stokab is owned by the municipality. The business model is depicted below (Fig. 5).

This model is similar to the private–wholesale model in terms of service provision. The expectation is that allowing several service providers to make use of the wireless infrastructure will enable innovation and affordable tariffs, while alleviating concerns over unfair competition. Particular motivations to opt for this model include: assuring control over the network and over future network investments (either exclusively by the municipality or a non-profit organization [Stockholm], or by a public–private partnership [Boston]), and ensuring a level playing field for all service providers. The roles and inputs borne by the public body are manifold, and public investment is relatively high, but service provision and marketing is left open to private players. The influence of public bodies on the initiatives can be rated as medium-to-high, as cities can set a range of criteria that service providers need to comply with, but are still dependent upon (and often uncertain about) the interest of a sufficient number of private players who have to make a business proposition to the citizens.

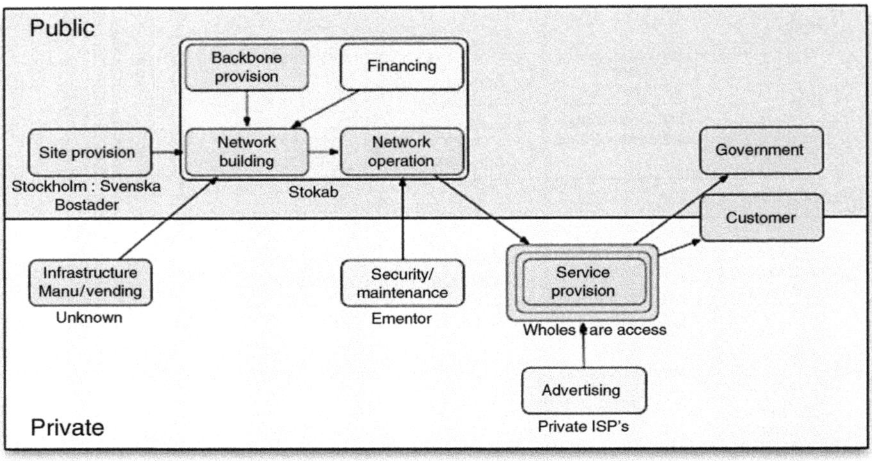

Fig. 5 Business model configuration of Stockholm case

Open Site Model

Paris (b) and Bologna are examples of what can be labeled an open site model. In the open site model, the city grants open access to a number of public sites or to its backbone network for whoever wishes to roll out a wireless network. The objective in this case is to have several private network builders and operators competing among each other. Given the fact that (some) infrastructure competition is aimed for, it is not likely that the municipality will use the concession or contract, linked to the use of public sites, to enforce a wholesale model at the level of service provision upon network owners. Private operators may of course still decide to adopt a wholesale model.

In this model, concerns over anti-competitive practices are minimal or non-existent. The role and input of public bodies is also minimal, and is limited to granting equal access to network operators. It can be expected that the influence of public authorities on prices and services will also be low. Both cases provide examples of this. In the Paris (b) case, the municipality has declared that it will open public sites to private actors that wish to roll out a wireless network, and that no (direct) public influence is sought after. By granting access to sites, the city merely hopes to better enable the deployment of private networks, and to stimulate competition in the broadband market. In Bologna, there have been attempts to combine this model with the requirement of limited free access for all citizens. However, as a result, only one private actor expressed an interest in network roll-out, and only one actor expressed an interest in service provision. The municipality is now exploring other models to follow up the pilot phase.

This does not mean that the open site model is not feasible. It is suitable in cases of high latent interest from private actors to enter the market. There is low direct public impact in terms of tariffs or geographical coverage, but there may be an

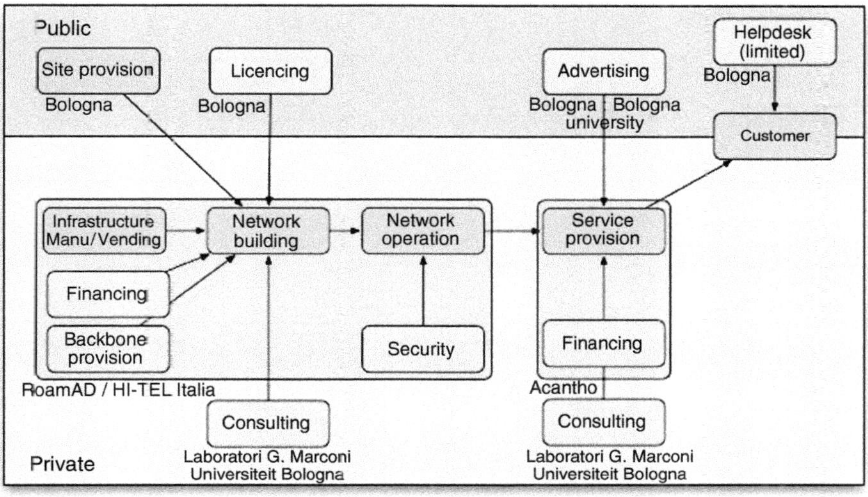

Fig. 6 Business model configuration of Bologna case

indirect impact through increased competition. The current business model of the Bologna case (i.e. with only one network owner and operator, and service provider, respectively) is depicted in (Fig. 6).

Community Model

In the community model, a group of individuals and organizations links up to form a wireless network. This can be disruptive to private as well as public models. Leiden and Turku are fairly vibrant examples of such a model. Several implementations are feasible. In Leiden, the network is built and operated by the wireless community itself. There is no specific ISP. In Turku, a virtual network layer is built upon existing private broadband networks. Members of the Turku Sparknet community open their routers (nodes) to all other members. The private company Openspark that provides the routers, is in essence the service provider of this virtual network layer. It authenticates and provides access to members (for free) and to guests who do not own a router (paid access). In other words, the community 'mandates' Openspark to be the private service provider. The overall business constellation of the Turku case is depicted below (Fig. 7).

In both cases, the municipality plays a similar role. Leiden as well as Turku have opened up the municipalities' own nodes for the virtual network, and have played a general facilitating role. The city of Turku also bought and distributed 500 additional access points. In general, both the inputs and the leverage of public bodies are quite low in this model and there seems no possibility for public bodies to influence the pace, scope and focus of the initiatives.

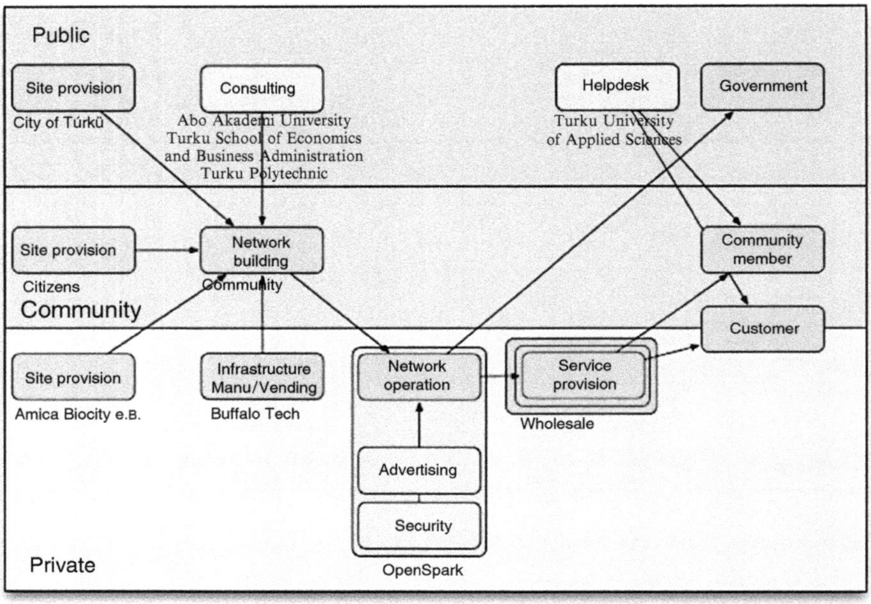

Fig. 7 Business model configuration of Turku case

Conclusion: Comparison of Public Inputs and Returns

Comparing the public inputs and leverage associated with each of the models, it is clear that cities around the world experiment with and adopt various 'business models' to achieve often similar returns. However, Table 2 illustrates that a number of clear links between public inputs and (envisaged) leverage or influence exist.

The table shows that the majority of initiatives aims to limit public involvement. Important considerations in favor of this include the desire to minimize public spending, and concerns over unfair competition. It is a recurring subject of debate whether there is a sufficient degree of market failure in the (wireless) broadband access market to justify any public involvement at all, whether any public gains are sufficiently high compared to the cost, and whether WiFi is mature enough as a technology to be used for city-wide coverage. Criticisms targeted e.g. at the market definitions and the business models used by municipalities are being raised by established mobile and fixed network operators and even by new entrants betting on new technologies (e.g. low-cost WiFi mesh, or WiMAX). It is outside the scope of this paper to provide an analysis of the relevant regulatory and legal frameworks, and of specific local circumstances. Incidentally, our cases did refer to rules being tightened in this regard in the US. Both in the US and EU cases, it appears that there is considerable uncertainty over what level of public involvement is permitted and what not.

Table 2 Public inputs and leverage in wireless cities

City	Input	Description	Leverage	Description
1. Private–private model				
Bristol	Low	– Site provision – Co-financing of pilot	Low/medium	– City has the right to offer municipal services within walled garden environment – Limited period of free Internet, financed by advertisements
Cardiff	Low	– Site rental	Low/medium	– Limited number of free accounts for city employees – City collects rental fee
Paris (a)	Very high	– Full network financing – Site provision – Outsourcing of network operation and service provision	Very high	– Outsourcing contract – Free access to hotspots for all citizens
Westminster	Very high	– Full network financing – Site provision – Outsourcing of network operation and site provision	Very high	– Outsourcing contract – Only dedicated services for municipality are offered
2. Private–wholesale model				
Philadelphia	Low	– Site rental – Exclusive license for 10 years – City as 'anchor tenant'	Medium/high	– Wholesale offering – License and rental fees – Limited coverage requirements – Price cap on wholesale tariff – Low subscription rate for socially disadvantaged households – "Free hotspots" at limited number of strategic locations – Number of free accounts for city employees
Portland	Low	– Site rental – Non-exclusive license for 10 years – City as 'anchor tenant'	Medium/high	– Wholesale offering – License and rental fees – Free, advertisement-based basic service, next to paid service – Preferential service for municipal services

(continued)

Table 2 (continued)

City	Input	Description	Leverage	Description
Sacramento	Low/medium	– Site provision – Access to city backbone network for backhaul – License for 5 years – City as 'anchor tenant'	Medium/high	– Initially, free subscriptions for all were demanded by city; now, this is being re-examined – Current plans involve limited basic free service and subsidies for socially disadvantaged accounts – Free access for schools – Preferential service for municipal services
San Francisco	Low	– License for 10 years	Medium/high	– Free, advertisement-based basic service – Negotiations on financial return for city
3. Public–public model				
St. Cloud	Full	– Fully public financing, ownership and operation of the network	Full	– Full control over coverage, services – Completely free access
4. Public–wholesale model				
Boston	Medium	– Site provision – Set-up of non-profit organization for building network and making wholesale offering to service providers – Limited co-financing by city	Medium?	– Not known as project is still in information phase
Stockholm	High	– Site provision – Building network and making wholesale offering to service providers through non-profit organization	Medium?	– Not known as project is still in information phase
5. Open site model				
Bologna	Low	– Site provision to multiple actors	Low?	– In the pilot phase, a limited free access service was demanded by the city. It is recognized that this requirement is probably 'untenable' after the pilot phase within the present model
Paris (b)	Low	– Site provision to multiple actors	Low	– Stimulus for competition
6. Community model				
Leiden	Low	– Site provision – Subsidy of one specific application	Low	– Some influence on topology of network by integration of city's own nodes
Turku	Low	– Site provision – Provision of additional access points	Low	– Some influence on topology of network by integration of city's own nodes and additional access points

If a municipality does choose to finance infrastructure roll-out, it may opt for the private–private (outsourcing), public–public or public–wholesale models. Outsourcing with enforcement of market-oriented prices may alleviate unfair competition concerns, but if there is already a competitive (fixed or mobile) broadband offering, such a model, just as the public–public model, may still be controversial. In cases of existing public (fixed) broadband investment, and doubts over the willingness to invest of private actors, a public–wholesale model may be suitable. It offers control over geographical scope, scale of investments and quality, and may limit concerns of unfair competition because it is open to several service providers.

In case the municipality does not have sufficient resources to finance network roll-out, the open site and community models offer 'low input-low influence' models that can be adopted. While offering minimal direct public impact in terms of pace, scope and focus, they may nevertheless yield indirect advantages that still (partly) fulfill public objectives. The private–private model with no or with limited public funding seems to be most suited for smaller, targeted projects.

For large initiatives, the private–wholesale model (often however with one or two preferential service providers) may yield the most optimal trade-off between public inputs and influence. It is adopted by most of the recent, large and high-profile wireless city initiatives in the US. Key elements of this model are site provision by the municipality, combined with some form of licensing. Usually, and increasingly, the city also acts as an anchor tenant and is very intensively involved in negotiations with the private network owner and one or more initial service providers prior to the launch. Returns are in the form of license and rental fees, guaranteed geographical coverage, cheap access for municipal employees, cheap or (limited) free access for disadvantaged groups or for citizens in general. That there is a serious limit to what cities can expect in terms of public returns, is illustrated by the Sacramento case, where the selected operator withdrew from negotiations because of public demands that were judged excessive, and by the scaled back activities of EarthLink in municipal WiFi in several US cities. In spite of this, the model appears to be continuously popular because of the relatively high public influence on tariffs and services, the limitation of public responsibilities and risks after the start-up phase, and less prominent concerns over unfair competition. However, given the recent setbacks it remains an open question under which specific conditions this model is viable, and whether it can be successfully exported to an European context and to less high-profile cities.

References

Bar F, Park N (2006) Municipal Wi-Fi Networks: The Goals, Practices, and Policy Implications of the US Case. Communications & Strategies 61 (1): 107–126.

CDG (2005) Something in the Air. Government on the Go Through Community-Wide Wireless. Centre for Digital Government, Washington, DC.

Gibbons J, Ruth S (2006) Municipal Wi-Fi: Big Wave or Wipeout? IEEE Internet Computing 10 (3): 66–71.

Gillett S (2006) Municipal Wireless Broadband: Hype or Harbinger? Southern California Law Review 79: 561–594.

Kramer R, Lopez A, Koonen A (2006) Municipal broadband access networks in The Netherlands – three successful cases, and how New Europe may benefit. In: Proceedings of the 1st International Conference on Access Networks, Athens, Greece.

Lehr W, Sirbu M, Gillett S (2006) Wireless Is Changing the Policy Calculus for Municipal Broadband. Government Information Quarterly 23 (3–4): 435–453.

Osborne S (2000) Public–Private Partnerships: Theory and Practice. Routledge, London, 348 p.

Shamp S (2004) WiFi Clouds and Zones: A Survey of Municipal Wireless Initiatives. Mobile Multimedia Consortium Paper, University of Georgia.

Van Audenhove L, Ballon P, Poel M, Staelens T, Van Laer T, Baelen D (2006) Urbizone: Internationale Best Practice van Wireless City Networks. SMIT Report, Brussels, Belgium.

Managing Communications Firms in the New Unpredictable Environments: Watch the Movies

Patricia H. Longstaff

Abstract This paper looks at a puzzle facing many industries: How can they survive and thrive in rapidly changing and unpredictable environments? The ideas in the paper are intended to be useful to many industries and firms. The movie industry is used as an example to test the applicability of some new work being done on resilience in unpredictable systems. Previous work has already shown that the U.S. movie industry (often referred to as Hollywood) is one of the least predictable industries in the world (in terms of which movies will be big hits) and it suffers some giant failures every year. Yet it is also one of the most successful and stable industries in the world, accounting for a very large share of U.S. exports every year. What works for the movie moguls may work for other organizations that find themselves in unpredictable environments. Movie revenues follow a power law distribution in each year. The reasons for this and the implications for strategy in similar environments are examined, including the implications of treating communications firms as complex adaptive systems. Rules of thumb are distilled that may be helpful for other industries and firms. These rules appear to be consistent with several widely accepted management theories, making it appear more likely that the rules from other systems will be relevant in developing resilience in business systems.

Introduction

Anyone who manages an organization will tell you that the number of "surprises" they deal with is growing all the time. In fact, most of us seem to keep busy just managing the crisis *du jour*. How do you manage an organization or make strategic plans in an environment where you can't predict what will happen tomorrow to

P.H. Longstaff
Center for Information Policy Research, Harvard University, S.I. Newhouse School of Public Communications, Syracuse University
e-mail: phlongst@syr.edu

B. Preissl et al. (eds.), *Telecommunication Markets*, Contributions to Economics,
DOI: 10.1007/978-3-7908-2082-9_20, © Springer Physica-Verlag HD 2009

your most important resources? There seems to be some tantalizing clues in current work being done on the concept of "resilience" – the ability of a system to bounce back after a bad surprise. It is being studied in systems as different as the human immune system and natural ecosystems. While resilient systems have many differences they seem to have a few things in common and it is those common threads that offer some hope for managing under uncertainty.

Managing Uncertainty?

Uncertainty is not new and it is not always unwelcome. Surprises often present both a danger and an opportunity for business organizations. Once you acknowledge it, uncertainty can become an offensive or defensive strategy. You can use it to disrupt your competitors and to protect your assets. But often it is not a weapon at all, but a fact of life in complex, evolving environments. Scott Snook of the Harvard Business School has taken an in-depth look at a tragic friendly-fire accident in the immediate aftermath of the Persian Gulf War in which two U.S. fighter planes shot down a U.S. helicopter. He asks why nobody predicted the problems that led to this accident before it happened. He concludes:

> Part of the answer lies in our inherent limitations as information processors. Part of the answer lies in our linear deterministic approach to causality. Part of the answer lies in the inherent unpredictability of events in complex organizations. (2000, p. 204).

Surprises happen. Both good things and bad things should surprise us when they happen but we should not be surprised *that* they happen. To the extent that surprises will affect investments, operations, or personnel, planning for the survival of those assets becomes a logical focus for modern business strategy. This often means building flexibility and diversity into the options available for critical resources. But as Snook points out, diverse adaptations that make life easier locally have tragic results when those local adaptations are unknown to outsiders who expect the Standard Operating Procedure (SOP) response. Snook calls this type of local adaptation "practical drift" and suggests that it cannot be eliminated, but it can be acknowledged as a potential problem.

Accepting uncertainty is not easy. It is much easier to believe that if you just had the right information you could predict surprises before they happen and build a fool-proof strategy to deal with them. A better way to manage uncertainty is to do so deliberately and strategically. Just because things are complex and uncertain, it does not mean they are unmanageable or ungovernable. The management just takes different forms and makes different assumptions. Where we had come to expect certainty, we are now (sometimes reluctantly) accepting the necessity of dealing with uncertainty.

This is not a new idea. The role of surprise in business management was first noted by Frank Knight in 1921 and echoed by John Maynard Keynes in 1936. Both rejected the possibility that the variables acting on a modern business can be measured and put into a universal mathematical formula or probability calculation. Knight reasoned that uncertainty was common in business systems because surprise was so common, even in the face of much calculation of probabilities. He believed these calculations were not helpful because the business world is ever-changing and no two transactions are exactly alike. Therefore, it is not possible to get enough examples of independent actions necessary for the Law of Large Numbers (which says that even where individual things happen in a random way, if you put enough of these events together in a reasonably stable environment you can make predictions that have a sporting chance of being accurate).

Burton Klein was one of the first economists to note that firms facing high uncertainty (such as new technology) must be managed differently.

> Highly adaptive organizations are required if the best use is to be made of a technology. .. However, the more structured and predictable firms become, the less adaptive they are likely to be. Generally speaking, highly structured organizations are inefficient when dealing with changes in their environments. (1977, pp. 56–57)

Nobody knows unpredictability like Hollywood. So the movie industry seems a good place to test some new work being done on resilience in unpredictable, networked systems. This new work is being done in many disciplines (including such diverse fields as mathematics, political science, ecology, and business management) and has revealed the possibility that complex, unpredictable systems, particularly those that operate as a network, seem to have some things in common. And, while you can't predict the individual events in these systems, you may be able to make some predictions about the kinds of strategies that will make them more resilient to failure. Previous work has already shown that the U.S. movie industry (often referred to as Hollywood) has some of the least predictable products in the world (in terms of which movies will be big hits) and it suffers some giant failures every year (DeVany and Walls 1996). Yet it is also one of the most stable and successful industries in the world, accounting for a very large share of U.S. exports every year (Litwak 1986; Vogel 2001). What works for the movie moguls may work for other organizations that find themselves in unpredictable environments.

It's not that people haven't made attempts to find the right formula for success in Hollywood movies, and things like sequels have some predictive power, but even the tried and true can be flops and that goal of predictable investment remains elusive. There are still surprise hits and surprise flops, and there are many more flops than hits. The conventional wisdom in the industry is that only one movie in ten ever makes a profit. This ratio is somewhat misleading because Hollywood accounting methods for "profit" are unusual, but the general idea that most films do not make money seems to be correct.

Every experienced Hollywood executive can give many examples of projects they did not predict correctly. These executives have been disappointed by many theories about box office success – theories put forward by both the creative people in the business (directors, writers, etc.) and by business analysts with sophisticated statistical analysis tools. Hollywood moguls are likely to roll their eyes and cut off any presentation that tries to posit predictability in their business. It is not easy for them to explain this unpredictability to investors and the corporate boards who now own most studios. But what they know from hard experience can be demonstrated with statistical analysis and some theories being developed in other hard-to-predict systems.

Anyone in the movie business will tell you that theirs is a very complex business, and it turns out that they are right in both the technical and the everyday use of that word. A recent paper that looked at budgets, revenues and profits from the 125 top budgeted films for the years 1996–2002 (Longstaff et al. 2004) confirms initial work examining more limited data (Sornette and Zajdenweber 1998) that block-buster movies follow a power law distribution in all years studied. Unlike a "normal" distribution, where there are a few events at the high and low end, with most events in the middle, a power law distribution indicates that extreme events are happening in a very few cases (blockbusters) but there are many more events (other movies) that are not successful and very few events in the middle range.

Movies and Power Laws

Below are two graphs based on 2002 movie data (Figs. 1 and 2). The graphs represent the budget and revenue distributions. Both indicate that the distributions are not bell-shaped normal curves but look more like a power law distribution. In statistics literature this is also characterized as a Gamma distribution (a power law distribution is part of the family of gamma distributions).

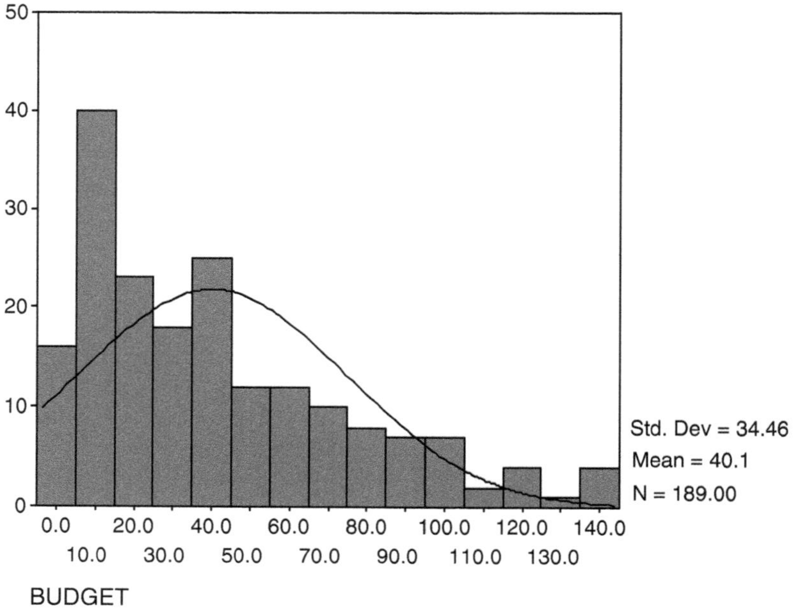

Fig. 1 Budget of hollywood movies 2002

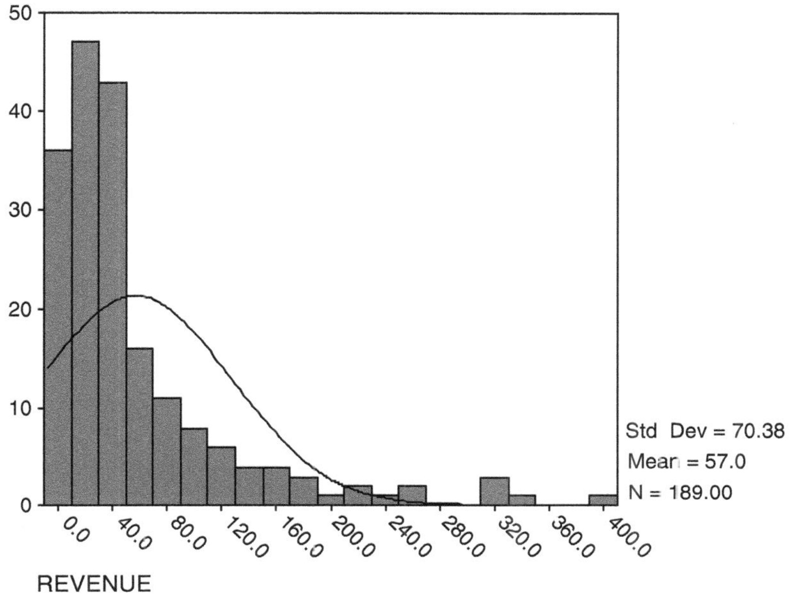

Fig. 2 Revenue of hollywood movies 2002

A power law distribution is often the signature of a "complex" and unpredictable system. A power law pattern is also said to be "the mathematical face of a special architecture, an architecture that is dominated by especially well-connected hubs" (Buchanan and Nexus 2002, p. 85). These systems are often subject to cascading behavior, where the response of the system is hypersensitive (because of its unstable structure) to individual events that have unpredictable effects on the entire system. This is usually illustrated with the image of a sand pile. Most of the next grains of sand to fall on the pile will stick to the pile, a few may cause a small landslide that collapses part of the structure, and a very few will cause a large landslide that collapses much of the structure (Bak 1996). There are some clues from these types of cascading, networked systems that can be useful for managers in an industry subject to a power law distribution – even if these clues do not promise predictability.

It must be noted that Hollywood movies are often parts of intellectual property "packages" that seek revenue not only from theatrical release, but from many other uses of the "brand" of the movie, including: home viewing (VHS and DVD), cable, ancillary products, and spin-off books, games, TV shows. In fact, in 2003, Americans spent far more on VHS and DVD rental and sales ($22.5 billion) than they did at the box office ($9.2 billion) and some movies expected to make most of their profits outside the theater (The Economist 2004). This section looks only at budgets for producing the movie itself and revenue from theaters.

Business as a Complex, Unpredictable System

What if the experienced Hollywood executives are right? What if managers in unpredictable environments let go of the idea that there is some magic management bullet or magic formula that means consistent success? Instead, these managers could assume that their businesses operate in an environment that is so complex that nobody can predict in all cases which product or strategy will be a blockbuster or even make a profit.

This does *not* mean that there are no incompetent business executives – their actions will always be one of the things that make this an unpredictable system. But it may be wise to change a strategy that assumes success can be "engineered" in advance by omniscient leadership. If predictability is impossible, then resilience in the face of failure seems imperative. And that is exactly what the movie industry seems to have achieved.

Systems are said to become "complex" when they have intricate interdependencies among their various parts and many variables operating at the same time. Examples of complex systems include the weather and the spread of disease in a population. Complex systems are also generally nonlinear. The effect of adding something to the system (an infected person or the air disturbed by a butterfly flapping its wings) may diffuse unevenly throughout the system because the other components of the system are not evenly distributed, or the force doing the distribution is not equally strong throughout the system. Think of throwing a handful of buttons on the floor and then connecting them in various ways: some are connected by heavy string, magnets connect some, and others are connected only by dotted lines on the floor. All the red buttons are connected to each other and some of the red buttons are connected to blue buttons. Most (but not all) of the blue buttons are connected to one yellow button while all of the red buttons are connected to another yellow button. The group of buttons is sitting on top of an active earthquake area. Could you predict what will happen to any one of the blue buttons if an earthquake hit its vicinity or someone pulled the string at one of the yellow buttons? (Based on Kaufman 1995, pp. 55–58)

Complex systems have another surprising property: adding an element that can be duplicated to the system may cause a shift in the total system that is much greater than the amount added. For example, sending a rumor about a company via e-mail to a friend in that company adds only one piece of information to that company's information system. But, because many agents (employees) in the company are connected via e-mail, the piece of information multiplies in the system as each employee sends it to many others. The phenomenon is typical in systems that are interconnected in a network.

Some complex systems are *adaptive* or are said to *evolve* when individual agents operate independently in response to forces in their environments via feedback. In some systems the agents can "learn" from one another when some agents obtain more resources and their actions are copied by other agents. In systems where the change is not learnable in the current generation by other agents (for example, the change is a mutation in an organism's genetic structure) it can become prevalent in succeeding generations if the change makes agents who have it more successful, enabling them to leave more offspring (this is evolution by natural selection). For example, a mouse with better hearing is more likely to survive the presence of foxes in her environment and will leave more offspring than other mice. Over many generations these better-hearing offspring will also leave more offspring and gradually the number of mice without the acute hearing will decline.

Management theorists have begun to use these ideas about adaptation, complexity, and unpredictability (Axelrod and Cohen 1999; Stacey et al. 2000; Schwartz 2003) although theorists like Alfred Chandler have been thinking about firms in turbulent times for quite a while (Chandler 1962; Perrow 1984).

In what would become one of the more influential business books of the late twentieth century, Peter Senge suggested that businesses must learn to adapt to change by creating "learning organizations" (1990). But he knew it wouldn't be easy. Business and other human endeavors are also systems. They, too, are bound by invisible fabrics of interrelated actions, which often take years to fully play out their effects on each other. Since we are part of that lacework ourselves, it's doubly hard to see the whole pattern of change.

Senge set out to destroy "the illusion" that the world is created by separate, unrelated forces and to develop understanding of *dynamic complexity* where cause and effect "are not close in time and space and obvious interventions do not produce the expected outcome." Subsequent writers, such as Robert Louis Flood, have expanded on this idea, expanded the evidence against predictability in complex business situations, and warned of the consequences for assuming that these processes are capable of being controlled.

An 'A caused B' rationality is a source of much frustration and torment in people's lives. If a difficult situation arises at work, then an "A causes B" mentality sets up a witch-hunt for the person or people who caused the problem (Flood 1999; Shook 2000).

And yet, many people in organizations that operate in complex systems continue to operate with this mentality. They believe that if they just look harder they will find the right formula for success. If their situation is truly complex and unpredictable, this quest will be doomed in many cases and it will take energy away from finding a strategy for resilience in the face of inevitable failures.

The Movie Business as a Complex System

In this section we look at how the movie industry fits current ideas about complex, unpredictable systems, but readers from other industries will want to compare the movie business to their own to see where the similarities will offer important insights.

It is certainly true that all industries are becoming more complex as they become more interconnected and the forces working on them become more global than local, but the movie industry exhibits the classic characteristics of complex systems more than most (Longstaff 2003). What makes the movie business complex? It has intricate interdependencies, many variables, nonlinear inputs, and adaptation.

Intricate Interdependencies

Like the button system described above, some of the connections in the movie business are strong and generally ongoing (directors and stars to their agents, producers to distributors) but most of the connections are weak because movies (at least since the demise of the studio system) are made as individual projects with people and firms attaching themselves to a project for a limited period of time. During the project's life all the players depend on one another and failure of one will affect the entire project. Many of these projects are connected to other projects where the participants are working simultaneously, have worked, or will work soon. Innovations or problems in one can spread to all of them rather quickly. At the level of the communications sector, all the communications industries (movies, cable, TV, print, etc.) are increasingly linked together by their need to compete for several scarce resources, principally the time, attention and money of consumers. Indeed, some have predicted that they will all "converge" into one industry. Although convergence is not *a fait accompli*, it is undeniable that increased competition has made all the formerly distinct industries look hungrily at each other's customers and in that sense they are now "linked" in ways they were not before. At the same time, each firm is linked to many other systems such as equipment and content suppliers as well as many layers of government. In addition, globalization links many more of these industries and firms to each other, making the system even more complex (Longstaff 2002). Thus any new movie will not just compete with other movies but with all the other options that potential moviegoers have for their free time. On the other hand, movie producers depend on other media to distribute advertising and marketing.

Wide Variety of Variables

The success of any particular firm or particular industry depends on a wide variety of variables, a few of which the firms or industries have some control over but many of which they have little or no control over. In the last 20 years the movie business has been buffeted by changes in the technologies they depend on for both production and distribution. Globalization and changes in U.S. demographics have made their audience much more diverse and increased the variables that operate on each film. Other uncontrollable variables for film producers include: the other movies in theaters at the same time, the "mood" of the public (if they are nervous about the economy or a terrorist attack they may not go to see movies that are not uplifting), and the weather on opening weekend (good weather may make more people choose to be outdoors). Some variables can be controlled (stars, budget, release date, etc.) but, as the research noted above and the experience of many producers show, manipulation of these variables does not give consistently predictable results.

Nonlinear Effects and Cascades

When forces changing the system do not add up in a simple systemwide manner, we say they are nonlinear. Adding something to the system may mean it changes by more than the amount added. This is particularly true for the success of a particular film since it depends heavily on recommendations of critics and "word of mouth" marketing. One popular national critic can have a great impact on the number of people who decide to see a movie, as can many well-connected local moviegoers (DeVany and Walls 1996).

There is a growing body of work that looks at systems that *cascade* at unpredictable points. Things such as epidemics and fads are examples of cascades in a system. In some systems, a cascade is the *tipping point* of the system – something that moves it from one state into another (Gladwell 2002). In fact there may be two tipping points in many networked systems. The first is when a system develops enough connections so that local islands of connections merge into a larger network where large cascades are possible. As the network increases, the cascades become larger and more likely until the *second* tipping point where the cascades become smaller and rarer because there is *too much connectivity*. The second tipping point is said to be a dilution effect and it happens when the individuals in the network are connected to so many people that any one of the connections does not have a great enough influence. Each individual is said to "tip" (decide to see a movie, for example) when a certain *fraction* of its neighbors makes this choice, not a certain *number* of them. If I have more people to whom I look for clues on movies or fads, it will take more of them to get me to join the bandwagon (Watts 2003; Strogatz 2003; Buchanan and Nexus 2002).

At first glance, this information about the operation of these cascading networked systems seems to indicate that there may be some way to figure out the best place to nudge them in order to make them tip your way – to make your movie the one that beats the odds to become a hit. But, while it may be important to know the amount of connectivity of a simple system in order to predict the likelihood of tipping, it is important to remember that if the system is adaptive and has many variables working on it, the likelihood of predicting a cascade that will tip the system is not high. This is true particularly when other people also have this information and are trying to nudge the system their way.

Multiple-Directional and Multi-Velocity Trajectories

The environment into which any particular film is born is complex at both the macro and micro levels. At the level of the communications sector, the growth rates or velocity of growth for the various industries that compete with movies are clearly not the same. The movie business (like print and broadcast) shows some signs of being a "mature" business that cannot look forward to much growth in its current markets. But it competes (for customer time and attention) with much faster-growing industries in cable, satellite, and gaming. This does not mean that these more mature industries will die, but it means that any financial opportunities that depend on rapid growth will be less available to them – even as they feel compelled to compete with the faster-growing industries. The system is made even more complex by a divergence in time frames: as the industries in the communication sector evolve faster to keep up with changes in their environments, other processes (policy making, business formation) move relatively more slowly and have difficulty keeping up with the changes.

Adaptation

Movie projects exist in a highly competitive environment with individual films living and dying in a relatively short period of time. Since the people working on films are highly interconnected, the ideas about why certain projects died early (or were never born) travel quickly through the system and others adapt their strategies based on their view of what works and what doesn't. The fact that there is often disagreement about what made a project fail means that this adaptation is not uniform and not predictable. DeVany and Walls note that "contingency rich" contracts help the movie business to maintain an ability to adapt to changing demand conditions, particularly contracts between distributors and exhibitors that allow everyone to take advantage of a movie that is becoming a hit while limiting the exposure of distributors who have signed up for a flop.

So how does the movie industry manage to thrive and prosper when it is so unpredictable? The answer seems to be that this industry has (without anybody planning it) developed resilience to the failures it deals with every day.

Dealing With Unpredictability Through Resilience

When confronted with a danger, the best strategy may be to develop a robust system that will keep the danger out or act as a buffer to keep the system from being affected by it. This is a good strategy if you know the nature of the danger(s) you are likely to face. If you are likely to be threatened by an army with spears and swords, building a wall around your city will act as a buffer and stop harm to those within. While there will be some temporary defensive measures put in place in case of attack, sooner or later the city will return to normal. But building a wall (literally or figuratively) may also keep out other things (like water and food to the besieged city) and keep those who are protected from taking advantage of opportunities outside the protected area.

When an individual or group (species, business organization) must operate in an environment where resources and dangers are unpredictable, one strategy for survival is to develop *resilience* (Gunderson 2003; Folke et al. 1998). Since they don't know what dangers are out there, they can't develop a single strategy for robustness – walls would not be appropriate because they would deprive the individuals of access to other resources they might need for unknown dangers. If a system with unpredictable danger/opportunities is attacked or deprived of resources, it must be able to bounce back to its previous state. Since they don't know what the dangers will be, they often don't know what will work to make the danger go away. In these cases adaptability and access to a variety of resources seem to be key for developing resilience.

In systems with unpredictable environments the members of the group try many things and hope that some of them will be able to survive whatever challenges or surprises they encounter. For example, birds often lay many eggs, but only a few hatch and fewer still survive to maturity. Businesses often develop many products in prototype form but abandon them if they do not meet specific targets. Thus, both *a limited investment in large numbers of the same thing* and *diversity in the things tried* can be tools for resilience. It should be noted at the outset that survival using these strategies does not necessarily involve anything even close to *stability* in the short-term fortunes of the individuals of that species or firms in that industrial sector. This strategy requires a willingness to accept many failures and/or to deal with the same challenge in different ways. This *should not* be interpreted to mean that some of the many eggs laid or the many different products developed are somehow inherently "fitter" than the others. All of the eggs were alike and all of the products had many unknowns in their development. A few eggs and a few products survive because they were lucky enough to find themselves in exactly the right environment for them to thrive. No one could have predicted which eggs or which products would be so lucky. In a more predictable environment, it may be possible to select the fittest eggs or products for extra support (as animals in more predictable or less dangerous environments often do by giving extra care to the healthiest infants).

The movie business would seem to be a perfect example of resilience in action. As noted above, movie producers operate in an environment with many unpredictable variables. The production part of the industry is not concentrated into large or permanent organizations, but consists of many independent operators who come together for specific movies and then move on to the next project. The industry is constantly experimenting with new ideas, looking for the next hit. And for every movie that is released there are many more projects that did not reach maturity. They were shelved or abandoned when it became clear that it was impossible to put together the right resources. Other movies were made but never released in theaters because it was determined that this would not be profitable, and they are released in secondary markets such as DVD or sold to cable. Instability at the level of the individual movie does not mean that there is instability at the level of the industry. Even though there are many failures, the industry survives because of a few big successes. This is made possible, in part by the adaptive contracts noted above and by the industry practice of *contingent compensation*. The people in the business (actors, directors, suppliers) make some money for their contributions and may be entitled to more (sometimes *much* more) if the movie becomes a hit. If the movie does not turn a profit it is only the investors who gambled on net profits who walk away with nothing. But, like those who bet on other unpredictable events such as horse races (or who lay many eggs or develop many projects), these investors take their losses as just one try in a bigger game and wait for a big hit.

In recent years, many businesses have been told that they can build resilience by developing "a broad portfolio of breakout experiments with the necessary capital and talent" from which there will be winners and losers. "Most experiments *will* fail. The issue is not how many times you fail, but the value of your successes when compared with your failures" (Hamel and Valikangas 2003). The movie industry seems to be a good example of this advice in action, but a view of resiliency on a larger time scale is often difficult for investors and executives from more traditional businesses where profits and losses are calculated on a quarterly basis.

Resilience Through Scales of Operation

In systems that operate at more than one scale, resiliency may operate at each scale and across the scales. There might be different time scales or different size scales at work in the same system. For example, in the human body, the immune system acts first at a local scale to confront an infection by sending a variety of forms of immune cells (within-scale resiliency through diversity or redundancy). But if this response is not successful, the system responds by "scaling up" its response and inducing fever. When similar *functions* (even if not similar *mechanisms*) operate across-scales it makes the system more resilient because they are redundant – if one fails the other will go into action.

In the movie business, most challenges are dealt with first at the individual film level. If the challenge is too big for that level or persists in spite of attempts at that level, the industry scales up through several organizations that represent the interests of types of players (producers, directors, actors, technicians, etc.) and if a challenge that threatens the industry cannot be dealt with at that level, the industry comes together in temporary coalitions of these organizations.

Resilience Through Loose Coupling and Slow Scales

Many authors have noted that slower parts of systems act as resilience mechanisms for the faster parts because they can "remember" how to handle certain surprises. In return the faster parts of the system give the slower parts information about changes taking place and allow it to adapt at its own time scale. In some cases, when the slower parts do not have this information they are liable to drastic, cascading change when the changes reach a critical level, particularly when the system has become tightly coupled. The very connectedness or sameness that makes it efficient can amplify internal weaknesses or external shocks. This has been seen in many systems.

> When the system is reaching the limits to its conservative growth, it becomes increasingly brittle and its accumulated capital is ready to fuel rapid structural changes. The system is very stable, but that stability derives from a web of interacting connections. When this tightly connected system is disrupted, the disruption can spread quickly, destabilizing the entire system. The specific nature and timing of the collapse-initiating disturbance determines, within some bounds, the future trajectory of the system. Therefore, this brittle state presents the opportunity for a change at a small scale to cascade rapidly through a system and bring about its rapid transformation. This is the "revolt of the slave variable." (Gunderson et al. 2002, 12–13, citing Diener and Poston, 1984)

In the movie business, this cascading change at the slow level was seen in the relatively abrupt collapse of the studio system. Under this system, individual players were bound by contract (tightly coupled) to particular studios. A variety of surprises caused the highly connected studio system to collapse and break into the loosely coupled industry organization we see today.

The importance of the relative *strength* of the connections in a system is increasingly seen as critical to the analysis of the system's behavior. In fact, it is a fairly good predictor of the stability and resilience of any group. Strong coupling within or between organizations would be predicted if there is a high level of resources reliably available, the system changes rapidly, and influence spreads quickly in the system (Weick and Sutcliffe 2001). The movie business appears to be both loosely coupled in some respects and tightly coupled in others. The level of resources (the audiences) is not high or reliable and the major parts of the system change slowly. But information can spread rapidly due to the connected nature of the business.

If the individuals are truly *tightly coupled*, any disturbance in the system will affect them all and if any of them fails, there is a much greater likelihood that others will fail with them. Think of a team of horses that is hitched to a wagon. This aspect of tight coupling has very interesting implications for organizations seeking efficiency through developing economies of scale. One of the seldom-acknowledged tradeoffs for doing the same thing many times in a highly connected way is that this tight coupling will inevitably create an unstable situation if redundancy is taken out of the system to lower costs. For example, we all know that you can produce individual widgets cheaper if you build a lot of them in the same way, using processes that are very closely tied to the processes of suppliers and customers. This kind of efficiency remains the Holy Grail for many firms that must compete with firms that have cheaper factors of production (e.g., lower labor costs). As attractive as this goal is, it comes with a dark side. If a critical supplier, customer, or piece of technology is removed, with no redundant source, the entire system is vulnerable to collapse. Loosely connected systems with built-in redundancy seem to be the most secure from this type of danger.

Loosely coupled systems are those where the components have weak enough links that they can ignore small perturbations in the system. The components of a loosely coupled system are said to have more independence from the system than tightly coupled components since they can maintain their equilibrium or stability even when other parts of the system are affected by a change in the environment. The components of loosely coupled systems are also better at responding to local changes in the environment since any change they make does not require the whole system to respond. Thus, if innovation or localized response to particular problems or opportunities in an unpredictable environment were the goal, then loosely coupled systems would have the best chance to find new answers and to develop resilience. A more tightly coupled system could lead to premature convergence on a solution since all the components would be responding more or less in unison.

Once again, when we look at the movie business through this lens, we see resilience in both the tight and loose coupling. This industry is made up of many small firms and individuals with loose ties when they are not working on a project but very strong ties once they begin a project. They must coordinate schedules and work together closely to meet the budget and the production schedule. At the level of the project, anything that affects the contributions of one will affect everyone. But outside of a particular project, a failure of one has almost no effect on the others. If one of them finds a good solution to a problem several others may try it, but it will not be adopted by the entire industry unless it has shown that it works in many situations OR unless there is a need for uniformity.

When tight coupling is needed in the movie business, the individuals and firms look to the higher level of organization provided by the professional organizations such as the Motion Picture Producers Association of America (MPAA). The relative weakness of these organizations has implications for the adaptations of the larger industry because this process can take place only as fast as the most loosely coupled component can (or is willing to) move. The temporary or partial independence of one or more components in the movie business will slow down (or change) the process. This is an important insight for executives who must make predictions about the ability of a merged organization to develop "synergies" that result in higher profits or lower costs. For example, if one unit of the newly merged company is a film production company (they are famous for being loosely coupled internally) and it must be coupled with a telephone company (they are equally famous for tight internal coupling) the result may be a slower adaptation process than the telephone culture is accustomed to. If the telephone firm becomes unstable after it has become more tightly coupled to the film company, the latter will become unstable as well. At this point the film company is likely to seek a more loosely coupled relationship – or even a break in the relationship.

The movie business may have been able to survive many failing movies because of its ability to organize many weak links into a network of influence. Networks have become the subject of much study in the search for resilience.

Networks, Power Laws, and Resilience

Note: Readers who are familiar with network theories may want to skip this section, but readers who have little or no knowledge of these relatively new ideas are urged read on because the conclusions of the paper rely, in part, on this section.

There is a new (and growing) body of work that looks at the connections between things that function as a network. Network Science gained popularity as the "small world" problem and, more recently, the "Kevin Bacon game." The former is the puzzle of why most people on the earth seem to be separated from one another by only six other people, or six degrees of separation. The latter uses movie actor Kevin Bacon and his connection to other people in the film industry to test the degrees of separation between them. The game proves that Hollywood is a very connected place and that every actor is connected to every other one by about four steps. If the game counted connections to agents and producers the connections might be even smaller (Watts 2003, pp. 92–100).

This new research on networked systems was originally done in a branch of mathematics known as graph theory. It is now being examined by many disciplines including political science, biology, sociology, and computer science. In some of the networks studied, the distribution of things in the network (e.g., wealth, Web links) follows a power law and the place of any particular thing in that distribution was difficult to predict. As noted above, networks that exhibit a power law distribution are characterized by a continuously decreasing curve, showing many small things existing with a few large ones (many people with small amounts of money and a few with a large amount, many Web sites with a few links and a few with many links). This is in contrast to systems where the distribution follows the typical bell curve with a few things at either end of the spectrum (a few small things at one end and a few large things at the other) but most of the things clustered in the middle. We have demonstrated in this paper that each year successful movies follow a power law distribution and it is quite clear that the business has many characteristics of networking.

There is some indication that networks following a power law develop differently from other systems. They seem to grow one node at a time (as in one Web page at a time, one person at a time) and some nodes will have preferential connections because the more connections they have the more they will get. These superconnected nodes are called "keystones" or "hubs." For example, some nodes become hubs when they are connected to more often by others because they were the first to fill a connection role or because they have more resources to devote to connections. Thus, if you are the first Ebay-type connection Web site, or you are a very large company such as Microsoft (with perceived resources to devote to connection) you are more likely to become superconnected. In these systems the connected tend to get more connected, not necessarily because they are better but because they were first or bigger to start with (Barabasi 2002).

Some networks have what is known as a "scale-free" topology. That is, there are many small nodes that connect to a few larger nodes that in turn connect to still larger nodes in a hierarchical configuration. The lower level nodes have no way to connect to other nodes in the system except through their local hub. Imagine the telephone network. You cannot connect directly to anyone except through your local exchange, which acts as a hub for your area. Unfortunately, we are now discovering that this type of network will often

> perform terribly under conditions of failure. For the same reason they are vulnerable to congestion-related failure (because they are too centralized), if any of the hierarchy's top nodes do fail, they will isolate large chunks of the network from each other. It is here that connectivity at all scales really comes into its own, for in multiscale networks there is no

> longer any "critical" nodes whose loss would disable the network by disconnecting it. And because they are designed to be decentralized not only at the level of teams but also at larger scales, they can survive bigger failures. (Watts 2003, p. 285)

Multiscale networks allow nodes to connect across scales, without requiring them to go through a hierarchical routing system. While this may not be the most efficient configuration, it does make the network *robust*, that is, allows it to survive failures because taking out the one hub that you connect to (or those that it connects to) will not deny you access to the whole system. The movie business shows many attributes of a multiscale network because there is no formal hierarchy and most people have connections to people in many parts of the business. An actor will have connections to producers, directors, technicians, and just about everyone else. And all of them will have similar connections. Thus, if one of a person's connections to the business fails they will not be separated from the business and will still have access to the resources of the network of people and resources that they need.

In some networks the "winner takes all" when one node has all the connections and there is one giant hub with many nodes (Watts 2003). This happens when nodes can choose which hub they will use to connect to the system and they will choose the hub that gives them the most connections. The more connections a hub has the more likely it will be chosen and eventually the system will tip and all will choose the most connected hub. A similar thing happens when people must choose a technology to connect to the system. They will choose the one that gives them the most connections to things in the system. The battle between VHS and Betamax was an example of this. The system tipped to VHS when consumers perceived that it gave them access to more movies.

In fact, there is strong evidence that the Internet is a winner-take-all network and only a few sites will have superconnections (Barabasi 2002). Any time you add a hub to a *random* network of individuals or groups (they are not connected except at random) you are likely to get this "aristocratic" (the rich get richer) configuration where power and scarce resources are drawn to the spot with the most resources. These networks where superconnected hubs form are often very efficient and robust at lower levels because destroying any of the less connected nodes will have little impact on the system. But this strength is also their Achilles' heel because destroying a superconnected hub can destroy the entire network. Any firm that becomes a superconnected hub for the sector becomes both an opportunity for efficiency and a danger because it can bring everyone down with it. The movie business would seem not to be this type of network because there are many people who serve as connectors and a "winner" that controls all the connections in the business has not emerged.

As in other complex systems, the work being done on networks also indicates that the strength of the ties between things is critical for understanding (if not always predicting) the operation of the networked systems. There is good evidence that weak ties (or loose couplings) are often more important than strong ones when dealing with a new opportunity or problem. If two firms are strongly linked (or tightly coupled) to each other they are probably also strongly linked to each other's links so what happens to one will affect all of them. Strong links work very efficiently as long as the firms (individually or collectively) don't face unique challenges or encounter new opportunities. If something unexpected happens, it will be the weaker links of each firm that will be bridges to other systems with other resources or ideas that can be used when they face a new problem (Buchanan and Nexus 2002, Chapter 2). Thus, the long-term stability of a system (or a firm) may actually increase if it has many weak ties – even if this means it is less predictable or less efficient in the short term. This has led to speculation that a balance between the need for stability and diversity is called for and the appropriate strength will depend on the number of connections available. "...the superconnected few should be linked to others mostly by weak links, while those with few links to others should be connected by strong links" (Buchanan and Nexus 2002, p. 149).

Resilience in Networks

Resilience in networks is also increased by diversity and redundancy. A variety of weak links for the superconnected hubs requires some diversity in the system because if everyone is the same it will be difficult to develop the variety of links that would be useful in case something unpredicted happens. Thus, a loss of too many of these weak links will have serious implications for the resilience of the system. As we have noted, the movie business is made up of many different types of people who are connected to one another through weak links to highly connected individuals. If these hub individuals connected to only certain types of people they would not be able to pull together all the resources needed to put a project together, and they would require connections to other hubs. The diversity of their links also makes them more likely to be able to take advantage of unexpected opportunities or deal with potential disasters.

Resilience is also improved in a network if hubs have some functional redundancy. If they are unexpectedly removed from the system there is something or some function that will perform their role in the system. In Hollywood this redundancy of hubs is accomplished by allowing many different professionals to serve as connectors: producers, agents and lawyers all often perform the task of putting people and deals together.

While this work on resilience in complex networked systems is still in the formative stages, Duncan Watts, one of the original researchers in this area, has this to say

> Already we can understand that connected, distributed systems, from power grids to business firms to even entire economies, are *both* more vulnerable and more robust than populations of isolated entities. If two individuals are connected by a short chain of influences, then what happens to one *may* affect the other even if they are completely unaware of each other. If the influence is damaging, then each is more vulnerable than they would be if they were alone. On the other hand, if they can find each other through that same chain, or if they are both embedded in some mutually reinforcing web of relations with other individuals, then each may be capable of weathering a greater storm than they would be by themselves. (Watts 2003, p. 303)

Most people with experience in the movie business would probably agree with these sentiments, even if they don't understand the science behind them. People who make movies must work in an environment that is complex, ever changing, and where what worked last year will not work next year. The causes of success and failure remain elusive. And yet the movie business is clearly successful. It may be an example to all industries that must operate in similar situations. It is possible that the goal for managers in complex, networked systems that exhibit a power law distribution may not be better predictions, but more loose connections across all scales of the business. In this kind of unpredictable environment, resilience may be the greatest (or only) success.

Conclusions

What does all this mean for business strategy? Can this information be used to make better decisions? Yes. As long as "better" does not mean "predictable" – particularly in a business with unreliable access to resources, intricate interdependencies, many variables, nonlinear inputs, and adaptation.

Even if they can't predict the forces working on their organizations, managers may be able to make their organizations more resilient. But beware! The insight provided by the study of these systems seems to indicate that if resilience is the goal, bigger is not better and increased efficiency (by eliminating redundancy or tight coupling with customers/suppliers) has serious consequences. These are not always welcome ideas.

Previous attempts at improving complex systems have met with some spectacular failures. James Scott has documented some of these well intended but disastrous policies and believes they argue for smaller, more diverse and more flexible institutions or strategies that can adapt with the system they are trying to improve.

> The intervention of scientific forestry, freehold tenure, planned cities, collective farms, ujamaa villages, and industrial agriculture, for all their ingeniousness, represented fairly simple interventions into enormously complex natural and social systems. After being abstracted from systems whose interactions defied a total accounting, a few elements were made the basis for an imposed order. At best, the new order was fragile and vulnerable, sustained by improvisations not foreseen by its originators. At worst, it wreaked untold damage in shattered lives, a damaged ecosystem, and fractured or impoverished societies.
>
> …we gain in ease of appropriation and in immediate productivity [with increased economies of scale], but at the cost of more maintenance expenses and less "redundancy, resiliency, and stability." If the environmental challenges faced by such a system are both modest and predictable, then certain simplification might also be relatively stable.
>
> …Even in huge organizations, diversity pays dividends in stability and resilience…. Much has been made of the rather complex family firms in Emilia-Romagna, Italy, which have thrived for generations in an extremely competitive world textile market by virtue of mutuality,

> adaptability, and a highly skilled and committed workforce.... These firms and the dense, diverse societies upon which they depend, have increasingly seemed less like archaic survivals and more like forms of enterprise ideally suited to postindustrial capitalism. (Scott 1998, pp. 352–354)

Does the movie business fit the pattern that is beginning to emerge in the study of other complex networked systems that must operate in unpredictable environments? In all of the ways examined here, the answer is yes. While acknowledging that this work is only in its infant stages, it is possible to formulate some ideas that warrant much closer attention and some possible strategies for organizations that can't wait until the definitive votes are cast. Many organizations in unpredictable environments may find that they accomplish their goals not by building organizations (and the rules that govern them) based on predictions, but based on adapting to the unpredictable and remaining resilient in the face of multiple failures. It is acknowledged that this will be a tough sell to those who have come to believe in the inevitability of finding the right formula for success. But sooner or later we may have to give up on trying to turn lead into gold and get down to the business of watching the world change while we change with it. Organizations looking for a model could do worse than studying the movie business.

Since all businesses live in different environments there is no one-size-fits-all plan for developing resilience. Not all businesses have the potential for the big hits that makes the system resilient in the face of many failures. But it will be worth any manager's time to look the ideas presented here in order to discover which ones might work for them. Business managers will recognize that several things on this list are consistent with well known ideas in management science. This is probably additional evidence that these ideas from other disciplines are not irrelevant to business.

1. Decide if your business (your industry and/or your firm) is unpredictable by nature or if it is just in an unpredictable phase of its development and stability can be expected to return. There may be a temporary unpredictability in resource availability or customer demand. You would then want a resilience strategy that allows you to bounce back when things return to normal and won't burden the enterprise with some complex coping mechanism that is no longer needed. If there has been a fundamental (or structural) change in the environment so that this unpredictability will be with you for the foreseeable future you will want to examine fundamental processes, particularly those that are tightly coupled with unpredictable resources.
2. Acknowledge the unpredictability (either permanent or temporary) and how this changes the definitions of success for the industry, the firms in it, and individuals in those firms. Expand the time scale for measuring success and acknowledge the goal of resilience. This may be the most difficult thing to do. It involves changing the expectations of all stakeholders, particularly investors and employees. It means acknowledging risk in specific terms and not punishing failure that is caused by unpredictable complexity.
3. Throw many seeds and acknowledge that there will be failures. Either throw many of the same kinds of seed in many environments and see where they grow, or, throw many different kinds of seeds and see which ones thrive in a particular

environment. Once again, this means acknowledging risk and making failure part of the cost of doing business, not a personal and professional death knell for those involved with the seeds that do not grow.

4. Set up feedback mechanisms that allow you to distinguish success and failures as early as possible. If the system is likely to have tipping points, decide how you will know them and how you will rush resources to (or from) the tipping system. This is especially important if the system is tightly coupled and subject t cascading failure or runaway success. For many companies (and government organizations) this is very difficult because their cultures punish anything that looks like failure. It's not surprising that nobody signals that things are failing (or not growing) because to do so would be professional suicide and jeopardize one's bonus.
5. Set up mechanisms that give you maximum adaptability and allow you to invest in success and cut losses on flops. This might include adaptive contracts and contingency compensation. This may mean fewer long-term contracts (for both employment and other resources) in "seedling" enterprises or products. It will also mean having resources ready to throw at any seedling that looks like it's about to take off. This may mean keeping a resource reserve for this purpose (but they will not be making maximum return while waiting in reserve) or having resources that can be quickly redeployed from other projects (those without long-term commitments).
6. Make sure there are weak links to resources outside your business, ones that can be used when the unpredictable happens. These links should be weak enough that they are not be affected by problems in either business. You might consider using a variety of suppliers or paying a supplier to be on "standby." Strong coupling should be used where efficiency is critical, but acknowledge the trade-off in stability and resilience for that function – strong coupling to resources prone to surprise would be limited to functions that have redundancy (back-up that will take over the function). If you need electric power to perform critical functions (cooling, running computers, etc.) you might have a strong link to an electric power supplier that gives you quantity discounts, but only if you have auxiliary generators or standby power suppliers that can step in to perform the function if the power is cut off by a terrorist attack.
7. Make sure that people who have to deal with the unexpected have links to many resources. These links should be on multiple scales and not require going through a hierarchy in times of opportunity or danger. If the person running your plant needs to get electric power quickly they should not have to go up the chain of command to get permission and they should have direct access to the persons who can deliver what they need, regardless of whether they are at similar levels of the organizations chart. If the night supervisor has to call the VP at the standby supplier the considerations of hierarchy should not get in the way.
8. Make sure that important functions have redundancy, preferably at different scales so that if the function is damaged at one scale it can be picked up at another. This is especially true for any function that acts as a hub. If communication services

are centralized there should be redundant communication services at the local level.

9. Hubs should have diverse connections that allow the people who use them to access many different types of resources – even ones they don't use very often. People and functions where things come together (such as manufacturing) need to have access to many things that they *might need* and even to things that nobody thinks they will ever need. This allows them to fix the problem without having to fix the supply chain first.
10. Stay in touch with both the slow and fast parts of the business. The part that changes slowly is just as important as the part that responds to every change in the environment. "The Fast proposes but the Slow disposes." There is great value in the people who have been performing a function for many years. They should never be treated as failures. Likewise, new ideas should not be dismissed without real consideration – particularly in light of feedback mechanisms set up to alert the organization about changes in the environment. Many of us do not notice changes in the environment until it is too late to respond with considerable effort, slowing down resilience.

While these ideas are applicable both to established firms and to startups, each group will have special advantages when building resilience. Startups have the advantage of being able to build hubs, establish links of various strengths, and create "ramp up" and "get out" trigger points from the outset, thereby avoiding any legacy networks and locked-in systems that will be difficult to adapt quickly enough to meet important surprises. Established firms and industries have the advantage of deeper understanding of the slow scales for their systems and the resources for building in redundancy. But all firms must get better at surviving surprises in an ever more connected and complex world.

References

Axelrod R, Cohen MD (1999) Harnessing Complexity: Organizational Implications of a Scientific Frontier, New York: The Free Press.

Bak P (1996) How Nature Works. Oxford: Oxford University Press.

Barabasi A-L (2002) Linked: The New Science of Networks. Cambridge, MA: Perseus, Chapter 6.

Buchanan M (2002) Nexus: Small Worlds and the Groundbreaking Science of Networks. New York/London: W.W. Norton.

Chandler A (1962) Strategies and Structure: Chapters in the History of American Industrial Enterprise. Cambridge, MA: MIT Press.

DeVany A, Walls WD (November 1996) Bose-Einstein Dynamics and Adaptive Contracting in the Motion Picture Industry. The Economic Journal, Vol. 106, No. 439, pp. 1493–1514.

Flood RL (1999) Rethinking the Fifth Discipline: Learning Within the Unknowable. London/New York: Routledge, p. 84.

Folke C, Berkes F, Colding J (1998) Ecological Practices and Social Mechanisms for Building Resilience and Sustainability. In: Folke C, Berkes F (eds.), Linking Social and Ecological Systems: Management Practices and Social Mechanisms for Building Resilience. Cambridge: Cambridge University Press.

Gladwell M (2002) The Tipping Point: How Little Things Can Make a Big Difference. Boston, MA: Little, Brown.

Gunderson L H (2003) Adaptive Dancing: Interactions Between Social Resilience and Ecological Crises. Navigating Social-Ecological Systems: Building Resilience for Complexity and Change. Cambridge/New York: Cambridge University Press.

Hamel G, Valikangas L (2003) The Quest for Resilience. Harvard Business Review, September, pp. 52–63, at p. 54.

Kaufman S (1995) At Home in the Universe: The Search for the Laws of Self Organization and Complexity. New York: Oxford University Press, pp. 55–58.

Klein, Burton (1977) Dynamic Economics. Cambridge, Mass: Harvard University Press.

Litwak M (1986) Reel Power: The Struggle for Influence and Success in the New Hollywood. New York: William Morrow.

Longstaff PH (2002) The Communications Toolkit: How to Build or Regulate Any Communications Industry. Cambridge, MA: MIT Press, Chapters 7 and 8.

Longstaff PH (July 2003) The Puzzle of Competition in the Communications Sector: Can Complex Systems be Regulated or Managed? Program for Information Resources Policy, Harvard University.

Longstaff PH, Velu R, Obar J (July 2004) Resilience for Industries in Unpredictable Environments: You Ought To Be Like Movies, Program for Information Resources Policy, Harvard University.

Perrow C (1984) Normal Accidents: Living with High-Risk Technologies, Princeton, NJ: Princeton University Press.

Schwartz P (2003) Inevitable Surprises: Thinking Ahead in a Time of Turbulence, New York: Gotham.

Scott JC (1998) Seeing Like a State: How Certain Schemes to Improve the Human Condition Have Failed. New Haven, CT/London: Yale University Press, pp. 352–354.

Senge P (1990) The Fifth Discipline: The Art and Practice of Learning Organizations. New York: Doubleday, p. 7.

Snook S (2000) Friendly Fire: The Accidental Shootdown of U.S. Black Hawks over Northern Iraq. Princeton, NJ: Princeton University Press.

Sornette D, Zajdenweber D (1998) Economic Returns of Research: The Pareto Law and Its Implications. Los Alamos e-print (cond-mat/9809366), September 27.

Stacey RD, Griffin D and Shaw P (2000) Complexity and Management: Fad or Radical Challenge to Systems Thinking? London/New York: Routledge.

Strogatz S (2003) Sync: The Emerging Science of Spontaneous Order. New York: Hyperion.

The Economist (2004) Romancing the Disc. February 7, pp. 57–58.

Vogel HL (2001) Entertainment Industry Economics, Fifth edition. Cambridge/New York: Cambridge University Press.

Watts DJ (2003) Six Degrees: The Science of a Connected Age. New York/London: W.W. Norton, pp. 92–100.

Weick KE, Sutcliffe KM (2001) Managing the Unexpected: Assuring High Performance in an Age of Complexity. San Francisco, CA: Jossey-Bass.

Shareholder Wealth Effects of Mergers and Acquisitions in the Telecommunications Industry

Olaf Rieck and Canh Thang Doan

Abstract In the past 10 years, Telecoms operators frequently engaged in M&As with the objectives to grow bigger and achieve higher earnings. However, the question arises whether and under what conditions they have really been able to achieve these objectives. The aim of this research is to investigate M&As in the telecoms industry and analyze the conditions under which such M&As can be considered successful. In doing so, we employ the event study method, which traces immediate market reactions to M&A announcements and the corresponding shareholder value effects. We first apply the event study method to our full set of M&As to obtain a broad picture of the shareholder value effects of M&As in the telecoms industry. We then proceed by investigating specifically (1) whether the shareholder value effects depend on the degree to which M&As are driven by a service diversification strategy; and (2) whether the shareholder value effects depend on the M&As being driven by an international diversification strategy.

Introduction

In recent history, corporate consolidation through Mergers and Acquisitions (M&As)[1] has reshaped a number of the world's key industries, such as automobile (Daimler-Benz and Chrysler), banking/insurance (UniCredito and HypoVereinsbank), airlines

O. Rieck (✉) and C.T. Doan
Nanyang Technological University, Singapore
e-mail: AORieck@ntu.edu.sg

[1] In this paper the terms mergers and acquisitions are used interchangeably to refer to transactions involving the combination of two independent market participants to form one or more commonly controlled entities. It should be noted that this study covers only mergers or controlling-stake acquisitions as classified by Hitt et al. (2005).

B. Preissl et al. (eds.), *Telecommunication Markets*, Contributions to Economics,
DOI: 10.1007/978-3-7908-2082-9_21, © Springer Physica-Verlag HD 2009

(Air France and KLM), and telecommunications (SBC and AT&T). A large body of literature in the fields of economics and strategic management has investigated the conditions under which M&As may be beneficial to the involved companies, the consumers, or society at large. One of the most widely used methods used this body of literature is the event study method, which looks at the companies' stock prices to assess the future benefits of M&A events. It has been widely observed that the release of important corporate information, such as M&A announcements, may trigger significant reactions of stock prices. When the market learns about these kinds of events, the involved stock price may immediately climb or fall. Hence, it appears that the market connects an event (e.g. a merger between two companies) with the prospect of future increases or decreases of cash-flow. Another way of putting this is that the stock price will immediately reflect "the good news" or "the bad news".

The stock price reactions illustrated in Table 1 could be purely random. But they could also be related to the release of information on the imminent M&A. If a useful interpretation of those observed returns is to be made, it must be contrasted against the overall market return for the same period. Any return that significantly exceeds the market return is to be considered abnormal (MacKinlay 1997). The type of news this study will focus on is the disclosure of M&A announcements of large telecommunication companies and the impacted stock price. The central question will be:

How does the release of information about a planned M&A abnormally affect the price of the underlying security?

Starting in the latter half of the twentieth century, social scientists began trying to explain why companies undertake M&As. In the Industrial Organization literature, the two most commonly identified reasons are efficiency gains and the strategic rationale (Neary 2004). Efficiency gains arise through economies of scale or scope which are due to the increased synergy between the involved firms. The strategic rationale follows the idea that M&As can alter the structure of the market, which in turn affects the firm's profits. Other motives for M&As include reduction of risk and change in organizational focus. It has been widely observed that an acquisition strategy is sometimes applied because of the uncertainty in the competitive landscape, as in the case of the telecommunications industry. If an industry undergoes dramatic

Table 1 Acquirer's stock price changes on the day of M&A announcement (Calculated from data extracted from Yahoo! Finance)

Acquirer	Target	Impact on Stock Price
Daimler-Benz	Chrysler	+8.38%
Time Warner	AOL	+9.40%
Vodafone	Mannesmann	+6.77%
UniCredito	HypoVereinsbank	+4.90%
Air France	KLM	+4.77%

changes or is in a state of uncertainty, a firm may make an acquisition in an attempt to diversify its product line or diversify geographically in order to spread risk. By doing so, it shifts its core business into different markets due to anticipated volatility within its core markets (Hitt et al. 2005).

In the case of the telecommunications industry, one of these uncertainties is caused by deregulation, another one is caused by technological change. Many countries have removed important legal and regulatory barriers for more competition and privatization. The removal of these barriers has opened the way for increased M&A activities within and across national boundaries and telecommunications industry segments.

Arguably, technological change has induced continued uncertainty about the future of this industry. The new "Everything over IP" paradigm could mean big changes for the industry's fundamental players (Zadeh 2004). Voice-over-IP (VoIP) will offer new opportunities to alternative carriers and present big challenges to traditional telecommunications companies. It is a disruptive technology[2] that will fundamentally change the traditional public-switched telephone network market (PSTN). The dominance of VoIP seems inevitable, but the broad-based transition to VoIP may take years, depending on the level of broadband penetration, regulatory conditions and attractive pricing of VoIP services (Punja 2005).

To deal with uncertainty in the light of deregulation and technological change, telecommunication giants are using M&As as an important growth avenue. Over the past decade, a wave of M&As could be observed, especially on the US and European telecommunications markets (Warf 2003). These M&As involve both large and small companies in a variety of different and similar industry segments (Warf 2003; Wilcox et al. 2001). Edward Whitacre, former CEO of SBC, puts this development in his own words after he announced that SBC would acquire AT&T in a $16 billion-transaction:

> The communications industry is undergoing a profound transformation as it transitions to unified, IP-based networks capable of delivering a host of integrated services. To manage this evolution, customers need a partner with the resources to provide new service platforms and product sets, while maintaining world-class reliability and security. This merger creates that company.(Source: AT&T News Release on January 31, 2005)

The remaining sections are organized as follows. Section "Development of Hypotheses" will discuss findings of previous literature in order to formulate my hypotheses. These will be tested using the event study methodology. This methodology will be discussed and described in detail in section "Theory and Methodology". The results of the empirical study will then be presented in section "Empirical Results". Discussion and limitations of to the findings will be discussed in section "Discussion and Conclusion" which will also conclude.

---

[2] The term refers to a technology or innovation, which radically transforms markets, creates entire new markets, or destroys existing markets for other (often older) technologies.

Development of Hypotheses

M&A in the Telecommunications Industry

According to Hitt et al. (2005), the strategic management process calls for an M&A strategy to increase the firm's strategic competitiveness as well as its returns to the shareholders. An M&A strategy shall only be used when the acquirer will be able to increase its economic value through ownership and use of the acquired firm's assets (Selden and Colvin 2003).

Thompson et al. (2005) argues that the main motivation for M&As is the maximization of profits and efficiency gain through a good combination with a new business. Because every firm possesses unique capabilities and resources, a good combination of two firms can be a unique match that maximizes the value of the resulting firm due to synergies. These synergies would not be available to these firms if they operated independently. Hence, the value of the target to the acquirer receives an extra premium on top of the second highest bid.[3] The benefits that make the firms pay such a premium include: (1) Economies of scale, which allow the combined firm to be more cost-efficient and profitable. Telecommunication firms are particularly keen on consolidation because of the high fixed costs and the need to spread these costs across a large customer base. (2) Superior pricing power due to the reduced competition and the higher market share result in higher profits. In telecommunications markets, where typically only few companies serve a given country market, any M&A induced market concentration may result in great increases of pricing power. (3) Benefits from different functional competencies, as would be the case when a telecommunications operator acquires an internet service provider.

Studies conducted by Park et al. (2002) and Ferris and Park (2001) find that in the telecommunications industry context M&A announcements generally show negative abnormal returns. Long-term integration costs, differences in corporate culture as well as agency-problems are possible explanations. However, Park examines 42 events between 1997 and 2000, whereas this study covers a much larger sample size with 88 events between 1998 and 2006. Ferris and Park (2001) investigate U.S. telecommunication firms in the years before 1993, focusing on long term value effects. On the other hand, potential benefits of M&As have been confirmed by a majority of other previous event studies. Wilcox, Chang and Grover (2001) examine short-term M&A announcement effects in the US telecommunication industry. They find that M&As are generally perceived positively by investors and therefore increases the market value of the acquirer. Another event study due to Uhlenbruck, Hitt and Semadeni (2006) shows that M&As involving Internet firms also show value-creating results for the shareholder. Therefore, choosing M&As as an avenue for growth may turn out to be a good strategy.

[3]The second highest bid would represent the market price of target firm.

In the light of the above review of previous empirical and theoretical literature, we therefore posit:

H1: M&As in the telecommunication industry generate positive abnormal returns.

Firm Diversification

A natural way to refine the empirical analysis of M&As is to look at the extent to which M&As can be seen as part of a diversification strategy, and what specific kind of diversification strategy they represent. In general, firms find it often easier to follow a focused strategy, staying within in the industry in which the firm is currently operating, than to diversify into an unknown industry in which they lack experience (Hitt et al. 2005). However, in the current global environment, a diversified firm can exploit strategic advantages that are generally not available to their undiversified competitors. Diversification is generally defined as the extent to which firms operate in different industries simultaneously and has proven to have a statistically significant influence on the value of the firm (Jose et al. 1986). To simplify the array of options with regard to M&A strategies, this study distinguishes between *conglomerate mergers* (broad firm diversification) and *non-conglomerate mergers* (focused/related diversification).[4]

There are several reasons why companies diversify. First, the formerly focused firm wants to spread the risk of market contraction and is forced to diversify because its products do not show further growth opportunities any more. Second, the outcome of a diversification strategy can result in growth and profitability as well as synergies with the firm's current activities, giving it a better reception in capital markets as the company increases in size. On the other hand however, diversification is generally riskier than a focused strategy and requires the most careful investigation on the company's own strength and weaknesses as well as on the target market itself. An empirical study conducted by Berry (1971) on US firms between 1960 and 1965 argues that conglomerate mergers in general show poorer performance than the average firms. Berry suggests that the firm's lack of experience in skills and techniques in the newly diversified industry may be one possible reason. Therefore, the more unrelated the diversification the more risky it is for the firm. Moreover, diversification may require redistribution of human and of financial resources that can cause the company to loose its focus, commitment and sustained investments in its core products.

Salter (1979) and a recent study by Moeller and Schlingemann (2005) found that acquirer returns are significantly positively associated with a deal between two firms in the same two-digit SIC code (non-conglomerate merger). In a more detailed analysis, Varadarajan (1986) extended the SIC-classification by defining

[4] Refer to section "Data" for diversification measures employed in this study.

broad-spectrum diversification (BSD) the mean narrow-spectrum diversification (MNSD). His findings suggests that firms pursuing a strategy of low BSD and high MNSD financially outperform those pursuing a strategy of high BSD and low MNSD.[5] In other words, firms of related diversification on average outperformed firms with unrelated diversification strategies. In fact, according to Bettis and Mahajan (1985), related diversification achieves a comparably better return performance than any other strategy. Arguments such as synergy, focus, competencies, and economies of scale may serve as explanations for these findings. In an event study on M&A announcement effects among telecommunication firms, Wilcox et al. (2001) found that M&As where the acquirers employ a more focused strategy (i.e. non-conglomerate) tend to experience greater market value increases upon the merger announcement. Based upon the above discussion, we posit:

H2a: *M&A activities classified as non-conglomerate mergers will generate positive abnormal returns.*

H2b: *Average abnormal returns for conglomerate mergers are significantly lower than for non-conglomerate mergers.*

International Diversification

Cross-Border M&A

Another dimension of a firm's diversification strategy is its internationalization strategy. Acquisitions made between companies with headquarters in different countries are called cross-border acquisitions. In recent years, the portion of cross-border M&A has steadily increased (Breedon and Fornasari n.d.). An extensive survey of senior executives[6] revealed that the majority of recent acquisitions have been cross-border. Although international diversification has captured considerably less attention in the finance literature than domestic diversification, in reality the former is more common. This is also reflected in the sample of this study (53 cross border vs. 35 domestic M&As). Domestic acquisitions are generally easier to execute and present less risk, but globalization is inevitable. Acquisitions are often made to overcome entry barriers (Hitt et al. 2005). Especially in the context of telecommunications, a cross-border merger definitely provides a more viable option than organic growth in the foreign geographic market. Firstly, building up an own network infrastructure is often more costly and time-consuming. Secondly, the acquirer can often continue to use the brand name and avoid costly marketing efforts in penetrating the market. For instance, when Vodafone (UK) took over Germany's Mannesmann, it could

[5] BSD is diversification into a different two-digit SIC industry. MNSD is diversification within the same two digit SIC industry but into a different four-digit SIC subcategory (Varadarajan 1986).

[6] According to Economist Intelligence Unit 2006 Global M&A Survey.

continue to provide mobile services under the "D2 privat" label. In fact, this was the largest cross-border M&A in the telecommunications industry with a price tag of 172 billion USD to date. This deal turned Vodafone instantly into the second largest mobile provider in Germany.

Due to the relaxation of regulation in many industries, the number of cross-border M&A activities among the European Union members generally continues to increase (Brakman et al. 2005). Some market analysts believe that this is due to the fact that many European corporations face stiffening competition and have reached saturation point in their domestic markets (Sarkar et al. 1999). Increased international competition is affecting in the telecommunications industry as well. With the erosion of trade restrictions and other regulatory barriers, the revenues from cross-border telecommunications services, such as telephony, data transmission, and entertainment offerings, have rapidly grown. This growth has been paralleled by an increase of the number of M&As among firms headquartered in different countries.

A number of studies on announcement effects of cross-border acquisitions come to the conclusion that shareholders experience significantly positive wealth gains. Eun et al. (1996) examined the effect of foreign acquisitions of US firms on the wealth of acquirer and target shareholders. The result shows that cross-border acquisitions are significantly wealth-creating for acquirers. Corhay and Rad (2000), using a sample of Western European cross-border acquisitions, also found weak empirical evidence that cross-border acquisitions are generally wealth-creating corporate activities. An event study on a large sample of Canadian firms shows that cross-border bidders earn post announcement abnormal returns that are significantly higher than domestic bidders (Tebourbi 2005). Cummins and Weiss (2004) examined the performance of cross-border transactions of European insurance companies. They found that cross-border transactions were value-neutral for acquirers, whereas domestic transactions led to significant value loss for acquirers. Finally, Morosini, Shane and Singh (1998) test whether foreignness enhances cross-border acquisition performance. They employ Hofstede's (1980) four cultural dimensions and its quantitative scores to measure national culture distance as a proxy for foreignness and find that there is a positive association between national cultural distance and cross-border acquisition performance. We therefore posit:

H3a: Cross-border M&A activities of telecommunication firms will result in positive abnormal returns for the acquirer within the event period.

H3b: Average abnormal returns for domestic M&A activities are significantly lower than for cross-border M&A activities.

Emerging Markets

A further dimension of a firm's diversification strategy concerns the question whether the firm should diversify into emerging markets or not. The World Bank defines emerging market as one with low-to-middle per capita income but potentially dynamic and rapidly growing economy where companies can seek lucrative opportunities for medium to long-term investments. While it is not easy to make an exact list of emerging markets, the best guides tend to be investment information sources (such as The Economist) or market index makers (such as Morgan Stanley Capital International). The MSCI Emerging Markets Index consisted of 25 emerging markets. Among them are countries in Latin America (e.g. Argentina, Brazil, Mexico), and Asian countries (China, India, Indonesia, Malaysia, Taiwan, Thailand, Russia) but also some Eastern European and African nations. A complete listing of all emerging countries can be found in the Appendix.

While developed countries remain the major targets for cross-border acquisitions, emerging economies have also been earmarked as important targets of incoming cross-border M&As since the 1990s. Companies that are racing for global leadership have to consider competing in emerging markets. The business risks in these countries are considerable but the opportunities for growth are huge, as their economies develop and living standards climb towards levels in the industrialized world (Thompson et al. 2005). Chari et al. (2004) examined shareholder value gains from developed-market acquisitions of emerging-market targets between 1988 and 2003. They found that joint returns for developed-market acquirers and targets are significantly higher when M&A transactions in emerging markets are announced.

Unsurprisingly, telecommunication firms rushed to seize opportunities in emerging markets, to reap first-mover advantages and to satisfy the growing demand for telecom services (Sarkar et al. 1999). Mobile operators especially, have become increasingly dominant in markets with poor fixed-line coverage. In these countries mobile phones are a substitute for traditional basic fixed services and extend access to population groups such as the urban poor and rural users. Mobile services are now considerably more affordable, both in start-up costs as well as in monthly recurring costs for those consumer groups. Vodafone, for instance, seized the potential growth opportunities offered by emerging markets when they acquired Telsim of Turkey. The mobile operator recognized that emerging markets were growing at three times the rate of developed markets. In light of this, Vodafone has also made investments in India, Romania, and South Africa. In the light of the above discussion, we therefore posit:

H4a: *Cross-border M&A activities into emerging markets will result in positive average abnormal returns.*

H4b: *Average abnormal returns for cross-border M&As into developed markets are significantly lower than for cross-border M&As into emerging markets.*

Dual Diversification

Prior research has focused on the causes and consequences of firm diversification. Only recently more attention was given to international diversification. However, much less attention has been paid to international diversification and its interaction with firm diversification. The formulation of the all previous hypotheses concerns only a single strategy (either service diversification or geographical diversification).

To study whether it is also beneficial for the firm when of both strategies are combined, a sub sample of events which are characterized as both non-conglomerate and cross-border will be tested. We investigate whether both strategies in combination affect each other negatively, or whether we can expect such combination to result in positive abnormal returns as well. So far there are only few studies which examine dual diversification (firm diversification in conjunction with international diversification). An empirical study of the media industry indicates that related diversification (non-conglomerate mergers) in conjunction with cross-border activities contribute to better financial performance than other combinations (Jung and Chan-Olmsted 2005). To test whether this is also valid for the telecommunications industry, our hypotheses are:

H5a: *Cross-border M&A activities which are non-conglomerate mergers will result in positive abnormal returns for the acquirer.*

H5b: *Average abnormal returns of non-conglomerate cross-border M&As are significantly higher than those of conglomerate domestic M&As.*

Theory and Methodology

Efficient Market Hypothesis

A capital market efficiently redistributes scarce resources from areas of surplus to particular areas of insufficiency, such as developing sectors. There are numerous studies in the economics and finance literature that have addressed the efficient markets hypothesis (EMH). In this study, paying some attention to the EMH is warranted because it provides the basis for the use of the event study methodology. The EMH assumes that (1) stocks are always in equilibrium, i.e. all available information is reflected in the total value of the capital market, and (2) that it is impossible for an investor to consistently "beat the market" (Brigham and Houston 2004). Market efficiency implies that stock prices incorporate all relevant information that is available to market traders. If this is true, then any financially relevant information that is newly revealed to investors will be instantaneously incorporated into stock prices (Fama 1998). Since release of news and information is unpredictable by nature, stock prices change in response to new information unpredictably. This means that all investment made have a zero net present value. As soon as there is any information indicating that a stock is under-priced, investors will immediately rush to buy the stock, bidding up its price to a reasonable level. Thus, only normal rates of return can be expected.

Financial theorists have discussed three forms of market efficiency. These three forms clarify further what is meant by the term "all available information". The *weak form* of the EMH suggests that all information contained in past price movements is fully reflected in current prices. Information about recent stock prices would be of no use in selecting stocks. This form of the hypothesis suggests that a trend analysis is meaningless. Historical stock price data are publicly available and costless to obtain. In other words, returns are unpredictable from past returns or other variables (Fama 1991). The *semi-strong form* of the EMH states that current market prices reflect all *publicly available* information. And again, if semi-strong form efficiency exists and investors have access to the publicly available information, it would be futile to study annual reports or any other published data because market prices would have adjusted to any good or bad news contained in such reports. The *strong form* of the EMH states that current market prices reflect all information pertaining to the company, including information available only to company insiders. Under this form of the EMH it is impossible for anyone to earn abnormal returns from the stock market. Studies of the semi-strong form of the EMH can be regarded as tests of the speed of adjustment of prices to new information. The leading research tool in this area is the event study method (Dimson and Mussavian 1998).

The Event Study Research Method

There have been numerous event studies undertaken in all kinds of areas, foremost in the field of finance and strategic management. For instance, Subramani and Walden (2001) study the returns to shareholders in firms engaging in e-commerce. Johnson et al. (2005) examine the impact of ratings of board directors by the business press on stockholders' wealth. Another area in which event studies have been widely used is in the evaluation of M&As. For instance, Wilcox et al. (2001) analyze M&A events in the telecommunications industry by testing the impact of near diversification, far diversification, and the size of the firms on the shareholder value. Uhlenbruck et al. (2006) focus on acquisitions of Internet firms and the potential for the transfer of scarce resources in a resource-based view.

The event study method is based on the assumption that capital markets are efficient such as to estimate the impact of new information on anticipated future profits of the firms. The core assumption of the event-study methodology is that if information communicated to the market contains any useful and surprising content an abnormal return will occur. In a capital market with semi-strong efficiency one can assesses the impact of the event in question on the market value of the company by calculating the abnormal return – the difference between the actual post-event return and the return expected in the absence of the event (MacKinlay 1997). McWilliams and Siegel (1997) gave a good reason for conducting event studies: "The event study method has become popular because it obviates the need to analyze accounting-based measures of profit, which have been criticized because they are often not very good indicators of the true performance of firms". Therefore, it is expected that event studies will continue to be a valuable and widely used tool in economics and finance.

According to (MacKinlay 1997) an event study can be roughly categorized into the following five steps:

1. Identifying of the events of interest and defining the event window size
2. Selection of the sample set of firms to include in the analysis
3. Prediction of a "normal" return during the event window in the absence of the event
4. Calculation of the abnormal return within the event window, where the abnormal return is defined as the difference between the actual and predicted returns
5. Testing whether the abnormal return is statistically different from zero

Kothari and Warner (2004) recommend MacKinlay's work as a standard reference for understanding and conducting event studies. Many recent event studies are based on the MacKinlay's outline (Wilcox et al. 2001; Subramani et al. 2001; Park et al. 2004). In this spirit, this study follows the methodology outlined by MacKinlay (1997). Also, in this study, Eventus,[7]which operates under SAS, was used to test the market reaction to the announcement of M&As.

1. Identifying of the events of interest and defining the event window size

The events of interest to this event study are M&A announcements of major telecommunication companies that are listed in the US-stock exchanges or European stock exchanges. More specifically, we are looking at the earliest announcements of the planned M&A in the media. The significance of an event can be identified by examining its impact on the firm's stock price. To accomplish this, the researcher defines a period of days over which the impact of the event will be measured. This period is known as the event window (denoted as L_2). The size length of the event window has to be justified (Fama 1998).

This study examines two symmetric event windows: a 3-day (1 day prior to the event day and 1 day after the event day) and a 5-day (−2, +2) event window. These window lengths are appropriate to capture any news that might have leaked shortly before the official announcement was made and also considers any short-term stock price reactions linked to the event after the announcement. Some researchers may decide to take longer event windows of 21-, 41-days or even longer for similar event studies. An examination of 500 published event studies in academic press revealed that short-term event studies deliver quite reliable results while long-term event-studies underlie some serious limitations (Kothari and Warner 2004). One disadvantage of using longer windows is that other, unrelated events may be confounded with the event of interest. If other relevant events occurred during the event window, it is hard to isolate the impact of one particular event. In addition, it is difficult to reconcile the assumption of efficient markets with the use of long event windows. The use of very long windows in many management studies implies that some researchers do not believe that the effects of events are quickly incorporated into stock prices. This can be seen as a violation of the assumption of

[7]Eventus™ is software which is designed for the specific purpose of event studies. Eventus™ is widely used in financial and economic research. It was developed by the US-based company Cowan Research LC – URL: http://www.eventstudy.com

market efficiency which is vital to the event study methodology (McWilliams and Siegel 1997). However, it may be reasonable sometimes to assume that information is revealed to investors slowly over a period of time. Therefore, some studies do apply longer event window sizes.

The estimation window is the control period proceeding the event period. In this study, the estimation window (denoted as L_1) for all events ends 12 days before the event and extends back to 120 days prior to the event date (Fig. 1). Estimation periods generally end before the event of interest so that the returns during the event window will not influence the model parameters.

2. Selection of the sample set of firms to include in the analysis

Next we have to find the companies to be examined. There are three steps to follow in order to generate a sample: (1) the definition of the population; (2) the specification of a sample frame; and (3) the design of a method to draw events from the sample frame (Fahrmeir and Tutz 2002). As mentioned above, telecommunication companies that are listed in (at least) one of the major US or European stock exchanges were defined as the target for this study. The population consists of all M&A announcements released by the companies that are listed in the above stock exchanges. In practice, the population list of M&A announcements is generated by searching for specific words in the title of media and news releases. This procedure will be discussed more thoroughly below.

The sampling frame is the fraction of the population of interest. Simple sampling is used throughout this study: all M&A announcements made by US- or European-listed telecommunication companies are considered equal in all aspect.

3. Prediction of the "normal" return within the event window

If the impact of M&A announcements on stock returns is to be examined, a measure of what shall be the "normal" return for the given stock is required. 120 days of historical stock data will be used for each event and the event window was defined to be 5 days in length. These 120 days are enough to calculate valid estimators needed for the event-study model (MacKinlay 1997). There are a number of approaches to calculate the normal return of a given stock. Four methods are usually used in past event studies: the *constant mean return model*, the *market model*, the *control portfolio return model* and the *risk-adjusted return model*. To discuss all these statistical models would go beyond the scope of this study. For firm-specific events, such as M&A announcements, an appropriate choice for such a measure is the market model (Fama 1998).

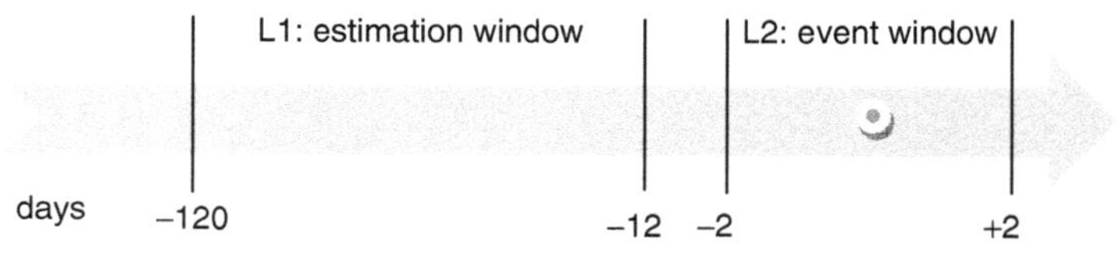

Fig. 1 Estimation and event window on a timeline

4. Calculation of the abnormal return

The market model event-study methodology is a statistical model which relates the return of any given security to the return of the market index. It is the most commonly used model in event studies. In case of this study, the Dow Jones Sector Titans – Telecommunications Sector index was used as the market index, as detailed below. To predict each firm's market model, daily returns were used to estimate a regression equation over the estimation period. The underlying securities are assumed to be independently and jointly normally distributed and shall be identically-distributed through time (MacKinlay 1997). For any company *i*, the market model is specified as

$$R_{i\tau} = \alpha_i + \beta_i R_{m\tau} + \varepsilon_{i\tau} \quad (1)$$

where $R_{i\tau}$ is the return of security *i* and R_{mr} is the rate of return of the market portfolio in period τ. $\varepsilon_{i\tau}$ is the zero-mean disturbance term. α_i and β_i are firm specific parameters of the market model. The market model assumes that in the absence of the event, the relationship between the returns of firm *i* and returns of the market index remains unchanged and the expected value of the disturbance term $\varepsilon_{i\tau}$ is zero. Using this approach the resulting regression coefficients and the firm's actual daily returns were then used to compute abnormal returns for each firm over each day of the event window period. The sample abnormal return $AR_{i\tau}$ on the event day τ is calculated for the *ith* firm by subtracting the predicted return of the market model from its observed return:

$$AR_{i\tau} = R_{i\tau} - (\hat{\alpha}_i + \hat{\beta}_i R_{m\tau}) \quad (2)$$

where the coefficients $\hat{\alpha}_i$ and $\hat{\beta}_i$ are ordinary least squares estimates of α_i and β_i The Cumulative Abnormal Return (CAR) for firm *i* over the event period $\tau 1$ to $\tau 2$ is then calculated as follows:

$$CAR_\tau(\tau 1, \tau 2) = \sum_{\tau=\tau 1}^{\tau 2} AR_{i\tau} \quad (3)$$

where ($\tau 1$, $\tau 2$) is the event window interval; and all other terms as previously defined. The abnormal returns represent the extent to which actual realized returns on any of the event days deviate from the returns that was expected based on the estimated firm-specific market model. In this sense, the abnormal returns can be seen as prediction errors ($\varepsilon_{i\tau}$). Following MacKinlay (1997), the variance of daily abnormal returns is known and can be calculated as:

$$\sigma^2(AR_{i\tau}) = \sigma^2_{\varepsilon i} + \frac{1}{L_1}\left(1 + \frac{(R_{m\tau} - \hat{\mu}_m)^2}{\hat{\sigma}^2_m}\right) \quad (4)$$

where $\sigma_{\varepsilon i}^2$ is the disturbance variance (i.e. the abnormal returns) from the market model and L_1 represents the length of the estimation period over which the market model residuals were estimated for firm i. The variance $\sigma^2(AR_{i\tau})$ in Equation (4) has two components: a disturbance term $\sigma_{\varepsilon i}^2$, which was estimated from the market model residuals, and a sampling error term. Thus, under the condition that the number of days in the estimation period L_1 is sufficiently large (greater than 30 days), the variance in abnormal returns converges to $\sigma_{\varepsilon i}^2$ and therefore it can be assumed that the daily abnormal returns are jointly normal with a zero conditional mean, that is, $AR_{i\tau} \sim N(0,\sigma_{\varepsilon i}^2)$ (MacKinlay 1997).

Furthermore, because the distribution of abnormal returns for all N securities can be assumed to be independent and normally distributed, the individual securities' abnormal returns can be aggregated within any given time period. The average abnormal return and the variance of average abnormal returns across all N securities in a given time period are computed as follows:

$$\overline{AR}_\tau = \frac{1}{N}\sum_{i=1}^{N} AR_{i\tau} \tag{5}$$

$$\operatorname{var}\left(\overline{AR}_\tau\right) = \frac{1}{N^2}\sum_{i=1}^{N} \sigma_{\varepsilon i} \tag{6}$$

where N equals the number of events; and all other terms are as previously defined. As expected, the average abnormal return is also normally distributed with a zero conditional mean and a conditional variance given by Equation (6). Then the average abnormal returns and the variance in the CAR can then be aggregated over the event window ($\tau 1$, $\tau 2$) as follows:

$$\overline{CAR}(\tau 1, \tau 2) = \sum_{\tau=\tau 1}^{\tau 2} \overline{AR}_\tau \tag{7}$$

$$\operatorname{var}\left(\overline{CAR}(\tau 1, \tau 2)\right) = \sum_{\tau=\tau 1}^{\tau 2} \operatorname{var}\left(\overline{AR}_\tau\right) \tag{8}$$

Analogously as above, inferences about the mean cumulative abnormal returns can be drawn using the following relationship (MacKinlay 1997):

$$\overline{CAR}\,(\tau 1, \tau 2) \sim N\left(0, \operatorname{var}\left(\overline{CAR}(\tau 1, \tau 2)\right)\right) \tag{9}$$

The impact on the cumulative abnormal returns across the securities in the portfolio follows a standard Z-score statistic.

5. Testing whether the abnormal return is statistically different from zero

(a) The Patell Z-Test

This study uses the standardized residual Z-test suggested by Patell (1976) to assess the statistical significance of the abnormal returns over the event interval. The literature also refers to the Patell test as a standardized abnormal return test or a test assuming cross-sectional independence (Cowan 2002). According to Patell (1976) the abnormal return for each security gets standardized by dividing it by the security's own estimate of variance. This standardization helps to make sure that no single firm in the sample dominates the results of the analysis and helps improve the power of the test statistics. Therefore several additional adjustments in measuring abnormal returns have to be made. First the abnormal returns for firm *i* on day *t* are standardized by the estimated standard deviation. The *standardized abnormal return* is defined as:

$$SAR_{it} = \frac{AR_{it}}{\sigma_{\varepsilon i}} \tag{10}$$

SAR_{it} follows a Student's t distribution with (L_1–2) degrees of freedom. Given a sufficiently large sample, the Central Limit Theorem says that the distribution of the sum of independent and identically distributed variables will be approximately normal, regardless of the underlying distribution. This rule would therefore also apply to an aggregation of SAR_{it}: Just like in Equation (5), the standardized abnormal returns in period *t* will be aggregated across the *N* firms to obtain the cumulative standardized abnormal return:

$$CSAR_t = \sum_{i=1}^{N} SAR_{it} \tag{11}$$

The expected value of $CSAR_t$ is zero. The variance of $CSAR_t$ is known to be

$$Q_t = \sum_{i=1}^{N} \frac{L_i - 2}{L_i - 4} \tag{12}$$

where L_i is the is the number of non-missing trading day returns in the *L*-day period $\tau 1$ through $\tau 2$ for firm *i*. In an analogous fashion as done in Equation (7), $CSAR_t$ are then accumulated over the event period (see Equations (13) and (14)) to produce the test statistic:

$$Z_{\tau 1,\tau 2} = \frac{1}{\sqrt{N}} \sum_{i=1}^{N} Z^i_{\tau 1,\tau 2} \tag{13}$$

where

$$Z^i_{\tau 1,\tau 2} = \frac{1}{\sqrt{Q^i_{\tau 1,\tau 2}}} \sum_{\tau=\tau 1}^{\tau 2} SAR_{it} \tag{14}$$

and

$$Q^i_{\tau 1,\tau 2} = (\tau 1 + \tau 2 + 1)\frac{L_i - 2}{L_i - 4} \tag{15}$$

Under the condition that $Z_{t1t2}{}^i$ is cross-sectional independent and some other conditions (see Patell 1976), $Z_{t1t2}{}^i$ follows the standard normal distribution under the null hypothesis (Central Limit Theorem). This Z-statistic is then used to test the significance of no abnormal return during the event window period.

(b) The Generalized Sign-Test

As discrete daily stock returns typically might not have a normal distribution in all cases, a nonparametric test can be used in conjunction with the parametric test in an event study, thus avoiding the dependence on normality of return distributions (Cowan 1992). This is to verify that the results of parametric tests are not dominated by outliers. It compares the proportion of positive abnormal returns around an event day to the proportion from the estimation period. The basis of the generalized sign test is that, under the null hypothesis, the fraction of positive returns is the same as in the estimation period. The actual test uses the normal approximation to the binomial distribution (Cowan 2002). For example, if 45% of market returns are positive in the estimation period, while 60% of firms have positive market returns on the event day "−1", the test reports whether the difference between 60% and 45% is statistically significant. Cowan reports the test to be well specified in general samples from NYSE, AMEX and NASDAQ stocks. In this event study, Cowan's (1992) generalized sign test is employed to provide a check of the robustness of conclusions based on the (parametric) Patell-test.

(c) The Paired t-test

A paired t-test is used to compare two population means. Here it is used to determine whether there is a significant difference between the *average values of event-window abnormal returns.*[8] This averaged difference is calculated based on the paired differences between the two values of each event-window day. The test statistics is calculated as:

$$T = \frac{\sqrt{N}}{\sigma}\left(\overline{CAR_1} - \overline{CAR_2}\right) \qquad (16)$$

where the numerator represents the mean difference, σ_t^2 is the sample standard deviation, N is the sample size and T is a Student-t quartile with n−1 degrees of freedom.

Data

The two inputs that are required for the event-study model are the events themselves, in this case M&A announcements, and historical stock price data (security prices and the reference index). They were both gathered from databases.

In order to explore the effects of M&As in stock prices, this research limits its scope to companies that are either listed at one of the major European stock exchanges (London, Paris, Frankfurt, Madrid, Amsterdam) or on a US stock exchange

[8] Represents the average value of cumulative abnormal returns over all event window days.

(NYSE, NASDAQ). The reason is that most of the telecommunication operators are listed on at least one of these stock exchanges. A large sample size is vital to reduce the sampling error which in turn also increases the reliability of the statistical results (Cohen 1988). Furthermore, it is generally known that these exchanges provide a high public confidence due to the high standards and listing requirements. Thus, using only major exchanges also increases the price stability of the securities – an important prerequisite for the event study method. Our hypotheses are tested on a sample of M&As completed by publicly traded acquirers between 1998 and 2006. Telecommunication operators were selected from the Thomson One Banker database indexes by searching for companies within the following industries: companies with primary SIC Code 4813 (Telephone Communications except Radiotelephone), 4812 (Radiotelephone Communications) or 4842 (Cable and Other Pay Television Services). The results were limited to companies that are (at the time of the event) listed at a major US or European stock. Next, with a list of 56 potential acquirers, the M&A events which are associated with every one of these carriers were separately retrieved from the Highbeamä Research database.[9] The following search on Highbeamä using a set of relevant search terms was performed and the earliest dates announcing the event was recorded:

- Search in article title only:
  - *{company name from list}* AND *buy* OR *acquires* OR *bid* OR *merge* OR *takeover* OR *merger* OR *acquisition* OR *buys* OR *merges* OR *merging* OR *acquiring*
- Search in following sources:
  - *"Newspapers"* (Business Wire, Associated Press, PR Newswire, Reuters, Wallstreet Journal, Washington Post)
  - Dates
  - Between *Jan 1, 1998 to 31 Oct 2006*

Events that were identified using these criteria were consolidated into a master list with duplicates removed. The preliminary sample frame had 493 M&A events. Out of this, an acquisition that resulted in a controlling stake for the acquirer, i.e. greater than 50% of the stakes are chosen. This means that acquisitions giving the acquirer lower than 50% in stakes were not considered. It is true that that in practice acquirers may de facto control the target even though they hold less than 50% of the shares. On the other hand, it is also true that owning only a minor stake, say 10% of the shares, could hardly result in a significant degree of control of the target. Determining whether or not an acquisition resulted in a controlling stake is difficult without using a great deal of insider knowledge and subjectivity. We therefore chose to apply the 50% rule outlined above, full-well acknowledging the practical limitations of our approach.

---

[9] HighBeam™ Research is an online research engine which sorts free, paid, and proprietary online articles and databases published in the past 20 years. It is a tool for serious business, education, and personal research – URL: http://www.highbeam.com.

Moreover, only M&A announcements containing accurate and detailed information about date of announcement, partner and transaction value were included. Last, only those events with at least 120 days of historic stock data[10] available were selected. Moreover, to avoid possible confounding effects within the event window, a number of M&A events were selectively omitted. Excluded events are those coinciding with other major firm-specific events that in turn might affect the stock price such as earnings alliance announcements, earnings, large investment decisions or new product introduction (McWilliams and Siegel 1997). Confounding events were identified using Highbeam Research functions. It displays all company news within a 4-days-range of the specific date. After meeting all these criteria, the final sample contains 88 M&A announcements of 37 telecommunication firms (see Table 6 in the Appendix). The distribution of events on the calendar years clearly reflects the M&A waves during 2004/2005 and 1999/2000 (see Fig. 2 below). The event date was again cross-checked with Factiva database and all relevant event data was then condensed into a master list. Once the M&A announcements were isolated and the event window was defined, the "normal" (expected) returns for that window needed to be estimated. This was done by using historical stock price data (adjusted closing prices) for all acquiring companies listed in the master list.

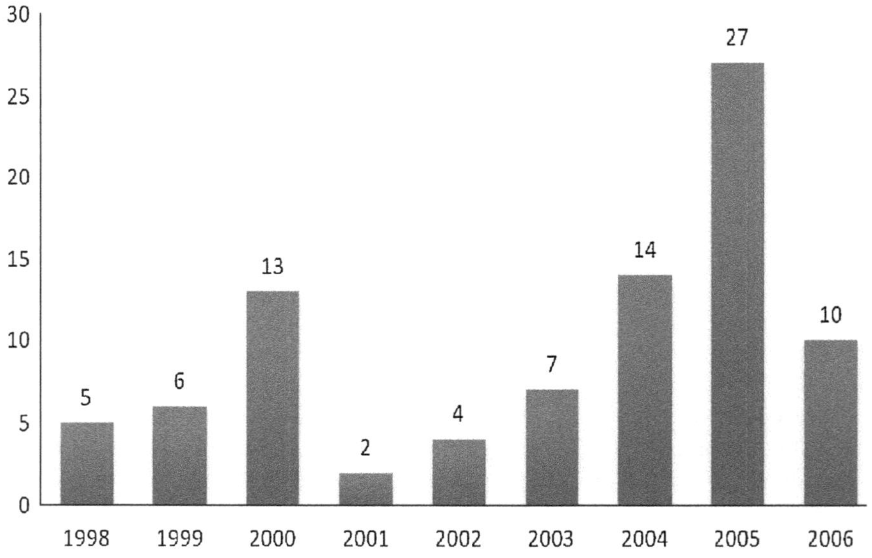

Fig. 2 Distribution of events during the sample frame 1998–2006

[10] Retrievable through Yahoo! Finance or CRSP database.

For each event, 132 days of historic stock data before the event date and 10 days of stock data after the event date were downloaded through either the Center of Research on Security Prices database (CRSP) or the Yahoo! Finance database.

As this event study focuses on M&As in a global perspective, many non-US telecom carriers were included. Most of these carriers are also listed in the United States and stock data could be easily retrieved from CRSP. However, in terms of the traded volumes, the main site of trading for those securities is at the stock exchange in their own respective countries. For instance, Germany's largest telecommunication operator, Deutsche Telekom, is traded both at the XETRA in Frankfurt with an average trading volume of 26,902,100 whereas the company's stock at the NYSE has an average trading volume of only 613,188. Any sudden event such as M&A announcements would be first reflected on their home stock exchange and the short-term stock performance would reflect the shareholder gain much more accurately. Therefore, historical stock data of the respective companies were always retrieved from their "home" stock exchange.

To estimate the abnormal returns under the market model, an index needed to be determined. The market model is a statistical model which relates the return of any given security to the return of the market portfolio, i.e. an industry index (MacKinlay 1997). Since the sample consists of major telecommunication carriers, the Dow Jones Sector Titans – Telecommunications Sector (DJTTEL) was used as the reference index for the market model (see Table 5 in Appendix). The index consists of 29 securities of leading global telecommunications carriers that are traded at their respective "home" exchanges. 17 out of 37 acquiring companies in this study's sample belong to the DJTTEL as of November 2006. To measure the direction of diversification of the M&As, this study employs the SIC[11] classification as conducted by Berry (1971) and Ferris and Park (2001). Many industrial organization studies have used objective measures based on standard industrial classification (SIC) counts to capture the aspect of diversification (Ramanujam and Varadarajan 1989). The first number assigns a product to a very broad category. Each subsequent number distinguishes the product at a progressively finer level. The SIC classification has been widely used among economists to determine in which industry segments the company is operating. As all acquirers in this study are listed telecommunications companies, they all operate with the two-digit 48xx SIC-code. The acquirers strategy can be determined by comparing both the SIC codes of acquirer and the target. Telecommunications M&A occurring solely within the 48xx SIC code (i.e. both acquiring and target firms) are termed non-conglomerate mergers. M&A where the target has a SIC code other than 48xx is classified as a conglomerate merger (Ferris and Park 2001; Ramanujam and Varadarajan 1989).

---

[11] A standard industrial classification (SIC) code categorizes US business establishments based upon the type of business activity performed at its location. All fields of economic activity are included in this system including both manufacturing and nonmanufacturing operations. The system is governed by the Office of Statistical Standards.

Empirical Results

Tables 2 and 3 show the results from the event study. The first two columns of Table 2 present coefficients and significances for the *additive event-window abnormal returns*[12] in the respective event window (also referred to as "Mean Cumulative Abnormal Return"[13]). The abnormal returns are calculated using the market model estimated from 132 to 12 trading days prior to the event announcements. The Mean CARs (given in parentheses) represent the cumulative market model-adjusted abnormal returns over the relevant event window. The Z-statistics for the (−1, +1) and (−2, +2) event windows are based on the standardized abnormal return method according to Patell (1976). The test statistics for the nonparametric generalized sign test are reported in the last two columns, with the number of positive and negative CARs given in parentheses (Fig. 3).

Table 2 Cumulative abnormal returns

		Z-value for mean CAR		Gen. sign Z-value for positive: negative	
Sample type	N	(−1, +1)	(−2, +2)	(−1, +1)	(−2, +2)
Complete sample (*H1*)	88	2.234*	1.871*	2.522**	0.603
		(0.85%)	(0.86%)	(55:33)	(46:42)
Firm diversification					
Non-conglomerate merger (*H2a*)	58	1.478#	1.427#	2.944**	1.106
		(0.63%)	(0.61%)	(40:18)	(33:25)
Conglomerate merger	30	1.764*	1.219	0.226	−0.505
		(1.29%)	(1.19%)	(15:15)	(13:17)
International strategy					
Cross-border M&A (*H3a*)	53	3.930***	3.038**	3.033**	1.384#
		(1.80%)	(1.85%)	(37:16)	(31:22)
Domestic M&A	35	−1.251	−0.747	0.267	−0.747
		(−0.58%)	(−0.63%)	(18:17)	(15:20)
Emerging markets					
M&A in emerging market (*H4a*)	22	2.167*	1.202	3.044**	1.339#
		(1.06%)	(0.82%)	(18:4)	(14:8)
M&A in developed market	32	3.41***	2.998**	1.547#	0.839
		(2.29%)	(2.52%)	(20:12)	(18:14)
Dual diversification					
Cross-border and non-conglomerate (*H5a*)	38	2.415**	1.819*	2.992**	1.370#
		(1.22%)	(1.27%)	(28:10)	(23:15)
Domestic and conglomerate	18	−1.688*	−0.414	0.005	−0.466
		(−1.41%)	(−1.02%)	(9:9)	(8:10)

The symbols #,*,**, and *** denote statistical significance at the 10%, 5%, 1% and 0.1% levels, respectively, using a one-tail test.

[12] Represents the sum of the mean cumulative abnormal returns over all event window days.

[13] Diagrams for the mean cumulative abnormal returns for the complete sample were generated for the 5-, 10- and 20-day event window and can be found in the Appendix.

Table 3 Paired t-test results for Hypothesis 2b, 3b, 4b, 5b

Event window	Mean (non-conglomerate)	Mean (conglomerate)	Mean difference	t-score
CAR(−1, +1)	0.220	0.450	−0.230	−0.4401
CAR(−2, +2)	0.146	0.248	−0.102	−0.3436
Test result	Hypothesis 2b rejected			
Event window	Mean (cross-border)	Mean (domestic)	Mean difference	t-score
CAR(−1, +1)	0.620	−0.193	0.813	1.9648#
CAR(−2, +2)	0.380	−0.126	0.506	1.6693#
Test result	Hypothesis 3b accepted			
Event window	Mean (emerging market)	Mean (developed market)	Mean difference	t-score
CAR(−1, +1)	0.367	0.813	−0.447	−0.5781
CAR(−2, +2)	0.172	0.536	−0.364	−0.8281
Test result	Hypothesis 4b rejected			
Event window	Mean (cross-border non-conglomerate)	Mean (domestic conglomerate)	Mean difference	t-score
CAR(−1, +1)	0.420	0.100	0.320	2.2857#
CAR(−2, +2)	0.262	−0.046	0.308	3.9983**
Test result	Hypothesis 5b accepted			

The symbols #,*,**, and *** denote statistical significance at the 10%, 5%, 1% and 0.1% levels, respectively, using a one-tail test.

Table 3 shows the results of the Student's paired t-test for the b-hypotheses. Columns two and three show the *averaged event-window abnormal returns.* An averaged difference is calculated in column four, based on the paired differences between the two values of each event-window day. The t-score is presented in fourth column is based at four and two degrees of freedom for the 5- and 3-day event window respectively.

Complete Sample

The first row of Table 2 reports the results for the complete sample (H1). As can be seen from the table, there is significant support for Hypothesis 1 that M&A activities will in general have a positive impact on telecommunication firms participating in these activities. The mean CAR for both windows is approximately +0.85% and significant at the 5% level. This is being supported by the generalized sign test for the (−1, +1) event window.

Firm Diversification

According to Hypothesis 2a, non-conglomerate mergers will generate positive abnormal returns. CARs of +0.63% and +0.61% are reported respectively for both windows, which are weakly significant at the 10% level. The generalized sign test is statistically significant at the confidence 1% level, indicating that the result is highly robust against outliers. Therefore Hypothesis 2a is weakly supported.

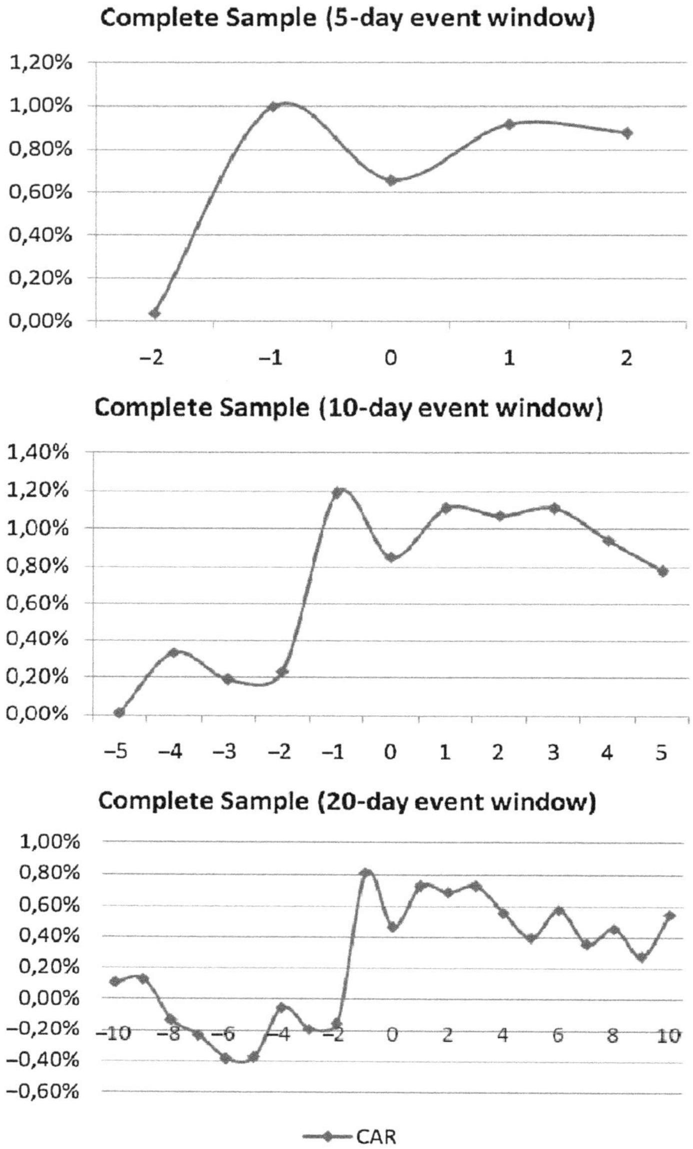

Fig. 3 CAR diagrams for complete sample

Hypothesis 2b states that conglomerate mergers will generate lower abnormal returns than non-conglomerate mergers. The CAR for the conglomerate mergers is twice as high (+1.29%) and is significantly positive at the 5% level for the (−1, +1) window. The sign test shows no significance which indicates that the gains might be due to outliners in the sub-sample. Also, no statistically significant evidence was found in the (−2, +2) window. The t-score for the paired t-test that measures the difference for both CAR means ($\Delta = NonConglom.CAR - Conglom.CAR$) is negative but not significant. Hence, there is no evidence that conglomerate mergers show lower abnormal returns than non-conglomerate mergers. Hypothesis 2b is rejected.

International Strategy

Hypothesis 3a suggests that cross-border M&A activities will result in positive abnormal returns. For both event windows, cross-border acquirers gained 1.80% and 1.85% (significant at the 0.1% and 1% level, respectively). The generalized sign tests for the number of positive and negative CARs are significant at the 1% and 10% level, respectively. These results indicate that the use cross-border M&A by telecommunication firms are perceived as a value-generating strategy. Hypothesis 3a is therefore supported.

Hypothesis 3b states that domestic M&A activities are less favorable than cross-border M&As. The Patell-test on the CARs results show that there is an insignificant wealth loss for domestic acquirers of –0.58% and –0.63%, respectively, indicating that CARs are not statistically different from zero. The paired t-test measures the difference of the cumulative abnormal return mean values.

($\Delta = NonConglom.CAR - Conglom.CAR$) and determines a possible significance. Here the t-score is positive and shows a weak significant difference of 0.81% and 0.50% at the 10% significance level for both windows. Cross-border M&A are hence to a certain extent adding more value to the firm than domestic M&A. Thus, these results provide weak support for Hypothesis 3b.

Emerging Markets

Hypothesis 4a states that cross-border M&A activities into emerging markets will result in positive abnormal returns. For the (–1, +1) window, the results show statistically significant, positive abnormal returns of +1.06%. This is backed by the generalized sign test which is significant at the 1% level. The examination of the (–2, +2) window results in positive abnormal returns but not statistically different from zero. Overall, the results show that cross-border M&A are beneficial to the foreign acquirer and consistent with Hypothesis 4a.

Hypothesis 4b postits that cross-border M&A activities of a firm into developed markets are less attractive for the stakeholders than cross-border M&A activities into emerging markets. As for the results of the Patell-test, transactions into developed markets result in positive gains for both event windows of +2.29% and +2.52% respectively (at the 0.1% and 1% significance level). Although CAR levels are higher than the ones observed for emerging markets (+1.06%), no inference can be made about the difference of both results. The paired t-test shows no statistical significance. In other words, both sub samples show significant positive abnormal returns while no conclusions can be made on the difference. Therefore, Hypothesis 4b is rejected.

Dual Diversification

Hypothesis 5a states that non-conglomerate cross-border M&A activities will result in positive abnormal returns for the acquirer. The wealth gains for cross-border M&A are +1.22% and +1.27% for the respective windows and statistically significant at the 5% level. The generalized sign test is also significant for both windows at the 1% and 10% significance level respectively. These results confirm Hypothesis 5a.

According to Hypothesis 5b, cross-border & non-conglomerate mergers generate significantly higher CARs than domestic & conglomerate mergers. The Patell-test for all domestic & conglomerate mergers indicate negative CARs in both event windows (−1.41% and −1.02%) which are significant at the 5% level for the (−1, +1) window. However, the generalized sign test for the number of positive and negative CARs is not significant for both windows. The paired t-test showed that the difference

$$(\Delta = CrossBorder\ \&\ NonConglom.CAR - Domestic\ \&\ Conglom.CAR)$$

is positive and significant at the 10% and 1% levels. This indicates that cross-border & non-conglomerate mergers do perform better than domestic & conglomerate mergers. Therefore, Hypothesis 5b cannot be rejected.

Discussion and Conclusion

Limitations

Like any other empirical study, this study is also subject to limitations, which should not remain unmentioned. For instance, in conjunction with the cross-border M&As, this study did not examine the prior international exposure of the firm that is making acquisitions outside its home market. In other words, an acquisition announced by a firm which is already strongly diversified in other geographic markets might not have the same effect as an announcement made by a firm that makes its first move outside its home market. However, the sample used in this event study consists of large telecommunications operators which are relatively similar in their international exposure. Therefore in any case, we should not expect this to be a major problem.

In contrast to some other studies, the SIC-based diversification measurement method employed in this study does not consider the relative size of the firms' various SIC operations. This limitation applies for very large enterprises which are classified in several SIC industries (such as Deutsche Telekom) and where the primary SIC industry is not necessarily the strongest business any more. Nonetheless, this study assumes the filed SIC to be the main business of the firm.

It is also uncertain whether a two-dimensional categorization (conglomerate merger/non-conglomerate merger) is suitable to characterize product diversification within the telecommunications industry, which is technologically evolving faster than any other industry. A more elaborate SIC measurement according to

Ramanujam and Varadarajan (1989) makes no sense within this research context as this study is examining M&A valuation effects of telecommunication operators only. There were no other quantitative measurement methods available to date. A thorough investigation by setting up a qualitative measurement and assessing every individual firm could be considered in future research.

Discussion

The results are consistent with previous event studies in showing that M&As in the telecommunication industry generally result in significant gains in the market values of the acquirer. Therefore, it can be concluded that the market is generally optimistic with regards to the potential of telecom carriers to add value in this industry. The highly competitive marketplace in the telecom sector means that high returns are no longer guaranteed for big telecom firms. Telecommunications networks have typically high fixed costs but comparably low marginal costs. As a result, the potential for economies of scale and scope remains enormous in this industry. All rival operators are racing to grow fast to reap those benefits. Investors may have realized that long-term growth depends on capital being diverted to productive purposes. However, reasons for cases where firms do not show positive gains after an M&A announcement can be that it is not always easy for a company to achieve synergies and to reap scale and scope. High integration costs and differences in corporate culture are reasons why M&A fail to add value. This shows investors' skepticism about the likelihood that the acquirer will be able to realize these synergies required to justify the premium paid (Selden and Colvin 2003).

Besides the positive abnormal returns derived from M&As of telecommunication operators, four other conclusions can be drawn from the empirical results:

1. The results weakly suggest that M&A involving firms that operate in the same or related industry are wealth-creating. However, such firms do not necessarily perform better than telecom firms following a broader diversification strategy through M&A.
2. Telecommunication operators engaging in cross-border M&A generally experience positive abnormal returns and outperform firms that expand domestically.
3. Cross-border M&A activities in emerging markets do add value to the acquirer but do not result in significantly higher abnormal returns for the acquirer than it is the case for M&As into developed markets.
4. Cross-border non-conglomerate mergers are found to add value to the acquiring telecommunications operator, while no significant stock reactions are found for acquirers engaging in domestic and conglomerate mergers.

As summarized under (1): Non-conglomerate mergers appear to be value-adding for telecommunications firms. On the other hand, conglomerate mergers do not experience a significant value reduction, either. Against expectations, the samples with conglomerate mergers show higher mean CARs than those with non-conglomerate mergers, but it could not be shown that the difference is significant. However, a broader diversification strategy might indeed provide a more predictable and reliable revenue stream. For instance, the ongoing modernization of packet-switched networks and the better congestion control due to bandwidth management results in much more efficient network utilization. Some telecom carriers with excess network capacity may be more cost-efficient across industries. Moreover, a large portfolio is proven to handle cyclical downturns in one business segment better as cyclical upturn in another business is likely to occur. This is currently already true for many telecom incumbents worldwide which struggle to maintain competitiveness in their former monopoly markets as local authorities have liberalized telecommunications to allow more competition. Firms in deregulated industries often want to provide bundles of products. This also applies to the telecommunications industry, especially in the context of today's technological convergence (Pernet 2006). Market analysts[14] suggest that consumers are demanding for bundled digital services. As a result, "triple play" providers are converging television (SIC 7812), Internet (SIC 7375) and telecommunication technologies (SIC 48xx). Large telecommunication providers started to see this trend and slowly back away from their pure-play strategy on which they relied on for the past decades. This means that operators must become active on formerly unrelated markets. In order to pick up the technology, incumbent firms may have develop their own technologies or acquire existing firms that are unrelated to telecommunications. In most cases, operators diversify through M&A. Therefore, it remains to be seen whether the results under (1) will show a significant shift in favor of a broad diversification strategy among telecom operators in the years to come.

Since they are operating under stiff competition in their home markets, it seems logical for large telecommunication firms to diversify geographically in cross-border M&A as stated under (2). However, with deregulation, and numerous competitors leveling the playing field, returns are no longer guaranteed. Cross-border M&As become essential as telecommunications companies operating without a global network will not be able to meet the growing demands of its customers in international data communication services (Park et al. 2002). The fact that most telecom M&As are cross-border shows that telecommunication operators look for new markets beyond national borders. This is an apparent attempt to expand the global market power as growth on their home markets is slowly but surely stagnating. Firm size and global reach are the new competitive edge, and is probably the reason for greater efficiency within this industry during the merger waves in 2004 and 2005. This study found that an increase in geographic diversification is positively and significantly associated with acquirer returns. The results reflect only a weak significance for a better performance than domestic M&A but they confirm a positive tendency of abnormal returns which are significantly high in magnitude for this sample and are consistent with previous studies.

According to (3), we could not find any evidence for the hypothesis that M&As into emerging markets add more value to the acquirer then M&A into developed

[14] According to DATAMONITOR: Fixed Line Telecoms in Europe (June 2006).

Table 4 Telecom connections in developed and emerging markets (Gartner Dataquest [September 2006])

	2005	2010	CAGR[15] 2005–2010	Share 2005	Share 2010
Fixed-line connections					
Emerging markets	686,168.9	789,620.5	2.8%	55.7%	60.0%
Developed markets	546,474.5	527,193.8	−0.7%	44.3%	40.0%
Total	1,232,643.4	1,316,814.3	1.3%	100.0%	100.0%
Mobile connections					
Emerging markets	1,364,589.0	2,675,386.7	14.4%	62.7%	72.7%
Developed markets	811,229.0	1,005,847.1	4.4%	37.3%	27.3%
Total	2,175,818.0	3,681,233.8	11.1%	100.0%	100.0%

markets. In fact, average CARs for developed market entrants show higher magnitudes than those of acquirers who operate in emerging markets and these results are also highly significant. The reason why the mean CAR is not significantly higher for mergers into emerging markets might be: First, the low disposable income of potential customers; Second, bad coverage and penetration of communications services and third, the lack of deregulation to enable more competition. But according to a Gartner press release,[16] emerging markets account for more than half of the world's total telecom connections in 2005. This portion will be expected to grow up to 69% by 2010 (see Table 4). The rapid expansion of the telecommunications industry in emerging markets is due to under-penetration and also due to the fact that communication-enabled services are crucial for overall development of these countries. Therfore, it might be worthwhile examining this issue again in future studies to find out whether telecom M&As into emerging markets add significantly more value to the acquirer than M&As into developed countries.

Final Remarks

Over the last decade, the telecommunications industry has shown most exciting developments. The ongoing dynamics in this industry, the profound changes in the institutional structure, the number of cross-border M&A and the strategic importance of this industry for modern information-societies enhanced the relevance on this industry for research. With this study we hope to have contributed to research on M&As by showing some useful empirical evidence to better understand the telecommunications industry. It can be basis for further research on this field and might also be helpful for research on different industries that share similar structure and conditions as telecommunications.

[15] Compound Annual Growth Rate.

[16] Gartner: Emerging Markets Hold the Key to Future Telecoms (September 2006).

Appendix

Table 5 Composition of the DJTTEL

Country	Exchange	Name
GB	LON	Vodafone Group PLC[a]
US	NYSE	AT&T Inc[a]
US	NYSE	Verizon Communications Inc[a]
ES	MCE	Telefonica S.A[a]
US	NYSE	BellSouth Corp[a]
DE	XTR	Deutsche Telekom AG[a]
US	NYSE	Sprint Nextel Corp[a]
GB	LON	BT Group PLC[a]
FR	PAR	France Telecom[a]
MX	MEX	America Movil S.A. de C.V[a]
HK	HON	China Mobile Ltd.
IT	MIL	Telecom Italia S.p.A.[a]
JP	TSE	NTT DoCoMo Inc
NL	AMS	Royal KPN N.V.[a]
JP	TSE	Nippon Telegraph & Telephone Corp.[a]
JP	TSE	KDDI Corp.
US	NYSE	Alltel Corp.
CA	TOR	BCE Inc.
ZA	JOH	MTN Group Ltd.
SE	STO	TeliaSonera AB[a]
NO	OSL	Telenor ASA[a]
US	NYSE	Qwest Communications Int. Inc.[a]
PT	LIS	Portugal Telecom SGPS S/A
AU	ASX	Telstra Corp. Ltd.[a]
SG	SGP	Singapore Telecommunications Ltd.
MX	MEX	Telefonos de Mexico S.A.[a]
KR	KSE	SK Telecom Co. Ltd.
TW	TWS	Chunghwa Telecom Co. Ltd.
KR	KSE	KT Corp.
CH	VTX	Swisscom AG[a]

[a]Security is part of the sample in this event study.

Table 6 List of events used for this study

Date	Acquirer	Target
1998–01–05	SBC	SNET
1998–05–11	SBC	Ameritech
1998–06–24	AT&T Corp.	TCI
1998–07–28	Bell Atlantic	GTE
1998–12–08	AT&T Corp.	IBM Global Data
1999–01–15	Vodafone	Airtouch
1999–08–06	Deutsche Telekom	One2One
1999–11–01	AT&T Corp.	Netstream

(continued)

Table 6 (continued)

Date	Acquirer	Target
1999–11–02	Qwest	US West
1999–11–17	Deutsche Telekom	Siris
1999–12–10	KPN	E-Plus
2000–01–27	France Telecom	Global One
2000–02–03	Vodafone	Mannesmann
2000–03–27	Deutsche Telekom	Debis
2000–05–07	NTT	Verio
2000–05–16	Telefonica	Lycos
2000–05–30	France Telecom	Orange
2000–06–05	Telefonica	mediaWays
2000–07–13	Deutsche Telekom	Slovenske Telekomunikacie
2000–07–17	BellSouth	Cocelco
2000–07–24	Deutsche Telekom	Voicestream
2000–08–28	Deutsche Telekom	Powertel
2000–12–18	KPN	Orange KPN
2000–12–21	Vodafone	Eircell
2001–11–14	Telstra	Clear Communications
2001–12–20	Comcast	AT&T Broadband
2002–03–26	Telia	Sonera
2002–05–06	Telefonica	Pegaso
2002–06–30	Telstra	Hong Kong CSL
2002–11–05	MTS	UMC
2003–02–17	Tele2	British Alpha Telecom
2003–03–01	America Movil	BCP Nordeste
2003–03–08	Telefonica	BellSouth Latin America
2003–08–11	Vodafone	Singlepoint
2003–08–29	America Movil	BCP Wireless
2003–09–05	Deutsche Telekom	Polska Telefonia Cyfrowa
2003–10–24	Telmex	AT&T Latin America
2004–02–23	France Telecom	Wanadoo
2004–03–17	Telmex	Embratel
2004–03–25	Comcast	TechTV
2004–04–07	Telstra	Kaz Group
2004–05–19	Telefonica	Telefónica CTC Chile
2004–06–03	Golden Telecom	Buzton
2004–07–08	TeliaSonera	Orange Denmark
2004–09–09	Deutsche Telekom	T-online
2004–11–01	TDC	Song Networks
2004–11–05	NTT	inter-touch
2004–11–08	British Telecom	Infonet
2004–12–03	British Telecom	Albacom
2004–12–07	Telecom Italia	TIM
2004–12–15	Sprint	Nextel
2005–06–27	MTS	Barash Communications
2005–06–28	KPN	Telfort BV
2005–07–11	Sprint Nextel	US Unwired
2005–07–27	France Telecom	Amena
2005–08–16	Cable & Wireless	Energis
2005–08–30	Sprint Nextel	Gulf Coast Wireless

(continued)

Table 6 (continued)

Date	Acquirer	Target
2005–09–29	Belgacom	Telindus
2005–10–03	NTL	Telewest
2005–10–31	Comcast	Susquehanna Communications
2005–10–31	Telenor	Vodafone Sweden
2005–10–31	Telefonica	O2 Group
2005–11–07	TeliaSonera	Chess
2005–11–29	Magyar Telekom	Orbitel
2005–12–13	Vodafone	Telsim
2005–12–14	British Telecom	Altanet
2005–12–21	Sprint Nextel	Nextel Partners
2006–03–06	AT&T Inc.	BellSouth
2006–04–04	NTL	Virgin Mobile UK
2006–04–27	British Telecom	Dabs.com
2006–05–01	Swisscom	Cybernet
2006–05–16	TeliaSonera	NexGenTel
2006–06–20	Telefonica	Be
2006–09–15	KPN	Tiscali Netherland
2006–09–18	Telecom Italia	AOL Germany
2006–10–25	British Telecom	Counterpane
2006–11–16	British Telecom	Plusnet

Table 7 List of emerging markets (as of June 2006)

Composition of the MSCI emerging markets index	
Argentina	Malaysia
Brazil	Mexico
Chile	Morocco
China	Pakistan
Colombia	Peru
Czech Republic	Philippines
Hungary	Poland
Egypt	Russia
India	South Africa
Indonesia	Taiwan
Israel	Thailand
Jordan	Turkey
Korea	

References

Berry C (1971) Corporate Growth and Diversification. Journal of Law and Economics, 14(2): 371–383.
Bettis R, Mahajan V (1985) Risk/Return Performance of Diversified Firms. Management Science, 31(7): 785–799
Brakman S, Garretsen H, Van Marrewijk C (2005) Cross-Border Mergers and Acquisitions. CESifo Working Paper Series No. 1602.
Breedon F, Fornasari F (2006) FX impact of cross-border M&A. Bank for International Settlements. Resource Document. http://www.bis.org/publ/bispap02n.pdf. Cited November 18, 2006.
Brigham E, Houston J (2004) Fundamentals of Financial Management. Thomson: South-Western.
Chari, A Ouimet Tesar L (2004) Acquirer Gains in Emerging Markets. NBER Working Paper No. 10872.
Cohen J (1988) Statistical Power Analysis for the Behavioral Sciences. Hillsdale, NJ: Erlbaum.
Corhay A, Rad A (2000) International Acquisitions and Shareholder Wealth Evidence from The Netherlands. International Review of Financial Analysis, 9(2): 163–174.
Cowan A (1992) Nonparametric Event Study Tests. Review of Quantitative Finance and Accounting, 2: 343–358.
Cowan A (2002) Eventus 7.0 User's Guide (revised version). Ames, Iowa: Cowan Research LC.
Cummins D, Weiss M (2004) Consolidation in the European Insurance Industry: Do Mergers and Acquisitions Create Value for Shareholders? SSRN Working Paper.
Dimson E, Mussavian M (1998) A Brief History of Market Efficiency. European Financial Management, 4(1): 91–193.
Eun C, Kolodny R, Scheraga C (1996) Cross-Border Acquisitions and Shareholder Wealth: Tests of the Synergy and Internalization Hypotheses. Journal of Banking & Finance, 20(9): 1559–1582.
Fahrmeir L, Tutz G (2002) Statistik – Der Weg zur Datenanalyse. Berlin: Springer.
Fama E (1991) Efficient Capital Markets: II. The Journal of Finance, 46(5), 1575–1617.
Fama E (1998) Market Efficiency, Long-Term Returns, and Behavioral Finance. Journal of Financial Economics, 49: 283–306.
Ferris S, Park K (2001) How Different Is the Long-Run Performance of Mergers in the Telecommunications Industry? SSRN Working Paper.
Hitt M, Ireland R, Hoskisson R (2005) Strategic Management: Competitiveness and Globalization (Concepts) 6th ed. Mason, OH: Thomson: South-Western.
Hofstede G (1980) Cultural Consequences: International Differences in Work-Related Values. Beverly Hills, CA: Sage.
Johnson J, Ellstrand A, Dalton D, Dalton C (2005) The influence of the financial press on stockholder wealth. Strategic Management Journal, 26(5), 461–471.
Jung J, Chan-Olmsted S (2005) Impacts of Media Conglomerates Dual Diversification on Financial Performance. Journal of Media Economics, 18(3), 183–202.
Kothari S, Warner J (2004) Econometrics of Event Studies. Handbook of Corporate Finance, Vol. A. Elsevier/North Holland.
MacKinlay A C (1997). Event Studies in Economics and Finance. Journal of Economic Literature, 35(1), 13–39.
McWilliams A, Siegel D (1997) Event Studies in Management Research Theoretical and Empirical Issues. The Academy of Management Journal, 40(3), 626–657.
Moeller S, Schlingemann F (2005) Global Diversification and Bidder Gains: A Comparison Between Cross-Border and Domestic Acquisitions. Journal of Banking & Finance, 29(3): 533–564.
Morosini P, Shane S, Singh H (1998). National Cultural Distance and Cross-Border Acquisition Performance. Journal of International Business Studies, 29(1), 137–158.

Neary P (2004). Cross-Border Mergers as Instruments of Comparative Advantage. University College Dublin/CEPR.
Park M, Yang D, Nam C, Ha Y (2002) Mergers and Acquisitions in the Telecommunications Industry: Myths and Reality. ETRI Journal, 24(1), 56–68.
Park N, Mezias J, Song J (2004) A Resource-Based View of Strategic Alliances and Firm Value in the Electronic Marketplace. Journal of Management, 30(1), 7–27.
Patell J (1976) Corporate Forecasts of Earnings Per Share and Stock Price Behavior. Journal of Accounting Research, 14, 246–276.
Pernet S (2006) Bundles and Range Strategies: The Case of Telecom Operators. Communications & Strategies, 63, 19–31.
Punja P (2005) Market Focus: Voice Over Broadband Will Force Big Changes for Carriers. Gartner.
Ramanujam V, Varadarajan P (1989) Report on Corporate Diversification: A Synthesis. Strategic Management Journal, 10, 523–551.
Salter M (1979) Diversification Through Acquisition: Strategies for Creating Economic Value. New York: Free Press.
Sarkar M, Cavusgil S, Aulakh P (1999) International Expansion of Telecommunication Carriers: The Influence of Market Structure, Network Characteristics, and Entry Imperfections. Journal of International Business Studies, 30(2): 361–381.
Selden L, Colvin G (2003) M&A Needn't Be a Loser's Game. Harvard Business Review, 81(3): 70–73.
Subramani M, Walden E (2001) The Impact of E-Commerce Announcements on the Market Value of Firms. Information Systems Research, 12(2): 135–154.
Tebourbi I (2005) Bidder's Shareholder Wealth Effects of Canadian Cross-Border and Domestic Acquisitions – The Role of Corporate Governance Differences. CEREG – Universitè Paris Dauphine.
Thompson A, Strickland A, Gamble J (2005) Crafting and Executing Strategy. Boston: McGraw-Hill Irwin.
Uhlenbruck K, Hitt M, Semadeni M (2006) Market Value Effects of Acquisitions Involving Internet Firms: A Resource-Based Analysis. Strategic Management Journal, 27: 899–913.
Varadarajan P (2003) Product Diversity and Firm Performance: An Empirical Investigation. Journal of Marketing, 50(3): 43–57.
Warf B (2003) Mergers and Acquisitions in the Telecommunications Industry. Growth and Change, 34(3): 321–344.
Wilcox H D, Chang K, Grover V (2001) Valuation of Mergers and Acquisitions in the Telecommunications Industry: A Study on Diversification and Firm Size. Information & Management, 38: 459–471.
Zadeh A (2004). Why Your Fate Rests at the Edge: The Convergence Value Chain Is Largely About Exploiting the Multiservice Edge. America's Network. August 1, 2004.

Part VI
Emerging Markets

Next Generation Networks: The Demand Side Issues

James Alleman and Paul Rappoport

Abstract The demand for next generation networks (NGN) for communications has mostly focused on the trend in technology. Recognizing that communications is a derived demand, we look at the demand for telecommunications services and then overlay these forecasts on the existing information and communications technology (ICT) infrastructure. We focus on the consumers rather than the technologies. We note that what consumers demand is for communications: The communications may be fixed, mobile, interactive, or unidirectional. With the technology and the move to IP protocol, all of these features can be handled in a few devices and networks – maybe only one. We provide an assessment of the forecast of market trends, and their implications for the regulator. The relevant demand elasticities are nearly unitary. Each of these factors alone implies that the market structure will be monopolist or an oligopoly at best. But amplified in combination, The need for clear, certain regulation of this segment of the ICT sector is an absolute necessity. Demand elasticities must be understood and factored into the consideration of the policy alternatives.

Overview

The demand for next generation networks (NGN) for communications has mostly focused on the trend in technology and the efficiencies of moving to networks driven by IP protocols. In this chapter, recognizing that communications is a derived demand, we look at the demand for telecommunications services and then overlay these forecasts on the existing information and communications technology (ICT) infrastructure.

We define communications in the broadest sense; the demand is derived in that consumers and businesses are mostly indifferent to the technology so long as it

---

J. Alleman and P. Rappoport
University of Colorado – Boulder
e-mail: James.Alleman@Colorado.edu

B. Preissl et al. (eds.), *Telecommunication Markets*, Contributions to Economics, DOI: 10.1007/978-3-7908-2082-9_22, © Springer Physica-Verlag HD 2009

functions. From this perspective, the suppliers of communication services, the telephone companies, cable companies, broadband providers, wireless providers, and more recently the video providers care about delivery protocols and investment requirements since these have a direct impact on the profits of the companies.

If the platform is indifferent, then what drives consumer choice is the importance of the availability and characteristics of the alternative communications services, along with price and quality constraints. The regulator, on the other hand, is influenced by both the demand- and supply-side considerations. Approaching the development of NGN from the demand-side allows us to add to the assessment of ICT and to consider additional directions or insights that might assist regulators and policy makers. They will be better able to evaluate policy options in a broader, *gestalt* manner; separate from the pleading of the vested interests of a specific sector or technology.

The chapter will also address pricing issues associated with bundling of services (triple-and quadruple-play).

Convergence

The issue of convergence in the telecommunications industry has been discussed for some time. Many definition exist (Bauer 2005), but for our purposes we will not deal with the technology or the protocols in general except to note that future communication networks will be radically different from those existing today. They will be broadband platforms on which applications will provide services to consumers and businesses. These networks of networks will look different from today's service specific platforms and, hence, the regulatory and policy issues will be complex and challenging. Unresolved questions include: (1) the pricing of both the wholesale and retail services, (2) degree of competition, (3) the level and quality of innovation and investment, (4) open or closed architecture, (5) secure interoperability between and among networks, and many more that we cannot even anticipate. Approaching the NGN from the demand side does not eliminate these questions, but, we feel, provides clearer insight how to address many of these issues.

Consumer Focus

Our focus is on the consumers rather than the technologies in this chapter. Consumers demand is for communications: one-way (video, traffic reports, location, etc.), two-way (traditional voice, e-mail, etc.), symmetrical (traditional voice calls, e-mails, etc.) or asymmetrical (file downloads, movies, etc.). The communications may be fixed, mobile, interactive, or unidirectional. With the technology and the move to IP protocol, all of these features can be handled in a few devices and networks – maybe only one. The full integration and convergence of the networks is only beginning to emerge. A recent discussion in Businessweek (2006) underscores this rush to convergence: "The future of wireless is to become the focal point for the fusion of consumer electronics, entertainment and telecommunications. The way the complex systems of technology will deliver the future will be a shift in both technology and the way it is applied."

Organization of Chapter

This chapter is organized as follows: The next section describes and briefly characterizes the demand for information and communications technology (ICT) services. The chapter provides a backdrop for assessing the forecast of market trends, and their implications for the regulator. The third section briefly addresses some of the policy issues to place them in the context of the pricing issues of a converged ICT sector. The final section concludes with recommendations and suggestions for future research.

Future Demand and the Market for Communications

The Derived Demand for Communications

We pose the following question: is there a point where the growth in ICT products and services runs into a wall – that wall being an income constraint. If there is a budget constraint, then real revenue growth in ICT services can only occur when (1) prices (or costs) fall or when (2) there is an increase in the percentage of a households income devoted to these services. Real revenue growth is not the same thing as product and service substitution such as the substitution of MP3 music files for music CDs. Clearly the magnitude of own and cross-price elasticities need to be considered when assessing the future of ICT. This point of view stands in stark contrast to those who see demand ever increasing due to the convergence of communication, entertainment and data services. If there is a binding budget constraint, the opportunities for revenue growth to pay for deployment of advanced networks are likely to depend less on consumer demand and more on regulatory edict.

Over the last few years there has clearly been an explosion of wireless subscribers. There are now more wireless subscribers than landline subscribers. There has been exponential growth in the internet both in terms of users and usage. These trends have led to optimistic extrapolations of increased demand for services that can be provided over multiple platforms. The technology required to deliver these services is available. But if the demand for these services is a derived demand, then the factors that stimulate demand require scrutiny. Foremost among these is a consumer's ability and willingness to pay for these services. In the end, it comes down to price and income. Not simply the price of these new services, but that price relative to the price of food, health care, shelter and transportation.

This section is organized as follows. First we examine the share of income devoted to communications and entertainment. We then examine the evidence on the demand for advanced products and services. We then focus on the competitive environment. Finally, we suggest policy implications.

Expenditure Share

To put this discussion in perspective, consider the share of income spent on telecommunications. Figure 1 displays the share of expenditure devoted to traditional telecommunications from 1981 through 2005 for the United States.[1] The data suggests a declining share after 2000. This decline appears to be a function of lower long distance prices and declines in second line penetrations.

Figure 2 expands the analysis and tracks the average monthly expenditure for local, long distance and wireless services for the United States (FCC 2006).

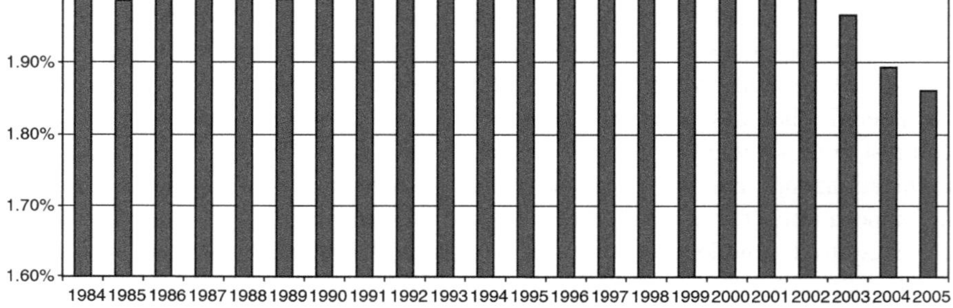

Fig. 1 Share of expenditures for telecom (United States households) (From BLS 2006)

[1] Includes local and long distance expenditures only. Figures derived from BLS (2006) Consumer Expenditure data.

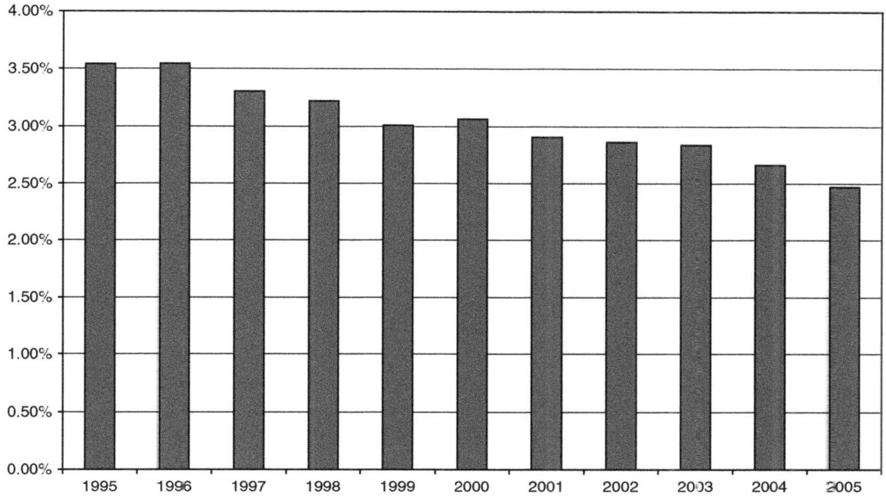

Fig. 2 Expenditure share of local, LD and wireless communications (United States households) (From FCC 2006)

Again we note a declining trend in the expenditure share.

The point is to suggest that households spend across a number of categories – housing, food, transportation, healthcare and so forth. If prices (costs) increase in these areas, given a budget constraint, then spending might decline in other areas such as communications.

Convergence suggests that the revenue "pot-of-gold" of household expenditures might better be gauged by focusing on voice, video, and data services and usage including entertainment.

Figure 3 displays the share of income allocated to entertainment. In this analysis, entertainment is defined as fees and admissions, television and sound equipment and services. Looking at the current entertainment share provides an upper bound to the services that could possibly be supplied in a connected environment – music, streaming video, over the air TV, games and other subscription services.

We note that the share of income spent on entertainment has been relatively flat over the last decade.

This suggests that the current fixation with the growth of internet content services should be viewed more in terms of substitution among alternative channels than as an increase in growth in the medium under consideration.

International Comparisons

How do expenditure shares for communications compare with other countries? A recent study looked at variations of expenditures on communications in developing

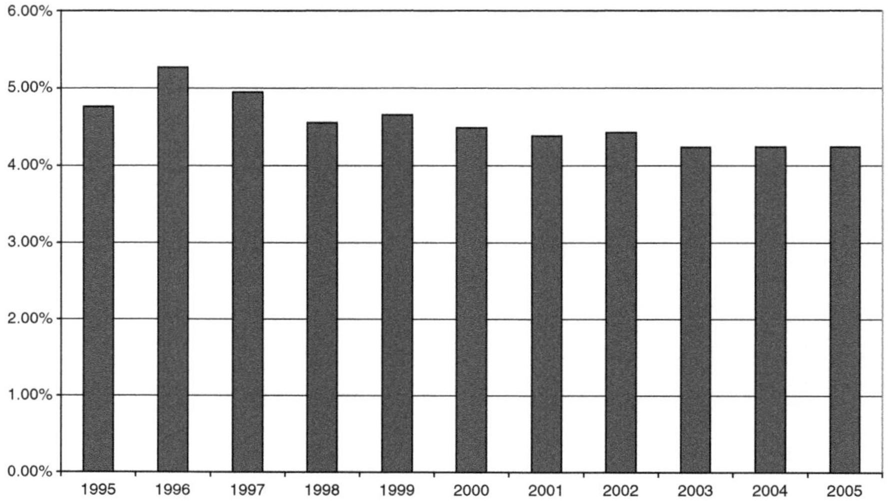

Fig. 3 Entertainment share (United States households) (From FCC 2006)

countries (Ureta 2007). The following table summaries the study's findings regarding expenditure share (Table 1).

Table 2 looks at information, communications and technology expenditures as a per cent of GDP. The rankings provide a different view of expenditure share over the period 2003 to 2004.[2]

Figure 4 shows the communications share across selected countries. The range of expenditure across countries is noteworthy. Part of the variation may be due to the definition of communication expenditure by country and the collection of statistics used to measure expenditure share. However, there is no adequate way to check for consistency in definition and computation. Nonetheless, if we look at the chart in terms of relative positioning, we note that the share of spending on communications in the US is on the low end. Why the large variation? Aside from China and Brazil, which are outliers in this review, why does South Korea have a communication share over twice that of the US? Keep in mind that we are looking at the share of overall consumer expenditures devoted to communications. A higher communication's share must imply a lower share in another consumption category. See Rappoport et al. (2007) for an elaboration of cross-country comparisons.

Music Downloads

Consider the music industry. Figure 4 displays the nominal value of the US sound recording industry. Sales are flat in nominal dollars and declining in terms of real dollars (RIAA

[2]ICT expenditure share was derived from Tables 5.10 and 5.11 of the World Bank (2006).

Table 1 Expenditure shares on communications (From Ureta 2005)

Country	Time period	Communication share	Entertainment share
Albania	2002/2003	3.8%	
Mexico	2000	3.4%	1.7%
Nepal	2002/2003	1.6%	
South Africa	2000	1.4%	

Table 2 ICT expenditure share 2003–2004 for selected countries (From World Bank 2006)

Country	ICT share of GDP 2003	ICT share of GDP 2004
United States	7.1	9.0
Austria	5.3	5.1
Belgium	5.5	5.3
China	5.3	4.4
Czech Republic	6.6	6.0
Ecuador	3.7	3.6
France	5.9	5.6
Germany	5.7	5.5
Greece	4.3	4.2
Italy	4.1	4.0
Japan	7.4	7.6
Nigeria	13.0	–
Paraguay	9.0	–
Poland	4.5	4.3
South Africa	8.0	7.3
Turkey	7.3	6.9
United Kingdom	7.3	6.9
Zimbabwe	11.8	16.0

2006). Much of the decline, however, is in traditional channels and media (full length CDs, DVDs, cassettes, vinyl LP). Figure 5 gives the Recording Industry Association of America's (RIAA) estimate of the share of music accounted for by downloads.

This figure clearly underscores the substitution in terms of the preferred or growing importance of the internet as a channel for delivery. The popularity of MP3 files is due in part to the increased level of choice – downloading singles, creation of custom play lists and so forth. However, perhaps the most significant factor is price. The rapid growth in MP3 downloads suggests that demand for MP3 downloads is elastic and that there are large cross-price elasticities.

Does the number of installed MP3 players stimulate MP3 downloads? US shipments of MP3 players will have reached 19 millions by 2006 (*InformationWeek* 2005). To some analysts, this represents a critical mass meaning that MP3 players have achieved a sustainable level to support continued growth. But what if the price of a downloaded song increased from $0.99 to $2.99? If demand for downloaded songs is price elastic, we would quickly observe the impact on the popularity of downloads.[3]

[3]There has been speculation of a move to tiered pricing. Consider the following from Spolsky (2005) "Here's the dream world for the EMI Group, Sony/BMG, etc.: there are two prices for songs on iTunes, say, $2.49 and $0.99. All the new releases come out at $2.49. Some classic rock (Sweet Home Alabama) is at $2.49. Unwanted, old, crap, like, say, Brandy (You're A Fine Girl) … would be deliberately priced at $0.99."

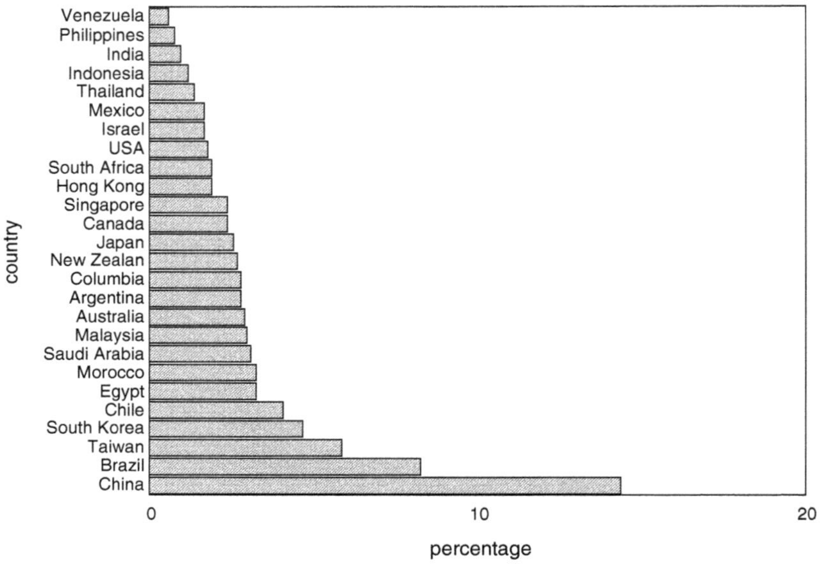

Fig. 4 Communications share (From Euromonitor International 2005)

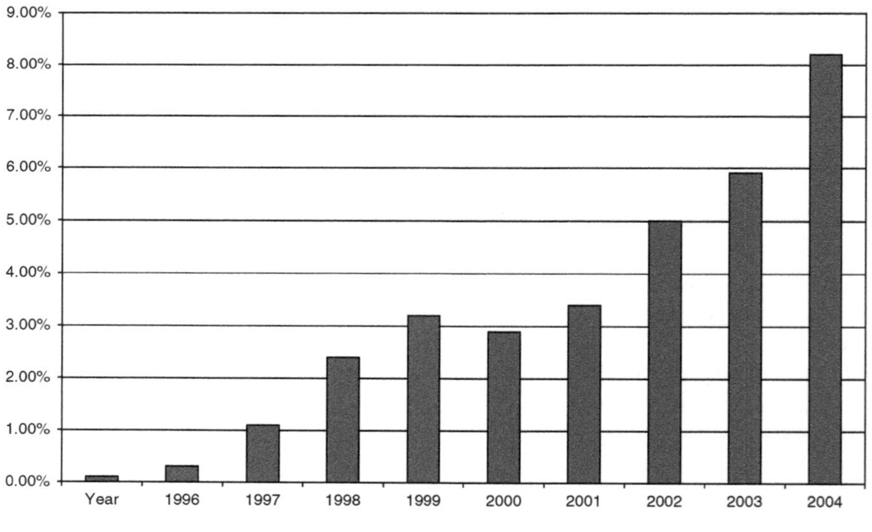

Fig. 5 Internet share of music channels (From RIAA 2005)

Note that the push to consolidate devices is thus a function of price, or more directly of a consumer's willingness to pay. The success of a device that is a combination of a camera, PDA, mobile phone and a platform that also serves as a music and video hub will be dependent on the value each of these functions bring. If prices drop for MP3 files (and streaming video files), then we can expect to see rapid growth in this type of device use; witness the popularity of the recently released Apple iPhone in the United States. Our contention is that the popularity of these next generation "Everything in a box" devices will wane if the price of content steadily increases.

Best Practice VoIP

A number of researchers have pointed to the continuing growth in VoIP as a harbinger of the future. Discounted flat rate telephony has been around for a number of years. In the US, best practice Voice-over-IP (VoIP) (no quality of service guarantees) is best represented by Vonage. This market is small and getting smaller. The overall impact of Skype and other "free" voice services (for example, Google), while expected to be significant, represent substitution of access lines. Since the demand for VoIP services at a price of zero (0) produces no revenue, one might ask whether next generation services (online applications, gaming, home networking, entertainment, telematics, ...) that are expected to be bundled with VoIP will command a premium price or a price that covers fixed outlays.[4]

In the early part of this decade, as the technology for offering high quality telephone services as an application over the internet VoIP became available, services raised the level of expectations that usage would greatly expand due to primarily to the lower cost of communications. VoIP was seen as a threat to the incumbent telephone operators. However, none of the pure VoIP in the United States companies has yet to turn a profit. Indeed, recently a major VoIP provider in the United States, SunRocket, went out of business (SunRocket 2007). The expected savings disappeared as long distance rates declined and as incumbent phone companies countered with flat rate unlimited calling plans. Quality of service continues to plague best practice VoIP providers.

The market for best practice VoIP in the United States was over estimated.[5,6] In a 2006 *Computerworld* survey, "... respondents ranked VoIP third among the technologies that didn't live up to their expectations in 2005" (Computer World 2006).

Whereas best practice VoIP providers have not seen significant growth in subscribers, Cable VoIP growth has been substantial and has accounted for the continued decline in overall telephone access lines – along with mobile services. In this case, the subscriber, in most cases, already has the broadband service and

[4]See for example a presentation by Nissen (2005).

[5]Best practice refers to services that have no quality of service guarantees. These are services provided over the public internet. These services are distinguished from IP-based voice services offered by cable companies over their private networks.

[6]IDC (2006) forecast saw VoIP growing to 44 million households by 2010.

the VoIP is an inexpensive add-on or as part of a triple play – cable television, broadband data, and VoIP services – moving from the incumbent telephone company to the cable provider. There are significant regulatory issues associated with VoIP. The most significant debate centers on the notion of 'net neutrality' Net neutrality embodies the notion that broadband customers should be able to subscribe to any service with no intervention by broadband providers. That is to say that customers would not have to pay more for higher quality of service (higher priority, faster service) when they visit content-rich web site such as Movie Link or Google. Without net neutrality, users who want higher priority delivery of packets would pay a premium.[7] Facility-based providers (telephone companies and cable companies) would have a real competitive advantage over best practice VoIP providers such as Vonage since they could require additional fees from these providers. Indeed, in the United States, at least some telephone companies did not allowed VoIP calls on their network, although they later reversed this position under regulatory pressure. Pricing of priority services is just another possibility of limiting VoIP take-up.

With respect to pricing of VoIP services, the telephone companies have been lowering their flat rate usage packages so that the differential between their prices and the prices of companies such as Vonage have diminished. When prices decline and demand is inelastic, overall revenue declines. At current prices of $25, the demand for best practice VoIP is inelastic.[8]

In contrast, the opportunity for VoIP in the developing world does promise to be a viable threat to incumbents due to different economic structures in these countries, for example see http://www.iworldservices.com/.

IPTV

The term IPTV covers a number of services, from streaming video to a multimedia personal computer to the provision of video services over fiber to the household. As with VoIP, IPTV was seen as an application that was expected to stimulate demand for bandwidth, and thus the demand for increased investment in networks and infrastructure.

For the most part in the US, IPTV is seen as the provision of video services by Verizon through its Fiber Optical Service (FiOS) service. FiOS provides the ability for Verizon to offer voice, data and video services. The market for FiOS services is not surprisingly a function of willingness and ability to pay. An open question is the reasonableness of the assumptions underlying Verizon's push into fiber. According to some analysts Verizon's estimates of customers signing on to their services is overly optimistic.

---

[7]See the special issue of Communications & Strategies (2007) in particular Bauer (2007) for a discussion of the elements of this controversy.

[8]See Rappoport et al. (2004), Table 1.

The reason – as with many services in the ICT space, competition tends to depress prices and lower prices lead to lower than expected revenues.[9]

These next generation applications provide some useful insights into the nature of demand. First, competition tends to lower prices. Second, in the case of IPTV, payback periods for the investment in infrastructure may be longer than assumed. "Past a certain point, the 'build it and they will come' strategy is not acceptable without subscribers paying a monthly fee that supports further investment in the outside plant." Broadbandproerties.com (2006, p. 56). The danger for the telephone companies such as Verizon is the cable company's ability to readily cut prices.[10] Third, in the VoIP market, facility-based players have significant technical and market advantages over non-facility based providers such as Vonage. For these and other next generation services, price matters.

Demand

How do these charts compare with forecasts that suggest increased demand for voice, video and data services and hence the increased demand for products that enable on to use these services anywhere and anytime?

We offer these observations:

First, the demand for advanced services coupled with broadband access is currently elastic. We should expect downward prices as these markets become more competitive. At some point demand will become inelastic with a concomitant decline in revenue.[11]

Second, there appears to be a movement towards bundling for price rather than bundling for quality thus reinforcing this downward pressure on price.[12]

Estimates of demand have been seriously overestimated.

What is the range of demand elasticities?

- Estimated elasticities for VoIP services are –1.0 at a price of $30 and –0.7 at a price of $25.
- For local and long distance services demand elasticities have been estimated in the range of –0.2 to –0.55.

---

[9]See for example the financial assessment in the September 2006 issue of Broadband Properties (2006).

[10]Case in point is Cox Communication's triple-play offer in Omaha of $69 per month for voice, video and data.

[11]Recall that when the demand is elastic (a value between negative infinitely and negative one), a decrease in price will increase revenue; however, when it is inelastic (a value between negative one and zero), a price decrease in prices will reduce revenues.

[12]Bundling for price implies that the price of the bundle is less that the sum of the individual components of the bundle. If the willingness-to-pay for the bundle is greater than the individual prices of the components, then the bundle "adds" value.

- Estimated elasticities for broadband access range from –0.8 to –1.5 (Rappoport et al. 2003).
- For mobile internet access estimated elasticities range from –1.0 and –2.0 (Rappoport et al. 2004).
- For IPTV, estimated elasticities are –1.02 for video only (Rappoport et al. 2004). Elasticities for IPTV seen as the triple-play are displayed in Table 1 below.

The following table displays the estimated elasticity for triple-play services at various prices (Rappoport et al. 2006) (Table 3). The table underscores the notion that the demand for the triple-play is clearly a function of a household's willingness to pay (price). The price elasticity of demand for a triple-play bundle is around –1.1 at a price of $80 (Rappoport et al. 2004).

We suspect that most of the services (and products) that are part of the scenario that sees explosive growth of web-based services and mobile applications are price elastic. For example, at a price of $60, the estimated elasticity for IPTV plus additional content (online content, two-way interactivity) is –1.06.[13]

When demands is elastic it is difficult to follow a policy for increasing prices. The competitive nature of the marketplace will thwart policies aimed at increasing prices. However, as the market for services matures bringing with it lower price elasticities, the ability to continually cut prices will diminish. It is at this point where prospective business models run into trouble. At this point, lower prices coupled with expanding services leads to lower revenues. For the regulator this poses a dilemma. Lower prices encourage increased usage. However, additional investment for next generation platforms requires a sustainable and predictable revenue stream. The investment required will require significant sunk costs. If the regulatory rules are not clear, this investment is problematic. However if the investment is made and the oligopoly market structure this will create is not recognized, it could result in economic rents and a reduction in welfare.

Table 3 Demand for triple-play services (From Rappoport et al. 2005)

Price	Elasticity
$40	–0.23
$50	–0.46
$60	–0.69
$70	–0.89
$80	–1.10
$90	–1.37
$100	–1.45
$110	–1.72
$120	–1.83
$130	–2.01

[13]Unpublished estimates derived from the CENTRIS omnibus survey.

Substitution

Does access to the internet result in expanded growth and sales? Or, does the internet merely reallocate the source of sales? It is clear from the RIAA data that overall sales for sound recordings are flat (even taking into account the estimated $1 billion in illegal music downloads).[14] This suggests that there is substitution from traditional channels to the internet as a channel.

Figure 5 provides additional evidence that the more likely scenario across most retail channels is substitution. Since the growth in total retail sales is less than the growth in e-commerce sales, e-commerce does not appear to stimulate total retail sales. Rather, e-commerce is growing due to convenience, choice and perhaps, current tax policy and is a substitute for on-site retail sales. Table 4 displays total retail sales and the percent of those sales associated with e-commerce (US Census 2006).

Mobile Telephony

Perhaps the most forward looking analysis of the future of mobile communications is a paper authored by Forge et al. (2005). In their conclusion, these authors expect there will be progressively more services developed around specific content and applications migrating to the wireless platform. They project scenarios where the price of services falls as the costs of providing those services drop. However, it is an open question whether such an assumption works for the provision of content. This is different from a consumer's willingness to pay for mobile access. The pricing of select services and content will typically be viewed as an add-on to existing access.

However, this may change. Google committed to offer up to $4.6 billion if the FCC will offer an "open system" – applications, devices, services and networks for the spectrum auctions in the 700 MHz band (Google 2007). While not accepted by the FCC, it indicates that pressure is on to change the traditional mobile structure and ultimately may serve to break the lock the traditional mobile carriers have on

Table 4 U.S. e-commerce (sales and in millions of USD) (From BLS 2005)

Year	Retail sales	E-commerce	Retail growth	Growth e-commerce	Share of total
2000	2,988,756	27,765			0.93%
2001	3,067,725	34,517	2.64%	24.32%	1.13%
2002	3,134,322	45,001	2.17%	30.37%	1.44%
2003	3,265,477	56,644	4.18%	25.87%	1.73%
2004	3,477,308	70,906	6.49%	25.18%	2.04%

[14]Most likely representing a biased view of the music market.

their customers and customers' access to content. Moreover, it would have broader implications for spectrum auctions.

Competition

That increased levels of competition tend to lead to lower prices is not surprising. What might be surprising is the magnitude of both own and cross-price elasticities. Consider the market for the triple-play bundle – voice, data and video. Purchased separately, the average price of these three services is $160. Competition to capture the triple-play market has resulted in initial prices as low as $69.

As markets become more competitive, we see significant price concessions. Consider the market for data services in the US. For most of the US, households have two choices for broadband services, DSL or cable modem access to the internet. Households will have other data options in the future as wireless and satellite data services appear. For now, however, consider the pricing of broadband. Table 5 display the average broadband price for areas where both DSL and cable modem services are available and where only one service (DSL or cable modem) is available.[15]

International Comparisons

The penetration of IPTV for selected countries is displayed in Fig. 6.

The incidence of personal computers and broadband internet access for selected countries are displayed in Table 6 (World Bank 2006).

The growth of next generation products and services is both a function of price and availability. From a cross-country perspective, the demand for broadband access is the key indicator for future growth.

Table 5 Price of broadband service (From Rappoport et al. 2005)

		DSL	
		Available	Not available
Cable modem	Available	$35.00	$45.00
	Not available	$43.00	

[15]Derived from the CENTRIS household survey. See www.centris.com

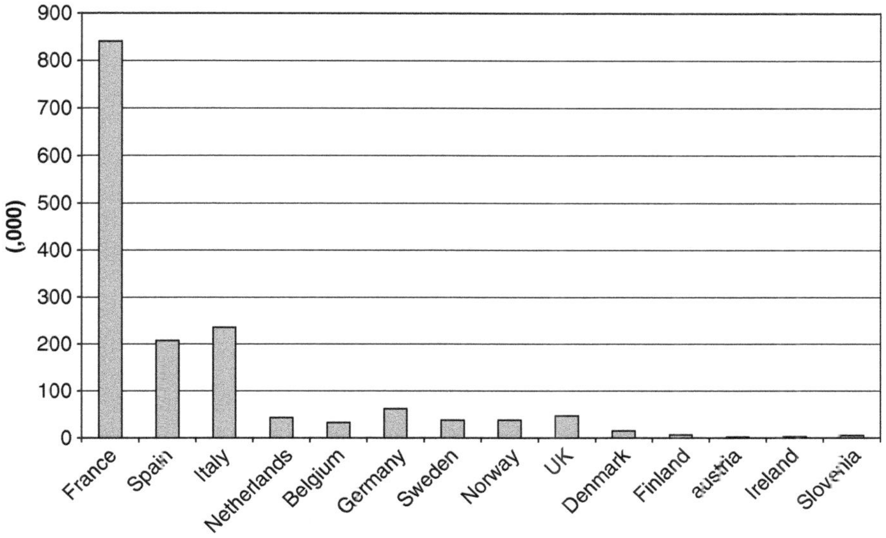

Fig. 6 IPTV subscribers. Source: IDC (2006)

Table 6 Personal computers and broadband access (From World Bank 2006)

Country	Personal computers per 1,000, 2004	Broadband access per 1,000, 2004
United States	749	129.0
Austria	418	101.3
Belgium	348	155.4
China	41	16.5
Czech Republic	240	16.5
Ecuador	56	–
France	487	108.1
Germany	561	83.7
Italy	315	81.7
Japan	542	145.8
Nigeria	7	–
Paraguay	59	0.1
Poland	193	32.7
South Africa	82	1.3
Turkey	52	0.8
United Kingdom	599	103.0
Zimbabwe	77	0.4

[16]See e.g. Taylor (1994).

Elasticities and Regulation

For the most part, the demand for traditional telecommunication services is inelastic.[16] Basic access to the network was deemed a necessity.[17] In the highly competitive environment created after the passage of the Telecommunications Act of 1996, competition resulted in expansion of services and lower prices. The impact associated with lower prices was the inability of competitive local exchange carriers to compete. Lower prices in an inelastic world lead to lower revenues.

For new services such as mobile and broadband access, estimates of demand are elastic. Thus, a reduction in price not only leads to a stimulus in the number of customers, it also leads to increased revenue. If we take the typical linear representation of a demand curve, at some point, with declining prices, "demand" becomes inelastic. At this point, further expansion occurs only with lower revenue. Current estimates of broadband demand suggest demand is close to the inelastic range. Strategies that lead to lower prices have been justified in terms of increasing market share. However, unless costs associated with network infrastructure drop faster, such expansion in a competitive environment is self-defeating. At best the market consolidates with only a few participants left.

The path of next generation services is likely to follow the same script. Demand is initially quite elastic, prices drop and eventually demand becomes inelastic. This appears to be the scenario for the triple-play package of voice, video and data. It also appears to be the path for subscription services (music, gaming, news).

If we include cross-price elasticities into the mix, the demand forecast for new products and services becomes even more problematical. As noted earlier, substantial cross-price elasticities lead to substitution not overall expansion. In a more general sense, a binding income constraint serves as a reminder that assumptions inherent in a business plan may be too optimistic. One only has to review the brief history of Iridium to learn this lesson.

In a world where advanced network architectures are being touted, if they are increasingly expensive, then developers of those networks will have no choice but to turn to regulation to set price floors, reduce entry incentives such as unbundled network elements (UNEs) in the United States. Witness the increased investment in the United States when the UNE requirement was lifted (Waverman 2006; Crandall 2005). Thus, "[n]ew entrants such as Vonage, Skype, Google, and Yahoo [which] have high disruptive potential but remain disadvantaged without their own access networks" (Bauer 2005).

---

[17]Although deemed necessary and highly inelastic price elasticity, most countries have imposed universal service obligations on the incumbent carriers, requiring them to support prices below costs and/or provide service in unprofitable areas. See Alleman and Rappoport (2000) for a discussion of the inefficiency of this policy. This leads to other problems as we discuss in the next section.

Conclusion

The Policymakers' Dilemma

While the technologist/policy makers may prefer one market structure outcome over another, what the consumer is interested in is communications – simple, easy-to-use, cost effective and available on demand. These needs are not always satisfied in the current market environment. Currently, they must be satisfied with multiple networks and devices. Business and households now have fixed telephones, mobile phone (many times more than one for a household), a broadband connection which could be satellite, cable, DSL, WiFi, or WiMax, and Blackberries. Are consumers indifferent to technology and the protocols to communicate? Does a consumer's desire to "communicate" transcend any one platform? Voice is not a unique form of communication; e-mail, facsimiles, video phones, and self-generated content are all means to communicate. For the next generation of consumers, simplicity, availability and access are required. To satisfy these consumers, the diversity of communications has significantly expanded. From this perspective, consumer demand is the driver of change.

The surge in mobile subscribers and the number of households that have internet access should be viewed carefully as at best a trend in unconstrained demand. Low access prices, the availability of content at zero or low prices has fueled this growth. There is substantial evidence that demand for new forms of access and usage are elastic. There is also evidence that cross-price elasticities are substantial. Coupled with the reality of a budget constraint and changes in household demographics, forecasts of continued growth must be tempered by the reality of rising prices for access and content. Otherwise, there is the possibility of over investment and the possibility of financial difficulties for firms that provide infrastructure.

An example is satellite radio: a microcosm of the industry. Satellite radio – a subscription service that is a substitute for over-the-air radio – is analogous to the fixed line access issue. Even with millions of subscribers and hundred's of million dollars in funding, the two satellite radio providers in the United States – Sirius Satellite Radio and XM Satellite Radio struggled and ultimately merged. Although satellite radio is growing faster than any other consumer product except for the iPod, both the supply-side and demand side realities have been recognized (Taub 2007). "'When you have two companies in the same industry, we have a similar cost structure. Clearly, a merger makes sense from an investor's point of view to reduce costs, and to have a better return,' said David Frear, the chief financial officer for Sirius." (Taub 2007, p. B1) This case illustrates Shapiro and Varian's maxim that the "technologies change, the laws of economics do not" (1999, pp. 1–2) and is a microcosm of the conclusion with respect to the ICT sector – limited demand, nearly unitary-elastic demand coupled with large economies to scale and scope, and sunk costs leaves little room to maneuver in the ICT sector.

Synopsis

Our conclusion based principally on the demand-side considerations is that dynamics markets must be considered. Demand elasticities must be understood and scale, scope, and sunk cost must be factored into the consideration of the policy alternatives. As we have shown, the relevant demand elasticities are nearly unitary and the last kilometer has significant scale and scope economies, as well as non-trivial sunk costs. Each of these factors alone implies that the market structure will be monopolist or an oligopoly at best. But amplified in combination, Thus the need for clear, certain regulation of this segment of the ICT sector is an absolute necessity. Hence, interconnection prices have to be controlled by the regulatory authority. And must reflect the "correct prices" which account for the opportunity cost of making the sunk investment.

Coupling the demand- and supply-side consequences of the economic facts on the ground places policy maker and regulators in a delicate position – they have a precarious balancing act to ensure the wellbeing of the ICT sector.

References

Alleman J, Rappoport P (2000) Universal Service: The Poverty of Policy. University of Colorado Law Review 71 (4): 849–878

Bauer J (2005) Bundling, Differentiation, Alliances and Mergers: Convergence Strategies in U.S. Communication Markets. Communications & Strategy 60 (4): 59–83

Bauer J (2007) Network Neutrality. International Journal of Communication 1 (2007): 531–547

Broadbandproperties.com (2006) September

Businessweek (2006) http://app.businessweek.com. Cited 23 December 2006

Bureau of Labor Statistics (BLS) (2006) Consumer Expenditure data. www.bls.gov/cex/home.htm

Computer World (2006) www.computerworld.com/managementtopics/management/story/0,10801,107306,00.html. Cited 10 January 2007

Crandall R W (2005) Competition and Chaos: U.S. Telecommunications Since the 1996 Telecom Act. Brooking Institute Press, Washington, DC

Euromonitor International (2005) World Consumer Income and Expenditure Patterns. Euromonitor International, London

Federal Communications Commission (FCC) (2006) Trends in Telephone Service. www.fcc.gov/wcb/iatd/trends.html. Cited 23 December 2006

Forge S, Blackman C, Bohlin E (2005) The Demand for Future Mobile Communications Markets and Services in Europe, Technical Report EUR 21673 EN, April. http://fms.jrc.es/documents/FMS%20FINAL%20REPORT.pdf. Cited 23 December 2006

Google (2007) Google Intends to Bid in Spectrum Auction If FCC Adopts Consumer Choice and Competition Requirements. http://www.google.com/intl/en/press/pressrel/20070720_wireless.html. Cited 31July 2007

IDC (2006). www.idc.com/getdoc.jsp?containerld = prus20211306. Cited 5 January 2007

InformationWeek (2005) MP3 Players Reaching 'Critical Mass', 13 April. http://informationweek.com/story/showArticle.jhtml?articleID = 160900450. Cited 23 December 2006

Nissen K (2005) The Coming Revolution in Voice Communication Services. In-Stat http://www.instat.com/events/asia/promos/asia_nissen_21367.pdf. Cited 30 December 2006

Rappoport P, Alleman J, Taylor L, Kridel D (2003) Willingness to pay and the demand for broadband services. In: Shampine A (ed.), Down to the Wire: Studies in the Diffusion and Regulation of Telecommunication Technologies. Nova Science

Rappoport P, Alleman J, Taylor L, Kridel D (2005) The Demand for Voice over IP: An Econometric Analysis Using Survey Data on Willingness-to-Pay. Telektronikk 4

Rappoport P, Alleman J, Taylor L (2006) IPTV – Telecom Provision of Video Services: An Econometric Assessment. Proceedings of Sixteenth Biennial Conference of the ITS, Beijing, 12–16 June

Rappoport P, Kridel D, Taylor L, Duffy-Deno K, Alleman J (2003) Forecasting the Demand for Internet Services, The International Handbook of Telecommunications Economics, Volume II: Emerging Telecommunications Networks, Madden, G (ed.), Edward Elgar Publishers, Cheltenham, UK, pp. 55–72

Shapiro C, Varian H R (1999) Information Rules, A Strategic Guide to Network Economics. Harvard Business School Press, Boston, MA

Spolsky J (2007) Price as Signal. http://www.joelonsoftware.com/items/2005/11/18.html. Cited 23 December 2006

SunRocket (2007) Critical Notice to SunRocket Customers. http://www.sunrocket.com/. Cited 31 July 2007

Taylor L D (1994) Telecommunications Demand in Theory and Practice. Kluwer, Dordrecht

Taub E A (2007) Loaded with Personalities, Now Satellite Radio May Try a Merger. New York Times, 1 January, p. B1. http://www.nytimes.com/2007/01/01/technology/01satellite.html?ref = business. Cited 1 January 2007

Ureta S (2005) Variations in Expenditure on Communications in Developing Countries: A synthesis of the evidence from Albania, Mexico, Nepal and South Africa (2000-2003). In Diversifying Participation in Network Development: Case studies and research from WDR Research Cycle 3 (editors A.K. Mahan and W.H. Melody) Available at http://www.regulateonline.org/component/option.comdocman/task,docview/gid,19/Itemid,40/#page=49

Ureta S (2007) Variations on expenditure on communications in developing countries. November

US Census Bureau E-Stats (2006) 25 May. http://www.census.gov/eos/www/papers/2004/2004reportfinal.pdf. Cited 23 December 2006

Waverman L (2006) The challenges of a digital world and the need for a new regulatory paradigm. In: Richards E, Foster R, Kiedrowski T (eds.), Communications: The Next Decade. The UK Office of Communications, London: 158–175

World Bank (2006) Information Technology. http://web.worldbank.org/WBSITE/EXTERNAL/DATASTATISTICS/0,contentMDK :20394827~menuPK :1192. Cited 10 January 2007

Technical, Business and Policy Challenges of Mobile Television*

Johannes M. Bauer, Imsook Ha, and Dan Saugstrup

Abstract Mobile TV is considered an important component of the future portfolio of mobile services. It illustrates many of the challenges of developing sustainable business models for advanced mobile services. The value net of mobile TV requires coordination and collaboration among a larger number of market players than mobile voice services. Using South Korea, a mobile TV pioneer, and the United States as rich case studies the paper examines how technology, policy, and business decisions jointly influence the mobile TV market. Whereas there does not seem to be a "best" market design, we identify choices that tend to impede the development of mobile TV. South Korea has been able to accelerate the early introduction of mobile TV but the US market-led approach seems to provide more flexibility to experiment and discover viable business models.

Introduction

Confronted with stagnating or declining voice revenues, mobile operators are looking to data services to reverse the trend. Finding a sustainable business model is complicated by the more differentiated value net of mobile data services. Not only are many more market players involved (e.g., content providers, application providers, portals, device manufacturers), coordination among them requires more complicated arrangements than in mobile voice services (Maitland et al. 2002). In that market, standardization was a sufficient instrument to achieve coordination among the fewer players. In contrast, mobile data services necessitate the negotiation of business

J.M. Bauer (✉), I. Ha, and D. Saugstrup
Department of Telecommunication, Information Studies, and Media and Co-Director,
Quello Center for Telecommunication Management and Law at Michigan State University
e-mail: bauerj@msu.edu

*An earlier version of this paper was published in *Communications of the Association for Information Systems* in October 2007. We appreciate permission to reuse some of the material.

B. Preissl et al. (eds.), *Telecommunication Markets*, Contributions to Economics,
DOI: 10.1007/978-3-7908-2082-9_23, © Springer Physica-Verlag HD 2009

aspects such as revenue sharing and security features that support financial transactions. They also require a closer coordination between more complex network platforms and devices. Moreover, compared to voice mobile, data services face a challenging market environment as they often have to compete with already established online substitutes.

Mobile TV is regarded as one important component of the future portfolio of mobile services, even though it will not contribute a large share of mobile revenues. Juniper Research (2007), for example, estimates global mobile TV revenues to $16 billion by 2011. The company expects that slightly less than $12 billion will be based on mobile broadcasting while the remainder will be streamed via the existing mobile communication networks.

Nonetheless, countries and service providers have made great strides to deploy mobile TV, probably recognizing its function as a facilitator of the migration to mobile data services in general. After considerable success in the deployment of early mobile data services, South Korea also took a lead in mobile TV services. In comparison, the U.S. pursued a more gradual, market-driven approach to mobile TV. A comparative analysis of the two markets not only sheds light on mobile TV but also on the challenges faced by advanced mobile service design in general.

This chapter has two main goals: (1) to develop a conceptual framework of sufficient analytical depth and flexibility to explain the development of advanced mobile services across countries and regions and (2) to examine the experience of South Korea and the US to draw lessons for the future design of mobile TV markets. The section of the chapter titled "Factors Influencing Mobile TV" briefly presents the key elements of a conceptual framework capable of organizing study of the factors influencing the sustainability of business models in advanced IT services. The section titled "Mobile TV in South Korea" discusses the development and present state of mobile TV in South Korea and "Mobile TV in the United States" the situation in the US. "Lessons and Outlook," finally, contrasts the experience of the two nations and extracts main conclusions for the future development of mobile TV in other countries and regions. All information is current as of May 2008.

Factors Influencing Mobile TV

Several technological platforms are available to configure mobile TV (see Bauer et al. 2007; Tadayoni and Henten 2006, for more details). The two principal solutions are streamed (or "in-band") mobile TV and broadcast (or "out-of-band") mobile TV. In both groups, several standards exist or are in development. In the case of streamed video, the signal is delivered using the existing cellular network, either in single streams to individual users ("unicast") or in a more efficient shared model to several users simultaneously ("multicast"). Existing video-capable handsets may be used but the costs of bandwidth may be high, especially in unicast mode. For this reason, streamed mobile video may be more attractive for asynchronous non-real time delivery or for the viewing of short video clips rather than live mobile TV.

In contrast, broadcast mobile video is delivered using a separate dedicated multicast frequency. Terrestrial mobile TV is typically configured in the VHF or UHF bands or in the L-Band (1.452–1.492 GHz). To receive broadcast mobile TV, new handsets with an additional broadcast receiver are required. Major competing solutions are terrestrial and satellite-based digital mobile broadcasting (T-DMB, S-DMB) developed and promoted in South Korea, terrestrial digital video broadcasting to handhelds (DVB-H), and Qualcomm's Media Forward Link Only (MediaFLO).

Which solution offers the most promising business proposition depends on factors internal to the service providers but also on conditions that are to a large degree beyond their control or even influence. Among the internal factors are the resource base and core competencies of the service provider, the attitude toward risk, and the overall competitive strategy. External factors include the market environment, the available technology, and the public policy framework. After a brief look at the emerging mobile television value net, this section looks at these factors and how their co-evolution determines success or failure of specific solutions to mobile television.

The Emerging Mobile Value Net

In the mobile voice environment with its relatively simple technology, coordination along the value chain could be effectively achieved by standardization (Maitland et al. 2002). The new environment features a larger number of players, including handset manufacturers, network operators (cellular and broadcast), content providers (including specialized producers of mobile content, broadcasters and cable programming networks), mobile portals, application providers, and service providers. Coordination among these participants in the value net goes beyond technical issues and also requires agreement on mutually compatible business models (e.g., forms of revenue and risk sharing). The new environment also enables more differentiated pricing options (e.g., advertising financed services, flat fees, multi-part prices, service-specific direct payments, and hybrids).

Moreover, market uncertainty is generally higher, raising complicated issues with regard to the synchronization of innovation and investment. This may lead to dynamic inconsistency problems, for example, if network operators proclaim to wait for equipment manufacturers to bring handsets to the market before network specifications are finalized whereas handset manufacturers adopt the opposite stance. Standardization is important but not sufficient to synchronize and integrate these technology and business aspects.

The specific coordination challenges vary with the mobile TV platform and the business model envisioned by the main players. For example, in-band solutions require that the mobile network operator arrange for the availability of terminals capable of receiving video; produce or acquire mobile content directly or in alliance with content providers; and package pricing and service plans that meet consumer needs. Mobile broadcasting requires, in addition, investment in network infrastructure. In this latter approach, it is not necessarily the cellular network operator who organizes the market. It could also be an equipment manufacturer, a content provider, or a broadcaster. In an extreme case, the role of the cellular operator could be relegated to providing upstream communication in return for a share in the overall revenues. The relative negotiating power of the mobile network operator and other players depends on the public policy rules that govern terminal certification, licensing, and access to content. In the U.S., certification is essentially done by the network operators, thus giving them a stronger say in the mobile broadcasting market compared to nations where a common standard has been adopted.

The tentative experience with mobile Internet access seems to indicate that these coordination tasks might be achieved more effectively if one company organizes the market players. For example, i-mode in Japan was in part successful due to the integrating role played by NTT DoCoMo. Likewise, Nate in South Korea benefited from the efforts of SK Communications, which coordinated handset manufacturers, network operation, and content provision. Both companies relied on the kiosk system to collect service revenues, at least for a select group of affiliated content providers, thus greatly reducing transaction costs for users. It is reasonable to assume that mobile TV will likewise have a higher chance of success where these coordination and integration tasks will be solved more efficiently, either by one of the major players or by independent system integrators.

Co-evolutionary Dynamics

Of particular interest is how factors, which are largely under the discretion of the mobile network operators, interact with those that are in part or fully external. It is this interaction that influences the performance of the mobile TV market segment, including the rate and level of diffusion of services, the mix of wireless technologies and services available, the speed of innovation in wireless services including the transition to new generations of wireless services, and prices charged for services. Key aspects that interact with the business strategic decisions of firms are technology (both the installed base and available alternative technologies), the legal and regulatory rules governing the provision of wireless services, and the market environment (both on the supply and on the demand side). These factors are linked by many feedbacks, which create strong dynamic interdependence (see Fig. 1).

Co-evolutionary approaches are an innovative way of modeling such interdependencies. Two or more units in such a system are said to co-evolve, if they have a persistent significant impact on each other (Murmann 2003). Although earlier approaches did not use the notion of co-evolution, basic aspects of this framework are visible in theories of learning organizations (Senge 1990), evolutionary models of innovation (Mowery and Nelson 1999), and system models for management and policy (Sterman 2000). A main difference to these earlier theories is that co-evolutionary models explicate the process of dynamic change in more detail by building upon recent developments in evolutionary research.

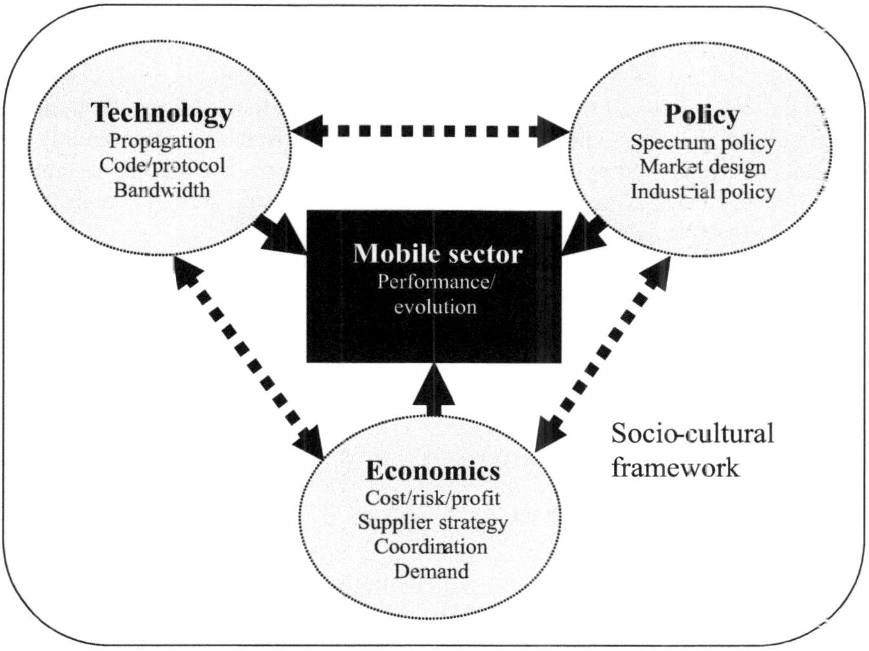

Fig. 1 Co-evolutionary dynamics

Applied to the context of mobile communication services, it is first important to identify what the relevant subsystems of the larger socio-technical system are that co-evolve. There is no straightforward deductive way to determine these components. Rather, a choice has to be based on a detailed understanding of the history of the industry and its present structure. The ability to observe the development of the mobile industry in different, somewhat isolated, regions assists in addressing this issue as the effects of alternative institutional arrangements may be observed. Depending on the specific question, a co-evolutionary model might distinguish between fewer or more subsystems. To simplify the analysis, we differentiate three realms: technology, policy, and economic aspects (supply and demand-side conditions as well as business strategy). The model could be further differentiated, for example, by unpacking the realm of firms to identify the different participants, including equipment manufacturers, network service providers, application and service providers, and portals. A key insight from co-evolutionary models is that events in one area affect (in anticipated and unexpected ways) developments in related areas but they do not fully determine them.

For example, spectrum policy choices (e.g., the amount of spectrum allocated, the assignment mechanism utilized – auctions, beauty contests, and number of licenses) have implications for the financial position of service providers and will, in turn, affect their investment and service pricing choices. Where suppliers are constrained by market parameters, such as consumers' willingness to pay and the competitive situation – and thus cannot simply roll higher license acquisition costs forward into prices – the upfront sunk cost will have real effects. For example, sunk costs may reduce the willingness of network service providers to share revenues with application service providers, hence delay content production and consequently the diffusion of service. Other policy choices, such as the specific market design, the ability for incumbents to use existing spectrum for new services, or roll-out obligations, will also have effects.

Likewise, the existing technology platforms will constrain the ability of service providers to deploy new features and services. Business decisions, on the other hand, will affect subsequent policy choices and the technology base. Jointly with other factors they shape the evolution and overall performance of the industry. The next two sections will illustrate these issues using South Korea and the United States as cases in point.

Mobile TV in South Korea

South Korea introduced the world's first handset-based mobile TV service in May 2005 (Lee and Kwak 2005). Mobile TV is another service in which the country has become a pioneer and global leader. The country illustrates the co-evolution of technology, policy, and business choices but also the risky and possibly problematic choices that can be made along the way. Apart from its own merits, understanding the development of mobile TV in South Korea may therefore be instructive for other countries as well.

Mobile TV Infrastructure

Mobile TV in South Korea is delivered via S-DMB and T-DMB infrastructures (see Fig. 2). In the first case, a satellite broadcasting center uplinks multimedia content via a Ku-band (12–13 GHz) signal. Content is downlinked to mobile devices using an S-band signal (2.630–2.655 GHz), allocated to DMB. The S-band is well-suited for broadcasting to small handset antennas as power output is not limited by international regulations. T-DMB uses the VHF band III and the L-band, which had already been set aside for Digital Audio Broadcasting (DAB). Multiplexed T-DMB uses only 1.5–1.7 MHz and is hence much easier to accommodate than the 6–8 MHz needed by DVB-H. To cover areas not reached by the S-DMB or T-DMB signals, a gap filler system of repeaters is used. Recently, T-DMB operators have introduced two-way data services through a return channel on the mobile cellular network, enabling more advanced services than one-way video and audio broadcasting. Such upgrades that combine broadcasting and mobile technology include MPEG-4 BIFS, middleware, traffic and travel information service technology, disaster broadcasting technology, and conditional access technology (DMB Portal 2008).

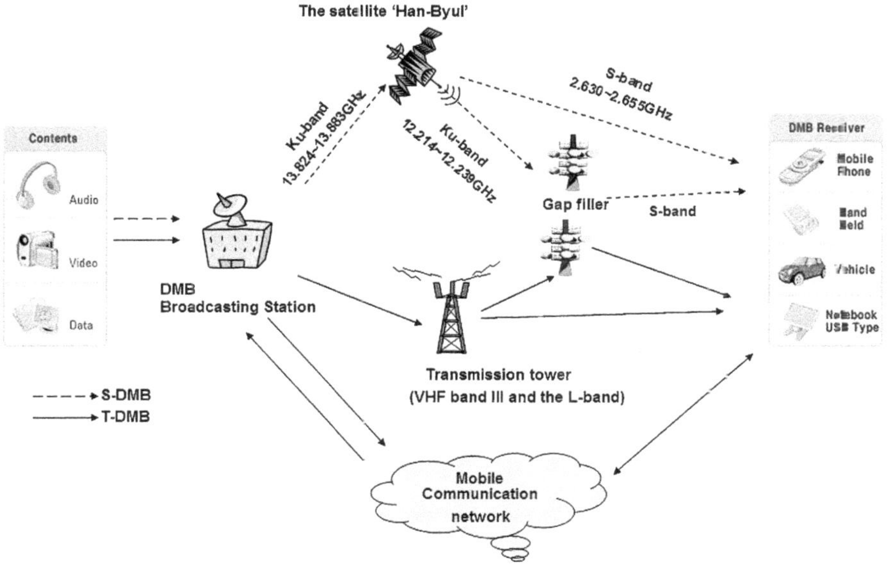

Fig. 2 DMB infrastructures in South Korea

Like Japan, South Korea uses System-E as the S-DMB standard, which relies on code division multiplexing (CDM) similar to the CDMA technology employed in its cellular networks. This technology is particularly suited to the country's mountainous terrain. Moreover, with this choice Korea aimed at prolonging the competitive advantages it had achieved in components and technologies used in CDMA. T-DMB uses the Eureka 147 standard, which is backward compatible and allows using the DAB network and frequencies (OECD 2007). When making these technology adoption decisions, policy-makers considered business as well as technology aspects. Criteria such as the cost effectiveness of the infrastructure, equipment, standards, and the relative competitive position of the mobile industry weighed in. Thus, policy decisions were closely linked to and co-evolved with technology and firm strategy.

Mobile TV Policy

Public policy decisions shaped mobile TV in several ways. Important industrial policy decisions were made by the Ministry of Information and Communication (MIC). Regulatory authority is vested in the Korean Broadcasting Commission (KBC) dealing with issues such as licensing, spectrum management, and competition among the service providers. In its early stages, Korean mobile TV has suffered from conflicts among players due to lack of a clear regulatory model for the converging broadcasting and communications services.

A mismatch between business and policy decisions was visible in the process of *licensing* mobile TV. During the early stages of S-DMB development, SK Telecom (SKT), the leading Korean mobile service provider, who sought to boost ARPU in the face of a nearly saturated cellular market, exerted a major push. SKT signed a Memorandum of Understanding with Mobile Broadcasting Corporation (MBCo), the Japanese DMB service provider backed by Toshiba, to cooperate in the development of DMB. MIC insisted on an early realization of S-DMB to establish a first mover position in the global mobile TV market. In September 2001, it initiated international registration at the International Telecommunication Union (ITU) of a satellite service corresponding to SKT's application. In contrast, KBC postponed its approval of S-DMB, reasoning that demand was insufficient and that fair competition with T-DMB needed to be ascertained beforehand.

Service was not launched until after the passage of the Broadcasting Law in March 2004, which suggested rapid introduction of S-DMB. KBC issued a request for proposals and despite the fact that there was only one application — TU Media, the joint venture between SKT and MBCo — KBC organized an evaluation committee of scholars, lawyers, experts in broadcasting and communication, and citizen groups. The committee selected TU Media as S-DMB service provider on December 14, 2004 after careful examination of the business plan, hearings and on-site evaluation. At that time, TU media had already invested substantial capital in launching a satellite ($97 million), the installation of gap fillers ($230 million) and the establishment of a DMB broadcasting center ($60 million) (SKT 2007). TU Media started a pilot service in January 2005 and launched its full broadcasting service in May 2005. Applications for T-DMB were invited in January 2005. In March 2005, six licenses were awarded to terrestrial broadcasters (KBS, MBC, SBS, CBS, YTN DMB, and KMMB). Commercial T-DMB service was launched in December of the same year. Shin (2006) argues that this process implies that Korean DMB has been developed by a leading mobile provider's technology push-to-market and government initiative rather than by market-pull.

Another important aspect of market design are the rules governing the *retransmission* of terrestrial TV programs via S-DMB (and possibly vice versa). A conflict of interest exists because terrestrial broadcasters, including KBS, MBC, and SBS, which were all given T-DMB licenses, directly compete with S-DMB. In its push for mandatory retransmission rights, TU Media invoked a combination of business and public interest reasons. It argued that, first, terrestrial programs are public property and that viewers want to and should be able to watch these channels on S-DMB. Second, TU Media argued that retransmission should be allowed to secure fair competition between T-DMB and S-DMB. Third, it claimed that refusal to retransmit would depress the profitability of S-DMB with negative repercussions on program providers and manufacturers (TU Media 2005). KBC left the issue to voluntary contractual agreements between providers thus giving terrestrial broadcasters relatively strong negotiating power. The first and only agreement was signed in 2007 between TU Media and MBC. This eventual contract may have come too late to reverse the stagnation of subscriber growth experienced by TU media since mid-2007. During the same period, T-DMB continued to grow indicating that S-DMB has lost ground relative to its competitors. From the point of view of TU Media this matter is an example of failed regulatory policy and counteracts other efforts by the Korean government to promote DMB.

A third important policy dispute was whether or not to allow T-DMB, like S-DMB, to offer *pay service*. The fees for S-DMB contribute to the costs of the satellite infrastructure as well as the gap fillers. T-DMB providers argued that, at least initially, advertising revenues would be too fragile and volatile to cover the costs of gap fillers and other start-up expenses. T-DMB service providers wanted to offer premium for pay services tailored to the different tastes of users in addition to free basic services. This two-tier business model was to generate additional revenues for T-DMB operators (DMB Portal 2008). However, the video and audio services of T-DMB are considered an expansion of free over-the-air broadcasting for home viewers. The utilized very high frequency (VHF) channels are regarded as a public asset (OECD 2007). Thus, although more than 7 million handsets, laptops, car navigators and other devices equipped with T-DMB receivers are in use, T-DMB has remained in financial distress, possibly due to these constraints on its pricing. KBC's approach, it could be reasoned, has contributed to the difficulties of the T-DMB service providers in finding sustainable solutions to recover their cost.

Business Aspects of Mobile TV

The evolution of the mobile TV market is also influenced by factors on the supply and demand side of the market (which are, in turn, shaped by policy decisions and technology). S-DMB and T-DMB have very different cost structures. According to MIC, S-DMB required a total investment of $500–800 million. As T-DMB mainly had to invest in a gap filler infrastructure, its costs are lower, estimated in the range of $50–80 million. The cost of content needs to be added to the investment expenses. Like in other information businesses most of the cost is an upfront fixed, possibly sunk, cost, whereas the incremental costs of serving additional users is very low.

From a consumer perspective, several important differences exist between the services. T-DMB is free of charge but offers only a lower number of terrestrial TV channels. S-DMB is priced from $6 to $13 per month and offers a higher number of channels. The business decisions of service providers in both platforms are highly interdependent and co-evolving. In South Korea, the large number of commuters, attitudes that favor new technologies and devices, and public policies in support of new technologies, created generally favorable market conditions. Nonetheless, market demand for S-DMB lags behind initial expectations. By spring 2008, the number of subscribers for S-DMB had reached approximately 1.3 million, well below the 2.2 million subscribers needed to meet the short-term break-even target. TU Media expects 6.6 million subscribers by 2010 but there are many contingencies to this forecast.

One factor is the pricing policy of T-DMB operators, who would like to introduce some form of payment. Options include a monthly flat fee, per channel fees, or charges for specific content. Another factor is public policy, such as the rules governing retransmission of terrestrial programs via S-DMB. The issue will again surface when South Korea introduces other convergence services, as is already visible in conflicts regarding IPTV and Wibro. Presently, S-DMB and T-DMB are imperfect but nevertheless strongly interdependent competitors. How competition will shake out will depend on future technology developments, the future public policy regime, and business decisions. Given the present market structure and market conditions, some form of consolidation or at least the integration of mobile TV into a broader range of communication services seems unavoidable.

So far, despite prolonged efforts by government and service providers, mobile TV is not yet profitable as a stand-alone service. In the S-DMB service, the accumulated debt of operator TU Media was estimated to 270 billion won (approximately $260 million) by the end of 2007. Whereas T-DMB is supported by big broadcasting firms, which are financed better than TU Media, it has also been in financial trouble for years. At the beginning of 2008, thus, the South Korean mobile TV market is thriving in terms of technological solutions but is facing continued financial difficulties. Even if mobile TV will eventually prove financially viable, the South Korean experience highlights the challenges for similar services elsewhere. Its rapid rise in South Korea was strongly supported by concerted government measures, which set technology standards, allocated spectrum and insisted on a free terrestrial service to promote uptake and to kick-start the market. As governments in other countries and regions are designing their mobile TV policy, some may choose a similar proactive industrial policy approach and others will opt for a more hands-off style. Whatever the details, policy decisions will inevitably affect the mobile TV market.

Mobile TV in the United States

In terms of mass commercialization, mobile TV in the US is still in its infancy. Several technologies that can deliver mobile TV service coexist in the US market place. Public policy has chosen a hands-off approach, leaving technology choices and service models to market players. A big challenge for service providers is the development of sustainable business models, in particular finding the proper balance between unicast and multicast streamed as well as broadcast mobile TV.

Mobile TV Infrastructure

Streamed mobile TV, using the existing 2.5G and 3G networks, is the most common form in which mobile video is presently delivered to customers in the US. The most prevalent solution is unicast technology, which provides individual streams to users. A typical example is MobiTV. If a user indicates that he or she wants to access mobile TV, a dedicated stream is sent to the handset via the cellular network. MobiTV is available in the US through Sprint, AT&T, Alltel and several other regional carriers.

Service providers have recently begun to migrate and upgrade streamed mobile TV from the existing 2.5G/3G networks to faster 3.5G/4G networks. For example, Sprint Nextel plans to develop and deploy mobile TV service based on its mobile WiMAX network (brand named "Xohm"). WiMAX technology promises to deliver high speed broadband and video service. With expected speeds of 2–4 Mbps for downloads and 1–3 Mbps for uploads, Sprint's WiMAX network will offer a higher number of mobile TV channels in unicast mode overcoming capacity constraints of the older cellular network platforms (Xohm 2008).

In the long term, unless significant advances in signal compression are realized, it is likely that broadcast mobile TV will become the dominant platform because it utilizes spectrum more efficiently. Due to US spectrum allocation decisions, only one technological option for the broadcasting of mobile TV, MediaFLO (Forward Link Only), has been launched commercially. MediaFLO is an end-to-end mobile multimedia broadcast system that uses UHF channel 55 (716–722 MHz). It enables the simultaneous delivery of TV-quality content to millions of subscribers. The four main components of a FLO system are the Network Operation Center (which consists of a National Operations Center and one or more Local Operation Centers), FLO Transmitters, a 3G Network, and FLO-enabled devices (also known as MediaFLO handsets). Realtime content is fed into the Network Operations Center directly from content providers. Non-realtime content can also be received over the Internet. The content is reformatted into FLO packet streams and distributed over a single-frequency network. In the target market the FLO packets are converted to the FLO waveform and sent out to mobile devices. A 3G cellular network provides interactivity and facilitates user authorization (Qualcomm 2005).

According to developer Qualcomm, MediaFLO has several advantages over other technologies. Distinguishing features include high capacity, layered modulation and source coding, superior service with fewer resources, support for wide and local service, fast channel acquisition time, low network deployment costs and optimized power consumption. Consequently, MediaFLO allows lower cost operation while facilitating a higher-quality consumer experience (Qualcomm 2007). The listed advantages are somewhat diminished by potential technological disadvantages. DVB-H, while the less efficient platform and inferior technology, has nevertheless been adopted by a larger number of companies. Like in the case of GSM which was inferior to CDMA but captured a dominant share of the market, this may have beneficial effects on the costs, prices, and diversity of available equipment. Hence, DVB-H may be able to take better advantage of economies of scale and diversity than MediaFLO, which is currently only deployed in the (admittedly large) US market.

U.S. Mobile TV Policy

In the US, mobile TV evolves in an essentially unregulated market environment. The roots of this approach date back to the early 1990s, when the Federal Communications Commission (FCC) started to allow more flexible uses of spectrum.

Licensees were gradually enabled to use spectrum for an increasingly broad range of voice and data applications. As licensees were free to deploy more advanced services in their existing 2G spectrum, the FCC saw no need to formalize a 3G policy. Spectrum of particular use for mobile data services was assigned as market demand for such spectrum became more pressing. Mobile TV will probably benefit from the latest round of spectrum auctions, in which frequencies in the 700 MHz band, well-suited for broadcast mobile TV, were assigned. As some of these frequencies are still occupied by incumbent users, it will take a while to clear the spectrum.

This policy environment has allowed service providers to experiment with mobile video applications and mobile TV. Consumer response could be tested without risky sunk investment. On the other hand, the differentiated and uncoordinated introduction of mobile video by network operators may have had detrimental effects on the whole value net and slowed the diffusion of such services. Due to network effects and economies of scale, other participants in the network of value generation, most prominently content providers and equipment manufacturers, may have had lower incentives to invest and innovate as the user base evolved slower and more organically in response to market developments. Moreover, there are complementary relations between these providers. One would thus anticipate slower market growth than in countries and regions that have agreed on a common standard. It is difficult to assess a priori whether the positive effects of more market experimentation outweigh the potentially negative effects of slower diffusion. There remains a risk that a critical mass of users may not be reached and that both streamed and broadcast mobile TV will not reach a financially sustainable state.

Business Aspects of Mobile TV

In the US, many operators are already offering streamed mobile TV on a commercial basis. Verizon Wireless provides (non-realtime) mobile TV services under the brand *V-Cast* over its EV-DO network. Sprint offers *Sprint TV* (packaged video clips) and *Sprint TV Live* (realtime TV) over its EV-DO and PCS networks. AT&T Wireless streams mobile TV over its EDGE and HSDPA platforms. Both Sprint and AT&T get their content from MobiTV, a company that specializes in taking television feeds and sending them over cellular networks (see Table 1).

These commercial services transmit mobile TV over the wireless network in separate streams to every user. However, mobile networks were not designed for real-time TV services. Given the state of compression technology, it is a relatively inefficient use of limited network capacity. The unknown willingness of consumers to pay for mobile TV service puts an additional constraint on the economic viability of unicast streamed mobile TV. Mobile broadcast platforms, despite their potential downsides, promise to overcome these constraints. South Korean mobile operator SK Telecom is a case in point. Nine months after it had first launched video streaming, it decided to invest in a satellite-based platform to relieve the serious congestion of its mobile network.

In the US, several mobile operators are migrating toward broadcast platforms for mobile video. Verizon is delivering TV via Qualcomm's MediaFLO network, making

Table 1 Streamed and broadcast mobile TV in the US

	Verizon wireless	Sprint Nextel	AT&T
Service	V-Cast	Sprint TV MobiTV	Cingular Video MobiTV
Network	EV-DO	CDMA/EV-DO	EDGE
Service type	Unicast VOD (non-realtime)	Unicast VOD Unicast live TV (non-realtime/realtime)	Unicast VOD (non-realtime)
Pricing	Option 1: Pay-as-You-Use (data rate), $1.99 per megabyte Option 2: Monthly subscription, V-Cast Vpak: $15 per month	Monthly subscription $20 per month in addition to any other Sprint Nextel calling plan	Mobile TV Limited $13 per month; Mobile TV Basic $ 15 per month; Mobile TV Plus $30 per month

it the first in the US to distribute broadcast-quality programming without straining traditional cellular networks. MediaFLO is also collaborating with AT&T to launch a mobile broadcasting service. Verizon's V-Cast Mobile TV service features nine live television channels (CBS Mobile, Comedy Central, ESPN Mobile TV, FOX Mobile, MTV, MTV Tr3s, NBC 2GO, NBC News2GO, and Nickelodeon) and one radio network (ESPN Radio) in more than two dozen cities including Chicago, Orlando, Las Vegas, and Seattle. The $15 per month service fee includes all channels and parental controls. For an additional $10, customers can bundle the mobile TV service with streamed mobile video services.

AT&T introduced mobile broadcasting service in May 2008 with an initial roll-out plan to 58 cities (covering a total population of 129 million, about 43% of total US population). The company will also use the MediaFLO broadcasting infrastructure built by Qualcomm. Its channel line-up comprises ESPN Mobile TV, FOX Mobile, NBC 2Go, NBC News 2Go, Comedy Central, MTV, Nickelodeon, CBS Mobile, and CNN Mobile Live. In addition AT&T offers one exclusive Sony Pictures channel. The service is offered in three versions: Mobile TV Limited is priced at $13/month for four channels; Mobile TV Basic offers nine channels at $15/month; and Mobile TV Plus for $30/month also offers the premium Sony Pictures channel.

Unlike their counterparts in South Korea, American customers will thus be able to receive both streamed and broadcast mobile TV services, either on a stand-alone or bundled basis, from the same operator. With the right handset, users will be able to receive live TV content equivalent to that broadcast on the comparable over-the-air television or cable channel, or on-demand content. The latter may be stored either on the network or on the phone for later viewing. Both Verizon and AT&T use MediaFLO for live simulcast TV and their cellular networks to deliver on-demand mobile video. The most sustainable business model may well be a dual approach in which customers use both on-demand and live simulcast channels. It is also too early to tell whether broadcast technology will eventually be utilized to deliver both types of content or a hybrid solution will emerge in which broadcast technology will be used for simulcast TV and streaming for asynchronous viewing and short live clips. The combination of a mobile and a broadcasting network enables interactive content, which is widely considered key to further growth of mobile video and the ability of mobile operators to generate revenues from their mobile networks.

Lessons and Outlook

The mobile TV market is an interesting example of the design of advanced mobile communications services. Whereas certain aspects are unique to mobile television, it also offers more generic insights. The early stage of development prevents a statistical analysis. Therefore, this paper resorted to a qualitative case study approach focusing on in-depth analyses of the mobile TV markets in South Korea and the US. The observations were organized using a co-evolutionary conceptual framework. This approach recognizes the interdependence and mutual conditioning of technology, business, and policy choices. The overall trajectory of the mobile TV (as that of any other) market is shaped by and emerges from the interaction of these groups of factors.

In such a dynamically interacting system, it is difficult if not impossible to specify a priori a "best" way of designing a mobile market environment. Rather, several combinations of technology, business strategy, and public policy are feasible that yield acceptable performance outcomes. Different constellations will probably result in slightly diverging trajectories and over time characteristic trade-offs between approaches may become visible. For example, mobile TV may be adopted faster in countries that select a common technical standard. On the other hand, technological experimentation and hence the discovery of superior solutions may be encouraged if no common standard is chosen. However, not all possible constellations of technology, business strategy, and policy will yield acceptable performance. There are some combinations that will be roadblocks to the further evolution of the market. An obvious example is the failure of spectrum policy to make frequency bands available to operators. Our paper attempted to illuminate both the obstacles to the development of the mobile TV market as well as the tentative implications of the two different market designs in South Korea and the US.

In South Korea, first initiatives to develop mobile TV came from the government but industry quickly responded and took a lead. Government and industry attempted to build on the momentum of earlier mobile data services and to solidify a global leadership role. A decision was made in favor of a dual approach to deliver mobile television via satellite and terrestrial platforms. The specific policy choices (most importantly free terrestrial service, voluntary agreements to retransmit content) sent mixed signals to the market. On the one hand, it kick-started the market; on the other hand, the specific limitations placed on revenue models seem to have contributed to continuing financial difficulties. Despite rapid subscriber growth, neither T-DMB nor S-DMB has reached a customer base that is sufficient to break even.

In contrast, the US mobile TV sector is market-driven. Public policy has taken a back seat and left the initiative to private industry. Cellular service providers were not constrained in offering mobile video and TV services over their existing platforms. Moreover, at present mobile TV is not subject to most of the regulations affecting over-the-air broadcasting. Mobile operators have introduced streamed mobile TV and video and are presently broadening their offerings to broadcast mobile TV. This gradual approach has allowed them to keep sunk cost lower but it has probably also reduced market growth. The evolution of the market is, furthermore, constrained by past spectrum allocation decisions that limit the bands available for broadcast mobile TV. Moreover, as market forces sort out the most successful technological platform and services, positive feedback from network effects and complementarities are weakened compared to a market in which one common platform is introduced. Other things equal, the technological fragmentation will most likely also result in slower market growth. However, the more flexible US approach has allowed new forms of risk sharing. For example, the MediaFLO network is built by Qualcomm and its services sold to retail customers by mobile operators. It also allows mobile service providers to offer streamed and broadcast mobile TV on a stand-alone and bundled basis. This should improve their ability to price differentiate and hence strengthen revenues.

In sum, our examination detected in both market environments aspects that complicate the development of sustainable mobile TV services. Some, such as spectrum allocation decisions, are caused by policy and hence could, at least in principle, be overcome. Others have to do with customer expectations (e.g., free access to over-the-air TV) or technology and are not fully controllable by policy and business decisions. For example, if mobile handsets will support mobile Internet access, the customer base for mobile TV may be diluted as some users may stream from Internet sources. Market players would be affected differently by such a migration. As mobile operators, within the limits set by their competitive market environment, choose their price level and structure, they may be able to offer pricing packages that counter such customer migration. On the other hand the creators of mobile content would experience slower market growth but not have effective strategies to mitigate the impacts of such a development.

For countries that are still early in the process of designing the market environment for mobile video and TV the main lessons are to remove, as far as possible, known obstacles and bottlenecks and let, at least initially, market experiments flourish. As in any fast-paced industry, measures of industrial policy, such as mandating a technological platform, face high risks. They are not necessarily bound to fail but fine-tuning them is a great challenge. It might be wiser to facilitate market experiments and intervene in more limited ways where private decentralized decisions visibly yield undesirable outcomes.

References

Bauer JM, Ha I, Saugstrup D (2007) Mobile Television: Challenges of Advanced Service Design. Communications of the Association for Information Systems 20, 621–631.

DMB Portal (2008). See http://eng.t-dmb.org/ (last visited June 18, 2008).

Juniper Research (2007) Mobile TV: The Opportunity for Streamed and Broadcast Services, 2006–2011. Second edition, Basingstoke, England.

Lee SW, Kwak DK (2005) TV in Your Cell Phone: The Introduction of Digital Multimedia Broadcasting (DMB) in Korea. Paper presented at the 33rd Telecommunications Policy Research Conference (TPRC), Arlington, VA, September 23–25.

Maitland CF, Bauer JM, Westerveld R (2002) The European Market for Mobile Data: Evolving Value Chains and Industry Structures. Telecommunications Policy 26 (9–10), 485–504.

Mowery DC, Nelson RR (eds.) (1999) Sources of Industrial Leadership: Studies of Seven Industries. New York: Cambridge University Press.

Murmann JP (2003) Knowledge and Competitive Advantage: The Coevolution of Firms, Technology, and National Institutions. Cambridge: Cambridge University Press.

OECD (2007) Mobile Multiple Play: New Service Pricing and Policy Implications. Paris: Organisation for Economic Co-operation and Development.

Qualcomm (2005) MediaFLO. FLO Technology. See http://www.cdmatech.com/download_library/pdf/MediaFLO_brochure.pdf.

Qualcomm (2007) Mobile Media for the Masses: The MediaFLO System. San Diego, CA: Qualcomm, see http://www.docstoc.com/docs/1896357/MediaFlo-Product-Brochure.

Senge PM (1990) The Fifth Discipline: The Art and Practice of the Learning Organization. New York: Doubleday.

Shin DH (2006) Prospectus of Mobile TV: Another Bubble or Killer Application? Telematics and Informatics 23, 253–270.

SKT (2007) SK Telecom Annual Report. Seoul.

Sterman JD (2000) Business Dynamics: Systems Thinking and Modeling for a Complex World. Boston, MA: Irwin, McGraw Hill.

Tadayoni R, Henten A (2006) Business Models for Mobile TV. Paper presented at RIPE, Amsterdam, November 16–18.

TU Media (2005) Digital Multimedia Broadcasting (DMB) in Korea. Seoul.

Xohm (2008) See http://www.xohm.com/ (last visited June 18, 2008).

A Cross-Country Assessment of the Digital Divide*

Paul Rappoport, James Alleman, and Gary Madden

Abstract This paper considers the constraints on information and communications technology (ICT) services growth. The focus of this analysis is on the drivers of demand per se, and not service growth as a function of infrastructure deployment and service availability. Indeed, given the investment required to design and deploy next generation networks, this focus raises the possibility that once networks are in-place, the price may not be affordable to most households. The policy implications of that scenario include significant (publicly funded) subsidies or increased regulation. We examine the growth in telecommunication and data services by constructing a cross-country model of countrywide expenditure shares. The analysis compares growth rates by country against the share of income allocated to communication service to infer an upper limit on potential market saturation. The analysis provides an alternative perspective on the 'Digital Divide' by separating the demand for advanced services into socio-economic (prices, income, demographics) and policy (universal access, availability) factors.

Introduction

An often cited statistic associated with communication services in the United States is the share of household expenditure on telephone services. For the period 1981 through 2005, this share varied in a narrow range from 1.9% to 2.4% (BSL 2007). The limited variation in this share is interesting given that major changes occurred in

P. Rappoport(✉), J. Alleman, and G. Madden
Temple University
e-mail: paul.rappoport@temple.edu

*We are grateful to Lester Taylor for comments and suggestions on an earlier version of this paper. The usual disclaimer applies.

B. Preissl et al. (eds.), *Telecommunication Markets*, Contributions to Economics,
DOI: 10.1007/978-3-7908-2082-9_24, © Springer Physica-Verlag HD 2009

the organization of communication services markets, e.g., the breakup of AT&T, the passage of the Telecommunications Act of 1996, the emergence of cellular telephony, and the introduction of VoIP as a means to bypass traditional POTS service.

This paper examines national communication service expenditure shares cross country. This examination, allows several questions to be addressed. In particular, is the near constancy observed in US household expenditure shares observed in other countries? In addition, at a point in time, is variation in expenditure shares related to country size, rate of economic growth or stage of economic development? Additionally, do expenditure shares vary by the nature of national regulation and state of competition? Also, how do these shares correlate with the availability and growth of advanced services such as broadband? Finally, how do share movements correlate with national expenditure on information and communications technology (ICT)?

This paper uses consumer expenditure data collected from several sources (reported by Euromonitor International 2005; Bureau of Labor Statistics 2007) to analyze national expenditure share patterns through time. Primary data sources include the ITU World Telecommunication Development Report database, Global Insight, UNESCO Institute for Statistics, the IMF and Euromonitor's compilation of national statistics.

Literature Review

Several studies examine aspects of cross-country differentials in computer penetration and internet usage, and their impact on economic growth and investment. Chinn and Fairlie (2007) find the global digital divide is mostly explained by national income and demographic variables. Kiski and Pohjola (2002) model national ICT investment and report a high-income elasticity for ICT. Garcia-Murillo (2005) in examining international broadband deployment shows that broadband growth depends on income, personal computer penetration and density. Röller and Waverman (2001) infer telecommunication infrastructure and policy are drivers of economic development. Alleman et al. (1992) review literature concerning analysis of POTS infrastructure and economic growth in explicitly analyzing the relationship between economic growth and telephone development in 13 southern Africa countries. Parks and Barten (1973) provide the basis for the cross-country analysis of consumption patterns employed in this study. In particular, they focus on expenditure shares for food, clothing, housing, durables and other goods.

Internationally, there is substantial public policy interest in national growth rates for internet access, broadband deployment, mobile network subscription and ICT investment. While we examine these measures, the focus of the econometric analysis is aggregate ICT expenditure. In particular, we are concerned with how much of the variation in communications expenditure shares is explained by computer penetration and internet use? Also, how much of expenditure share variation is explained by income and demographics? Finally, and more importantly, how much of the variation is explained by these factors?

Expenditure Share

In this study the communications services expenditure share is an amalgam of traditional telephone service, wireless service and telecommunications equipment expenditures. Near constant expenditure share movements, in the absence of substantive quality change, indicates that the prices of communication services are stable relative to other category (e.g., food, housing, clothing, transportation) prices. When demand for communications services is inelastic, increasing price leads to greater expenditure (*inter alia*, a relatively higher expenditure share). Second, if the underlying service-mix changes then the net effect of these changes depends on the net price impact. For example, if the POTS price falls then more calls are made at lower this price implying a smaller expenditure share. In particular, international calling revenue declined even when the volume of international calls increased principally due to price falls.[1] However, variation in the communication share may also reflect the state of national economic development.

The paper next reviews national expenditure data. Two datasets are used for this purpose. The Euromonitor dataset contains information for 26 countries on expenditure and information on information goods (the most recent data are for 2004). A second dataset contains national data segmented the regions: North America, Eastern Europe, Western Europe, Latin America, Africa, Austrasia, and Asia, and these data are used to examine the constancy of communication expenditure shares.

Communication Share

Figure 1 displays the communications share by country.

Examination of the reported national expenditure share cross-country indicates that the US communications share variation is relatively low.[2] Further, aside from outliers China and Brazil, South Korea has a communication share twice that of the US. Figure 2 displays category expenditure shares for health and medical services, and communications. The variation in health are cross country suggests a partial explanation of the variation in communication services, with the US healthcare share substantially exceeding that of other nations.

From Fig. 3 we note that China, e.g., has higher expenditures for communications than in all other categories. In the US, healthcare expenditure represents the largest component of consumer expenditure.

Finally, Fig. 4 displays national consumer expenditure by category by communications expenditure.

---

[1] See for example the FCC's International Traffic Data (2005).

[2] Part of the variation may be due to different national definitions of communication expenditure or the statistics used to measure the expenditure share. At this time, there is no adequate way to check for consistency in definition and computation.

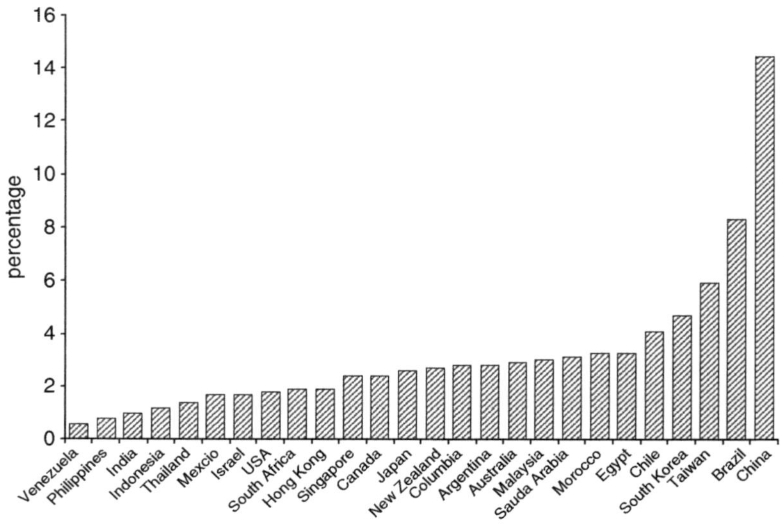

Fig. 1 Communication share 2004 (From Euromonitor International 2005)

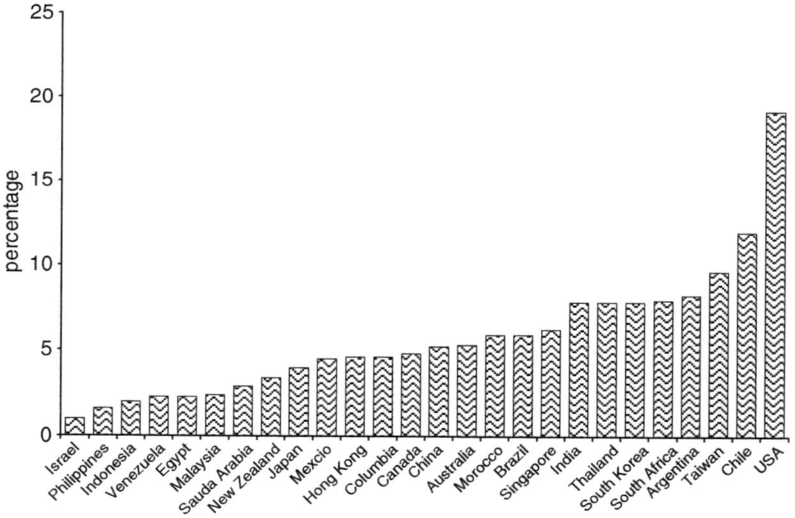

Fig. 2 Healthcare share 2004 (From Euromonitor International 2005)

Entertainment provides a proxy measure of communication service and applications (e.g., music downloads, streaming video and online gaming) demand. Increased demand for necessities (e.g., housing, food, energy, transportation, clothing and healthcare) impact on the communication share.

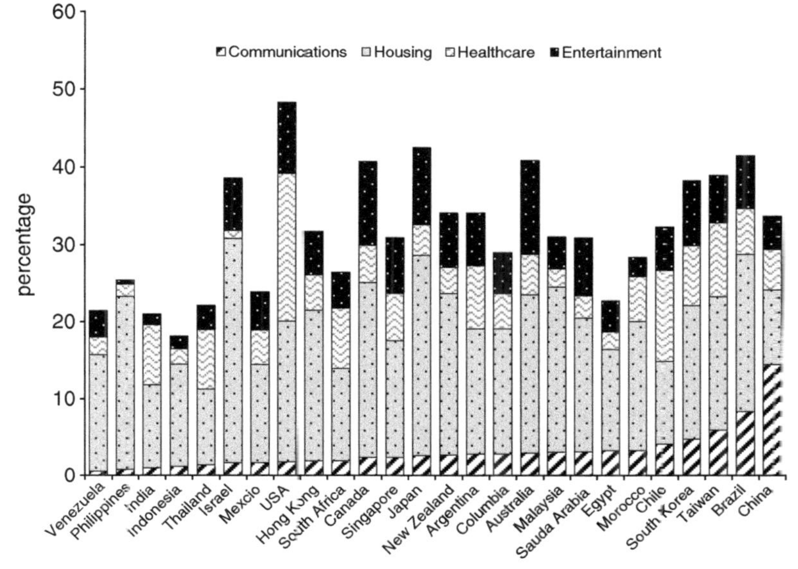

Fig. 3 Consumption shares (From Euromonitor International 2005)

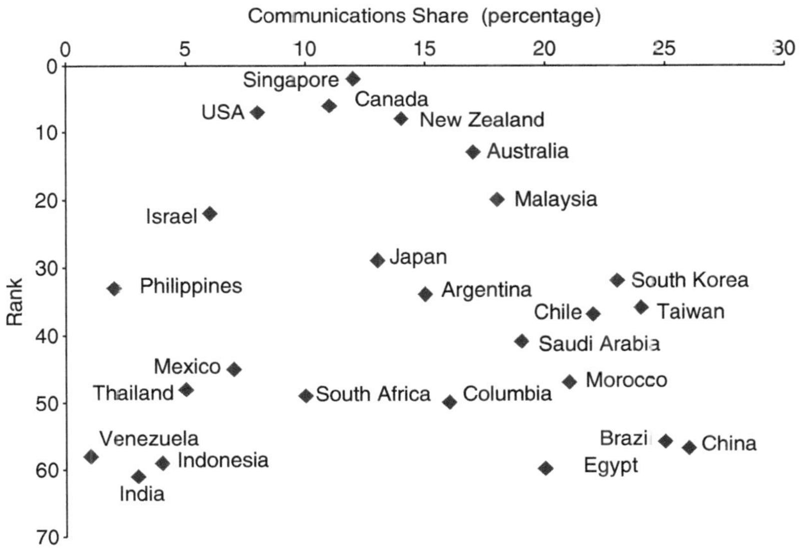

Fig. 4 Expenditure shares (From Euromonitor International 2005)

Explanations

A.T. Kearney (2006) publishes a Globalization Index that ranks nations across 12 dimensions.[3] Figure 5 lists Kearney's Globalization Index and national communication shares.

Figure 5 suggests an apparent positive relationship between the Index and national communications expenditure shares. In particular, nations with a low globalization score report lower communication expenditure shares. That is, developed countries invest more in ICT infrastructure. Additionally, developed nations rank highest for computers and the proportion of households that have internet service. This finding appears to suggest that 2–4% is a steady state communication share for developed countries.

Figure 6 displays the relationship between the Globalization Index and the entertainment expenditure share.

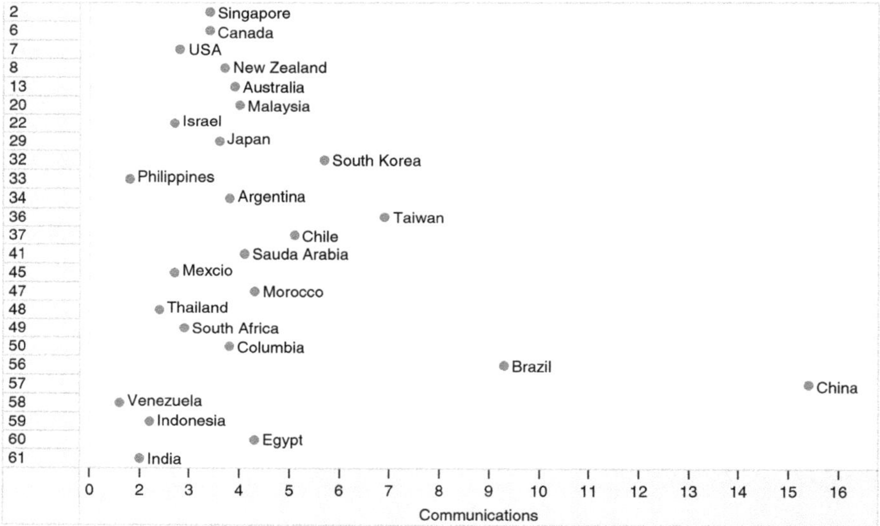

Fig. 5 Globalization index rank 2004 by communication share (From Euromonitor International 2005; A.T. Kearney/Foreign Policy 2006 and authors' calculations)

[3]The A.T. Kearney/Foreign Policy (2006) Globalization Index ranks 62 countries (representing 85% of the world population), based on 12 variables grouped in the categories: economic integration, personal contact, technological connectivity and political engagement. The Index attempts to quantify economic integration by combining data on trade and foreign direct investment. Technological connectedness is gauged by counting internet users and hosts, and secure servers. Political engagement is assessed by the number of selected international organizations and treaties that a nation signs, as well as national financial and personnel contributions to UN peacekeeping missions and levels of governmental transfers. Personal contact is charted by national international travel and tourism, international telephone traffic and cross-border transfers, including remittances.

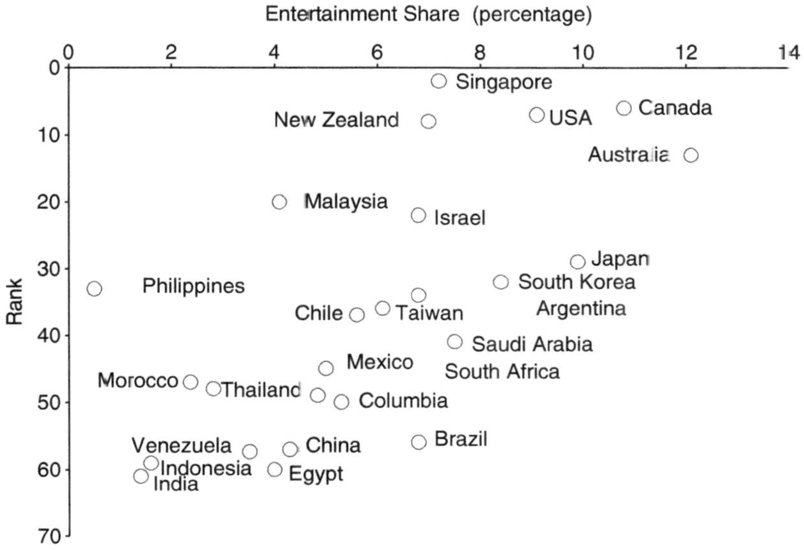

Fig. 6 Globalization index rank by entertainment share (From Euromonitor International 2005; A.T. Kearney/Foreign Policy 2006 and authors' calculations)

A positive relationship between globalization and entertainment shares is evident. A rationalization is that Globalization rank is correlated with income, which is an entertainment demand driver.

Figure 7 shows 2004 per capita ICT investment by communication share.

The horizontal and vertical grid lines represent average variable values. The upper-right quadrant identifies countries with high ICT investment and communication share. This quadrant includes Brazil and South Korea. Conversely, the upper-left quadrant contains high ICT investment and low communication share nations – more developed and mature countries are located in this space (ICT investment is an important category in Kearny's Globalization Index).

We note from Fig. 8 that national communications expenditure share are inversely related to GDP per capita. In general, the higher is national income, the smaller is the communications share. This finding is not necessarily in conflict with research that suggests information technology products and services are positively related to income. That is, the communications share is not a proxy for communication demand.

Figure 8 displays the relationship between the communication share and GDP per capita.

To illustrate this point, we identify a different relationship for the housing share and GDP. Figure 8 suggests a positive relationship between national income and housing expenditures.

Implicit in the computation of share is price. For the most part, housing (and healthcare) prices increased substantially more than the prices of communication services. A similar finding holds for the relationship between expenditures on healthcare and GDP per capita.

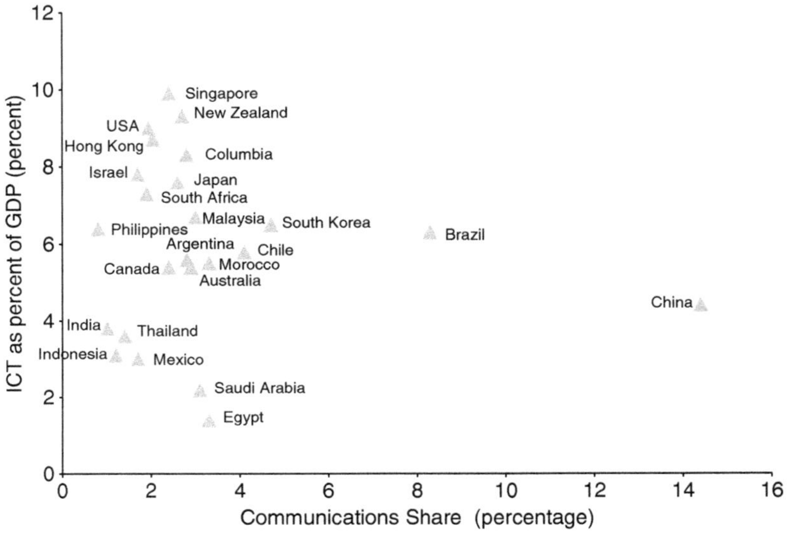

Fig. 7 ICT Expenditure per capita by communication share (From Euromonitor International 2005 and authors' calculations)

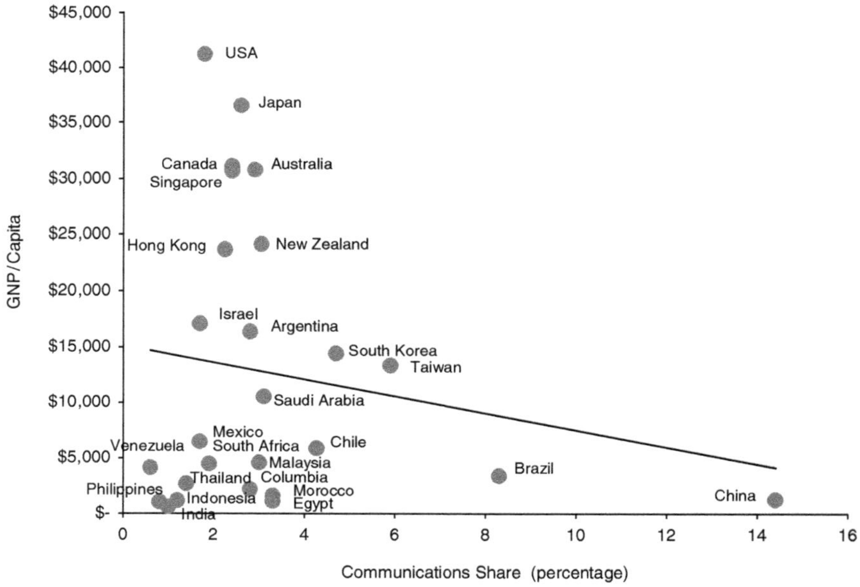

Fig. 8 Communications share by GDP per capita (From Euromonitor International 2005 and authors' calculations)

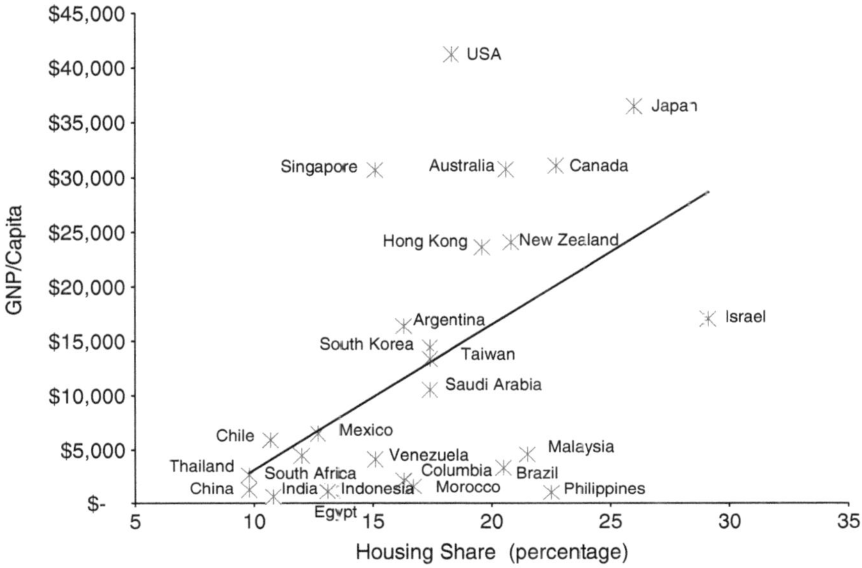

Fig. 9 Housing share by GDP per capita (From Euromonitor International 2005 and authors' calculations)

Communications Expenditures and the Information Age

Figure 10 displays mobile subscription per 1,000 inhabitants.

Figure 11 displays national mobile phone penetration by communication share. For the most part, the higher is mobile phone penetration, the lower are communication expenditures.

Excluding China and Brazil, internet subscription typically increases with the communications share. For 2004, no strong correlations are apparent for broadband or mobile subscription, ICT expenditure or personal computer penetration with national communication shares.

Temporal Communication Share Patterns

This section focuses on the relative stability of the communications expenditure share for the period 2000–2005. Of interest are the share growth rates, in particular the pattern of increased, constant or decreased national share growth during the period. Table 1 display the communications share growth rates by region.

Figure 12 lists internet subscription (subscribers per 1,000 households) and corresponding national communication share.

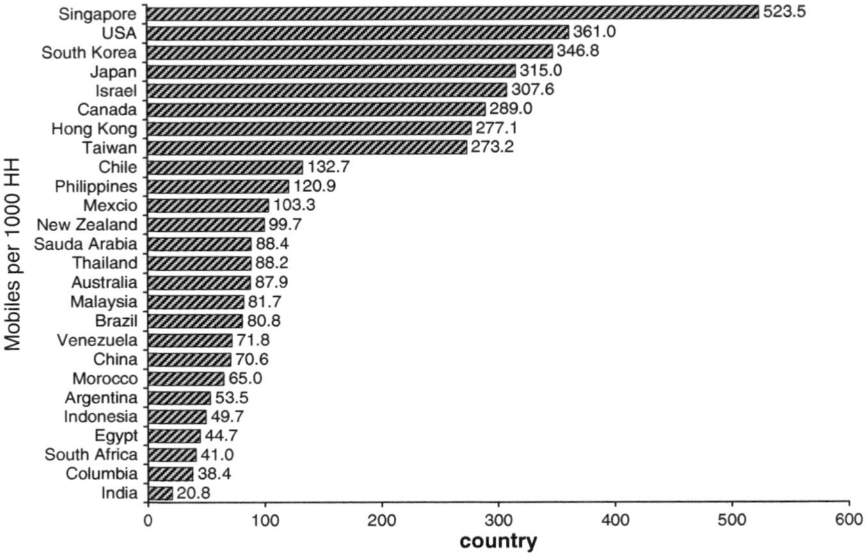

Fig. 10 National mobile subscription (From Euromonitor International 2005)

Table 1 Communications share growth rates by region

Country	Communications share growth rate
Africa	0.64 %
Australasia	1.73 %
Western Europe	1.99 %
Latin America	2.23 %
Eastern Europe	2.57 %
Asia Pacific	2.75 %
North America	3.52 %

Hong Kong's growth rate for this period is approximately 4%, which is four times the regional average (Fig. 13). PC China, which reports the largest share, experienced a decline during the period. More volatility is observed for Eastern Europe (Figs. 14–16).

Latin America experienced an increased growth of the share.

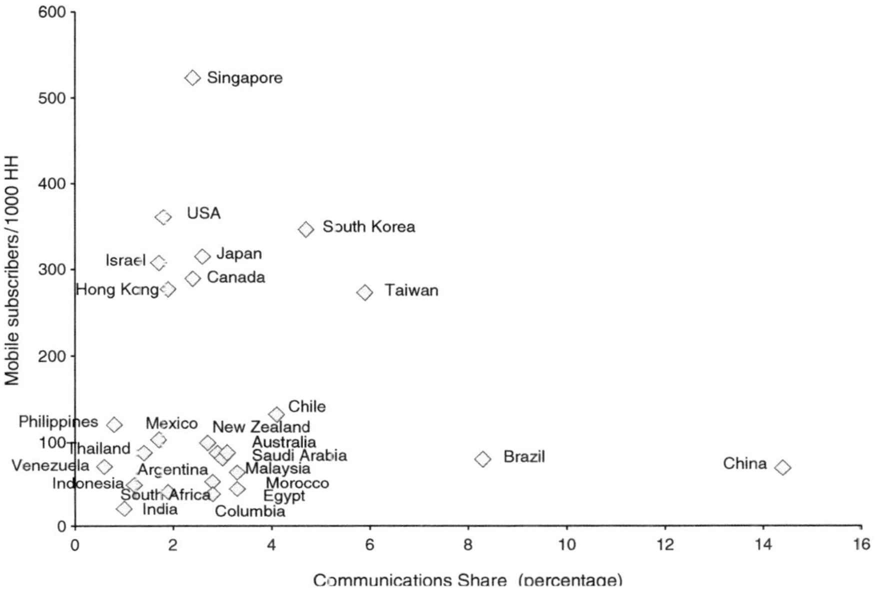

Fig. 11 Mobile subscription by communication share (From Euromonitor International 2005) Note the data did not allow the comparison of western European countries, which have high mobile penetration and high GDP per capital

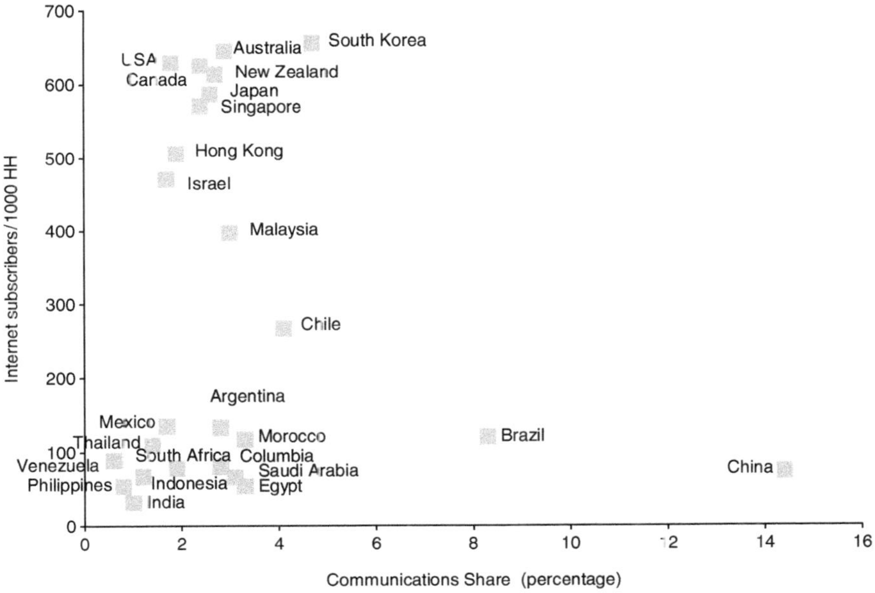

Fig. 12 Internet subscription by communication share (From Euromonitor International 2005 and authors' calculations)

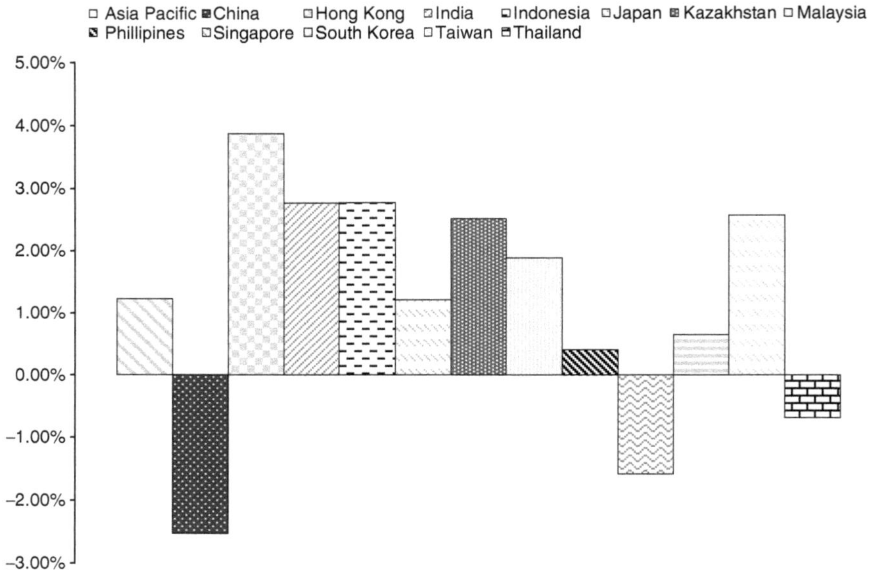

Fig. 13 Communications share average annual growth, Asia Pacific 2000–2005 2005 (From Euromonitor International)

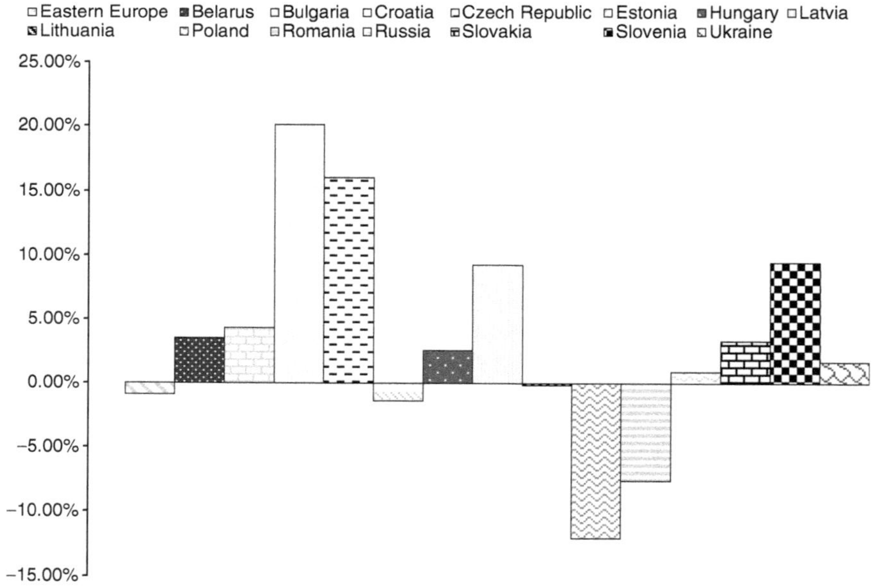

Fig. 14 Communications share average annual growth, Europe 2000–2005 (From Euromonitor International 2005)

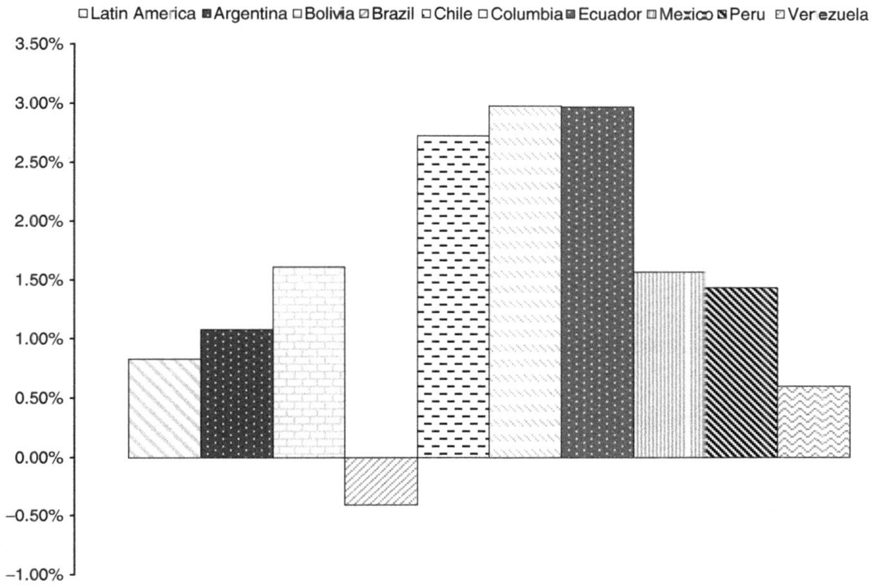

Fig. 15 Communications share average annual growth: Latin America 2000–2005 (From Euromonitor International 2005)

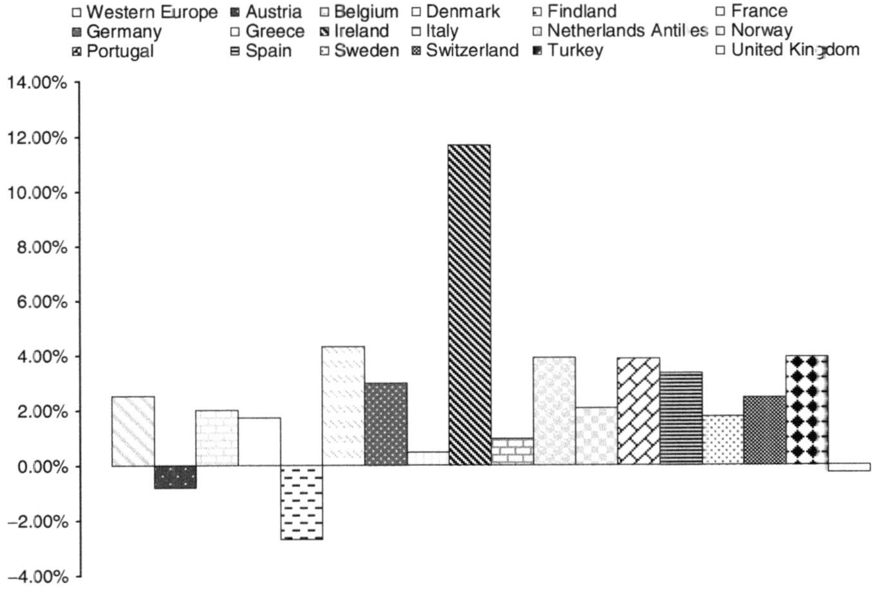

Fig. 16 Communications share average annual growth, Western Europe 2000–2005 (From Euromonitor International 2005 and authors' calculations)

Table 2 Regression coefficients (From Authors' calculations)

Variable	Sign	p-value
GDP per capita	+	0.02
Inflation rate	–	0.08
Broadband subscription	+	0.06
Mobile subscription	–	0.01
Globalization rank	+	0.03
PC ownership	+	0.10

Model of Communications Share

Following Parks and Barten (1973) we specify a linear expenditure system to analyze budget shares via the underlying utility function:

$$U = \sum_i \beta_i \log(x_i - \gamma_i) \quad (1)$$

where x is a vector of the commodities consumed. The budget constraint is given by:

$$y_i = p_i x_i = p_i \gamma_i + \beta_i (y - \Sigma_j \, p_j \gamma_j) \quad (2)$$

where β_i is the marginal propensity to consume the ith commodity and g_i is the threshold level for commodity i.

The corresponding share equations are:

$$w_i = p_i \theta_i + \beta_i (1 - \Sigma_j \, p_j \beta_j) \quad (3)$$

where w_i is the share of the ith good and θ_i is the threshold parameter.

The results from model estimation for communications share are:

The results in Table 2 indicate the communication share is larger for countries with a lower globalization rank. This finding is consistent with the inference that countries that are less open have state-run telecommunication monopolies and higher prices. Countries that embrace openness and competition tend to see communication prices fall. Further, broadband and PC ownership are positively related to the communications share; mobile subscription is negatively related to the share (Table 2).

Conclusion

We provide a cross-country analysis of the communication share of spending. We note that there are substantial differences in shares cross country. However, for most of the world, these shares are relatively stable through time. The analysis suggests that the communications share of developed countries provides a benchmark with which to predict the steady state level the share trends toward.

References

A.T. Kearney/Foreign Policy (Carnegie Endowment for International Peace) (2006) The Globalization Index. http://www.atkearney.com/shared_res/pdf/Globalization-Index_FP_Nov-Dec-06_S.pdf. Cited 29 July 2007.

Alleman J, Rappoport P, Taylor L, Mueller M, Greene P, Gerarity C (1992) Southern Africa Telecommunications/Economics Scoping Study, Task II, U.S. Agency for International Development Contract, October.

Bureau of Labor Statistics (BLS) (2007) Consumer Expenditure Survey. http://www.bls.gov/cex/home.htm.

Chinn M D, Fairlie R (2007) The determinants of the global digital divide: A cross-country analysis of computer and internet penetration. Oxford Economic Papers 59: 16–44.

Euromonitor International (2005) World Consumer Income and Expenditure Patterns. London: Euromonitor International.

Federal Communications Commission (FCC) (2005) International Traffic Data. http://hraunfoss.fcc.gov/edocs_public/attachmatch/DOC-272545A1.pdf. Cited 28 July 2007.

Kiski S, Pohjola M (2002) Cross-country diffusion of the Internet. Information Economics and Policy 14: 297–310.

Garcia-Murillo M (2005) International broadband deployment: The impact of unbundling. Communication & Strategies 57: 83–108.

Parks R, Barten A (1973) A cross-country comparison of the effects of prices, income and population composition on consumption patterns. Economic Journal 83: 834–852.

Röller L, Waverman L (2001) Telecommunications infrastructure and economic development: A simultaneous approach. American Economic Review 91: 909–923.

Russian Information and Communication Technology in a Global Context

Svetlana Petukhova and Margarita Strepetova

Abstract At present, the growth rate of Russia's high-tech sector outstrips significantly the average world indicators: in 2005, for example, information and communication technologies grew by 24.5%; however, the share of Russia's information sector in total production amounts to about 2% of GDP. The ICT sector in Russia is now in an early phase of development, it can learn from pioneer countries and their historical parallels, notably, countries of the Asian and Pacific region, which have managed, over a short period of time, to make great strides in this area. The rising impact of the Internet on the population's quality of life can be observed in areas such as health care, education, culture, everyday life, and work. In Russia the ICT sector shows three times higher growth rates than those of the national economy as a whole. Legislation on ICT has laid the foundation for market liberalization.

Introduction

The modern world faces a virtual environment of human interaction, characterized by the decreasing importance of geographical borders, undisclosed nationality of the participants and the possibility to anonymously access resources. It is based on the global computer network, the Internet, which already unites billions of people. This number is constantly growing and most likely in the next 5–7 years the majority of the world's population will have reliable access to computer networks.

Today the Internet provides new growth opportunities to both new and established companies and is a priority when it comes to competing globally, it is changing the scale of the global marketplace. Over 80% of top managers of international corporations agree that the Internet has altered the core mechanism of competition in many industries. The current phase where the Internet technology merges with international capital began when the global players started to adapt to the "new economy."

S. Petukhova and M. Strepetova (✉)
Institute of Economics, Russian Academy of Sciences
e-mail: petukhova@transecon.ru
e-mail: strepetova@transecon.ru

B. Preissl et al. (eds.), *Telecommunication Markets*, Contributions to Economics,
DOI: 10.1007/978-3-7908-2082-9_25, © Springer Physica-Verlag HD 2009

A decade-long development has brought the Russian business information market up to global standards, at least in terms of technology. However, a number of significant factors should be considered when talking about the information market in Russia. This includes an undeveloped market culture, low ability to pay of the majority of corporate and individual customers, lack of technological equipment. Small and medium businesses that constitute the majority of the customer base are unable to adequately invest into ICT goods and services under present economic conditions. Information distribution is also a major issue as the lack of transparency among business and banking sectors hampers the disclosure of information.

The factors that attract conventional businesses into the internet space include:

- Internet has become a part of the daily business life. The world cyberspace is a completely new and unexplored sphere of the economic activity. It provides unique opportunities for services both directly and through virtual models.
- Factors of space and time lose their vital importance. Economic activity obtains global nature.
- Transaction costs go down as the customers work directly with the producers without third parties. It is true even for the operations with the transnational structures enabling more companies to get involved into the business operations.
- There is a possibility to customize the products in the real time mode, fulfilling unique personal demands of the customers.
- All companies have similar interaction conditions. This enables small but aggressive companies to enter the market and compete with larger and well known companies.
- Electronic market mass media is a gratifying opportunity for advertising. It facilitates interactive marketing among potential customers.

Internet technology development and its integration into the world economy are responsible for the formation of the global markets. These markets – financial, commodity, and labor – are global not only from the point of view of geography, but also based on the number of the actors involved.

Russia in the Worldwide Information Space

Russia joined the integration process into the worldwide information space with a big delay. However, the importance of the issue was recognized, and a number of significant, crucial documents were adopted. So, in 2004 a "Concept for information technology use by the federal governmental bodies until the year 2010" (ratified with a resolution of the Russian Federal government from September 27, 2004) and a "Concept for regional informatisation until year 2010" has been developed (Russian Federal Government 2004). The first concept aims at building up an effective system for governmental ICT based services. The latter targets the effectiveness of the governance of regional socio-economic development and fosters the computer literacy of the population (Porokhovsky 2003).

As the Federal Target program "Electronic Russia (2002–2010)" comes to its end, the government will be one of the main consumers of ICT goods and services.

Some 2.6 billion US dollars are planned in the budget aiming to build up a modern ICT infrastructure within the public sector (Russian Federal Government 2002). It is considered to set up a Russian Investment Fund for ICT with some nominal capital of 52 million US dollars (Russian Federal Government 2004). This Fund will provide financing for innovation projects, as well as small enterprises in the field of ICT. Russian ICT Venture Fund belongs to another important supporting organization. The total value of the Fund will amount to 100 million US dollars with a public contribution of 75% and the rest provided by private investors (Russian Federal Government 2002). This Fund is already in operation, and it has resulted in the manufacturing high-tech products that are able to compete in both domestic and global markets.

As a result of the successful implementation of these programs, the sales volume of the ICT sector in Russia could reach some 40 billion US dollars by the year 2010 (more than five times as much as in 2003) (Russian Federal Government 2002, 2004). This will enable Russia to join the established world leaders in the field of ICT. It will also make it possible to raise employment in the sector up to 3.5 million people. This will amount to 5% of the working population and marks an increase of about 350% in comparison with 2004 (Russian Federal Government 2004). Already today, the growth rate of the Russian ICT sector is way above the world average: 24.5% and 20.0% in 2005 and 2006 respectively (Statistika Rossii), while the world market reached approximately 7% annually on average (Germany 3.5%; UK 5.1%; France 6%; China 11.6%; India 22.9%) (Doktorov 2006). Such a growth rate requires significant investment. Over the last years, investment into the Russian ICT sector has not exceeded 2.5% of the national value (Doctorov 2006). Investors are lured by the large companies, though the sector itself would rather benefit from small and medium enterprises.

Internet belongs to the most innovative fields of the ICT. Within 2006, more than 4.5 billion US dollars were placed into the communication sector by the Russian investors (Statistika Rossii). This is 7.1% more than the year before. Foreign investors have contributed with some 4.0 billion US dollars, which is 19.5% more than in 2005 (Statistika Rossii). The total ICT sector value in 2006 has reached according to the estimations 17.8 billion US dollars (Doktorov 2006).

Though the federal government shows that it is willing to achieve a break-through in the field of information technology, some serious difficulties in governmental support mechanisms hinder consistent development. The establishment of effective mechanisms for successful cooperation between the state and the business sectors is one of the main tasks. The tax advantage, for instance, plays a significant role in fostering the achievement of the established targets. A suggestion of special tax privileges for the ICT sector has raised harsh criticism from the Federal Taxation Authority. As the State Duma proposed the introduction of a unified 6% net profit tax for software exporters, the Federal Taxation Authority only agreed to reduce the single social tax rates and to free this target group from the value added tax. However, even this modest advantage will allow the ICT companies to reduce significantly their total tax payments. This example shows how the proclaimed ambitions of the Russian government are shattered against the reluctance of a particular governmental body. However, this attempt of the Taxation Authority to protect their

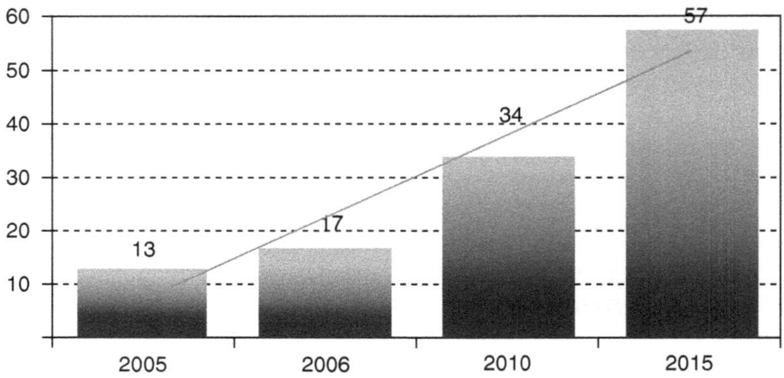

Fig. 1 Russian IT market, billion US dollars (From J'son & Partners. www.J'son&Partners/services/market _research/connect/)

own interests has its reasons. If the ICT sector is subject to tax privileges, this may lead to misuse and will therefore complicate the daily work of the responsible authorities. However, neither is the Federal Ministry for Finances willing to provide budget funds, nor is the Federal Ministry for Economic Development and Trade eager to offer resources from the so called Stabilization Fund in order to support ICT sector development (Fig. 1).

The Role of ICT for the Development of the Russian National Economy

It is a highly complicated task to measure the impact of the Internet on the overall economic development within the system of traditional economic and statistical indicators. It is well known that the economics of telecommunications is a specific field, where many practical activities are not reflected by the actual statistics. Within the sector itself, the intensive innovation processes are continuously progressing. Topology and network organization, intellectual compound of the equipment are changing dramatically. The technology renewal rate goes up (almost twice for each generation); the range of the offered services becomes wider. The effectiveness of ICT in general and of the Internet in particular, and their direct or indirect impact on structural change within the national economy over the last decade can be approached using the following criteria:

(a) The share of ICT in gross domestic product is one of the key indicators reflecting national information technology market development and its impact on the national economy. Among the countries that belong to the established leaders, this indicator exceeds 3%. In Russia it makes up some 1.4% which is typical for economies in transition (ICC Russia 2006). Direct impact of the global network on the main macroeconomic indicators has a tendency to grow. The share of communications in GDP is supposed to reach 10% in 2010 (RBC Daily 2005–2007).

Per capita consumption of ICT products in Russia is many times lower than in industrialised countries. Russia has a significant reserve of ‘postponed’ demand for ICT technology which will secure the overall high consumption rates for ICT products and services. The manufacturing, distribution and services sectors belong to the largest consumers of the Internet services in Russia. In the late 1990s their share accounted for more than 62% of all Internet traffic (ICC 2006). At present, there is a tendency for this share to go down, while the share of personal consumption, consumption by trade, distribution and services sectors grows rapidly. Expenditures on communications technology within the household budget reached in 2006 some 5% with a tendency for further growth. In 2008 this indicator should reach some 5.5% (Doktorov 2006; Russian Federal Government 2004). Within the manufacturing industry the following sectors proved to be the main consumers of ICT products: chemical, printing, forestry, paper, transport engineering, construction and transport.

The use of the Internet influences greatly the dynamics of e-commerce. Some 25% of the Internet users in Moscow are involved in e-commerce operations. The actual national volume of e-commerce is yet relatively small. It amounts at present to about 110–120 million US dollars annually, 70% of which falls on Moscow (National Technological Base 2002–2006). The scope of Internet penetration in trade operations amounts to approximately 10% of all firms though this figure grows rapidly. As the internet tariffs go down (50% by the year 2010), intensification of e-commerce is expected (RBC Daily).

(b) The analysis of intersectoral balances makes it possible to document the dynamics and to discover tendencies of structural change within Internet services consumption as well as ICT penetration in the national economy in general. A clear growth tendency is seen for the demand of private households and of the distribution and services sectors. The most important factors of influence here are technology development in the national economy and in the telecommunications sector, total value and distribution of the GDP, living standards and the size of the population.

The fact of the rising Internet use is reflected clearly by the Russian labour market. The demand for highly qualified personnel is going up. Particularly in the ICT market, the remuneration for qualified staff is growing rapidly which inevitably leads to structural change in the entire national labour market. In the first stage, ICT companies have only offered the endowment of firms with computers. Today, personal computers represent only one element of an information network or a built-in compound of a production line. According to the expert estimations, software services and networks development account for up to 30–60% of the total informatization budget of the large companies (National Technological Base 2002–2006).

(c) The profitability of the ICT sector is still high. Mobile internet services represent the main source of revenue and substantial revenue growth is expected from it in the future. In the first 6 months of 2006, the revenue from mobile

Internet services reached 380 million US dollars, a 73% increase compared with the same time-period of 2005. Its quota in the total revenue from value added services (VAS) accounted to 25% (Statistika Rossii).

In terms of cellular communications penetration, Russia has already reached the level of European countries. According to the national average, there are 84 mobile phones per 100 citizens (Statistika Rossii). In Moscow and St. Petersburg there are 131 and 114 mobile phones respectively per 100 citizens. The subscriber market in both capitals has already reached a certain saturation. The further growth of the penetration rate of cellular communications is based on its spread in the regions, given that at present, cellular networks are operating nationwide, but usage is unevenly distributed.

The number of mobile services subscribers (based on the number of SIM-cards issued) has reached 130 million, providing some 90% penetration rate (Statistika Rossii). A further growth of 7% is forecasted. A market quota of the alternative providers for fixed phone connections grew by 4% reaching 36%. IP-Telephony grew 30% (Statistika Rossii). The market segment for mobile Internet outpaces traditional Internet connections in its growth. This tendency is a result of abrupt raise of GPRS users throughout the country. At the moment, this group of customers is as large as 16 million people (including 8.5 million active users) (National Technological base 2002–2006).

There is an interesting tendency in the fixed phone connections segment. The growth rates here for the long-distance calls are slowing down. However, it is still too early to speak of market saturation as diffusion covers only the upper, richer level of consumers. The market potential for mobile services is not exhausted at all, and the marketplace for the long-distance calls is in its infancy (Fig. 2).

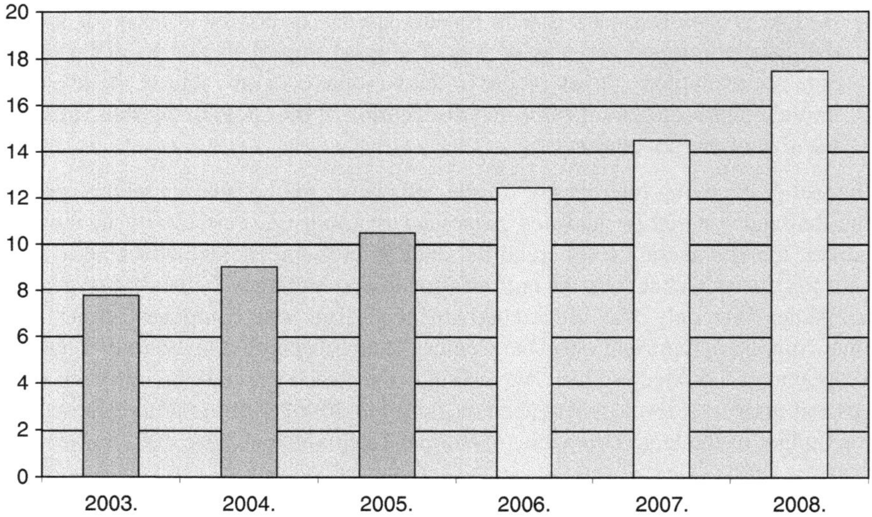

Fig. 2 Dynamics of Russian ICT market, billion US dollars (2006–2008 Forecast) (From ICC 2006)

Consistent policy in the field of national ICT production support plays an important role in accomplishment of the newly established targets. The market for electronic engineering and corresponding components whose value worldwide exceeded 1 trillion US dollars. According to the various expert estimations, the Russian market share accounts for some 0.1–0.3% (Giglavy et al. 2002). This fact puts under the spotlight two main goals for the national economy to reach. First, the internal market for Russian electronic engineering products must be restored as national producers do not have control over it at the moment. Secondly, Russian producers should enter global electronics and information technology markets with competitive products.

The role of the state as a customer is much more significant in Russia than in many other industrial countries: the sales within the public procurement procedures amount to 20% of the total Russian market for information technology. Meanwhile in the USA where the absolute expenditures on ICT are high, the state purchases account for only 7.5% of the national ICT market (ICC 2006). On the one hand, an active role of the national government provides an impulse for national computer engineering, information security means and other ICT production. On the other hand, informatization within the governmental bodies at present is bound to the process of building up information networks infrastructure. This stands for a predominance of hardware and software licences over the services. Until recently, the companies had faced significant difficulties in long-term projects realisation. The projects had to be divided in 1-year modules with tenders taking place every year. In the coming 3–5 years, the government might not only become the largest ICT hard- and software consumer, but also a major customer for ICT services. This will certainly positively influence the structure of the national markets for information technology.

The Role of Techno-Parks

From the strategic point of view, it is crucial for Russia to control the intellectual property rights of software and contents produced in the country. The Internet expansion depends first of all on technical, intellectual and financial resources which the country has at its disposal. The so called techno-parks have peaked in the last 2 years. The program for their launch and development was adopted by the government not so long ago. The term “techno-park” stands for a sort of a “factory” for small and medium enterprises with knowledge intensive business models. Their main purpose is to enable an effective transfer and commercialization of the technology created by the universities and research institutes. Universities and research institutes are to produce innovative technologies, and techno-parks take care of the small and medium enterprises.

The “Programme for techno-parks development until the year 2010” stipulates six locations in European and Asian parts of the country. It is planned to build up two techno-parks in the Moscow Region, in St. Petersburg, Nizhnij Novgorod and Novosibirsk – one techno-park in each city. This Federal Program has started in

2005 and is to become a large national project. The target here is to develop modern enterprises which will be able to create competitive products and therefore take over the leading positions within the global ICT market (Rynok 2005; Russian Federal Government 2004).

The programme draft makes provision for up to 123 billion roubles (4.3 billion US dollars) until the year 2010. Sources for financing include: federal budget (16%); individuals of the federation (12%); non-governmental sources (72%). By 2010 the total investments into Russian techno-parks will exceed 100 billion roubles (3.5 billion US dollars). The total production value of the Russian ICT sector by 2010 could reach 1 trillion roubles annually (35 billion US dollars). The realization of the programme will secure the rapid development of the information technologies within the Russian Federation as well as its growing contribution to the national economy. However, in case financial support fails, the country will remain dependent on imports with regard to ICT products and services. Also, the lucrative export opportunities appearing due to the growth of the global ICT outsourcing market (estimated at 140 billion US dollars in total until 2010) would be missed (National Technological Base 2002–2006).

The Future of the Internet in Russia

Internet penetration and development is limited by a number of factors, classified as follows:

- Internal factors (ICT market is hard to enter, lack of reliable information about the companies and market development, low consolidation and capitalization rate, etc.)
- Demand side factors (foreign hard- and software prevail in the ICT consumption; as a consequence, national ICT products and Internet services production remains low)
- Structural factors (national demographic situation, skill levels of ICT profes sionals)

As for the internal factors, more transparency is needed. Some years ago, mergers among market players have started, the foreign investments are lured and the first IPO of ICT companies were announced.

The structure of Russian ICT market (hard-, software and services) differs significantly from that of other industrial countries. The hardware share still accounts for more than two-thirds of the market, and services make up only 26% (data from 2005). This is a completely different trend in comparison with the EU countries. The majority of the hardware and components are being imported. Services, the key sector for future development are still insignificant in value. As a result of the international division of labour, the Russian ICT market is characterized by a low added value. Its volume in value terms comprises about 1.1% of the global ICT market and its produced added value is less than 0.4% (Doktorov 2006).

Internet regulation is another crucial issue to be solved in the near future. The use of Internet technologies in Russia is regulated by a number of the different laws.

The introduction of a special act is not considered. Concerns about social consequences of the Internet have led to sporadic regulation: communication services, intellectual property management, procurement procedures, advertisement, etc. Providing a legal basis for the public protection against harmful content in the public Internet belongs to the tasks of the government. In this regard, it is crucial to consider carefully the regulation spheres and to make it clear who is responsible for the cyber space regulation and in what way.

The key problem is, in our view, the professional training of the employees for the ICT market. At 4% annually the global ICT market shows the highest employment growth rate. Some 33% of the employees in this sector are working in the USA, 37% and 15% in Europe and Japan respectively (Giglavy et al. 2002). There is only 15% of the total number of ICT specialists that work outside the three regions mentioned above. Russia accounts for a small share of the global GDP in the field of ICT technology. Its share in the global ICT workforce market is, therefore, insignificant. This ratio mirrors the level of economic development among the countries, their technological progress, it hints at the degree of social well-being and sustainability of growth and their investment and social perspectives.

However, in Russia nationwide there are at least 200 higher education institutions, which is one-fifth of all education institutions (2). (Statistika Rossii 2007). Besides, similar to the other economic fields, there are employees that have degrees in other disciplines although employed in the ICT sector. This shows the particular demand for professionals with a degree in information technology. On the other hand, not all graduates of ICT departments work according to their specialization. In Moscow, roughly 70% of ICT graduates are employed in the ICT sector. In the regions, this figure is even lower and amounts to about 30% (Economica i vremya 2006–2007). The potential of high-skilled workers is not fully used. The weakness of the remuneration system for the professionals working in the knowledge based sectors is the main reason for this tendency.

The quick development of ICT in general and the Internet in particular reveals the crucial role of further professional training. The global ICT players such as Microsoft, IBM, Oracle, HP, Intel, Cisco Systems and others provide a platform for the training centers which are seen as the most effective tool in the moment. These centers offer the practical experience to the students assigning them real-life tasks introducing them to an innovation based economy. It is also a good opportunity for students to find a job according to their professional background in the future.

The accumulated potential value of Russian patents is estimated to be several trillion US dollars. However, these refer to already existing inventions and know-how which has never been implemented due to a lack of favorable conditions for intellectual property generation and use. The share of the non-material assets in the Russian industry is the lowest in Europe (along with the other CIS countries). The ICT sector development secures the rehabilitation of the intellectual potential. It is also an effective tool against the brain-drain, in particular within science and knowledge based sectors. Techno-parks and special economic areas are to secure fair working conditions for the highly educated specialists according to their professional potential.

Conclusions and Actual Tasks

Conclusions:

- Information and communication technology belongs to the most dynamic sectors of the Russian economy.
- The national government has achieved significant progress in supporting the ICT sector. Private–public partnerships are seen as a perspective model in which a particular governmental function is delegated to a private company or a particular national project is realized with private capital.
- The ICT sector is, more than any other, oriented towards innovation. Its development is seen as an important factor and an opportunity for the national economy to switch from being based on natural resources to a high-tech platform for other industries.

Actual tasks:

- The development of ICT needs governmental support, mainly through laws, tax policy, and the protection of intellectual property. More governmental support for the ICT development is needed. The average annual revenues for the venture funds world-wide amount to 20–30%. From the economic point of view, this is much more lucrative than investment returns offered by banks. From the economic point of view, it is more economically favourable for the government to invest in ICT rather than in other more traditional industries.
- More attention should be paid to the ICT professional education. By financing the ICT sector, the government supports personnel and intellectual potential in this field as well as the national science in general. To a large extent, governmental support helps solving the problem of brain-drain. At the moment, up to 30% of Microsoft software products are produced by Russian-speaking specialists (National Technological base 2002–2006).
- As far as the introduction and development of ICT and the Internet throughout the country is concerned, it is necessary to take into account a number of specific barriers. General difficulties typical for the economies in transition as well as the heterogeneity of the Russian population spread (large cities, regions and periphery) need to be considered. Those problems can put in danger the spread of the informational space throughout the country and its inclusion into the world economic space. Fundamental decisions on the organizational structural reform of companies, public organizations and governmental bodies are still to be taken.
- The experts report a paradoxical phenomenon: peak prices for oil and gas cease the ICT services development. The higher the revenues from raw materials are, the lower is the demand for ICT services from the side of the governmental sector. In the few preceding years, the state has recognized the necessity of the support for high-tech and education within the ICT sector. Finding an effective solution for this task has an absolute priority.

References

Doktorov B (2006) The Internet – New Russian Miracle. "Socio", Moscow.

Giglavy, A.V., Y.M. Gornostaev, V.I. Drozhzhinov i dr (2002). Sovershenstvovanie gosudarstvennogo upravleniya na osnove ego reorganizatsii i informatizatsii. Mirovoy opyt. Pod red. Drozhzhinov, V.I. Tsentr kompetentsii po elektronnomu pravitelstvu pri Amer. Torg. Palate v Rossii. M., Eko-Trends, 2002 (Giglavy A.V., Gornostaev Y.M., Drozhzhinov V.I. et al. (2002). Reorganization- and Informatization-Based Improvement of State Administration. World Experience. Ed. Drozhzhinov, V.I. E-Government Competence Center, American Chamber of Commerce in Russia. Moscow, Eko-Trends).

ICC (International Chamber of Commerce) (2006).

ICC Russia (2005) www.iccwbo.ru/about/about_icc/

National Technological Base, Federal Target Program for the Years 2002–2006.

Porokhovsky, A.A. Ekonomicheskaya teoriya v sovremennoy Rossii. Globalnye tendentsii i natsionalnye traditsii. M. 2003 (Porokhovsky, A.A. (2003). Economic Theory in Modern Russia. Global Trends and National Traditions, Moscow). pp. 57–64.

Russian Federal Government (2002) E-Russia (2002–2010) Federal Target Program (ratified with a resolution of the Russian Federal government No. 65, January 28th).

Russian Federal Government (2004) Concept for information technology use by the federal governmental bodies until the year 2010 (ratified with a resolution of the Russian Federal government No. 1244-p, September 27th).

Rynok (2005) IKT na poroge bolshikh peremen (Russia: ICT Market on the Verge of Big Changes). (April 8, 2005)

Statistika Rossii (Russia's Statistical Office) (2007) ICT Database Information and Publishing Center, 2007, pp. 282–293.

www.J'on@Partners/services/market_research/connect/

Part VII
New Perspectives on the Regulatory Framework

The Regulatory Framework for European Telecommunications Markets Between Subsidiarity and Centralization*

Justus Haucap

Abstract This paper discusses the advantages and disadvantages of centralizing regulatory competences in the European telecommunications sector. As is demonstrated, political economy suggests that an over- rather than an underregulation of telecommunications markets has to be expected. This tendency has been strengthened by the allocation of competences under the current regulatory framework which endows the European Commission with far reaching veto rights under the so-called article 7 procedure. In order to delimit the risk of overregulation through institutional safeguards, it should be easier for regulators to deregulate than to regulate a market. Current suggestions and ambitions by the Commission to extend its veto right or to establish a European regulator for telecommunications should be dismissed. Instead, we suggest limiting the Commission's veto rights to (a) markets, in which regulation creates significant cross-border externalities, and (b) cases where national regulators do not move into the direction of deregulation. If, however, a national regulator decides to deregulate a market, the Commission's ex ante veto right should be abandoned (but ex post intervention by the Commission still be possible) in order not to hamper deregulation which tends to be the desired result, in principle. In addition, we argue that the "insufficiency of competition law"-criterion of the three-criteria-test needs to be taken more seriously than has been so far by national regulators.

J. Haucap
University of Erlangen-Nuremberg, Department of Economics, Nuremberg, Germany
e-mail: justus.haucap@wiso.uni-erlangen.de

* This paper builds on previous work by Haucap and Kühling (2006, 2007, 2008), which resulted from a larger project undertaken for the German Ministry of Economics and Technology (see Baake et al. 2007).

B. Preissl et al. (eds.), *Telecommunication Markets*, Contributions to Economics,
DOI: 10.1007/978-3-7908-2082-9_26, © Springer Physica-Verlag HD 2009

Introduction

The liberalization of Europe's telecommunications markets over the last 10 years has been a major success story. There are probably few microeconomic reforms that have resulted in similar benefits for consumers over the last decade. Prices have come down, service levels have increased (e.g., through reduced waiting times), innovative services have been introduced, network quality has increased, and the overall level of consumption has grown enormously. Today, telecommunications services are cheaper and better than ever before.[1]

A main driver of these reforms has been the European Commission which facilitated the reforms in the EU Member States with its 1996 reform package. While most Member States (with the particular exemption of the UK) were not able to commit to a liberalization process on their own (even though the overall benefits for consumers were well known from the US experience), the European Commission managed to bind the Member States to reforms through a liberalization package.

After the initial reforms have been successfully introduced in 1998, the framework governing electronic communications markets was revised in 2003 when the current framework was implemented. It consists of five directives,[2] the Radio Spectrum Decision (676/2002/EC) and the so-called Market Recommendation, which contains the markets that are likely to be subject to ex-ante regulation and which have to be analyzed by national regulatory authorities (NRAs). This framework has been implemented by all 27 Member States by now, and it has governed competition in telecommunications markets for the last 5 years.

The 2003 reforms have shifted competences away from the Member States to the European Commission. Most importantly, the Commission now has a veto right concerning the NRA's market definition and market analysis, according to Article 7(4) of the Framework Directive.

Following the so-called 2006 review of the framework, the European Commission has tabled new proposals on how to further improve the regulatory framework in November 2007. The main elements of these proposals were (a) to establish a European Electronic Communications Markets Authority (EECMA), (b) to extend the Commission's veto rights also to the choice of remedies imposed by NRAs on regulated entities, (c) to either harmonize parts of the radio spectrum management or to shift more responsibilities for radio spectrum to the EECMA to be founded, (d) to further harmonize universal service requirements, user rights, data protection guidelines and security aspects, and (e) to introduce functional separation as a new remedy available to NRAs.

---

[1] For a summary of the reforms and achievements in Germany see Dewenter and Haucap (2004) and Haucap and Heimeshoff (2009).

[2] These five directives are the Framework Directive (2002/21/EC), the Access and Interconnection Directive (2002/19/EC), the Authorisation Directive (2002/20/EC), the Universal Service Directive (2002/22/EC) and the Privacy and Data Protection Directive (2002/58/EC).

Before the European Commission finally came out to propose the establishment of EECMA in November 2007, several other, more or less similar proposals were floated, such as the establishment of a "European FCC" or an "Enhanced European Regulators Group (ERG)". What all of these proposals had in common is the wish to further harmonize and centralize the regulatory framework across Europe. As mentioned above, the European Commission has also proposed, as a further alternative, that its veto rights in the regulatory process would be expanded. Under the current framework the Commission has veto rights only in the first two steps of the three-step regulatory procedure, i.e. the Commission can veto the market definition and the market analysis as conducted by NRA, but the Commission has no veto rights on the remedies imposed on the undertaking to be regulated by the NRA.

It may be interesting to note that the proposals of November 2007 already represent the second attempt of the European Commission to establish a central regulator at the European level. The first rather cautious attempt was undertaken in the context of the 1999 review when a report on the costs and benefits of the establishment of a European Regulatory Authority for Telecommunications was ordered. That study, not conducted by London Economics at that time, but by Cullen International and Eurostrategies (1999), concluded that the establishment of a European regulator was not warranted, as its costs would clearly outweigh the benefits. Maybe for the reason that the 1999 study has come out with a rather clear (and apparently undesired) picture on the costs of a European regulator exceeding its benefits, this time the European Commission has come forward with its proposals without commissioning another cost–benefit analysis.

Even though some attempts have been made to save the Commission's EECMA proposal at least in a modified fashion (such as the proposal by MEP Pilar de Castillo to establish a Body of European Regulators in Telecommunications, BERT, which has been adopted by the European Parliament's Industry Committee on 7 July 2008), it is unclear whether it is going to be implemented in the context of the current reforms. In fact, the European Parliament's Industry Committee has already watered down the Commission's proposals to further harmonize and/or centralize regulatory decisions on 7 July 2008, after the proposals had faced significant opposition from many Member States, the European Regulators Group and many industry sources. And even the EP's proposal to set up BERT as a mandatory body to be consulted by NRAs before decisions can be made may be stopped by the Council of Ministers. From a political perspective, the most drastic harmonization proposals may be shelved for the time being. However, given that the proposals have also resurfaced after the 1999 review, it is not unlikely that they will resurface again in a number of years. In fact, the EP's Industry Committee already announced that "BERT's mandate should be reviewed by 1 January 2014 to ascertain whether it needs to be extended".

Hence, we consider it still a worthwhile endeavor to analyze the pros and cons of further harmonization. For this purpose, we adopt a comparative institutional perspective to examine the benefits and costs of the various institutional arrangements proposed. First, this paper will analyze the costs and benefits of centralization of executive competences on a general level in section "The Benefits and Costs of Centralization" of this paper. In a second step, we will summarize some important particularities of telecommunications markets in section "Important Particularities of Telecommunications Markets", before we elaborate on the political economy of (de-)regulation in section "On the Political Economy of (De-)regulation". Building on this basis, we will examine and compare the details of the various policy proposals in section "A Brief Evaluation of Current Policy Proposals". Section "An Alternative Institutional Proposal" develops an alternative proposal for how to allocate veto rights between a central institution (such as the European Commission itself) and the NRAs. Section "Conclusions" draws conclusions.

The Benefits and Costs of Centralization

Benefits of Centralization

The economic theory of fiscal and legal federalism has identified a number of benefits and costs that the centralization of executive competences can bring along. The benefits largely result from:

1. The internalization of potential cross-border externalities of some particular policy and/or regulation
2. Economies of scale and transaction cost savings associated with centralization
3. Commitment values for policy makers at the lower (less central) level
4. Avoidance of a potential regulatory "race to the bottom"

The internalization of potential cross-border externalities of some particular policy and/or regulation refers to a situation where a policy or regulation in one Member State has positive or negative side effects on citizens, consumers and/or firms in at least one other Member State. The most illustrative example is probably climate change where emissions and climate policy in one country clearly affect the well-being of people in other countries. Similarly, in the area of competition policy a merger between two undertakings in one country may affect customers and competitors not only in the two firms' home country, but also in other countries if the relevant market is larger than one country. Therefore, it is sensible to shift responsibilities for competition policy from the national to the EU level, as far as the Common Market is concerned.

With respect to telecommunications markets, international roaming is a prime example for cross-border externalities of regulation. If Member State A caps prices for international roaming in its own territory the beneficiaries are foreign consumers, while domestic mobile operators suffer as do domestic consumers since – in response to a decrease in roaming revenues – other mobile telecommunications prices may increase or handset subsidies decrease due to what has been called the waterbed effect (see, e.g., Littlechild 2006). If, however, foreign consumers benefit while domestic parties suffer, no NRA is likely to be overly keen on the regulation of international roaming fees within its own country. In fact, of the 18 markets contained in the 2003 market recommendation, the market for international roaming was the only one that has not even been analyzed by one single NRA before the European Commission took over regulation of international roaming charges in 2007. Clearly, international roaming is a clear issue to be dealt with by the European Commission or some other central agency, as the cross-border externalities are obvious.

Economies of scale may result if one agency or bureaucracy can achieve certain objectives or perform some functions at a lower cost than if this task was executed by two or more agencies or bureaucracies. This may be the case if the duplication of certain tasks can be more easily avoided in a central agency, but not with local authorities. In that case, allocating competences to the central level can save resources and thereby reduce overall cost. Similarly, centralization may save transaction costs not only for the administration or agency itself, but also for the firms or citizens dealing with the relevant agency or bureaucracy (i.e., the advantage of "one-stop-shopping").

It is at best doubtful, however, whether such economies of scale and substantial transaction cost savings are generated through the proposed harmonization of telecommunications market regulation across EU Member States. As almost all telecommunications markets are, at least currently, national or even regional markets, the market analyses still have to be performed at the national level. That means, there are still 27 different market analyses and hardly any economies of scale in this respect. The current notification procedures under article 7 of the Framework Directive even increase the cost of regulation compared to a more localized regulatory system.

One also should keep in mind that the differences between the 27 national markets will in any case still lead to differences in the regulatory outcome. This in turn implies that any transaction cost savings for concerned multinational businesses will be rather small. While the European Commission argues that the costs of doing business will be reduced if the methodologies applied by different NRAs are more harmonized (or less "inconsistent" as the Commission says), it is just not plausible that the costs of doing business decrease, for example, if all NRAs used analytical cost models to determine mobile termination charges. In fact, the cost of doing business may be lower if some NRAs use much more simple methodologies.

Furthermore, one has to keep in mind that market structures and the availability of alternative infrastructures vary in parts heavily between Member States so that, even or, to be more precise, especially when the same methodologies are applied, the regulatory outcomes will differ between these Member States. This in turn implies that the cost of doing business will not necessarily be reduced for multinational telecommunications firms, as the remedies imposed on some providers in some countries (such as local loop unbundling or bitstream access) may not be imposed in other countries (because the market structure is different) which in turn means that business models feasible in the first group of countries will not be feasible in the other countries. Hence, the harmonization of regulatory methodologies will not lead to any reduction in the cost of doing business for multinational telecommunications operators, unless there is also a harmonization of regulatory outcomes. As we will argue below, the latter would be clearly inefficient, however, due to the significant differences in both consumer habits and attitudes and in existing infrastructures and market situations.

The third potential benefit of centralization lies in the commitment value that centralization may offer. At the local, regional or even national level, decision makers may find it difficult to commit to a policy measure such as the liberalization of a particular industry, as the public pressure from incumbent interest groups (namely firms and unions) may be too strong to resist for political decision makers and regulators as well. In such a situation it may be rational to voluntarily delegate decision rights to a higher level that is not so exposed to lobbying and public pressure. The European Commission, for example, is less likely to give in to public pressure from particular national interest groups, as the Commission's re-election chances do not really depend on one particular national interest group. This starkly contrasts with national or local politicians who may easier give in to such pressures for re-election concerns even if they know that a liberalization of the particular interest would be welfare enhancing. Hence, the European Commission is usually in a better position not only to initiate, but also to stick to welfare enhancing reforms such as the liberalization of network industries and other sectors.

Finally, it is sometimes argued that a lack of centralization or harmonization may lead to a race to the bottom between jurisdictions that compete for industries that bring jobs and tax revenues. The idea is that some standards, e.g. environmental regulations, may be inefficiently soft or low, as competing jurisdictions try to lure businesses to their location. In such a case, minimum standards or centralized decision making can prevent such a wasteful race to the bottom. Similarly, competing jurisdictions may be forced to engage in wasteful subsidy races to attract businesses. Again, some centralization or centralized control, such as the European prohibition of State aid, may be well justified and welfare enhancing. As we will see below, this argument is only a weak one when it comes to telecommunications markets though.

Before we discuss in greater detail, however, how these benefits of centralization apply to telecommunications markets in Europe, let us now first have a look at the other side of the medal. What are the costs associated with further centralization?

The Costs of Centralization

For an economist, there are two sides to everything. No benefit comes without a cost or, as Milton Friedman has famously put it, "there is no such thing as a free lunch". Indeed, there are five main disadvantages associated with a further centralization of decision rights. These disadvantages are due to:

1. Differences in consumer preferences and habits across jurisdictions ("one size does not ft all")
2. Differences in natural and historical endowments (such as infrastructure deployment)
3. Lower information costs at a local level
4. Different agency costs at local and central levels of decision making and
5. Differences in administrative costs

Differences in consumer preferences and habits imply that different regulations and policies may be appropriate for seemingly similar problems. For example, if consumer habits vary between countries, some countries may find it beneficial to establish certain user rights and consumer protection mechanisms while these regulations may be at best superfluous and at worst detrimental to consumer welfare in other countries with different patterns of consumption and user habits. If individual preferences differ strongly between jurisdictions, then a "one size fits all"-approach is most likely to be inefficient. Similarly, the attitude towards universal services may differ between Member States as the preferences with respect to universal services may well be different. While the constituency in some Member States may want to have some particular service available for everybody and may be willing to bear the cost of that, other Member States may not want to include this particular service as part of a universal service obligation. Hence, further harmonization with respect to user rights and universal service obligations does not appear to be justified, given that user preferences, habits and attitudes may be rather different in different Member States.

Similarly, while certain regulations (such as local-loop unbundling) may be beneficial in some countries without competing infrastructure, such a regulation may be rather harmful in other countries with infrastructure competition. "Horses for courses" is an English idiom that nicely catches what is meant here: What is suitable in one situation may not be suitable in another, rather different situation. Centralized decisions and harmonized solutions are inefficient in such a case.

Thirdly, in many situations institutions that are located closer to the problem at hand (such as the regulated industry) usually have lower costs of gathering the relevant information. The information costs are therefore usually lower if the responsible decision maker (agency) is located at the same level as the concerned market or problem.

Fourthly, agency costs are generally the higher the more distant the agents are from their principals. From the perspective of the new institutional economics, the regulator can be viewed as an agent of both the regulated industry and the concerned customers (Goldberg 1976; Crocker and Masten 1996). The regulator's task is to protect consumers against abuse of market power on the one hand and to protect investors against opportunism by the constituency on the other hand, given that investors usually have to undertake specific investments which result in sunk costs. According to this view, the regulator can be seen as a (independent) mediator which balances consumers' and investors' interests. The more distant a regulatory body now is from both consumers and regulated firms, the more difficult it is for the two parties to monitor the agent's activities. Hence, the risk increases that regulatory decisions will not serve the two parties' interest, but the regulator's self-interest, which results in increasing agency costs.

Finally, there may be additional administrative costs associated with centralization. The European Commission, for example, envisaged the new entity, EECMA; to consist of 120 additional staff.

Hence, as always in life, there are costs and benefits associated with centralization and decentralization as well. The costs of centralization can be phrased as the benefits of decentralization and vice versa. To determine the optimal balance in the allocation of competences between the European and the national level, the details of the concerned industry or policy area have to be analyzed, given the criteria outlined above. Such an analysis will be provided in the next section.

Important Particularities of Telecommunications Markets

Telecommunications markets have some important particularities which have to be kept in mind when analyzing the optimal balance of power between central and local levels of political or regulatory decision making.

First of all, almost all of the telecommunications services that may require regulation (due to their natural monopoly characteristics and the sunk costs involved with the underlying network infrastructure, see Baumol et al. 1982; Kruse 2002) are based on *local* infrastructure. This also implies that the regulated services in question are not tradable across borders, which in turn means that the risk of a regulatory race to the bottom is *comparatively* low in telecommunications. While it may be conceivable in theory that one NRA regulates its incumbent softer than other NRAs regulate "their" incumbents, it is still unlikely that this is (a) for strategic reasons and (b) would lead to more than insignificant cross-border effects. First of all, one has to understand that in telecommunications weak regulation at home does not provide any direct advantage in foreign markets, as the relevant services are not tradable. As the European Regulators Group (ERG) has correctly pointed out, "(…) electronic communications services provide an interesting counterexample in that the types of services which are typically heavily regulated (access and access-related products) can also only be sold in a single geographic location. It is not possible, for instance, to sell local loop unbundling across borders" (ERG 2006, p. 9). With tradable products, the home firm may have an advantage if it can lower its production costs (e.g., through state aid) and thereby increase its competitiveness vis-à-vis foreign rivals. This does not work with non-tradable services, however. Any advantage conferred upon a firm, can at best be indirect, e.g., through better access to financing or lower cost of capital. In this case, a much more direct way to improve a firm's finances would be to lower its taxes, change the deprecation or other accounting rules, reduce or take over part of the labor costs, etc. Harmonizing regulation while allowing tax and accounting rules, labor laws and other regulations to completely differ will not achieve any level playing field in any case, even if the risk of strategic regulatory favoritism were high (which it is not, of course).

In any case, a much better way to prevent strategic regulatory favoritism or to reduce its risk would be to strengthen the NRAs' political independence. This chance has been missed by the European Commission, however, as no proposals into this direction (strengthening regulatory independence) were made.

A second important particularity of telecommunications markets is the rapid technological change that characterizes this industry. Various media markets are converging from previously separated markets into a single one. There are signs of increased substitution between fixed and mobile telecommunications (see, e.g., Heimeshoff 2008), and the same holds for the relationship between cable TV and fixed-line telephony. The traditional monopoly of fixed-line telephony is coming under pressure from two sides: Mobile telecommunications on the one side and cable TV on the other. Fixed line telephony itself is rapidly changing as well with new broadband networks such as VDSL being deployed and VoIP gaining a growing foothold in the market. In such a highly dynamic industry one should honestly admit that it cannot be clear ex ante which particular form of regulation works best. The conflict between fostering service-based or infrastructure-based competition is only one example. Any economist or bureaucrat claiming to know precisely what regulation is really welfare maximizing in the long-run should be given a free copy of the collected works of Hayek, as ignoring the limits of our knowledge is a form of hubris that is dangerous for a free and open society.

Harmonizing regulation in a highly dynamic environment is akin to putting all eggs into one basket. As it is highly uncertain what regulation is welfare maximizing in the long-run, some regulatory diversity is highly beneficial in order to allow for learning processes. As the ERG (2006, p. 11) has put it, "(...) in fast moving and innovative environments, some diversity of approach can be positive, as it allows Member States to experiment or learn from each other." Instead of fragmentation and inconsistencies, the European Commission may therefore rather speak of pluralism and diversity in this context.

There are, however, two arguments that support some centralization of decision rights at the European level. Firstly, as already outlined above, there are clearly cross-border externalities associated with the regulation of international roaming charges. Hence, in the particular case of international roaming, centralization of regulatory competences appears to be warranted. The same holds true for other pan-European services as far as their regulation exhibits similar cross-border externalities. Very few pan-European services have been identified until now, however, the almost only exception being mobile telecommunications in airplanes.

Some coordination of radio spectrum allocation may also be warranted on the grounds of cross-border externalities, as their can be indirect externalities of the following kind: The development of new technologies and services may only be profitable if the potential market to be served is sufficiently large. Hence, if only one Member State or a limited number of countries makes spectrum available for certain technologies or services, this may not suffice to provide the necessary innovation incentives. A mutual interdependence between Member States' allocation decisions can exist so that the designation of frequency bands needs to be coordinated in order to achieve a critical mass to provide sufficient incentives for R&D activities related to a particular use of spectrum.

However, in contrast to international roaming where foreign consumers benefit (or suffer) while domestic firms and consumers suffer (or benefit), a similar conflict of interest between foreign and domestic interest groups does not exist with respect to spectrum allocation, which is first of all a coordination problem but not one of rent shifting between countries. Hence, some coordination mechanism should be sufficient to achieve efficient solutions with respect to spectrum allocation.

Apart from cross-border externalities, the commitment argument outlined above has certainly some merit as a second argument in favor of some degree of centralization. Especially at the beginning of a liberalization process the political resistance from both unions and the concerned industry is usually high. This argument lends less support to further centralization today, however, as the liberalization of telecommunications markets was initiated more than ten years ago. Even in the relatively new Member States, market liberalization processes have become irreversible by now. In fact, in most Member States the structure of the market for political influence or the equilibrium between different interest groups has shifted. While at the beginning of the liberalization process very few new entrants have existed at all and those that did exist were not very well organized to lobby the Government (through associations etc.), this situation has fundamentally changed now with new entrants being also well organized in the political arena. Therefore, it would have made sense to safeguard the liberalization process through some higher degree of centralization for, say, the first five years of the liberalization process and then decentralize executive competences from there on. Starting with a rather decentralized allocation of competences and then moving to a more centralized setting, however, is exactly the wrong path.

In summary, the arguments in favor of an increased degree of centralization are only applicable to a very limited extent. The issue of international roaming, which has been resolved by now – even though in a rather unsatisfactory, draconic fashion, lends support to some degree of centralization, which has been achieved already, however. The commitment argument, in contrast, would have lent support to more centralization in the early days of telecommunications liberalization, but not today where market liberalization is almost irreversible.

Overall, it testifies to an unflappable confidence to neglect the bulk of expert advice and demand further harmonization and centralization without having a single serious argument, as Commissioner Reding has done so constantly (see, e.g., Reding 2006, 2007).

As we will argue, however, there is an enhanced role for the European Commission in rolling back overregulation and achieving a better regulation for Europe, which is, of course, one of the prime objectives of the European Union. To fully derive this argument, let us first outline some of the political economy of (de-)regulating telecommunications markets.

On the Political Economy of (De-)regulation

Ideally regulators should only pursue those objectives that have been assigned to them by legislation. That reality conforms to this ideal is a pious hope, however. As Viscusi et al. (2000, p. 44) write in their classical textbook on economic regulation, "in theory, regulatory agencies serve to maximize the national interest subject to their legislative mandates (...) Such a characterization of regulatory objectives is, unfortunately, excessively naïve. There are a number of diverse factors that influence policy decisions, many of which have very little to do with these formal statements of purpose."

Hence, the positive theory of regulation assumes that regulators also pursue their own objectives, which consist in maximizing their own influence or power. Put differently, it is more attractive to direct or to work for a large and/or influential agency or authority than for a smaller and less important institution. Therefore, bureaucracies tend to grow and to expand their tasks and competences. People want to be important, and they are (or feel) the more important the more issues they are allowed to decide. While the desire to enhance one's competences also exists in private organizations, the inefficient growth of bureaucracies is a particular problem of the public sector. In contrast to private businesses, there is usually both a lack of product market competition and a lack of capital market pressure for public institutions while product market competition and capital market pressures discipline private firms in their behavior not to use their resources inefficiently.

Regarding the efficient regulation of telecommunications markets a particular problem lies in the incompatibility between the long-run policy objective to reduce regulation and to foster the development of effective competition on the one side and the desire of bureaucracies to grow (or to remain important, at least). This means that, in principle, market forces should be left to themselves in the long run and Government intervention be rolled back. This objective, however, requires a reduction of bureaucracy and Government intervention – a development which runs against the objectives of the bureaucrats employed in regulatory agencies. As deregulation necessarily implies a loss of importance for the regulatory agency, it is not likely that regulators will foster deregulation and certainly not deregulate too early. Instead, it is most likely that the deregulation process will move to slowly. This problem can be traced back to the inherent information asymmetry that results from the specialized expert knowledge that regulators necessarily have in order to regulate telecommunications markets. As a result, regulatory activities, which are in fact detrimental for overall welfare, can nevertheless be portrayed as welfare enhancing to the public, as politicians and consumers are unlikely to know better. Hence, one important task is to contain the bureaucracy's desire to expand its business by implementing adequate institutional barriers to overregulation.

Currently, the so-called three-criteria-test is used as the standard to decide whether markets need to be regulated or can be deregulated. The three-criteria-test consists of three questions, namely:

1. Are there non-temporary legal or structural barriers to entry?
2. Is there a long-term tendency towards effective competition?
3. Is competition law insufficient to address the competition concerns?

While the first two criteria are usually rigorously tested, the test of the third criterion often appears to be rather cursory (also see Monopolkommission 2008). The Bundesnetzagentur as the German regulator, for example, mostly uses more or less the same wording to justify why competition law is not sufficient to deal with competition problems in a particular market. A typical justification is the following

"The use of competition law alone would only allow for selective interventions. More detailed competencies are required to positively regulate matters. Furthermore, telecommunications law allows faster interventions [than competition law], as the regulator's decisions have to be executed immediately" [own translation, J.H.].

This reasoning is fairly general, however, and always applicable, as it is not specific to any particular market. A more market-specific justification should explain why fast interventions are necessary in a particular market (e.g., because the risk that anti-competitive behavior causes irreversible damages is especially high) or why other advantages of regulation are especially important for that particular market. A general statement about the differences between regulation and competition law cannot suffice, however, to prove the insufficiency of competition law to deal with a particular market. Without a detailed discussion of a market's particularities, competition law would either be always sufficient or never sufficient to address competition concern.[3] Hence, the tendency to almost always declare ex ante regulation as superior to competition law strengthens the tendency to rather deregulate too late than too early. Overall, the three-criteria-test does not provide a sufficient barrier to overcome the inherent problem of overregulation in telecommunications markets.

In telecommunications markets, overregulation is especially problematic, however, as the welfare losses from over- and underregulation are not distributed symmetrically. Quite generally, if an innovation or investment is not undertaken as a consequence of overregulation in dynamic industries, the entire market rent that could have been realized otherwise is lost. If, in contrast, prices are too high due to underregulation, "only" an allocative efficiency loss results which is confined to the Harberger triangle (also see Baake et al. 2007). Hence, the costs associated with the two different types of errors (type-I and type-II-error or false positives and false negatives) are very different. As a consequence, more emphasis should be placed on preventing overregulation. In case of doubt, it is better to deregulate too early than to regulate too early or too long. The difference in the welfare cost of over- and underregulation should be reflected in the institutional setup of the regulatory system. Put differently, it must be made more difficult for regulators to regulate than to deregulate.

A Brief Evaluation of Current Policy Proposals

The current reform proposals should be evaluated against this background. As mentioned above, as cross-border externalities are, at least for the time being, largely restricted to international roaming and as commitment problems mainly existed in the early phases of the liberalization process, but not in 2010 (i.e., 15 years later), the case for establishing a European regulatory agency is weak.

[3] For a more detailed, comprehensive critique of the application of the 3-criteria-test see Möschel (2007) and Never and Preissl (2008).

In fact, as a result of 1999 review the Commission identified the same "problem" as during the current review, noting that the "inconsistent application of certain provisions of telecommunications legislation is hindering the development of effective competition and the deployment of pan-European services" (European Commission 1999, p. ix). It also noted on the same page, however, that "the Commission is not persuaded that a regulatory body at Community level would currently add sufficient value to justify the likely costs. The Commission therefore does not propose to establish a European Regulatory Authority for communications services at this stage."

The European Commission, therefore, concluded "that the creation of a European Regulatory Authority would not provide sufficient added value to justify the likely costs. In addition, it could lead to duplication of responsibilities, resulting in more rather than less regulation. The issues identified that might be better dealt with at EU level can be addressed through adaptation and improvement of existing structures."

Furthermore, the Commission explicitly acknowledged that "it would be disproportionate to establish a new Community institution to address the limited number of issues that might be better undertaken at Community – rather than at national – level. There would be considerable costs of setting up a new regulatory body at European level, in view of all the associated political, legal, technical, economic and linguistic skills that would be required for it to carry out its task effectively across the Community. These costs do not just relate to the administrative costs related to the Agency itself, but the wider cost to the economy as a whole of adding another layer of administration. The issues on which dissatisfaction have been expressed (for example in interconnection, licensing, competition, consumer protection, frequency management and numbering assignment) do not appear to justify the establishment of a new agency" (European Commission 1999, pp. 9–10).

What has been correct in 1999 is also correct today. The difference is, of course, that for the 2006 review the European Commission has not even bothered to estimate or quantify the cost and benefit of further harmonization. Similarly, the scope of the supposed problem of inconsistency has not been quantified at all. This is rather troublesome given that Kiesewetter (2007, p. 61) concludes his detailed cross-country comparison of remedies applied by NRAs in European telecommunications markets as follows: "In wholesale markets there is an increasingly uniform picture concerning the ex ante remedies imposed upon firms with significant market power. This is especially true for unbundled access lines and broadband access at the wholesale level." In other words, the perceived inconsistency problem hardly exists. Again, any further centralization of the sort that has been proposed is not warranted.

The observation that the European Commission aims at expanding its competencies without any economic need or objective justification is consistent with the recent analysis of Bolkestein and Gerken (2007). Their analysis provides four important insights:

1. The European Union is not successful in preventing protectionist measures by various Member States. The Common Market is only being insufficiently protected by the European Commission.
2. The Common Market is developing into a densely regulated area. The Commission is not working towards deregulation, but towards harmonization ion high levels.

Bolkestein and Gerken (2007) even see a shift of paradigm in the Commission's philosophy away from competitive deregulation towards harmonization, while different national traditions are neglected and lumped together.

3. The Common Market is increasingly overloaded with social policy objectives.
4. The Common Market is also instrumentalized to pursue other objectives outside the Commission's areas of competency. The "harmonization competency" provided for in article 95 of the EC Treaty was used "to establish EU regulations in areas where the EU does not have any competencies". According to Bolkestein and Gerken (2007) these measures are portrayed as if they were safeguarding the Common Market even though the Common Market is often not at all or at best marginally affected. Harmonization is pursued under the disguise of protecting the Common Market against artificial distortions. Bolkestein and Gerken (2007) even speak of a "perversion of the Common Market mandate".

Especially the fourth hypothesis is supported by our analysis of the current reform proposals. It is not the Common Market that is protected by the proposed measures, but the European Commission's self-interest, while any sound economic justification is completely missing. As Bolkestein and Gerken (2007) note the tragedy is "that, as a frustration relief, European politics are driving economic regulations forward in areas which do not need to be regulated on a European level while national egoisms prevent European policies where they are really necessary" [own translation, J.H.].

Indeed neither the current nor the proposed allocation of competences is well suited to address the risk of overregulation. Quite in contrast, NRAs and European Commission may compete for influence so that market deregulation is slowed down while additional markets may be subjected to regulation too easily, given the political economy of regulation. Hence, the question arises whether some alternative allocation of competences may be more promising to address the risk of overregulation.

An Alternative Institutional Proposal

In order to effectively avoid overregulation the institutional framework should be designed so as to make it more difficult for regulators to implement new and additional regulations than to undertake deregulatory measures. This is of particular importance in telecommunications markets which are characterized by rapid technological change and the continuous introduction of new services so that significant welfare gains result from investment and innovation.

Therefore, we have developed the following proposal to re-allocate executive competences between the European Commission and NRAs (see Haucap and Kühling 2006, 2007):

The European Commission should have ex ante veto rights in all three steps of the market regulation procedure if there are significant cross-border externalities of regulation. The standard of "significant cross-border externalities" needs to be stronger than the current "potential effects on trade" standard. Significant cross-border externalities of regulation can be expected if the prices for consumers of one Member State are likely to be affected by regulatory measures of another Member State.

In the absence of significant cross-border externalities of regulation, the European Commission is allocated a *one-way deregulation competence* for all three steps of the market regulation procedure for measures proposed by an NRA that do not lead to a reduction of regulation compared to the status quo.

If, in contrast, a proposed measure leads to a reduction in regulation, the European Commission should not have any ex ante veto right. In this case, the usual ex post control mechanisms remain.

With the particular exception of regulations that have significant cross-border effects this allocation of competences results in an exclusive deregulation competence of the European Commission, as the Commission can only impose a veto if regulations are not removed. If, in contrast, an NRA intends to deregulate a market, the Commission would not have a veto right but had to refer to ex post measures of intervention. We can also summarize our proposal in the following table (Table 1):

To decide whether the intensity of regulation is deceasing or not, the various remedies should be ranked ex ante according to the intensity of intervention associated with them. This may be done in a separate EC directive.

Furthermore, one has to keep in mind that even in those cases where the European Commission would not have an ex ante veto right according to our proposal it would still have the following ex post intervention rights, as it has today:

1. Treaty violation proceedings
2. Competition proceedings under articles 81 and 82 of the EC Treaty, as has happened, for example, against Telefonica in 2007 or Deutsche Telekom in 2003 and
3. State aid proceedings under articles 87 to 89 of the EC Treaty

That means that the European Commission does not have to sit still if Member States are rolling back their initial market liberalization. If there are indeed distortions of the Common Market the Commission has three potential ex post instruments at her hand.

In summary, our proposal guarantees through an intelligent institutional set-up that deregulatory measures are easier to implement than additional regulation. Hence, our proposal is a real contribution to achieving better regulation for Europe. A the same time, the proposal ensures that there is some scope left to "experiment" and to find the best regulation which can only be achieved through learning from different regulatory approaches. The ERG should therefore actively benchmark regulatory systems and outcomes to foster such a learning process.

Table 1 Proposed veto rights for the European Commission

Proposal by NRA...	Markets with significant cross-border externalities	Markets without significant cross-border externalities
Does not lead to deregulation	Ex-ante veto right of European Commission	Ex-ante veto right of European Commission
Leads to deregulation	Ex-ante veto right of European Commission	No ex-ante veto right of European Commission

Conclusions

Anybody looking for a detailed analysis of the costs and benefits of the European Commission's proposals within the documents tabled in the course of the 2006 review will be disappointed. This is true for the Commission's central and often repeated statement that there is a close relationship between investment and innovation on the one side and regulatory intervention on the other side,[4] and it also holds for the proposal to further centralize and/or harmonize regulation of telecommunications markets across the EU. The simple question "where is the problem?" has not been answered in any satisfactory, let alone empirical, way.

The common theme of the proposals is the wish to centralize competences. While the European Commission favors competition in almost all markets, it fails to see the benefit of some little pluralism and diversity (that is still left) of regulatory approaches. This is a worrisome tendency.

In contrast, this paper has proposed an allocation of competences that is based on the analysis of the costs and benefits of centralization and that is aiming at preventing overregulating telecommunications markets. The trick is to install asymmetric veto powers so that deregulation is made easier for NRAs than additional or more intense regulation.

In addition, the current three-criteria-test should be kept as the standard for de- and also re-regulating markets, but the third criterion ("insufficiency of competition law") needs to be taken more seriously than it is now.

References

Baake P, Haucap J, Kühling J, Loetz S, Wey C (2007) Effiziente Regulierung dynamischer Märkte, Law and Economics of International Telecommunications Vol. 57. Nomos, Baden-Baden.

Baumol WJ, Panzar JC, Willig RD (1982) Contestable Markets and the Theory of Industry Structure. Harcourt Brace, New York.

Bolkestein F, Gerken L (2007) Protektionismus und Regulierungswut: Wie die Zukunftsfähigkeit der Europäischen Union gleich von zwei Seiten gefährdet wird. Handelsblatt 58 (22-3-2007), p. 6.

Crocker KJ, Masten SE (1996) Regulation and Administered Contracts Revisited: Lessons from Transaction-Cost Economics for Public Utility Regulation. Journal of Regulatory Economics 9: 5–39.

Cullen International & Eurostrategies (1999) The Possible Added Value of a European Regulatory Authority for Telecommunications, Brussels-Luxembourg.

Dewenter R, Haucap J (2004) Die Liberalisierung der Telekommunikationsbranche in Deutschland: Bisherige Erfolge und weiterer Handlungsbedarf. Zeitschrift für Wirtschaftspolitik 53: 374–393.

European Commission (1999) Towards a New Framework for Electronic Communications Infrastructure and Associated Services: The 1999 Communications Review, COM (1999) 539, Brussels.

European Parliament (2008) Telecoms Package: EU-Wide Spectrum Management for Full Benefits of Wireless Services, Press Release, Brussels, 8 July 2008.

---

[4]For an acute critique see Gerpott (2008).

European Regulators Group (ERG) (2006) Effective Harmonisation Within the European Electronic Communications Sector: A Consultation by ERG, ERG (06) 68, ERG, Brussels.
Gerpott T (2008) Kommentar zu Impact of the Regulatory Framework on Investment Across Europe. In Picot A (ed.), Die Effektivität der Telekommunikationsregulierung in Europa – Befunde und Perspektiven. Springer, Berlin, pp. 27–30.
Goldberg V (1976) Regulation and Administered Contracts. Bell Journal of Economics 7: 427–448.
Haucap J, Heimeshoff U (2009) Zehn Jahre Liberalisierung in der Telekommunikation: Erfolge und Zukunftsaussichten. Forthcoming in Jens U, Romahn H (ed.), Macht und Moral: Netzindustrien in der Diskussion. Metropolis Verlag, Marburg.
Haucap J, Kühling J (2006) Eine effiziente vertikale Verteilung der Exekutivkompetenzen bei der Regulierung von Telekommunikationsmärkten in Europa. Zeitschrift für Wirtschaftspolitik 55: 324–356.
Haucap J, Kühling J (2007) Zur Reform der Telekommunikationsregulierung: Brauchen wir wirklich noch mehr Zentralisierung? Wirtschaftsdienst 87: 664–671.
Haucap J, Kühling J (2008) Europäische Regulierung der Telekommunikation zwischen Zentralisierung und Wettbewerb. In: Picot A (ed.), Die Effektivität der Telekommunikationsregulierung in Europa – Befunde und Perspektiven. Springer, Berlin, pp. 55–80.
Heimeshoff U (2008) Fixed-Mobile Substitution in OECD Countries, Working Paper, University of Erlangen-Nuremberg.
Kiesewetter W (2007) Marktanalyse und Abhilfemaßnahmen nach dem EU-Regulierungsrahmen im Ländervergleich, WIK Diskussionsbeitrag Nr. 288, Bad Honnef.
Kruse J (2002) Deregulierung in netzbasierten Sektoren. In: Berg H (ed.) Deregulierung und Privatisierung: Gewolltes – Erreichtes – Versäumtes, Duncker & Humblot, Berlin, pp. 71–88.
Littlechild SC (2006) Mobile Termination Charges: Calling Party Pays Versus Receiving Party Pays. Telecommunications Policy 30: 242–277.
Monopolkommission (2008) 17. Hauptgutachten: Weniger Staat, mehr Wettbewerb: Gesundheitsmärkte und staatliche Beihilfen in der Wettbewerbsordnung, Bonn.
Möschel W (2007) Der 3-Kriterien-Test in der Telekommunikation. MultiMedia und Recht 2007: 343–346.
Never H, Preissl B (2008) The Three-Criteria Test and SMP: How to Get It Right? International Journal of Management and Network Economics 1: 100–127.
Reding V (2006) The Review 2006 of EU Telecom Rules: Strengthening Competition and Completing the Internal Market, Speech, Brussels, 27 June 2006, online at: http://europa.eu/rapid/pressReleasesAction.do?reference = SPEECH/06/422&format = HTML&aged = 0&language = EN&guiLanguage = en.
Reding V (2007) Towards a True Internal Market for Europe's Telecom Industry and Consumers: The Regulatory Challenges Ahead, Speech, Brussels, 15 February 2007.
Viscusi K, Vernon J, Harrington J (2000) The Economics of Regulation and Antitrust, 3rd edn., MIT-Press, Cambridge, MA.

Surveying Regulatory Regimes for EC Communications Law*

Maartje de Visser

Abstract This paper discusses what institutional model is best able to address identified deficiencies in the enforcement of the 2002 EC Electronic Communications Framework, ie a lack of consistency in the application of the legal rules and the lack of independence of the front-line institutions for the daily administration of EC law: the national regulatory authorities (NRAs). An examination of the three paradigm models otherwise available in European law reveals that the current 'network-based' model is basically sound. While it should be strengthened and supplemented, it should not be replaced. It is argued that it is time to move beyond these 'basic' questions of institutionalization to the more fundamental question of the constitutionalization of this model, through a debate on the legitimacy and accountability of its central construct: the European Regulators Group (ERG).

Introduction

Something is rotten in the state of electronic communications.

> There is a fragmentation of regulation across the 27 Member States, lack of independent regulators in several EU Member States, sometimes also a lack of properly resourced regulators, delays in applying remedies, as well as problems caused by inefficient remedies

Commissioner Reding laments in a recent speech.[1] What must be done about such deficiencies?

In the field of electronic communications, change is the State of the Art. First and foremost, change is fostered at the technological level, through the deployment

M. de Visser
Dr (Tilburg), Post-Doctoral Fellow, Tilburg Institute of Transnational and Comparative Law (TICOM)
e-mail: Maartje.deVisser@uvt.nl

*A more extended version is available as TILEC Discussion Paper 2007-028 at <http //www.tilburguniversity.nl/tilec/publications/discussionpapers/>. This article reflects the state of the law at 31 December 2007.

[1] SPEECH/06/795, 3.

B. Preissl et al. (eds.), *Telecommunication Markets*, Contributions to Economics,
DOI: 10.1007/978-3-7908-2082-9_27, © Springer Physica-Verlag HD 2009

of new networks and services. We also witness change in the substantive rules, adapting as they must do, to market and technological evolution.[2] Even at the institutional level, the relationship between the two dominant actors – Commission and national authorities – has developed in a dynamic manner. Initially the Commission was in the driver-seat as it used Art 86(3) EC to achieve liberalization.[3] The 1998 framework saw the rise of national authorities as the 'cornerstones' for enforcement.[4] Under the current regime the pendulum has swung back somewhat, evident in the Commission's powers under Art 7 of the Framework Directive.[5]

Reding's quote suggests that the time has come for another amendment to the institutional framework. This article considers, in Part I, the main criticism levelled at the current model. This is to flesh out what improvements are perceived as necessary by the Commission and market participants Having done so, we shall investigate available alternatives in Part II. We assess their strengths and weaknesses within our understanding of the need for change. We conclude with Part III, in which we discuss suggestions to improve on the existing institutional set-up.

Defining the Problem

A perusal of the Commission's Annual Implementation Reports[6] and the documents related to the revision process[7] reveals two main interlinking institutional deficiencies. The first relates to inconsistencies in the application and enforcement of the law. The inconsistency is believed to exist at two levels. On the one hand, national authorities are accused of choosing different remedies to address similar problems.[8] On the other hand, national courts apply heterogeneous standards when deciding whether to suspend an NRA decision pending appeal – in most cases influenced by their eagerness, or absence thereof, to grant such interim relief.[9] In both cases, the result is the same: the scope and degree of regulation in force at any point in time may differ from one Member State to the next.

---

[2] The first legislative measures in the sector challenged existing legal monopolies to bring about liberalization. The 1998 ONP framework was directed at the 'original sin' of the former incumbent to allow the development of a genuinely competitive market. The current 2002 framework is premised upon a fully liberalized and competitive sector. A recent example of the perceived need to adapt substantive rules to technological evolution is the discussion on next generation networks.

[3] Dir 88/301/EEC and Dir 90/388/EEC. Further: Larouche (2000).

[4] Commission (1999a, 9). In particular, they were in charge of applying the SMP regime.

[5] Dir 2002/21/EC.

[6] The Reports are available at DG INFSO's website.

[7] <http://ec.europa.eu/information_society/policy/ecomm/tomorrow/index_en.htm> accessed 9 July 2007.

[8] Commission (2006a, 2007a, b); SPEECH/06/795; SPEECH/06/442; SPEECH/07/86; Hogan & Hartson and Analysys 2006 and numerous responses to the Commission consultation process, accessible at the website of DF INFSO.

[9] E.g. Commission (2006b, 10).

The importance of consistency is intimately linked to the Internal Market. Most sectors of the economy now have a European, if not a global dimension. There are ever more pan-European undertakings, or at least undertakings operating in several Member States. Contradictory or incompatible decisions by national authorities or courts – inter-State as well as intra-State – complicate business life as compliance costs are augmented.[10] It leads to distortions of competition and predictability suffers, creating insecurity which enhances entrepreneurial risk.

Inconsistency knows various causes. Most common is a difference in regulatory capacities, resources and expertise.[11] The authority or court enforcing the law lacks the requisite personnel or financial support to do so or misinterprets the European rule due to a knowledge or experience deficit. This unintended consistency may be contrasted with intentional deviation. A nationalist outlook, political pressure or regulatory capture may induce the authority to deliberately misconstrue or misapply the European rule in favour of domestic interests or undertakings.

The current regime comprises three instruments to counter this unwanted behaviour.[12] Art 7 of the Framework Directive mandates that NRAs consult the Commission and their peers on draft decisions.[13] The notifying NRA must take the utmost account of any comments received. In some cases, the Commission holds a veto right over draft decisions concerning the first two steps of the SMP procedure – market definition and SMP analysis.[14] We note that in these two steps the NRAs are guided by the Commission Recommendation on relevant markets[15] and the Commission Guidelines on SMP.[16] Then there is the European Regulators Group (ERG).[17] Providing an interface between the NRAs and the Commission, the ERG is meant to contribute to the consistent application of the electronic communications rules in particular as regards the third stage of the SMP procedure – remedies.[18] Finally, Art 4 of the Framework Directive stipulates a right of appeal against NRA decisions.

---

[10] Larouche (2005).

[11] Majone (1996, 277).

[12] Of course, these are complemented by the generalist consistency tools laid down in Arts 234 and 226 EC, cf. Case C-478/93 Kingdom of the Netherlands v Commission [1995] ECR I-3081 [38].

[13] Art 7(3) Framework Directive.

[14] Art 7(4) Framework Directive. The Commission must consider that the measure would create a barrier to the single market or have serious doubts as to the measure's compatibility with EC law. Thus far, five veto decisions have been adopted: Commission Decision C(2004)527final in Cases FI/2003/0024 and FI/2003/0027, Commission Decision C(2004)3682final in Case FI/2004/0082, Commission Decision C(2004)4070final in Case AT/2004/0090, Commission Decision C(2005)1442final in Case DE/2005/0144 and Commission Decision C(2006)7300final in Cases PL/2006/0518 and PL/2006/0514.

[15] Commission Recommendation 2003/311/EC.

[16] Commission Guidelines 2002/C165/03.

[17] Commission Decision 2002/627/EC establishing the European Regulators Group as amended by Commission Decision 2004/3445/EC. The ERG's website can be found at <http://erg.ec.europa.eu/>.

[18] Art 7(2) Framework Directive. National courts are left out of this network for obvious constitutional reasons.

To ensure meaningful control, national courts are charged to take 'the merits of the case duly into account'.[19] The Article also specifies that NRA decisions must in principle remain intact pending the outcome of the appeal. On the one hand, systematic suspension would induce unnecessary delays in the implementation of regulatory decisions. On the other hand, the special expertise of the NRA and the extensive public consultation process that precedes most regulatory decisions[20] warrant a presumption of legality.

The second deficiency relates to the lack of independence of some NRAs.[21] Independence can be jeopardized through excessive government interference or through close ties with market participants. An example of the former is unlimited discretion for the executive to dismiss the head of the NRA.[22] The concentration of regulatory tasks and responsibilities relating to the control of the incumbent within the same Ministry serves as an example of the latter situation.[23] A lack of independence allows for decisions based on considerations other than those listed in the electronic communications framework. As with the inconsistency deficiency, market operators would be faced with different regulation in different Member States.

The need for independence is compelling. It guarantees the credibility of the regulatory function as it ensures that the authority can act without reference to political or other volatile interests.[24] Similarly, it operates as a barrier to regulatory capture. This means that regulatory outcomes will be more consistent and neutral as between different interests and over time.[25] This in turn will quite likely positively influence the legitimacy of the authority and its output as well as the degree of regulatory compliance. In relation to the causes of inconsistency outlined supra, we observe that the emphasis on independence is primarily directed against intentional misbehaviour.

Art 3 of the Framework Directive prescribes the legal and functional independence of NRAs from undertakings active in the eCommunications field.[26] Member States are hence barred from entrusting regulatory responsibilities to such undertakings. There must further be no possibility whatsoever for market participants to unduly

---

[19] Further: Art 10(7) Framework Directive and Art 2 Dir 2002/77/EC.

[20] Art 6 and Recital 15 Framework Directive and paras 144, 145 of the Commission Guidelines 2002/C165/03.

[21] E.g. Commission (2007b, 14); SPEECH/06/795.

[22] IP/06/1798, MEMO/06/487, IP/07/888, MEMO/07/255.

[23] IP/05/430, IP/05/875, IP/05/1296, MEMO/05/372, MEMO/05/242, MEMO/05/478, IP/06/464, IP/06/948, IP/06/1798, MEMO/06/158, MEMO/06/271, MEMO/06/487.

[24] Independence as a solution to the 'commitment problem' has been advocated strongly by Majone, e.g. Majone (1997, 2000, 2002). Baldwin and Cave (1999, 70) also note that independence allows the authority to develop a high level of expertise necessary to make decisions on complex questions.

[25] Maher (2004, 228).

[26] Art 3(2) first sentence Framework Directive.

influence, or interfere with, the regulatory work of NRAs.[27] This is commonly achieved through strict rules on conflicts of interests.[28] Under certain conditions, the law demands structural separation between NRAs and undertakings.[29] The scenario is that of a Member State which has maintained the former incumbent under public ownership or control. The State is then responsible for regulatory functions, through the NRA, and economic activities, through an active interest in the performance of the public undertaking(s).[30] In such a case, the State must ensure that the latter interest does not affect the exercise of regulatory responsibilities.

Assessing the Alternatives

We can roughly distinguish three models for the administration of EC law. Of course, many variations and combinations of the models are possible and indeed occur in practice. Our aim, however, is not to provide an exhaustive account of the workings of specific sectors such as the CAP or fisheries. Instead, we intend to offer a general survey of available alternatives to the current regime in electronic communications law with particular emphasis on their strengths and weaknesses in relation to the deficiencies identified.

Centralized Enforcement

When we talk about centralized enforcement, we refer to the situation where legislation elaborated at the European level is also applied and enforced by the EC institutions, in particular the Commission. Textbook examples are the regulation of anti-competitive conduct from the 1960s until the millennium,[31] State aid,[32] merger control[33] and trade law.[34]

---

[27] Also Case C-91/94 *Criminal Proceedings against Thierry Tranchant and Téléphone Store SARL*[1995] ECR I-3911, Joined Cases C-46/90 and C-93/91 *Procureur du Roi v Jean-Marie Lagauche and others* [1993] ECR I-5267, Case C-69/91 *Criminal Proceedings against Francine Gillon, née Decoster* [1993] ECR I-5335 and Case C-92/91 *Criminal Proceedings against Annick Neny, née Taillandier* [1993] ECR I-5383.

[28] E.g. §4 Gesetz über die Bundesnetzagentur (Germany), Art L.131 Code des postes et des communications électroniques (France), Art 4(1)(c) Wet OPTA (the Netherlands), Members' Code of Conduct (Ofcom), available at <http://www.ofcom.org.uk/about/csg/ofcom_board/code/> (accessed 9 July 2007).

[29] Arts 3(2) second sentence and 11(2) Framework Directive.

[30] Stevens and Valcke (2003, 169) submit that the trigger should be interpreted to encompass not only a majority, but also a minority interest in, or control over, an undertaking active in the eCommunications field.

[31] Reg 17/62, now replaced by Reg 1/2003. For a detailed examination of the origins of the competition rules, Goyder (2003).

[32] Arts 87-89 EC.

[33] Reg 139/2004.

[34] Arts 131-134 EC.

Regulation theorists accept that centralized governance is the most effective tool to bring about homogeneity in regulatory approaches.[35] A single enforcer may be expected to ensure internal consistency in its decision-making. In the European context, if that authority is the Commission, we can further assume that it will interpret, apply and enforce the law correctly, thereby eliminating unintentional inconsistency.

The Treaty stipulates that Commissioners must be independent.[36] Reality is however not as black-and-white as the law might lead us to believe. Admittedly, the Commission will overall be less susceptible to regulatory capture than an NRA might be. Although one could argue that it is certainly more cost-efficient for an undertaking to lobby a single authority as opposed to 27 different NRAs, it is not certain to what extent lobbying at the EC level would exert the desired effect. Indeed, capture requires a relationship of information asymmetry which makes the authority dependent on the information coming from a single, external source. At Community level, there is no single entity on the other side of the Commission: industry groups are not united or powerful enough given the divergence in the interests of their members, and individual firms are too small. The Commission can however be vulnerable to political pressure. On the one hand, the Commission is dependent on the cooperation of Member States to fulfil its mission, thereby opening up the possibility of undue leveraging and bargaining. This is mainly due to the doctrine of national procedural autonomy and the need for directives to be implemented to be effective. If the Commission were to become the sole enforcer of the electronic communications rules, it would in all likelihood remain dependent on the Member States for the supply of local information. On the other hand, veto decisions are made by the full Commission, allowing for pressure from other Directorates-General to take account of considerations other than those related to the electronic communications policy when deciding.

It is axiomatic that the success of the Commission is the *sine qua non* of centralized governance. The experience with competition law reveals however that this enforcement model puts a vast strain on the typically limited resources of the Community. The Competition Directorate-General had only a handful of senior officials during its first year of operations, gradually rising to about 20 by the end of the second year, and increasing steadily to 78 by April 1964 and to just over 100 by 10 years later.[37] Yet, as early as 1967 the Commission was faced with the daunting total of 37,450 cases that had accumulated since the entry into force of Regulation 17 just 4 year prior. Only 222 decisions were ever adopted.[38] These were often so far apart in time that their internal consistency was difficult to assess. More fundamentally, the inability to enforce the law with a high frequency has repercussions for its perceived effectiveness and, in the long-term, credibility.

---

[35] Baldwin and Cave (1999, ch. 5).

[36] Art 213(1) EC.

[37] Goyder (2003, 31).

[38] Forrester (2003).

Rules lacking in credibility rules will have a similar impact on the incentives of the business community to take entrepreneurial risks as inconsistently enforced rules. Centralized enforcement also suffers from political constraints. Member States are commonly hesitant to transfer enforcement competences to the Community level for sovereignty reasons, and invoke the principles of subsidiarity and national procedural autonomy in support.

Decentralized Enforcement

Centralized enforcement is however the exception. As a general rule, Member States are in charge of bridging the gap between general EC legislation and individual cases. They carry out the application and enforcement of EC law, for instance in the fields of customs, agricultural policy, banking and insurance. When the legislation takes the form of directives, Member States already take over from the EC level at a fairly high stage of generality, when they have to implement the directives in their respective systems. Furthermore, sometimes extra layers of legislation are issued at EC level to further develop the original legislation. The power to issue such extra legislation is usually delegated to the Commission, in a comitology setting.[39] The literature is split on the merits of the comitology system. The committees have been praised as

> a fruitful collaboration between [the Commission] services and those Member State administrations which are most often faced with having to apply, on the ground, the implementing measures adopted at Community level.[40]

Yet they are also critiqued for their obscurity, lack of accountability and transparency, the corporatist nature of their processes as well as the exclusion of Parliament from the system.

To the extent that there are no common rules in the matter[41] and subject to compliance with general obligations deriving from the EC Treaty,[42] the Member

[39]Art 202 third indent EC. Comitology committees are forums for discussions, consist of representatives from Member States and are chaired by the Commission. Consider Bergström (2005), Joerges and Vos (1999), Craig (2006), Andenas and Turk (2000).

[40]Joerges and Vos (1999, 53), Bergström (2005, 10).

[41]The Community lacks the competence to directly legislate on, let alone harmonize, national procedure. Indeed, Member States have strong national traditions in the field and will be loathe to accept too many intrusions, Schwarze (1996).

[42]Member States must adopt rules that are effective and equivalent to domestic laws, Case 33/76 *Rewe-Zentralfinanz eG and Rewe-Zentral AG v Landwirtschaftskammer für das Saarland* [1976] ECR 1989. For an example in telecommunications Case C-462/99 *Connect Austria Gesellschaft für Telekommunikation GmbH v Telekom-Control-Kommission and Mobilkom Austria AG* [2003] ECR I-5197. Also, they must behave loyally towards the Community institutions pursuant to Article 10 EC.

States act in accordance with their own constitutional traditions, procedural and substantive rules when administering Community law. This feature, coupled with the absence of a sustainable relationship between the national actors and with the Commission, exerts a clear negative impact on the consistent and independent application of the EC rules. Xénophon Yataganas argues this succinctly,

> There is an institutional vacuum between EU legislators and the implementation of European laws by the national authorities at the Member State level. The absence of adequate features of conflict resolution and an unequal expertise and independence of the national regulators further undermines the efficiency of the system. The lack of a European administration infrastructure makes cooperation between the national administrations essentially depend on their mutual trust and loyalty. In the perspective of enlargement, this situation becomes clearly insufficient to ensure the credibility and legitimacy of the European rule-making processes.[43]

A certain contribution to consistency is made by Arts 234 and 226 EC and the threat of Member State liability.[44] These suffer from a number of drawbacks however. For infringement proceedings to commence, the Commission must first detect and collect evidence of the distortive behaviour, which is far from easy for the non-blatant cases. It must then decide whether it wishes to use its resources to actually prosecute the matter – a decision which is heavily influenced by political reasons relating to the dependence of the Commission on the Member States in the adoption and execution of EC law. As far as preliminary references are concerned, national courts might be overly confident in asserting that they understand the Community rules and that there is hence no need for a reference. Furthermore, all three procedures are slow and cumbersome, sporadic and operate on an ex post basis.

Yet decentralized enforcement also exhibits more positive features. It yields access to a regulatory capacity far greater than that available to the Community. Also, it brings the benefits of proximity, flexibility and diversification. Policies are executed at the same level as where the beneficiaries of that policy, or those subject to it, are located. Member States are able to adopt solutions that match local preferences. In particular, they can experiment with rules, processes and enforcement allowing the emergence of 'better' solutions than those currently in place. Finally, the default nature of decentralized enforcement model renders this the politically

---

[43]2003. This is referred to by Nicolaides et al. (2003) as 'the implementation deficit' and by Majone (2000, 279) as 'the institutional deficit'. In a way, we can of course also qualify the main problem of centralized governance as an implementation/institutional deficit. After all, the insufficiency of resources also resulted in defective implementation of the relevant legal rules. Keeping in with normal European parlance however, we will reserve references to the implementation deficit to discussions on decentralized governance.

[44]Cases C-6 & 9/90 *Andrea Francovich and Danila Bonifaci and others v Italian Republic* [1991] ECR I-5357, Cases C-46/93 & C-48/93 *Brasserie du Pêcheur SA v Bundesrepublik Deutschland and The Queen v Secretary of State for Transport, ex parte: Factortame Ltd and others* [1996] ECR I-1029, Case C-224/01 *Gerhard Köbler v Republik Österreich* [2003] ECR I-10239, Case C-173/03 *Traghetti del Mediterraneo SpA v Repubblica italiana* [2006] ECR I-5577.

most acceptable model – a feature that must not be underestimated in a time when observance of the subsidiarity principle is taken ever more seriously.[45]

The Agency Model

The agency model holds the middle between centralized and decentralized enforcement. While the Member States retain an important role in the enforcement stage, there is a heavier EC-level presence through agencies. The precise extent of this presence is dependent on the type of agency involved. We identify three broad categories.[46] The first, information agencies, itself comprises two categories. Some agencies collect, analyze and disseminate information relating to their specific policy area.[47] Others also create and coordinate expert networks.[48] Expert networks comprise National Focal or Reference Points, required to cooperate with the agencies and, at national level, coordinate the activities related to the agencies' work programme. The second, management agencies, assist the Commission in the management of EU programmes inter alia through the execution of budgetary implementation tasks. The third, regulatory agencies, are in some way involved in regulating economic and social policies, for instance by monitoring implementation of the relevant regulatory framework.[49]

Agencies generally function under the authority of an administrative, governing or management board, which lays down the general guidelines and adopts the work programme of the agency. Agency boards commonly include one or two representatives from each Member State and one or several Commission representations.[50] Each member of the board has one vote, and the norm is to require a two-third majority for board decisions. It is clear that Member States can exert strong national influence through the agency board. The executive director is responsible for oversight of the day-to-day work of the agency, drawing up the work programme, implementing that programme and preparing the agency's annual report. He is appointed either by the board on a proposal from the Commission or by the Commission on the basis of candidates suggested by the board.

---

[45] Also Scott and Trubek (2002).

[46] The typology followed here derives from Vos (2003, 119). Other taxonomies are proposed by inter alia Commission (2002), Craig (2006, 154ff) and Geradin and Petit (2005).

[47] E.g. CEDEFOP, EUROFOND, ETF.

[48] E.g. EEA, EU-OSHA, EMCDDA.

[49] E.g. OHIM, CPVO, EMEA, EFSA.

[50] They may also include members appointed by the European Parliament or representatives of the social partners or other relevant stakeholder groups. These members commonly do not have the right to vote.

The contribution of the agency model to consistency surpasses that made by the decentralized model, but remains below the level achieved under centralized enforcement. On the one hand, the agency structure allows for regular contact between the Community and the national level as well as – albeit on a smaller scale – between the Member States. This helps to mitigate unintentional inconsistency and might make it more difficult for intentional inconsistency to be practiced for fear of detection and punishment.[51] On the other hand, the model is characterized by a multitude of actors, which is shown to compromise absolute consistency.[52] There are no specific control mechanisms and faulty behaviour must accordingly be sanctioned under Arts 234 or 226 EC, with their attendant weaknesses. Also, in certain cases, the addition of another layer of administration – the agency – might raise search and administration costs to a level where consistency suffers.

The agency model could exert some noteworthy positive effects on the independence of national actors when compared to decentralized enforcement. The links between the two governance levels may provide incentives for national actors to adhere to Commission and peer beliefs and hence make them less prone to undue political or market interference.[53] As the agency model facilitates the use of experts, it contributes to a reduction in information asymmetries between the market and the administration. With national actors less dependent on the market for information, their independence should be strengthened.

However, the agency model suffers from legal-political limitations. Following the infamous judgment in *Meroni*,[54] the Treaty is read to prevent the creation of fully-fledged regulatory agencies, i.e. agencies with real decision-making competences that would function as a centralized enforcer, but mainly staffed with Member State representatives.

The Commission continues to reaffirm that *Meroni* prevents it from proposing 'meaningful' agencies.[55] A cynic's view would be that the Commission simply uses the *acquis communautaire* to avert a reduction in its own powers. The Commission advocates the 'unity and integrity of the executive function' which it locates in the Commission and its President.[56] The executive function includes all that occurs after the passage of primary regulations and directives. According to Paul Craig, real agencies with discretionary powers would challenge the unity and integrity of the Commission's executive function.[57]

---

[51] Here we must not think solely of the threat of infringement proceedings or perhaps even Member State liability, but also – and arguably primarily – of a loss of face towards other Member States or the Commission.

[52] Cf. Baldwin and Cave (1999).

[53] Further the text between (n 89) and (n 90).

[54] Case 9/56 *Meroni & Co, Industrie Metallurgiche SpA v High Authority* [1957] ECR 133.

[55] Commission (2001, 2002). There are however tensions within the Commission regarding the topic, with some members wishing to move beyond *Meroni* and create true regulatory agencies, cf. Majone (2002).

[56] Commission (2002, 1).

[57] Craig (2006, 163).

The Commission's sentiments are echoed by the Member States. They too fear a loss of control. The tradition of decentralized enforcement means that Member States will not look kindly upon what they perceive as a transfer of their 'prerogatory' competences and role in the enforcement stage to a Community body. According to Majone, the lack of a European tradition of regulation by independent agencies further fuels Member States' reluctance: why should they grant European agencies powers that they were unwilling to delegate to domestic institutions?[58]

Comparing and Contrasting the Alternatives

Applying our findings to the electronic communications sector, we note the following. It is clear that the decentralized model is inappropriate as it would exacerbate the consistency and independence problems. This was in fact the institutional structure under the 1998 framework and discarded precisely for those reasons.[59] This leaves centralized or agency enforcement. Both models, it has been shown, have apparent credentials to address – and in the case of centralized enforcement: eradicate – the identified deficiencies. Yet, we also saw that political realities act against the actual implementation of either of these models.

As regards centralized enforcement, the Treaty itself endows the Commission with independent enforcement powers in competition law and State aid.[60] In the former case, Member States believed that the rules would play a marginal role and would not be seriously enforced.[61] They were hence not opposed to a strong role for the supranational Community institutions. In the latter case, the rules are addressed to the Member States and allowing 'self-regulation' would in all likelihood not lend the desired results. Matters are radically different in the electronic communications field. Member States were in accord that telecommunications belonged to the 'hard core' of their national sovereignty[62] and the Treaty was accordingly silent on this sector.[63] We further recall the strenuous resistance against the liberalization policy of the

[58] Majone (1997, 3).

[59] Eurostrategies/Cullen International (1999), Forrester et al. (1996); the Commission's *Annual Implementation Reports*.

[60] Arts 85(1) and 87–89 EC respectively.

[61] In large part because competition rules and institutions in the then six Member States were in a primitive state, Gerber (1994, 103), Goyder (2003, 28).

[62] The importance of telecommunications for the economic development of national sovereign states required, or so it was believed, political control. The need to ensure complete national coverage at equal and affordable prices was used as an additional argument.

[63] In 1992, the Treaty on European Union (TEU) introduced a chapter on Trans-European networks (TENs). Article 154 EC states that the Community has the task of contributing to '[T]he establishment and development of trans-European networks in the areas of (..) telecommunications (..) infrastructures'. This has however never taken off in the field of telecommunications.

Commission under Article 86(3) EC as epitomized in the well-documented challenges against the first two directives promulgated under that provision.[64]

Turning to the agency model, the European Parliament has strenuously campaigned for a European Regulatory Authority for Telecommunications. At its instigation, two of the Telecommunications Directives comprising the 1998 framework mandated the Commission to investigate the added value of such a creature.[65] The Commission's opinion, set forth in the 1999 Review, broadly followed the study it had commissioned on this issue.[66] It considered that 'the creation of a European regulatory authority would not provide sufficient added value to justify the likely costs'.[67] The outcome was much to the satisfaction of the Council, who had, for reasons outlined earlier, initially rejected the Parliament amendments. In its reform proposals, released on 13 November 2007, the Commission now foresees the creation of the European Electronic Communications Market Authority (EECMA) in replacement of the ERG.[68] Whether the Authority will actually see the light of day is far from certain however. Cyprus, the Czech Republic, Germany, Malta, the Netherlands, Poland, Spain and Slovakia have sent a letter to Commissioner Reding, in which they object to the creation of the Authority. They opine that, given differing national circumstances in the various Member States, the existing institutional structure should remain in place. Leaving aside the factual correctness of these assertions, we must recognise the underlying deeply ingrained beliefs about competence questions as well as the more pragmatic concern that Member States have often spent considerable time and resources putting in place a national enforcement infrastructure and will understandably be loathe to see their efforts rendered relatively meaningless. On the assumption that their voting behaviour in the Council accords with the sentiments expressed in this letter, these eight Member States constitute a blocking minority, meaning that the proposal would fail.

Sustaining the Status Quo

If the above analysis is correct, it means the continuation of the status quo. We will refer to this regulatory model as 'network-based enforcement'. This name is derived from the close relationship between the Commission and the NRAs, which are conceived as partners in a legal framework of rights and obligation.

[64] Case C-202/88 *France v Commission (Terminal Equipment)* [1991] ECR I-1223 and Joined Cases C-271/90, C-281/90 and C-289/90 *Spain v Commission* [1992] ECR I-5833, directed at Dir 88/301/EEC and Dir 90/388/EEC respectively.

[65] Art 8 Dir 97/51 and Art 22 of Directive 97/33.

[66] Eurostrategies/Cullen International 1999. This study followed two earlier reports dealing at least partially with the issue of the establishment of an ERA: Forrester et al. (1996), especially 51-82 and NERA/Denton Hall (1997).

[67] Commission (1999b, 9).

[68] Commission (2007c). The proposal closely follows Commission (2005).

In particular, it reflects the key position accorded to the ERG for structuring relations between the Community and the NRAs as well as between the NRAs. A cursory glance reveals clear positive features inherent in network-based enforcement. As with the agency structure, it strikes a balance between the uniformity characteristic of centralized enforcement – through the special control powers bestowed upon the Commission under Article 7 – and the subsidiarity benefits intrinsic in decentralized enforcement – by establishing national authorities as the point of reference for the daily administration of the electronic communications rules. The hierarchical notification and veto mechanism, the ERG's Common Positions as well as the good social working relations between the NRAs bolstered by frequent interaction (either face-to-face at ERG meetings or through the ERG intranet) almost certainly allows for even greater consistency than the agency model would. Crucially, unlike the agency structure, network-based enforcement is politically acceptable because it leaves both the Commission and the Member States with a feeling of control over the enforcement process. For the Commission, the special control mechanisms recognize its position as guardian of the Treaty. For the Member States, retaining responsibility for the day-to-day enforcement of the Community rules recognizes the national procedural autonomy and subsidiarity doctrines. This is not to say that there is but one construction of the network model. Other famous examples include the European Central Banks System (ECBS),[69] the Committee of European Securities Regulators (CESR)[70] and the European Competition Network (ECN).[71]

However, where does that leave the identified consistency and independence deficiencies? It is proper to begin by setting out the recent initiatives of the ERG to address inconsistencies.[72] To assist in the SMP procedure, it will develop case studies of regulatory 'best practices'[73] as well as adopt more Common Positions in identified priority areas. The ERG has also identified NRAs with relevant experience and knowledge in relation to particular regulatory issues or areas which will make themselves available to other ERG member for practical advice ('knowledge centres'). Further, it has agreed to the automatic establishment of Art 7 expert groups to advise affected NRAs whose notifications have entered a Phase II procedure, or in respect of which the Commission proposes to issue a serious doubts letter. The Expert Group would provide the affected NRA with a full expert analysis and enable it to amend a notification.[74] Finally, the ERG undertakes to monitor NRAs' compliance with Common Positions.

---

[69] <http://www.ecb.int/ecb/orga/escb/html/index.en.html>.

[70] <http://www.cesr-eu.org/index.php?page = home&mac = 0&id=>.

[71] <http://ec.europa.eu/comm/competition/ecn/index_en.html>.

[72] Annex 1 to the ERG advice to the Commission in the context of the Review. See the entire process of correspondence between the Commission and the ERG (accessible through the DG INFSO and ERG website) for other institutional proposals.

[73] These are aimed in particular in fleshing out the 'proper' application of ERG(06)33.

[74] The ERG gives the example of the Bundesnetzagentur who requested such a group for its leased line market notification. The group's recommendation was for the Bundesnetzagentur to withdraw its notification. NRAs can also seek informal peer review of their analysis prior to finalization and notification.

The Commission's proposal to extend its veto power to cover remedies must be evaluated in the light of these recent ERG initiatives and the premise behind network-based enforcement.[75] The Commission should explicitly consider why these initiatives should be complemented by an extended veto power and the costs this option would entail, in terms of Commission resources as well as the damage this could do to the good working relationships between the NRAs and the Commission. Here, it must also be emphasized that absolute consistency will, as under the decentralized or agency model, not be achievable. Nor is this desirable: as with technologies, policy innovation arguably requires a certain degree of regulatory emulation between the NRAs.

Another proposal could be to make the Commission an ERG member. The Commission currently provides the Secretariat of the Group and is able to attend and participate in its meetings.[76] As regards the collaboration between the Commission and the NRAs, the 2004 ERG Report notes that

> A very positive evolution was the increasing and deepening cooperation with the Commission services. It has become visible that the EC and the NRAs in the ERG are partners, often with the same objectives.

And that

> As can clearly be seen from ERG documents, the cooperation between the ERG and the Commission Services was very productive. Many issues and work items were discussed extensively between the ERG and the Commission.[77]

Membership in the ERG would acknowledge this practice. For case-based interactions, the ERG could develop – more so than is currently the case and in line with its recent initiatives – into the natural forum for discussions with the Commission and between the NRAs in the context of the Art 7 procedure.[78] For general policy work, Commission membership could add more clout to ERG output – again in keeping with the new approach. In addition, the NRAs would still have access to a forum where they might meet without Commission presence: the Independent Regulators Group (IRG).[79] Making the Commission an ERG member in this respect

---

[75] Commission (2006c, 9) and accompanying Staff Working Document. For a more elaborate version of this argument, Larouche and De Visser (2006).

[76] Art 4 fourth al. ERG Decision; ERG (03)07 Arts 1.4 and 5.6.

[77] ERG (05)16 1. This much is also evident from the language used in the Conclusions of ERG Plenary Meetings, referring e.g. to 'complementarity of activities' or 'that there will be close coordination with Commission services'. In addition, the Commission reports its efforts on the same matters as NRAs are dealing with and gives information on its activities, COCOM meetings and policy proposals.

[78] Akin to what the ECN is for the Commission and the national competition authorities (NCAs) for the notification of draft decision and possible ousting of jurisdiction under Art 11 Reg 1/2003. Consider the ERG's comments in Annex 1 of its advice to the Commission (n 72) 2.

[79] <http://irgis.anacom.pt/site/en/irg.asp>, ERG(06)03. The IRG is an unofficial forum of NRA heads, established in 1997 and used for informal strategic discussions which do not involve the Commission.

helps differentiate the two bodies, as they currently display great overlap in terms of mandate, chairmanship and work programme.[80]

The current position of national courts leaves ample room for improvement. The electronic communications regime conceives these actors as a control instrument vis-à-vis the NRAs; and the legislature therefore, it seems, saw no reason to include specific consistency devices addressed to the courts. This is however misconceived. On the one hand, in certain situations national courts act on a par with the NRAs, rendering consistency tools immediately relevant.[81] On the other hand, it is of little help to synchronize the decisions of the NRA, only to see their actions wrongly undone by national courts. The formalistic case law under the 1998 regime and the overly frequent suspension of NRA decisions by some courts or the difference in standards applied, attests that the threat to consistency at the judicial review stage is not a mere theoretical teaser.

Inspiration may be drawn from the current competition law model. This regime knows three special consistency tools.[82] First, national courts are given the express competence to ask the Commission for information or assistance in the face of uncertainty in the application of the law. Secondly, they must notify their judgments applying the EC competition rules to the Commission, who enters these into a publicly accessible database. Thirdly, the Commission and the national competition authorities may make amicus curiae submissions to the national court. These formal measures are complemented by an informal one: the Association of European Competition Law Judges. All four mechanisms deserve to be seriously examined for possible parallel application in the electronic communications regime.

It may be objected that the first tool is already available pursuant to Art 10 EC[83] and that the second tool is operated on a voluntary basis.[84] It can also be argued that experience to date under the competition regime shows that none of these mechanisms is applied very often. Both claims are true. Nevertheless, not all courts will be aware of their power under Art 10 or the existence of the possibility of notification and consultation of the database. Formal inclusion in the law does offer this awareness. Indeed, consistency tools have a clear educational value. They alert national courts that they find themselves within the realm of European law and the

[80] For instance, the IRG could issue separate documents where the views of the NRAs and the Commission do not align or take over the advisory tasks of the ERG.

[81] E.g. under dispute settlement, Arts 20 and 21 Framework Directive or where competitors institute damage actions for an undertaking's failure to comply with an NRA decision.

[82] Art 15 Reg 1/2003, Commission Notice on the cooperation between the Commission and national courts.

[83] C-2/88 *Criminal Proceedings against JJ Zwartveld and Others* [1990] ECR I-3365 [17]-[22]. Examples of assistance mentioned by the Court include the production of documents and having Commission officials give evidence in national proceedings.

[84] The non-confidential versions of those judgments that have been voluntarily submitted can be found at.<http://ec.europa.eu/information_society/policy/ecomm/article_7.2/national_judiciaries/index_en.htm>. In terms of scope and accessibility, contrast with those sent under the competition regime:<http://ec.europa.eu/comm/competition/antitrust/national_courts/index_en.html>

concomitant need for a Europe-friendly perspective.[85] National courts must appreciate the special institutional characteristics of the national authorities[86] and their relationship with the Commission through the notification and veto procedure. This facet, more so than its direct effects, is the real contribution of the specific tools to consistency. Here we must also note that, unlike national authorities, national courts are not grouped in a formal network, for obvious constitutional reasons. Information structures and distribution platforms for case-sharing thus acquire even greater importance.

If we accept this reasoning, then the introduction of special consistency tools should positively impact on courts' behaviour in relation to suspension of NRA decisions. Nevertheless, harmonization of the conditions which must be met before a decision may be suspended is a worthwhile exercise. The Commission has proposed just this in the revision documents through alignment with the case law of the ECJ.[87]

The contribution of the network structure to the independence of national authorities can be explained as follows. Imagine a scenario in which there are two authorities, A and B. In the absence of any links between A and B, we may expect a fair number of A's decisions to be influenced by national – political or economic – considerations. This is because A will be dependent on its government and/or the market for resources as information, advice, legitimacy and authority. Further, the immediacy of the gains of acceptance and reputation within A's national context outweigh the far more distant and uncertain costs of non-acceptance or punishment by B within the European context. Now suppose that A and B are both members of a transnational network, and under a duty to cooperate with each other. This will quite likely induce A to adopt EC-conform decisions independent of national considerations. First, resource dependency on the government and the market is largely replaced by resource dependency on B through these cooperation duties – particularly the Art 7 procedure. Secondly, if A were to adopt non-independent decisions, these would be detected by B (again through the Art 7 procedure) who could then punish A by denying it prestigious positions within the network – such as chairmanship of a working group or project team, or a position as a knowledge centre – or by limiting cooperation with it. In sum, network membership creates powerful incentives for a high degree of independence in decision-making.[88]

---

[85] The fact that the eCommunications regime is laid down in directives arguably exacerbates matters, as it obscures its Community origins and Internal Market imperatives.

[86] In particular, their broad and discretionary competences.

[87] Commission 2007d Art 4(1). A proposal to this effect was already made during the negotiations on the current regulatory framework: European Parliament, A5-0053/2001FINAL amendment 22, A5-0435/2001/FINAL amendment 27.

[88] Majone (1996, ch. 12) arrives at a similar conclusion, albeit from a different premise. His starting point is the 'commitment problem' of national governments towards regulatory policies, and the credibility of national agencies to address this problem. Majone argues that the credibility of these agencies, and their commitment to regulatory policies can be strengthened through teamwork.

These effects are usefully supplemented by the Commission's proposal for a provision outlawing the overturning of NRA decisions by the Minister or the issuance of (specific) instructions to the authority.[89] A similar proposal has been made at the level of content regulation during the revision of the Television Without Frontiers Directive.[90] Since some NRAs are also responsible for content regulation, alignment with this proposal at the network level is prudent indeed.

Conclusion

The utopian model of enforcement does not exist. This realism is a sobering call for tempered expectations by undertakings, institutions and the public. We saw that all available alternatives – centralized, decentralized and agency-based enforcement – incorporate trade-offs between positive and not-so-positive features. In the final analysis, the current network-based model should be strengthened and supplemented, but certainly not replaced.

Recent efforts of the ERG may be expected to make an important contribution to increasing the level of consistency. They may indirectly further enhance NRAs' independence. From the Commission's perspective, the best way forward would seem a closer association with the ERG and hence the NRAs on a footing of trust and equality – rather than increasing the hierarchical element in their relationship through an extended veto power.

Legislative attention should focus on two matters. The first concerns the role of national courts, where material consistency gains can be achieved. The second concerns the legitimacy and accountability of the ERG – in particular in view of its heightened status – to allow it to evolve into a mature structure in the eCommunications institutional regime in the years to come.[91]

Non-traditional bodies or institutions are, sooner or later, plagued by legitimacy and accountability questions. Think of the NRAs,[92] comitology committees[93] or the

[89] As with the conditions for suspension, suggestions to this effect were already made by Parliament in the negotiations on the current version of the Framework Directive, (n 87) amendment 10.

[90] Dir 89/522/EC as amended. The proposal can be found at Commission 2007e Art 23b(2).

[91] Admittedly, the Commission mentions some of these in its letter of request to the ERG (n 72), but seems to perceive their relevance only in relation to the far-reaching institutional scenario of transforming the ERG into some sort of regulatory agency. It is argued here that these questions are also relevant if the current state of institutional play remains in place.

[92] On the Continent, a number of legal systems do not readily admit that independent authorities be given wide, discretionary powers, usually for reasons of a constitutional nature. Under the 1998 framework, it could thus be noted that the status of NRAs was often dubious, that Ministries tended to keep important competences to themselves and that the court were overly strict in reviewing NRA decisions, focusing too much on competence, e.g. Eurostrategies/Cullen International Report 1999, Commission. *Annual Implementation Reports*.

[93] Consider the text between (n 39) and (n 41).

EC structure as a whole.[94] Reliance on the perceived or proven effectiveness of these new creatures is typically not enough to silence the critics. In relation to the ERG, we note that its relationship with the European Parliament is embryonic, with the Commission typically functioning as an intermediary, for instance when it comes to the submission of annual reports.[95] Further, while ERG Common Positions exert considerable practical impact on the work of the NRAs, and hence indirectly on the market parties to which the NRAs address their decisions reflecting these Common Positions, they are currently immune from judicial scrutiny given their soft law character.[96] It is imperative to adopt a pro-active approach to the legitimacy and political and judicial accountability of the ERG to allow it to devote its resources to where they matter: the efficient exercise of its responsibilities.

References

Andenas M, Turk A (eds.) (2000) Delegated Legislation and the Role of Committees in the EC. The Hague: Kluwer Law International.

Andersen S, Eliassen K (eds.) (1996) The European Union: How Democratic Is It? London: Sage.

Baldwin R, Cave M (1999) Understanding Regulation: Theory, Strategy and Practice. New York: Oxford University Press.

Bergström CF (2005) Comitology: Delegation of Powers in the European Union and the Committee System. Oxford: Oxford University Press.

European Commission (1999a) Fifth Annual Implementation Report. Directive 2002/21/EC of the European Parliament and of the Council of 7 March 2002 on a Common Regulatory Framework for Electronic Communications Networks and Services [2002] OJ L108/33.

European Commission (1999b) Towards a New Framework for Electronic Communications Infrastructure and Associated Services. (Communication) COM(1999)539, 10 November 1999.

European Commission (2001) European Governance. (White Paper) COM(01) 428 final, 21 July 2001.

European Commission (2002) The Operating Framework for the European Regulatory Agencies. 11 December 2002. (Communication) COM (02)718 final.

European Commission (2005) Draft Inter-institutional Agreement on the Operating Framework for the European Regulatory Agencies. COM(2005) 59 final, 25 February 2005.

European Commission (2006a) Consolidating the Internal Market for Electronic Communications. (Communication) COM(2006) 28 final, 6 February 2006.

European Commission (2006b) Eleventh Annual Implementation Report. COM(2006)68 final, 20 February 2006.

European Commission (2006c) Review of the EU Regulatory Framework for Electronic Communications Networks and Services. (Communication) COM(2006)334 final, 29 June 2006.

European Commission (2007a) 2nd Report on Consolidating the Internal Market for Electronic Communications. (Communication) COM(2007)401 final, 11 July 2007.

---

[94] E.g. Andersen and Eliassen (1996), Craig and Harlow (1998), Snyder (1996), Weiler (1999).

[95] Art 8 ERG Decision.

[96] A full legitimacy assessment of the ERG can be found in De Visser (2007).

European Commission (2007b) Twelfth Annual Implementation Report COM(2007) 155, 29 March 2007.
European Commission (2007c) Proposal for a Regulation of the European Parliament and of the Council Establishing the European Electronic Communications Market Authority. COM(2007)699 final, 13 November 2007.
European Commission (2007d) Proposal for a Directive of the European Parliament and of the Council Amending Directives 2002/21/EC on a Common Regulatory Framework for Electronic Communications Networks and Services; 2002/19/EC on Access to, and Interconnection of, Electronic Communications Networks and Services, and 2002/20/EC on the Authorization of Electronic Communications Networks and Services COM(2007)697 final, 13 November 2007.
European Commission (2007e) Communication Concerning the Common Position of the Council on the Adoption of a Proposal for a Directive of the European Parliament and of the Council Amending Council Directive 89/552/EEC on the Coordination of Certain Provisions Laid Down by Law, Regulation or Administrative Action in Member States Concerning the Pursuit of Television Broadcasting Activities (Audiovisual Media Services Directive) COM(2007) 639 final, 18 October 2007.
Commission Guidelines (2002) 2002/C165/03 of 9 July 2002 on Market Analysis and the Assessment of Significant Market Power Under the Community Regulatory Framework for Electronic Communications Networks and Services [2002] OJ C165/6.
Commission Notice (2004) Commission Notice on the Cooperation Between the Commission and the Courts of the EU Member States in the Application of Articles 81 and 82 EC [2004] OJ C101/54.
Commission Recommendation (2003) 2003/311/EC on Relevant Product and Service Markets Within the Electronic Communications Sector Susceptible to Ex Ante Regulation in Accordance with Directive 2002/21/EC of the European Parliament and of the Council on a Common Regulatory Framework for Electronic Communications Networks and Services [2003] OJ L114/45, 11 February 2003.
Craig P (2006) EU Administrative Law. Oxford: Oxford University Press.
Craig P, Harlow C (eds.) (1998) Lawmaking in the European Union. The Hague: Kluwer Law International.
De Visser M (2007) Revolution or Evolution: What Institutional Future for EC Communications Law? TILEC Discussion Paper 2007-028 http://www.tilburguniversity.nl/tilec/publications/discussionpapers/2007-028.pdf.
Commission Directive (1988) 88/301/EEC on Competition in the Markets in Telecommunications Terminal Equipment [1988] OJ L131/73, 16 May 1988.
Commission Directive (1990) 90/388/EEC on Competition in the Markets for Telecommunication Services [1990] OJ L192/10, 28 June 1988.
Commission Directive 2002/77/EC (2002) on Competition in the Markets for Electronic Communications Networks and Services (Consolidated Services Directive) [2002] OJ L249/21, 16 September 2002.
Commission Staff Working Document (2006) On the Review of the EU Regulatory Framework for Electronic Communications Networks and Services SEC(2006)816.
Council Directive 89/522/EEC (1989) of 3 October 1989 on the Coordination of Certain Provisions Laid Down by Law, Regulation or Administrative Action in Member States Concerning the Pursuit of Television Broadcasting Activities [1989] OJ L298/23 as Amended by Directive 97/36/EC of the European Parliament and of the Council of 30 June 1997 [1997] OJ L202/60.
Directive 97/33 (1997) of the European Parliament and of the Council on Interconnection in Telecommunications with Regard to Ensuring Universal Service and Interoperability Through Application of the Principles of Open Network Provision (ONP) OJ L199/32.
ERG(05)16 (2005) ERG Annual Report 2004.
ERG(06)03 (2006) Independent Regulators Group/European Regulators Group.
ERG(06)33 (2006) Revised ERG Common Position on the Approach to Appropriate Remedies in the ECNS Regulatory Framework.

European Parliament A5-0053/2001FINAL (2001) Report by the Committee on Industry, External Trade, Research and Energy on a Proposal for a Directive of the European Parliament and of the Council on a Common Regulatory Framework for Electronic Communications Networks and Services (COM(2000) 393 – C5-0428/2000 – 2000/0184(COD)) A5-0053/2001 FINAL, 7 February 2001.

European Parliament A5-0435/2001/FINAL (2001) Recommendation by the Committee on Industry, External Trade, Research and Energy for Second Reading on the Council Common Position for Adopting a Directive of the European Parliament and of the Council on a Common Regulatory Framework for Electronic Communications Networks and Services, A5-0453/2001 final, 4 December 2001.

Eurostrategies/Cullen International (1999) Draft Final Report on the Possible Added Value of a European Regulatory Authority for Telecommunications. Brussels/Luxembourg: ECSC-EC-EAEC.

Forrester I (2003) Modernisation of EC Competition Law. In Annual Proceedings of the Fordham Corporate Law Institute: EC Competition Law Reform, ed. Barry Hawk, ch 12. New York: Juris.

Forrester, Norall & Sutton (1996) The Institutional Framework for the Regulation of Telecommunications and the Application of the EC Competition Rules. Brussels/Luxembourg: ESSC-EC-EAEC.

Geradin D, Petit N (2005) The Development of Agencies at EU and National Levels: Conceptual Analysis and Proposals for Reform. Jean Monnet Working Paper No 1/04 http://www.jean-monnetprogram.org/papers/index.html.

Gerber D (1994) The Transformation of European Community Competition Law. Harvard International Law Journal 35: 97–147.

Goyder D (2003) EC Competition Law. Oxford: Oxford University Press, 4th edn.

Hogan & Hartson and Analysys (2006) Preparing the Next Steps in Regulation of Electronic Communications accessible at the website of DG INFSO.

Joerges C, Vos E (eds.) (1999) EU Committees: Social Regulation, Law and Politics. Oxford: Hart.

Larouche P (2000) Competition Law and Regulation in European Telecommunications. Hart: Oxford.

Larouche P (2005) Coordination of European and Member State Regulatory Policy – Horizontal, Vertical and Transversal Aspects. In: Geradin, D, Muñoz, R, and Petit, N (eds.), Regulation Through Agencies in the EU – A New Paradigm of European Governance, pp. 164–179. Cheltenham: Edward Elgar.

Larouche P, De Visser M (2006) The Triangular Relationship Between the Commission, NRAs and National Courts Revisited. Communications & Strategies 64: 125–145.

Maher I (2004) Networking Competition Authorities in the European Union: Diversity and Change. In: Ehlermann C-D, Atanasiu I (eds.), European Competition Law Annual 2002: Constructing the EU Network of Competition Authorities, pp. 223–236. Oxford: Hart.

Majone G (1996) Regulating Europe. London: Routledge.

Majone G (1997) From the Positive to the Regulatory State: Causes and Consequences of Changes in the Mode of Governance. Journal of Public Policy 17: 139–167.

Majone G (2000) The Credibility Crisis of Community Regulation. Journal of Common Market Studies 38: 273–302.

Majone G (2002) Delegation of Powers in a Mixed Polity. European Law Journal 8: 319–339.

NERA/Denton Hall (1997) Issues Associated with the Creation of a European Regulatory Authority for Telecommunications. Brussels/Luxembourg: ECSC-EC-EAEC.

Nicolaides P, Geveke A, Den Teuling A-M (2003) Improving Policy Implementation in an Enlarged European Union. Maastricht: EIPA.

Reg 17/62 (1962) Council Regulation (EEC) No 17 (First Regulation implementing Articles 85 and 86 of the Treaty) [1959–1962] OJ Spec Ed 87.

Reg 1/2003 (2003) Council Regulation No 1/2003/EC on the Implementation of the Rules on Competition Laid Down in Articles 81 and 82 of the Treaty [2003] OJ L1/1.

Reg 139/2004 (2004) Council Regulation (EC) No 139/2004 of 20 January 2004 on the Control of Concentrations Between Undertakings [2004] OJ L24/1.

Schwarze J (1996) The Europeanization of National Administrative Law. In: Jurgen (ed.), Administrative Law Under European Influence – On the Convergence of the Administrative Laws of the EU Member States. Baden-Baden: Nomos.

Scott J, Trubek D (2002) Mind the Gap: Law and New Approaches to Governance in the European Union. European Law Journal 8: 1–18.

Snyder F (1996) Constitutional Dimensions of European Economic Integration. The Hague: Kluwer Law International.

Reding V (2006a) The Review of the EU Telecom Rules: Strengthening Competition and Completing the Internal Market. SPEECH/06/442, 27 June 2006.

Reding V (2006b) Why We Need More Consistency in the Application of EU Telecom Rules. SPEECH/06/795, 11 December 2006.

Reding V (2007) Towards a True Internal Market for Europe's Telecom Industry and Consumers – The Regulatory Challenges Ahead. SPEECH/07/86, 15 February 2007.

Stevens D, Valcke P (2003) NRAs (and NCAs?): The Cornerstones for the Application of the New Framework – New Requirements, Tasks, Instruments and Cooperation Procedures. Communications & Strategies 50: 159–189.

Vos E (2003) Agencies and the European Union. In: Zwart T, Verhey L (eds.), Agencies in European and Comparative Law, pp. 113–147. Antwerp: Intersentia.

Weiler J (1999) The Constitution of Europe. Cambridge: Cambridge University Press.

Yataganas X (2003) Delegation of Regulatory Authority in the European Union. Jean Monnet Working Paper No 3/01 http://www.jeanmonnetprogram.org/papers/index.html.

Innovation and Regulation in the Digital Age: A Call for New Perspectives

Pierre-Jean Benghozi, Laurent Gille, and Alain Vallée

Abstract Relations between innovation and regulation are all but fluid and simple. When innovation, beyond just developing new techniques, means redefining the very framework for implementing and operating technologies, it often means breaking the rules, challenging them. In the digital economy, the way it overturns the regulation and rule setting system is particularly radical. Innovation is key to create competitive advantage in a highly dynamic sector such as information and communications technologies. Firms invest heavily in productive resources and take steps to protect their competitive advantage. Productive resources are either network and connection infrastructure or consumer control which is rarely seen as such. It could be consumer's attention or visits which requires massive investments in content, for example. It does include intermediation platform like search engine, programs, knowledge, entertainment and other immaterial products of the digital age. There is a need to rethink regulation on the basis of innovation and mobilization of these productive resources. The digital economy calls for a more holistic consideration of the link between innovation and mobilization of value on the one hand, and regulation on the other.

Introduction

For more than 3 decades, the ICT sector has been the scene of some of the most spectacular innovations, both in terms of their industrial and economic effects and their societal impact. Investments have flowed into the sector in massive volumes, particularly in mobile telecommunications networks; products and services have penetrated at an unprecedented rate; and the sector has experienced comprehensive reorganizations, particularly through the influence of new entrants.

P.-J. Benghozi (✉), L. Gille (✉), and A. Vallée (✉)
Economics and Management Research Centre (PREG),
Ecole Polytechnique, CNRS 1
e-mail: Pierre-Jean.Benghozi@polytechnique.edu
e-mail: laurent.gille@telecom-paristech.fr
e-mail: alain.vallee@telecom-paristech.fr

B. Preissl et al. (eds.), *Telecommunication Markets*, Contributions to Economics,
DOI: 10.1007/978-3-7908-2082-9_28, © Springer Physica-Verlag HD 2009

The substantial drop in the cost of equipment and related operating costs, reflected in lower prices for the services offered, has greatly facilitated the rollout and adoption of technologies. Furthermore, in the digital service industry, particularly content services, the modes of distributing and consuming these services seem to be prevailing over the conditions for producing them, and specific regulations on this content are having a profound impact on regulations for its "container." With the expansion of the Internet, companies – whether existing, emerging or from other sectors – experienced new economic and transaction models. They went so far as to adopt such unheard-of economic practices as providing products and services for free, in general content services: they saw this as an opportunity to secure market share, control consumers or generate indirect revenue in amounts greater than their direct revenue. Using ICTs, they built their organizations in such a way as to minimize fixed costs and break free of geographical limitations.

It could be said, therefore, that regulating the sector, far from impeding the development of innovations, actually served as something of a neutral arbitrator for technological developments, if not actually a breeding ground for them. In this context of rapid technological progress, opening markets to competition acted as a clear driving force: in particular, it allowed for market globalization and substantially redefined the shape of said markets. Many examples show that new entrants were frequently the ones to break with the past in ways made possible by technological advances. At the same time, the incumbents hold an opposing view, namely, that the constraints imposed by regulations curbed their potential for innovation and investment, leaving the newcomers to take the initiative.

Throughout these developments, there has been constant debate over the relationship between regulation and innovation. In the 1970s, arguments focused on the relative dynamic of 'rate of return' or 'price cap' regulations, whereas more recent discussion has centred on the neutrality of the Internet and the issue of what should happen to the radio frequencies freed up by the transfer to digital television and how they should be allocated.

Clearly, tension between systems of innovation and regulatory models is nothing new. It becomes critical, however, when it reaches the point where 'regulation holidays' seem to be the only solution to achieve a new level of network deployment.[1]It is therefore becoming urgent to look into these relationships more directly and comprehensively.

This paper intends to explore these areas of tension between innovation and regulation. It hypothesizes that the current forms of innovation and their related strategies are helping to radically redefine the traditional configurations of the very firms and markets on which the principles and practices of regulations are based. We support the view that observable tensions call for new forms of regulation, based more directly on analysis of the productive resources and structuring of the value chains in innovation systems for digital products and services.

[1]Even though a laissez-faire approach involving lack of regulation is often considered a method for management.

In the first section of this paper, we identify the sources of friction emerging between innovation and regulation. After discussing the observable areas of tension, we then describe the current characteristics of innovation in the industry and the principles of network economy at the root of these forms of regulation. We then discuss how such a possibility would revitalize traditional approaches.

Strained Relations Between Innovation and Regulation

The sources of tension between innovation and regulation are well known and varied. Indeed, there is an inherent incompatibility between the innovative approach and management based on established rules.[2]Innovating means not only developing new techniques, but more often redefining the very framework for implementing and operating technology: the most disconcerting innovators frequently become brilliant trail-blazers. This expanded vision of innovation, also characterized by an ability to "break the rules," defy them, circumvent them and judge success or failure, could even be described as the main strength of national innovation systems (Freeman 1987).

The fact remains that the unique trait of the innovation process in the digital economy is that it is particularly radical in the way it overturns, or at least circumvents, the system of regulations and rule-setting. There are several types of reason for this, including the conjunction between service production and management of large infrastructures, the size of investments made in network renewal and building a mass audience and the relationship between technical and usage innovations, among others.

Innovation Today

In the telecommunications sector, as in the rest of the industry, innovation has taken a central role as the main vehicle for competitive positions and consumer control (Henderson and Clark 1990). This development has taken the form of faster industrial dynamics and deep-level redefinitions of the way that firms and their organizations are configured. These characteristics of innovation, shared across the entire industry, explain the difficulties currently facing the information and communication technologies (ICT) sector.

Acceleration in the design process for products and services and the focus on being the first to market has forced all firms to abandon the traditional, linear model of innovation that moved from research to development and finally to implementation. This strategy has had several consequences. Today, innovative firms strive to

[2]See, for example, Schumpeter who has already discussed the notion of destructive creation.

account for use-related issues and limitations starting in the R&D phase, considering equipment, goods and services simultaneously. In so doing, they have succeeded in radically redefining the nature of their "products" (Midler 1995) by working from the standpoint of useful features rather than simply attempting to master technology or equipment. It is also important to note that this acceleration in the design process has substantially changed methods for protecting innovation by changing the procedures for standardization and interoperability.

In turn, these new modes of design led to a deep-level reorganization of partnerships and value chains by viewing innovations as industry systems and architectures (Jacobides et al. 2006): this has made the boundaries of firms and their competitive framework much more flexible, since competitive and cooperative relationships are developed simultaneously, and are constantly changing and conditional (Bengtsson and Kock 2000). Finally, firms have tried to be more systematic in identifying and capitalizing on all of their available productive skills and resources (Teece et al. 1997), whether tangible or intangible; this has caused them to completely reconfigure their strategic processes for vertical and horizontal integration.

Friction (and Contradictions) Between Competition and Innovation

This complete reworking of the types of innovation in the industry has served to increase the traditional difficulties of regulating a network economy via a basic focus on regulating competition: on the one hand, it changes the role of innovation in the market dynamic and the strategies adopted by firms, and on the other, it has a substantial impact on production processes within and between firms.

Structural tension

Stimulating Competition and Protecting Innovation

The traditional, substantial principles of economic regulation deal mainly with the conditions for exercising competition (dominant position, concentration, fair competition etc.). In these systems, innovation is considered, first and foremost, a driving force and a tool for creating a legitimate competitive advantage – an advantage that is legally protected, no less, if it is based on intellectual property. Developments in the digital economy, however, have demonstrated that innovation, according to its nature, the rate at which it propagates, its externalities and its scope, can create global positions of domination (Microsoft, Google etc.) that are sometimes unassailable and have even been the cause of legal proceedings. At the same time, the rules of competition prohibit any illegitimate or unfair competitive advantage. This is why, given their cost and the associated risk, the innovation and development of certain new infrastructures could only occur in the context of a "regulatory holiday."

Indeed, this was the only solution to ensure that entrepreneurs would invest when they were hesitating, or even giving up, in the face of insufficient return on investment in a regulatory environment that forced them to share the profits from those investments. This "regulatory holiday" is in itself already a radical departure from traditional regulation in contexts where innovation in networks calls for massive investments, whether for building infrastructure or creating an audience.

On top of this first example related to networks, a second source of tension can be found in the nature of digital services, and concerns intellectual protection of content (copyright). This protection aims to stimulate new ideas by guaranteeing the authors and producers of works the exclusive right to use them; at the same time, however, supporters of consumer protection are pushing for the broadest possible access to such works, justifying the practice of fair competition concerning all who might wish to use them.[3]

Innovation involves "securing" a productive resource, that is, having the "exclusive" right to use that resource. Regulation involves facilitating access for all to such productive resources so as to reduce any discrimination concerning those that might distort the mechanisms of competition. Since the principles of protection have not been redefined, the rise of digital content runs the risk of causing clashes, with no possible resolution, between owners claiming rights to intellectual property and users insisting on the right to competition: recent developments in the theory of essential facilities provide an illustration of this.

Reconciling Rapid Innovation with Time-Consuming Regulation

The regulations of competition are "static" regulations, in the sense that they are based solely on established situations (drawing on the current state of affairs, frequently even long after the fact).[4]When the pace of innovation picks up, regulators are accused of misjudging the speed required for the development of the economy. Indeed, by their very nature, regulations have a difficult time defining a framework to analyze the situation and anticipating the status of various stakeholders. Since their regulations are not dynamic (with forward-looking objectives, for example), regulators seem to lack the appropriate tools.

Simultaneously, however, the same economic stakeholders (and others, such as potential newcomers) demand transparency and stability so that they can invest and choose equipment for the long term. It should be added that the "informational" realm inhabited by the economic stakeholders (in terms of both supply and demand) – that is, all of the information and "signals" on which they base their decisions to act or consume – coincide only somewhat with the legal realm (that is, all of the

[3]This tension was particularly noticeable in France upon the adoption of the recent law on digital rights.

[4]In a discussion on innovation and competition, Joseph Stiglitz offers broader consideration of the issue of the vitality of competition and criticisms of traditional antitrust policies (Stiglitz 2002).

rules and institutions responsible for monitoring their application). Along with other issues, such as the perceived value of a product or service, "piracy" of songs illustrates this type of disconnect.

"Accelerators" Unique to the Digital Economy

Several factors unique to the operations of the digital economy even further intensify the contradictions in and the process for disputing regulations.

The Specific Importance of the Technological Dimension

The importance of technical issues, interoperability or the lack thereof, and shared standards has long since been recognized in the ICT sector. Today, these concerns also weigh on the content-based industries related to the sector.

As we have previously stated, the pace of technical change in content and on the networks is based on (and requires) massive investments, to cover infrastructure rollout and the expansion of applications and to anticipate and support their technological development. These investments are immense in terms of capacity, radical technological improvements, and improvements in productivity and features that build anticipation and speculation among economic stakeholders. The situation is nothing new, and other stakeholders have already pointed out that the same type of financial speculation marked all of the main innovations in infrastructure (railways, the automobile, radio transmission, the computer industry, telecommunications, the Internet etc.). However, beyond the economic consequences of these investments, competition and its regulation become especially difficult to evaluate when technology is continually emerging and evolving, because it suggests that certain choices can be made in advance. Still, the above observations on the role of innovation in redefining all industry architectures demonstrate that these expectations cannot be reduced to a mere extension of the current state of affairs.

Telecommunications infrastructures are frequently described as layered types of architecture. The lowest levels contain the bearer layer; then comes the transportation layer, with functionalities that transform the basic support service into an infrastructure capable of transporting top-level applications and formats coded to support traffic; on top of this comes middleware, containing commonly used functions; and at the highest level come applications that allow users to interact directly. Working from this type of layered approach with well-defined boundaries, regulations deem that open and fair competition can take place between all types of provider within each layer. Many authors are pushing for regulation of the Internet "by layer," respecting the architectural integrity of each technical layer in the various registries and functional components of the communications system. Their argument rests on the idea that regulatory intervention cutting across different layers runs up against disparities in the economic dynamics involved, inevitably creating imbalances between the objectives and the operational resources of the regulations.

The essential intuition behind this principle (Lessig 1999) is to allow each feature to be implemented at the level where it is required, masking the internal mechanisms of layers that are independent from one another.[5]

The shape taken by innovation in the digital industries, combined with developing value chains, offers serious cause for doubting this layered configuration. It could be said that, until now, history had demonstrated above all the ability for new offerers of technology to appear in the value chain.[6] This emergence tended to cause the value chain to rebalance, to the detriment of those offering content on distribution circuits that remained largely unchanged: the main aim of these offerers of technology was to ensure that content was supplied for the technology they offered.[7]From this perspective, the current situation, with its network economy, is substantially different. New entrants are no longer content to "integrate vertically" or to compete with those already established on the market. They are more thoroughly redefining the entire industry architecture underlying the production of goods and services. For example, the way DRM is standardized (collective and public or proprietary) may or may not eventually open the door to a strong control position for those stakeholders that have mastered the standards. Intrusion by those offering technologies is no longer merely leading to a rebalancing; it is also changing the structure of the business models and architecture of industries. Hinge developers even believe that issues of interoperability hinge on differences in the value chain.

Controlling access through digital platforms (decoders, set-top boxes, DRM) is therefore central to the industrial strategies of firms in the sector (Chantepie and LeDiberder 2004). These platforms, a technological bottleneck at the interface between the content industries and the technical media used to disseminate that content, are becoming the tool of choice for "internalizing network externalities": proprietary technical systems, locked content and so on are at the root of vertical and horizontal integration strategies alike.

The issue is not purely theoretical. The pervasive nature of information technologies means that they have the potential to trigger a huge explosion in the market: rebuilding the distribution channels that recent technological convergence helped to overshadow (satellite, mobile, Internet, fixed networks, radio etc.) and developing integration chains (from terminal to software) within homogenous sectors similar to those that can be found in video games (Sony vs. Nintendo vs. Microsoft, for example). In the area of mobile telecommunications services, the success or failure of the introduction of WAP (Wireless Access Protocol) illustrates the issue well. The formation of sector-specific consortiums, like the Open Mobile

[5]This holds true, for example, in the "end-to-end" principle, which aims to simplify the infrastructure by coming up with "dumb" networks and "intelligent" applications, thus setting up an implicit hierarchy between upper and lower layers.

[6]Philips, Sony.

[7]This held true for Sony, in particular, after it failed to impose the use of Betamax over JVC's VHS alternative.

Alliance in the mobile phone industry, can be considered a sign of this type of development. Faced with these changes, sector-specific and global regulations are walking a fine line between opening digital platforms for controlling access, network access (must carry/must offer) and access to content deemed "essential" by the regulating authority.

The Effects of Network Externalities

A well-known structural paradox among specialists in this area is that network externalities lead to the establishment of a monopoly, as the value of an application increases with its number of users. This widespread concentration trend is linked to economies of scale and is the key characteristic of a sector with high fixed costs. These economies of scale combine with positive consumer externalities (the "club" effect) (Curien and Moreau 2007; Bomsel 2007; Benghozi and Paris 2005) and are found primarily where there are fixed costs and risk: in distribution and promotion. In terms of content services, consumption is concentrated on a few "hits," despite the existence of a wider offering due to the "long tail" (Anderson 2006). *The Long Tail: Why the Future of Business Is Selling Less of More*, Hyperion Books (Anderson 2006) of products that initially struggle when faced with high distribution costs and low public demand (the effects of this long tail are felt some time afterward). The industrial structure of the sector and the unique position of the majors cannot be explained in other terms. Reaching critical mass first is the objective, even if it means subsidizing the first subscribers in a "winner takes all" strategy (Shapiro and Varian 1999).

Conversely, the "public" nature of content available on the Internet actually acts as a disincentive for economic stakeholders to invest. Faced with the difficulties of discriminating between users and limiting usage, these stakeholders are able to delay investment in establishing a new service or market if they are unsure that they will be able to go on to shape its development. This is the crux of the debate between search engines and content producers (notably the press) (Asbroeck and Cock 2007), and is the same challenge that network operators face with the "neutrality of the Internet" (Sirbu 2007).

Digitization thus changes the balance of power between the various parties in the production–distribution–consumption chain. The marginal cost of reproducing and communicating digital content is moving toward zero – content is becoming collective property – and "cost-free" duplication can even be carried out beyond the productive system (Gensollen and Bourreau 2004). While productivity gains had rarely, until very recently, been shared with consumers, the main effect of digitization is to suddenly give consumers access to a huge volume of low-cost – even free – content of every kind.

From the industry's point of view, digitization allows businesses to maximize economies of scale. Financing is still required to establish infrastructures and terminals, and this must be analyzed alongside the complementary items, network infrastructure, terminal equipment and digital content to understand how this financing can be made possible. It is a problem that telecommunication operators know particularly well, though without being able to find a satisfactory solution.

Such contradictions are particularly visible in the debates and strategies concerning questions regarding standards and interoperability today. The coupling of externalities and the interoperability of applications effectively call for standardization so as to reduce investments for suppliers as well as users (Lemley et al. 2003). Standardization reinforces the effects of a natural monopoly: the goods most widely adopted enjoy a far greater market share than their competitive advantage. However, DRM in music, for example, has shown that when there is no interoperability (due to competition or non-existent legislation), the consumer is faced with a battle of standards where the aim is to be first and enjoy the "winner takes all" effect. Thus, the imperfection of a "spontaneous norm" can also create an inefficient market when no natural monopoly exists, as is the case in the software industry.

The fact remains that trends toward concentration are closely monitored and regulated if they concern content services: what does this mean for diversity and plurality, something a healthy society should guarantee for its members? This is symbolically the primary objective of a regulation policy that controls concentrations, particularly in a domain where the diversity of content is seen as essential for consumers. However, with information technology, technological infrastructures do not necessarily give rise to a particular or determined industrial structure, at least at the level of services: small businesses exist alongside major ones in the same market, for example.

The "concentration and diversity of production" are not "invariably negatively correlated" (Menger 2006). A business with a monopoly can sometimes be more readily encouraged to offer more diverse cultural goods than companies that are in competition. Similarly, maintaining diversity is not an end in itself for competitive analysis (Perrot and Leclerc 2008; Ferrando et al. 2007), though it does allow any offering to be submitted to competitive mechanisms, i.e. to access the market. In terms of competitive mechanisms, the disappearance of an offering that has not found enough customers due to its high price or unsuitability is a good thing.

Parallel to this, regulators are challenged by the definition of diversity. It can be measured in three ways – variety, balance and disparity – while the range offered can be very different from that which is consumed. Nevertheless, it remains to be proved (in order to refine economic reasoning) that there is a real preference for diversity and that the consumer is really able to choose when faced with a plethora of offerings (Benhamou and Peltier 2007). A theory on the formation of taste also still needs to be produced. In a situation that is characterized on the Internet by an excess of offerings, the question posed seems more often one of distribution than one of diversity (Benghozi and Paris 2005). In this context of excess, cultural goods find themselves in competition when faced with consumers' limited "experience capacity," meaning that businesses will need to create "a desire to experience" in order to promote their offerings and differentiate themselves (Bomsel 2007). The scarcity has shifted from products to the technical and functional capacity to choose them, play them and incorporate them into one's world. Scarcity has shifted toward the cultural capital of users (Moreau and Peltier 2004), with the business model of cultural industries reorganizing according to these functions (Benghozi and Paris 2005). This merely represents the continuation of past moves toward privatization and the individualization of cultural consumption (Pasquier 2006), which has been accelerated by the emergence and spread of digital networks.

The Transformation of Value Chains

With the digitization of networks and content, a new network of content services is replacing the former. The stakes here are enormous.

The complete digitization of the production–distribution–consumption chain of content services has given rise to the development of new economic models and new forms of market, based most notably on free-of-charge offerings, financing via advertising, audience[8]and, beyond this audience, consumer access and control.[9] How effective are the traditional rules of competition? And, in terms of the functioning of competition, how can "free" services be competed with?

Beyond the effect of near-zero distribution costs, digitization also modifies, in terms of content, conditions for maximizing the value of rights. This is something that the product's creators clearly have a vested interest in, but in reality it is important for every link in the chain, particularly distributors. The chronological organization of markets is a traditional way of progressively reaching the various strata of payment consents (Bomsel 2007). How can this chronological organization now be controlled when the Internet has rendered not only costs, but also duplication and diffusion delays, non-existent?

The main problem of the Internet economy is therefore the absence of any evaluation by the market, as is the case for the off-line economy: managing competition and measuring consumers' disposition to pay in order to reduce external effects caused by the interconnection of networks.

What Does National Regulation Mean in the World of the Internet?

The distance between the "on-line" domain and the legal domain is widened further once the geographical dimension is considered: What is the significance of national regulation for a worldwide exchange network such as the Internet? Users' ability to cross borders between legal territories poses a range of legal challenges that are yet to be properly regulated and have even been swept under the carpet: this applies to customs duties, the sale of medicines, the control of personal data and the right of expression.

---

[8]The audience becomes a commodity and capital in itself, as non-profitable companies, or companies with uncertain future profits, have been able to sell themselves very highly on this basis (as seen with Skype).

[9]The recent interest in the theory of two-sided markets is a good illustration of this: see Bourreau and Sonnac (2006).

As a result of this situation, a number of authors maintain that transnational exchanges over the Internet could be exempted from governmental regulation. Some groups also promote, where necessary, the type of self-regulation that currently exists in community-based exchange structures (e.g. Open Source) as well as commercial ones (cf. eBay and YouTube). However, in this matter of efficiency/inefficiency of national regulations in the Internet context, it is important to point out the example of Internet portals in China where the big international Internet players have to take into account the political and economic dimension of the regulation. The key is to take into account both the political and economical dimensions of regulation. Is regulation destined to become a power struggle between multinational organizations and states? Will there be room for "negotiations" in the regulations? And does the Internet call for new areas of regulation and intervention?

Intrinsic Economic Contradictions

Reconciling Network and Content Economies

The digital (services) economy is at the conjunction of the network economy and that of content and information services.

In terms of content services, the digital economy is – like all creative domains – an "industry of extremes" in which some extraordinary successes make up for the far more numerous total failures. Consequently, businesses' actions and the organization of the sector are largely directed at managing this risk, whether by acquired monopolies/oligopolies (the majors) or conceded monopolies (authors' rights, copyright). The sharing of risk is the main economic question in the organization of vertical relationships along the value chain, with the cultural goods economy even more unpredictable than the value chain is long. Controlling distribution, in particular, is proving to be a key area, as this is where the risk is greatest. Businesses similarly employ brand strategies, along with prescription, awareness, loyalty, recommendation and sampling strategies, in order to minimize risk.

On the networks' side, components (protocols, languages etc.) and software can be combined to offer a range of different services: web pages, intranets, chat rooms, auctions, exchanges, search engines, information services and downloads. What's more, these services can be supplied (almost) instantaneously around the world. The result is a period of rapid innovation in an intensely competitive environment: open technologies and low costs encourage entrepreneurs to enter the market and compete, and it becomes difficult to maintain an advantage over the long term.

With regards competition, authors' rights, copyright, privacy and plurality of information regulation – in short, everything that contributes to the development of digital services – objectives and types of regulation are often differing and even contradictory on occasion, particularly between competition on one side and exclusivity on the other, whether the latter is linked to artists' rights, the protection of an innovation, a material investment (controlling productive resources) or an immaterial one (link with the consumer or future consumer).

The timing of the development of networks and services is also in question. Since telecommunications have existed, operators' motivation to invest in improving capacity has involved anticipation of demand: otherwise, why do it? Those offering services believe capacity should be increased without worrying about the services to be carried, as available capacity will always find a taker and limitations in network capacity have a harmful effect on service innovation. This old debate has now returned.[10]However, it is no longer merely about contesting the traditional logic of telecommunications networks: improving network capacity has always led to major developments in terms of services. The debate now is about contesting the exclusive forms of regulation that result from this, and their economic foundation: who takes the risk and how are they "compensated"? This is the key question in current regulatory debates such as that of the "regulation holidays" concerning the investment of telecommunications operators in fibre optics, and that of "Net neutrality" in the US. The debate is currently focused on the following issue: should conditions of access to the network and usage of it be "frozen" to allow services to develop freely, which the Internet's major players are demanding? Our analysis leads us rather to shift the perspective by questioning how the commercialization of services is or can be a way of financing networks, rather than the intermediary access service.[11]A change of focus such as this transforms the nature of the economic analysis by studying the status and real economic role of the access services: are they transforming from long-term to short-term? From paid to free-of-charge? From service to audience?

Taking Innovation into Account at the Heart of the Digital Economy

As we have seen above, the dynamic of the digital services industry relies profoundly on innovation, while regulation applies an outdated system to innovation. Innovation today (in ICT and other areas) doesn't just refer to the introduction of new technical supports to improve processes, infrastructures or terminals: it also concerns the development of new business designs and cognitive industrial systems in areas of major uncertainty, where business simultaneously develops its facilities, products, markets and audience (Hatchuel et al. 2007). These are revolutionary systems, as they simultaneously redefine the organization of business models, markets, goods/services, functionalities, forms of transaction and value chains (basically the rules of the game and companies' outlook on the world) by mobilizing resources such as networks, technology, applications, skills, marketing and sales (in a modular

[10]Why install FTTH if there is no market for video services? In such a case, ADSL largely suffices.

[11]For links between unbundling, access conditions and innovation, see Baranes and Bourreau (2002).

way that can be configured according to specific portfolios of resources). This radically new type of innovation explains the difficulties in drawing up strategies to defend stakeholders, such as regulation, by analyzing the technologies, markets and existing competitors. We are thus in a model of continual, major innovation, and not in a rhythm of marginal innovation or an imitation/improvement model of industrialization/lowering costs. The advantage of the "first mover" doubles the "winner takes all" effect by increasing the consequences: it is now impossible, for example, to imitate the Google experience, simply because it has become impossible to acquire its audience level (the grail of the Google business model) and thus its levels of revenue, investment and innovation. This calls for a much more flexible, progressive approach,[12]but one whose principles are yet to be established. These principles can be envisaged either in economic terms (refining models to define "optimal" situations) or operative ones (rethinking regulation and arbitration procedures).

The Way of Innovating Challenges Established Models

New technological developments require a new division of labour that allows people to develop new competitive advantages: new stakeholders can emerge in the added-value chain, while traditional stakeholders can broaden their activities. As innovation is constant and unstable in ICT, no economic model or competitive balance is unanimously accepted and shared: in such a context, each innovation represents a unique arrangement and calls for specific negotiations. As such, innovative IT applications bring about a new economic system: new markets, a new segmentation of these markets, new products, a new rhythm, new transactions and new prices. The boundaries between goods, services and market segments become fluid (see the example of the iPod). Powerful groups and small businesses coexist, technical structures overlap and are interdependent, and a new balance emerges between primary and secondary markets.

Furthermore, the nature of digital services and products is such that their adoption is determined by usage more than by people being aware of their requirements. The characteristics of the product or service are less important than its use. Products are complex, and their appropriateness for a specific service is difficult to evaluate. What's more, different products' and services' functionalities overlap, while one piece of technology may offer a range of functions. Everyone knows that most users do not fully exploit the capabilities of their computers, software or high-speed connection; the ostentatious nature and image of the consumption is just as important as suppliers' innovation strategies. The relevant way of innovating is through experimenting on the market (Greenstein 2007). This evolution depends on the development of new business models and transactions based on cost-free offerings or prescriptions modifying both the types of competition in the market and the way of evaluating companies' value (Borreau 2007;2005 Benghozi and Paris).

[12]In the style of Sun Tsu or Laozi rather than Clausewitz, to offer a traditional contrast.

The key question remains that of creating value to generate financial flows to supply the production–distribution–consumption chain, from consumers' readiness to pay to the remuneration of the innovators. The rise in power of business models built on free-of-charge offerings makes the mechanisms of value creation and reappropriation considerably more complex, whether the cost-free offerings are "imposed" by the consumer, who believes it "costs nothing" and thus should be free, or whether two-sided markets, built on audience and advertising revenues, are considered more efficient.

These ideas must thus be rethought in order to mobilize productive resources to generate more innovation without compromising markets' competitive functioning.

Rethinking Regulation on the Basis of Innovation and the Mobilization of Productive Resources

Freedom of enterprise, a fundamental principle of the market economy, is a bounded freedom, that is, a freedom circumscribed by so-called regulatory frameworks, which have gradually been developed to protect economic stakeholders from the "unfavourable" effects of free enterprise. Such rules protect employees (labour law), governments (tax law etc.), shareholders and creditors (financial law) and consumers, as well as partners and competitors, against the "abuses" that this freedom might entail: overpricing, decline in quality, lack of innovation and so on. The question that arises in connection with these rules involves where to set the limits and how to ensure a balance between conflicting interests.

Economics has identified the conditions under which markets allocate resources efficiently. If these conditions are not met, economists speak of market failures and recommend that the authorities establish rules and incentive mechanisms to bring market conditions closer to those of "perfect" markets. The authorities diagnose the "ills" affecting markets and apply "remedies" to restore satisfactory market functioning.

From the standpoint of businesses, however, the term "resources" has another meaning.[13]To start a productive activity, the entrepreneur must mobilize productive resources for the purpose of bringing products to market, using the techniques that characterize the activity concerned. We have just seen how different and much more complex the process of resource mobilization is in the digital economy. It therefore seems important to us, in analyzing regulation, to consider this notion of resources in a broader sense than the strictly economic one, as follows: all inputs to production,

[13]An entire strand of the literature, known as the "resource-based view," has grown on the basis of Wernerfelt's work. Wernerfelt postulates that the competitive advantages of firms are based on their ability to mobilize the resources at their disposal and deploy them in a firm-specific manner (Wernerfelt 1984).

including intangible resources (such as a brand), that give a product objective or subjective qualities that contribute to successful marketing of that product. From a regulatory standpoint, the advantage of adopting this perspective is that it regards the competitive strength of businesses as deriving more from control over scarce resources than from a given market positioning.[14]

Above and beyond the theoretical question of how to characterize such resources, much of the current economic debate has to do with the status granted to them: are they freely controlled[15]by the entrepreneur who wishes to mobilize them, or is control over productive resources "regulated" by a public authority? In other words, how does freedom of enterprise apply to resources that are necessary to the business activity? Clearly, more or less tight control over certain resources, i.e. more or less high barriers to competitors' access to these resources, may constitute what is euphemistically called a competitive advantage. Now, if a firm is to make a profit in the context of a competitive market, the economics of the firm dictates that it must have competitive advantages. Thus, the more possibilities there are of gaining control over certain productive resources, the greater will be the incentives for entrepreneurs. This dilemma – providing incentives for entrepreneurship while allowing a reasonable measure of control over those productive resources for which such control can be distinguished – leads to the formulation of sets of rules that lay down what is prohibited and what is permitted, as well as – increasingly nowadays – to the provision of incentives, which tend to encourage what is desirable and to constrain what is undesirable.

To Reward Innovation or to Encourage the Dissemination of Innovation?

This dilemma, which has been explored from all sides by economists, can easily be illustrated by an example that will also clarify the regulatory implications of our question. The example concerns the exploitation and appropriation of innovations. We noted above the tension of a general nature between protection of creative work and encouragement of the dissemination of ideas. In the case of ICT, this debate takes a sector-specific form in the conceptual conflict between protection of intellectual property and regulation of essential resources.

We have noted that innovation is the key to competitive advantage in a highly dynamic sector such as ICT. Firms therefore invest heavily in creative resources so as to market goods and services offering substantial benefits to consumers in terms of either price or functionality. To recoup their investment, firms take steps to

[14]Following Teece et al. (1997), these questions have continued to be debated (see e.g. Spannos and Lioukas 2001).

[15]The concept of control used here encompasses control over access to these resources, from their use to their appropriation.

ensure that they will, temporarily, have an advantage: the exclusive right to exploit this innovation, directly or indirectly (e.g. through licensing). Thus, they control a resource, an innovation (an "idea") over which the intellectual property regulations give them exclusive rights of use, if not a "monopoly." In recent decades, protection of intellectual property has been strengthened – in terms of the duration and coverage of protection and the contestability of resources – giving innovators still greater control.[16]It is characteristic of this process that where intellectual property is easy to copy – and the digitization of knowledge is making this increasingly easy – the protection is increasingly strong.

At the same time, however, the desire to encourage the spread of new goods and services, added to the creation of innovations, has brought many public authors and stakeholders to defend new appropriateness principles. This is the case for the Essential Facilities Doctrine, or EFD. This old doctrine[17] has been enjoying a revival in the United States and subsequently in Europe for infrastructure based activities.[18]Its principle states that if a entrepreneur owns a resource that is difficult to replicate, and that access to it has a strong influence on the existence of dependent competitive markets, the resource's owner must be compelled to open it up. This means it must be made available to third parties, often the owner's competitors, under reasonable economic conditions, and often for almost at-cost pricing. This doctrine is gradually spreading, and access to essential facilities is an increasingly common clause of competition regulations.

To summarize, wherever facilities are easy to duplicate, the tendency is to over-protectiveness, but wherever facilities are difficult to duplicate, the tendency is to demand substantial openness. Under these circumstances, sector regulation specialists in charge of regulating essential facilities often find it difficult to understand the arguments used in intellectual property law, while intellectual property specialists sometimes struggle with the concept of essential facilities. It seems to us that this once-marginal debate should now be opened to a wider audience. This would allow an economic doctrine to be determined based on the correct indicators, striking a balance between protecting and allowing access to productive resources, whatever their nature. The participants in digital economics are increasingly faced with conflicting regulation dynamics, and are justifiably calling for further clarification on these issues.

Which Productive Resources for Innovations?

In order to resolve the tension we have just seen, it is necessary to identify the productive resources of innovative companies more precisely. In particular, it seems

[16]The same trend can be observed in the case of copyright.

[17]It was first expressed in the United States in 1912, in a Supreme Court decision concerning railway networks.

[18]At the beginning of the 1990s the European Commission used for the first time this theory about harbor facilities in its decision Sealink II of December 21st, 1992 (Décis. Comm. IT, in December 21st, 1992, Sea Containers c. Stean Sealink, Europe in March, 1994, n 115, the Ob. L.I.).

to us absolutely imperative to explicitly measure the differences between the two types of resource we are studying. Creative resources involve the ability to conceive of, create and implement innovations of any kind. Essential facilities combine to form other productive resources and generally include network and connection infrastructures, such as ports, airports, train stations, television decoders and so on, and production infrastructures like stadiums, for instance, and natural sites. Both types are clearly identified in economic literature and regulations, but only in established frameworks. In our opinion, this range of productive resources should be extended to comprise intermediary "objects" that are sometimes considered intellectual property and sometimes "essential facilities." This is the case with intermediation platforms (Penard 2006) (with distribution, advisory or search engine capabilities), programs, knowledge, entertainment (sport rights, for example) and other immaterial products of the digital age like the rights associated with standards. This phenomenon is spreading to brands and identifiers, two absolute and unquestionable immaterial properties whose usage owners are careful to control and which are increasingly seen as essential facilities that can lead to bitter international disputes. Indeed, certain "brands" have such a market presence that they end up in the "public" domain; this is the case for examples like the yellow pages or free-call numbers, and it can also affect interoperational standards like Windows.

Another productive resource that is rarely seen as such, but strikes us as being equally material, is consumer control. Consumers cannot be owned, but they can be "captured"; they are often targeted with the aim of ensuring their loyalty toward those who are investing in their capture. In other words, consumers are the focus of "commercial" investments aimed at captivating, if not entirely capturing, them. As with any investment, protecting the product of the investment, in this case consumption is the object of increasing thought being similar to thought related to controlling and regulating this resource.[19]

This issue, well documented in the field of access providers, arises just as strongly for content providers, who nowadays mainly use attractive content to produce audiences they can resell to their advertisers. The productive resource in this case is the consumer's attention or visits, requiring massive investment in either content or, more commonly, the ability to generate traffic. Producing attention or visitor traffic is expensive: how can the investment be protected? In this case, customers, consumers and users can be considered a productive resource, too; even though they don't pay anything themselves, their attention or visits are billed to a third party, and they are often rewarded implicitly or explicitly for their attention or visits through the use, for example, of content offers and loyalty schemes.

Another productive resource concerns what are known as scarce resources, such as the radio-frequency spectrum, numbers, domain names and the public domain. These are scarce resources inasmuch as, for a given sum, they are not available in unlimited quantities. A public authority often owns them and the granting of access to these resources is becoming increasingly crucial. The current issue involves the

[19]This is no doubt what is at stake in the current debate regarding the "portability" of cell phone numbers and, more generally, of identifiers.

digital dividend, meaning the available spectrum resources freed by the digitization of audiovisual content. What access rights do various economic stakeholders have to these resources?

A Stake for Regulation: Planning the Composition and Availability of Productive Resources

The diversity of productive resources available to companies in an innovative context more generally poses the traditional question of regulation. If an economic stakeholder invests in the production of a resource that cannot be entirely duplicated by its competitors, what protection should be accorded to the exploitation of that resource? Conversely, if an economic stakeholder needs a competitor's resource for its productive activity and cannot recreate it under the same economic conditions, what access rights should be granted? In other words, what degree of control can an economic stakeholder legitimately demand for a resource that cannot be replicated?

Whether the purpose is to encourage the development of new goods or to ensure, for society, the quality of a service, the degree of control granted on a resource is always the result of an investment. The issue is therefore to guarantee the right to exploit a productive investment, meaning one that has led to the creation of a productive resource.

The issue has already been raised in these terms in areas where investments are only considered if an *ex-ante* operating monopoly is guaranteed, as in the case of mines and oilfields. In other areas, such as pharmaceuticals and artistic creation, the fundamental principle remains the same, but there is still some discussion about the actual particulars: the time span of the protection and defining elements of the innovation, for example.

More generally, even in areas traditionally removed from these issues, this is how productive investments are viewed nowadays. This is illustrated by the recent debates, particularly in France, over how to capitalize subscriber acquisition costs from an accounting point of view.

Naturally, the resources don't necessarily exist as separate entities. The stakeholders compose them, organize them and put them to use in the specific context of value chains. The issue of access rights is therefore linked to their appropriation, transformation and sharing by a number of stakeholders contributing to the collective creation of a product or service. It is no longer just necessary to identify competitive positions, but also to understand a resource's strategic situation in a value chain. Those who access the resource might want the chain to remain reasonably stable, but those who produce the resource may, on the contrary, want the chain to change in order to free the economic and strategic constraints imposed by regulated access. In this area, as in the end market, it is difficult to consider access conditions to a resource, meaning the constraints of its opening, without simultaneously considering how to control those who have access to it.

Regulations have always been led by supply rather than demand, in the sense that they much more bind supply than demand. Their focus is how the end points of a sector can have access to resources created earlier in the process, rather than the obligations to be imposed on those accessing the resource. Reversing this perspective raises the question of how a resource's investment risks should be shared. Is it possible for regulations to ensure that the significant risks involved in innovative disrupting technologies are financed, without also considering regulations that focus on contracts rather than access? The financing of very-high-speed networks currently poses the question of risk-sharing very clearly. Large systems like Galileo are therefore struggling to see the light of day. Without regulations that can take into account the construction of new productive resources by an entrepreneur and its potential customers, only massive public funding appears able to support and guarantee heavy, regular investment by spreading the risk to the entire community. The risk shared by participants in a value chain, defined by the contracts that bind them together, then becomes an essential point in resource regulation in an economic context whose production conditions (fixed costs) and consumption conditions (external factors) considerably increase the risk taken.

Conclusion

Shifting the balance of regulation more toward demand and with a central focus on the mobilization of productive resources will lead, in this context, to a reformulation of the more traditional question of discrimination in new terms, rather than on the basis of pricing. Can the holder of a productive resource that cannot strictly be replicated impose discriminatory terms of access on those who want access to it? For example, is the holder to be permitted or not to choose those allowed access to it? Can the holder set tariffs for access on a discriminatory basis? Can it make differences in the quality of the resources on offer? Can it discriminate in terms of risk-sharing? And so on. In other words, must the holder demonstrate "neutrality" in granting access to the resource, particularly between itself and third parties, or is it permitted to breach that neutrality?

On the other side, can an authority responsible for a regulatory mechanism discriminate between stakeholders, and, if so, on what terms? The issue of asymmetrical regulation is another facet of regulatory problematics that sectoral regulators have often had to deal with, taking account of the relative market weights of the economic stakeholders involved. Some rulings in case law[20]implicitly feel that minority stakeholders in a market can be permitted to engage in collusion: such practices enable them to cope with the discrimination applied by a dominant market player without forcing the latter to open up a resource.

[20]See for example Bronner vs Mediaprint/Austria.

The consequences of the regulation of resources are therefore not radically opposed to the effects of the regulation of market agreements: whether a resource can or cannot be replicated depends on what agreements are authorized. On the other hand, the change in standpoint does lead to a clearer definition of questions that are usually no more than incidental to traditional regulation:

- What are the economic terms of production and consumption of these resources? How efficient are the productive functions? What is their possible sub-additivity? What are the direct and indirect externalities? What is the degree of rivalry between products for consumption? And so on.
- What does it mean to say that a resource can be replicated or imitated? There is extensive case law in this area, but the extension of the concept of the productive resource makes it necessary to cross-correlate the rulings in such jurisprudence. Are the concepts of resource and supply mutually substitutable or can they at least be interlinked?
- How are the terms of access and discrimination relating to productive resources established?

These initial considerations are static in nature. There is, however, a need to look at the dynamic aspects. The protection of a resource can, on the face of it, be no more than temporary, and needs to be reviewed regularly. This is due not only to the very rapidly changing context of innovation but also to the phases of regulation, as such: contexts in which there is intense competition to win markets are routinely followed by situations in which there is cooperation and non-competition in the market. Regulatory frameworks regulate – with varying degrees of success – the restoration of competition between the monopolies that have been permitted.[21] The protection enjoyed by a stakeholder on the grounds of its investment in a productive resource should also be called into question periodically. This is usually, but not invariably, the case (see the granting of use of scarce resources). However, the issue is particularly pressing at present with regard to certain resources that have received no consideration, such as "consumer control." Taking the example of contracts signed between customers and service providers, what is the right frequency for the re-examination of their validity? In the case of a mobile telephone service, for how long can the subscriber, the cost of whose terminal has been heavily subsidized (constituting a genuine investment), be "locked" into the relationship with the provider?

The nature of the protection for a business investment in a customer base seen as a resource is currently a hotly debated issue. What is the right duration for such protection? On what terms can a customer redeem the investment or have it redeemed? What information is to be given to customers to enable them to assess for themselves the degree of loyalty they should show? On the other side, this raises the issue of the possibility of resale of resources whose "control" is likely or to which access has been granted: is such resale to be permitted, and, if so, on what

[21]See the comments of Joseph Stiglitz in Competition and Competitiveness in a New Economy (Stiglitz 2002).

contractual and economic terms? There are countless cases of this kind of resale: the selling of customer bases by self-employed professionals, audiences and visitor traffic and brands (franchises, for example); the sale of data on individuals and statistics; the so-called secondary market for spectrum, numbers, domain names, rights of access to State-owned property and business licences; and so on.

Conventionally, discrimination is considered to be weakened by easy resale, but significantly reinforced by possibilities for bundling market offerings. Joint consideration must be given to resale and bundling, since the latter provides opportunities for predation where market offerings are difficult to replicate. The problem is that much tough, because the aggregation of market offerings is particularly easy in the case of intangible productive resources. Evidence of this can be found, for example, in the extreme diversity of potential competitors in the financial services market: banks, insurance companies, large retail chains, airlines, telephone operators, ISPs and more.

Going beyond such questions – which are, when all is said and done, problems that fall within the usual, legitimate sphere of activity of economic stakeholders – the issue of resource-based regulation also calls for examination of the prohibited character of certain investments, aimed at controlling them. At present, the law condemns certain practices whose purpose is to control a resource (corruption, for example). There is nevertheless a need to look in a more systematic manner at the rules that should be applied and the definition of "proportional" incentives capable of guiding investor behaviour. The borders here are at present poorly defined. The rules and incentives that have been applied by means of regulatory mechanisms can be seen as difficult to manage due to their externalities and indirect consequences. Indeed, it is possible to see, as a response to this problem, the emergence of increasingly widespread forms of self-regulation, defined not only as collective regulation provided by the community formed by all the stakeholders involved (enterprises and users), but also as individual acceptance of ethical rules or meta-rules. In this regard, ICTs are following an evolutionary path already seen in the life and environmental sciences. Economic stakeholders tend – or should tend – to take certain social concerns of an ethical nature directly on board without prompting: sustainability, personal data, security, the control of automatic systems and so on. This form of ethics is probably likely to penetrate the economic sphere and lay down principles for the protection of and access to productive resources, but such principles are yet to be defined for the most part.

Finally, the issue of regulation leads us to raise the question of sanctions. What sanctions should be incurred by those who fail to obey the rules? What bias those who fail to respond to the incentives should suffer? Rules assume, in principle, the existence of mechanisms for sanctions or degrees of bias "proportional" to non-performance, but in an economic system whose stakeholders have widely differing strengths, the regulator will not necessarily be strongest, and the economic stakeholder subject to the regulation will not necessarily be the weaker party. Even the courts are often caught up in power plays: pressure may be exerted on those issuing legal rulings, and there can be difficulty in duly enforcing the rulings. This is a context in which the issues are no longer merely national, but international: regulation comes up against the links that exist between market regulation and industrial policy, and the debate is constantly being reshaped by current affairs: the latest developments in anti-trust proceedings or the regulation of mergers and acquisitions.

Regulation of the productive resources in play in the digital economy can therefore be seen to be extremely fragmented at present when one considers the range of resources involved: some appear to be excessively protected, others excessively open, and the ways in which they are treated are highly disparate. The digital economy calls for a more holistic consideration of the link between innovation and the structuring and mobilization of value, on the one hand, and regulation, on the other. Technological convergence, which is no more than a tighter interlocking of productive resources, necessitates such conceptual globalization, which must inevitably cast doubt on the current state of the art of regulation.

Acknowledgments This paper had been developed with the support of the Chair Innovation & Regulation of Digital Services, jointly created in 2007 by Orange, Ecole Polytechnique and TELECOM ParisTech.

References

Anderson C (2006) The Long Tail: Why the Future of Business Is Selling Less of More. New York: Hyperion

Asbroeck B V, Cock M (2007) Belgian Newspapers v Google News: 2–0. Journal of Intellectual Property Law & Practice 2(7): 463–466

Baranes E, Bourreau M (2002) An Economic Guide to Local Loop Unbundling. Communication & Strategies n°45 1st quarter

Benghozi P-J (2006) Mutations et articulations contemporaines des industries culturelles. Création et diversité au miroir des industries culturelles. X. Greffe. Paris: La Documentation Française: 129–152

Benghozi P-J, Paris T (2005) The Economics and Business Models of Prescription in the Internet. In: Brousseau E and Curien N (Eds.), Internet Economics. Cambridge University Press, Cambridge, UK

Bengtsson M, Kock S (2000) Co-opetition in Business Networks: To Cooperate and Compete Simultaneously. Industrial Marketing Management 19(5): 411–426

Benhamou F, Peltier S (2007). How Should Cultural Diversity Be Measured? An Application Using the French Publishing Industry. Journal of Cultural Economics 31 (2):85–107

Bomsel O (2007) Gratuit! Du déploiement de l'économie numérique. Paris: Gallimard

Bourreau M, Sonnac N (2006) Competition in Two Sided Markets (Special Issue). Communications & Strategies 61(1): 11–15

Chantepie P, LeDiberder A (2004) Révolution numérique et industries culturelles. Paris: La Découverte

Curien N, Moreau F (2007) The Convergence Between Contents and Access: Internalizing the Market Complementarity. Review of Network Economics 6(2): 161–174

Ferrando J, Gabszewicz J, et al. (2008) Intermarket Network Externalities and Competition: An Application to Media Industries. International Journal of Economic Theory 4(3): 357–379

Freeman C (1987) Technology Policy and Economic Performance: Lessons from Japan. London: Pinter

Gensollen M, Bourreau M (2004) Communautés d'expérience et concurrence entre sites de biens culturels. Revue d'Economie Politique 113

Greenstein S (2007) Economic Experiment and Neutrality in Internet Access. N. W. Paper
Hatchuel A, Le Masson P, and Weil B (2007) Building Innovation Capabilities. The Development of Design-Oriented Organizations, in Innovation, Science and Industrial Change, the Handbook of Research., ed. Jerald Hage and Marius Meeus. Oxford, Oxford University Press, 294–312
Henderson R M, Clark K B (1990) Architectural Innovation: The Econfiguration of Existing Product Technologies and the Failure of Established Firms. Administrative Science Quarterly 35: 9–30
Jacobides M G, Knudsen T, et al. (2006) Benefiting from Innovation: Value Creation, Value Appropriation and the Role of Industry Architectures. Research Policy 35 (8): 1200–1221
Lemley M A, Merges R P, et al. (2003) Intellectual Property in the New Technological Age, Aspen Law & Business, New York, 1112 p.
Lessig L (1999) Commons and Code. Intellectual Property Media and Entertainment Law Review 9: 405–415
Menger P M (2006) Artistic Labor Markets: Contignent Work, Excess Supply and Occupational Risk Management. Handbook of the Economics of Art and Culture, Elsevier 1(1), V. A. Ginburgh & D. Throsby (ed.), pp. 766–806.
Midler C (1995) "Projectification" of the Firm: The Renault Case. Scandinavian Journal of Management 11(4): 363–375
Moreau F, Peltier S (2004) Cultural diversity in the movie industry: a cross-national study. Journal of Media Economics, 17(2): 123–143
Pasquier D (2006) L'espace privé comme lieu de consommation culturelle. Création et diversité au miroir des industries culturelles. X. Greffe. Paris: La Documentation Française
Penard T (2006) Faut-il repenser la politique de la concurrence sur les marchés internet? Revue internationale de droit économique 1: 57–88
Perrot A, Leclerc JP (2008) Cinéma et concurrence, Paris, La documentation francaise
Shapiro C, Varian H R (1999) Information Rules: A Strategic Rules to the Network Economy. Cambridge, MA: Harvard Business School Press
Sirbu M (2007) What Is the Network Neutrality Debate About? A Workshop on Net Neutrality: American and European Perspectives. Innovation and Regulation in Digital Services Chair (innovation-regulation.eu). Paris, May 29, 2007
Spannos Y E, Lioukas S (2001) An Examination into the Causal Logic of rent Generation: Contrasting Porter's Competitive Strategy Framework and the Resource Based Perspective. Strategic Management Journal 22(10): 907–934
Stiglitz J (2002) Competition and Competitiveness in a New Eeconomy. Austrian Ministry for Economic Affairs and Labour. Economic Policy Center. Edited by Heinz Handler and Christina Burger. Vienna. July
Teece D J, Pisano G, et al. (1997) Dynamic Capabilities and Strategic Management. Strategic Management Journal 18(7): 509–533
Wernerfelt B (1984) A Resource-Based View of the Firm. Strategic Management Journal 5: 171–180

www.ingramcontent.com/pod-product-compliance
Ingram Content Group UK Ltd.
Pitfield, Milton Keynes, MK11 3LW, UK
UKHW021454240926
475688UK00009B/170
* 9 7 8 3 7 9 0 8 2 0 8 1 2 *